Leveraging AI for Innovative Sustainable Energy:

Solar, Wind and Green Hydrogen

Hind Hammouch
University Sidi Mohamed Ben Abdellah, Morocco

Laeeq Razzak Janjua
WSB University, Poland

IGI Global
Scientific Publishing
Publishing Tomorrow's Research Today

Vice President of Editorial	Melissa Wagner
Managing Editor of Acquisitions	Mikaela Felty
Managing Editor of Book Development	Jocelynn Hessler
Production Manager	Mike Brehm
Cover Design	Phillip Shickler

Published in the United States of America by
IGI Global Scientific Publishing
701 East Chocolate Avenue
Hershey, PA, 17033, USA
Tel: 717-533-8845
Fax: 717-533-7115
Website: https://www.igi-global.com E-mail: cust@igi-global.com

Library of Congress Cataloging-in-Publication Data

LCCN: 2024055703 (CIP Data Pending)
ISBN13: 9798337300450
EISBN13: 9798337300474

British Cataloguing in Publication Data
A Cataloguing in Publication record for this book is available from the British Library.

Table of Contents

Detailed Table of Contents

the transformative potential of AI while offering insights into overcoming barriers to its widespread adoption in renewable energy systems.

The urgency of considering climate change and working for carbon neutrality has accelerated cutting-edge models in energy generation and urban sustainability worldwide. This impetus to address climate change and the shift that cities demonstrate towards a more sustainable, carbon-neutral future has catalyzed city willingness towards solutions that deliver reduced emissions and improve the quality of our urban media. This provides a fantastic context for how smart cities can reduce greenhouse gas use by working both wind and solar renewable agents with very low, even zero, marginal costs. Wind turbines produce clean electricity. Solar energy is made up of photovoltaic panels that capture solar power. Renewable technologies in cities give them a tremendous opportunity to shift away from fossil fuels, cut carbon emissions, and drive international carbon neutrality goals. This chapter underlines the call for diverse pathways concerning intellectual property rights on renewable energy sources to build a sustainable and greener future.

Artificial Intelligence (AI) and Machine Learning (ML) are reshaping the frontiers of Very-Large-Scale Integration (VLSI) by revolutionizing energy management, vital for sustainable smart cities. VLSI, in which millions of transistors are combined onto a single chip, contributes significantly to the development of low-power electronic systems. These systems are further optimized by AI & ML techniques in terms of design, manufacturing to operations and help energy consumption from becoming bare minimal and their performance optimization. This chapter comprehensively explores the diverse aspects of artificial intelligence and machine learning in VLSI for energy management towards sustainability in smart cities.

Virtually every place doing business with us is in the throes of rapid change: smart cities and villages are replacing traditional communities; urbanization is transforming social systems that were once localized phenomena; sustainable growth, one integral tenet for achieving more equitable Sustainable Development

Artificial intelligence in energy systems has brought a significant improvement of efficiency, management and optimization making them dependable, there are also big concerns in the cybersecurity domain. A major worry is the risk of AI algorithms being influenced by data poisoning which is when bad actors introduce corrupted data to cause malfunctions on a system. AI-powered energy systems frequently connect thousands of IoT devices and every one of these represents a potential gateway to a system breach if not correctly protected. In addition, threats such as ransomware, phishing and Distributed Denial of Service (DDoS) attacks can compromise the integrity of energy management systems. Security of AI models and data, strong encryption protocols to ensure that these infra are secure from the cyber threat elites of today because of which we must build a real-time monitoring system too. Energy industry can prevent catastrophe by prioritizing cybersecurity.

The power of digital twins, virtual replicas of physical assets and systems that are revolutionizing the efficiency and performance is significant for optimization for renewable energy. Agriculture sector uses digital twins in the context of precision agriculture, fishing and reforestation, manufacturing can simulate machinery and robots predict maintenance breakdowns before they happen or calculate needs for supplies. Bringing new perspectives on cutting-edge technologies in the area of renewables it simulates wind turbine. With a real-time digital twin in place, energy operators can observe how the system behaves across diverse operating conditions, detect inefficiencies and work on data to make judicious decisions toward enhancing performance. In case of solar power systems, they monitor the performance efficiency of solar panels and predict probable failures which may result into loss from energy production.

Artificial intelligence (AI) utilizes machine learning algorithms to determine the optimal location for solar panels, predict energy demand, and manage battery storage more effectively, translating into higher energy production and lower running costs. Solar energy production AI automates towards maximum output with minimal operators, enhancing output and minimizing data entry errors as a result. AI-powered predictive maintenance improves the equipment lifetime, reducing unscheduled failures and downtime. Long-term economic impacts are job creation as AI enables new jobs such as energy analytics, automation, and smart grids. In addition, AI-based solar energy not only draws significant investments but also contributes to the growth of global markets and accelerates sustainable energy efforts in regions around the world. Governments embrace AI in energy policies that subsidize and incentivize adoption, rendering solar energy more economical and ubiquitous, paves the way for a resilient, AI-enabled green economy of the future.

The current decade is a turning point in the energy sector's transition, with the introduction of green energy along with the optimization of efficacy through machine learning and artificial intelligence (AI). As a result, competitive policies are needed to manage multifaceted tasks on one platform. When energy policies fail to accomplish both energy as well as climate targets over their entire lifecycle, the socioeconomic ramifications can be profound. These deficiencies are said to result from poor decision-making and insufficient incentives, which should encourage equity, equality, fairness, and inclusivity in energy policy and decision-making regarding projects. This chapter seeks to assess the many obstacles posed by the emergence of AI in the energy industry. Specifically, the study addresses (1) the development stage decision-making process, 2) the execution stage implementation management process, (3) integrating deep learning, machine learning and data science in the energy systems, and (4) the substantiality requirements of energy systems.

Environmental, social and governance (ESG) principles have become critical to corporate strategies across industries. However, ESG requires sector-specific approaches as industries face unique challenges. This chapter examines ESG applications in four sectors: finance, energy, fashion and food. In finance, ESG factors influence investment decisions and risk management with institutions evaluating companies' performance in addressing climate risks, regulatory challenges and social concerns. The energy sector, a major carbon emitter, prioritizes renewable energy transitions and emission reductions, alongside employee

the performance of wind turbines by optimizing accurate wind pattern forecasting and energy supply stabilization through intelligent grid integration via predictive analytics. Electrolysis processes are improved with AI, its energy efficiency is increased, and the green hydrogen is distributed and stored as well as possible. While barriers still exist in issues such as data privacy and computational demand, developments in edge computing and machine learning appear to bring about encouraging opportunities for moving past these barriers.

Sustainability is a global principle highlighting the interconnectedness of economic, social, and environmental dimensions. Sustainable development is a challenge not only for the EU but for the entire world. This chapter aims to examine the 27 EU Member States using cluster analysis. Five indicators were included in the examination and these represents all EU 27 sustainable development strategies. Indicators were compared with GDP and HDI, the two most relevant indicators of economic development and human well-being. The research posed the question: can we identify homogeneous groups of countries regarding sustainability among the 27 EU Member States? The analysis revealed that when clustering based on GDP, the Member States typically fell into two, three or four clusters. In contrast, when clustered according to the HDI, the number of clusters varied between two and four. Ultimately, four distinct clusters emerged: cutting-edge, fast-growing, prudently advanced, and tail-enders. The performance of Member States can be identified according to their positions within these clusters.

This research paper delves into the rapid progress of energy technologies for renewable sector in the distinctive context. The research analyses the important improvements across diverse sectors of renewable sector like encompassing solar, wind, biomass, hydropower, geothermal, and ocean energy. India's ambitious renewable energy vision and mission, as exemplified by the National Solar Mission and the Green Energy Corridor innovative techniques, have spurred significant growth in the sector. Similarly, it scrutinizes advancements in wind energy, biomass and bioenergy innovations, and the potential of geothermal and ocean energy sources in elaborate. Through previous scenario and an evaluation of the economic and environmental goods, this research gives a comprehensive view on India's renewable energy. It underscores the important role of technological advancement in propelling India's sustainable energy transition and stats the recommendations for policymakers, industry stakeholders, and researchers to further improve the country's renewable energy growth.

stakeholders, demystify complex technologies, and cultivate trust. By emphasizing tailored messaging, interactive tools, real-world examples, and continuous feedback, the chapter addresses the challenges of technical complexity and public skepticism. The proposed strategies seek to deepen understanding and foster proactive community engagement, ultimately enhancing resilience and safety against wildfires. These insights offer valuable guidance for researchers, practitioners, and policymakers in developing impactful education campaigns to promote the integration of AI and IoT in wildfire management.

Green hydrogen, which is produced via water electrolysis powered by renewable energy, has emerged as a favored solution to decarbonize sectors such as transportation, heavy industry and power generation. But things like high production cost, inefficiency, and a difficult operation still pose challenges for its widespread adoption. The transformative potential usually sits with artificial intelligence (AI). The analytics on data through AI algorithms means far better adjusted to live demands with solar and wind inputs for instance. Less energy is wasted and production costs fall. Predictive maintenance, backed with AI-driven algorithms, enables electrolysis systems to run continuously with less downtime, thereby reducing operational risks while ensuring better scalability.

Artificial intelligence (AI) is increasingly being integrated with green hydrogen production, so as to enable a seamless transition from conventionally suffocating energy sources. Water electrolysis, using renewable energies such as solar and wind energy as a power source, is the key technology of green hydrogen with considerable advantages towards decarbonizing energy-intensive industries and promoting global sustainability. Innovative application of AI is changing the way green hydrogen operates for efficiency and scalability by improving production processes saving energy in electrolysis making it cost effective. The AI and green hydrogen production benefits not only energy efficiency, but also serves to further the cause of climate action by weaning society off fossil fuels. And this clever, foresighted investment clears a path for a cleaner future which also adds one more way out of many possible ways that the world can reach its sustainability goals and move off carbon-neutral energy.

Preface

The world energy transition is reshaping globally as a result of both the sustainability agenda and AI. Exploring the frontier of intelligent systems, AI is transforming the world's approach to energy generation, storage and distribution. With climate change intensifying and an overdependence on fossil fuels having significant environmental consequences, AI-driven solutions create immense potential to optimize energy efficiency, lower carbon footprints and stabilize the grid. This book explores how AI is also used to predict energy needs and make smarter solar and wind farms, as well as support the mass rollout of green hydrogen as a clean alternative to fossil fuels. Powering Renewable Energy Systems Predictive analytics, intelligent grids, automation, and machine learning algorithms are changing renewable energy systems using AI-based innovations.

OVERVIEW OF CHAPTERS

Chapter 1: A Study of Drivers and Barriers of Green Supply Chain Management Practices in the Context of Developing Framework for SMEs

This chapter examines the driving forces and challenges in adopting Green Supply Chain Management (GSCM) by Small and Medium Enterprises (SMEs), which include government control, market requests, and financial constraints. The authors discuss both the barriers and opportunities that affect the successful integration of GSCM practices.

Chapter 2: AI Renaissance in Renewable Energy: A Path to Sustainability

Focusing on the integration of AI in renewable energy sectors, this chapter highlights AI technologies' role in addressing the intermittency of renewable resources, democratizing energy access, and fostering economic growth, all aligned with global sustainability goals like the Paris Agreement.

Chapter 3: Applying Internet of Things (IoT) for Green and Sustainable Future: Intellectual Property Rights (IPRs) Raising Carbon Neutrality for Smart-Sustainable Cities

The authors explore how smart cities can use IoT, wind, and solar energy to reduce emissions and drive carbon neutrality goals. This chapter also addresses the significance of intellectual property rights in building a greener future through sustainable energy solutions.

Chapter 4: Artificial Intelligence and Machine Learning in Digital and Analog VLSI: Sustainability and Energy Management in Smart Cities

This chapter investigates the role of AI and Machine Learning in Very-Large-Scale Integration (VLSI) for energy management, highlighting its contributions to energy-efficient design and manufacturing processes for sustainable smart cities.

Chapter 5: Assimilating Fuzzy Logic in Smart City and Village Development for Sustainable Futures with Green Energy Solutions

The authors discuss the application of fuzzy logic in the planning and decision-making for sustainable development in smart cities and villages, providing case studies and offering insights into the role of this approach in achieving Sustainable Development Goals (SDGs).

Chapter 6: Brand Advocacy and Marketing Strategies in Green Economy for Sustainable Future: Harnessing Hydrogen Power

This chapter explores the critical role of brand advocates and marketers in promoting hydrogen power as a clean energy solution. It highlights strategies for building consumer trust and educating stakeholders on hydrogen's benefits in the transition to a green economy.

Chapter 7: Building Public Trust in AI-Enhanced Energy Systems through Strategic Communication

The chapter explores how public trust is essential for the widespread adoption of AI-driven energy systems, advocating for strategic communication to address concerns like data privacy and ethical issues, while aligning technologies with societal values.

Chapter 8: Cybersecurity Challenges in AI-Driven Energy Systems: Current and Future Prospects concerning Ethical and Legal Provisions

Focusing on the cybersecurity challenges faced by AI-powered energy systems, this chapter highlights the risks of data poisoning, ransomware, and other cyber threats, stressing the need for robust security measures to protect energy infrastructure.

Chapter 9: Digital Twins in Renewable Energy Systems: Enhancing Efficiency Projecting Sustainability

The use of digital twins—virtual replicas of physical assets—is examined in this chapter for its potential to optimize renewable energy systems. Through real-time simulations, digital twins can predict inefficiencies and help improve the performance of wind and solar energy systems.

Chapter 10: Economic Implications of AI-Driven Sustainable Energy: Unlocking the Potential of Solar Energy

This chapter delves into the economic benefits of AI-powered solar energy systems, discussing the optimization of solar panel locations, energy management, predictive maintenance, and long-term impacts like job creation and economic growth.

Chapter 11: Energy Policy Driven by AI Towards Sustainable Future

The authors address the role of AI in shaping energy policy, exploring the complexities of managing energy systems with AI, and the need for policies that are equitable, fair, and inclusive in achieving sustainability goals.

Chapter 12: Enhancing AI-Driven Solutions for Virtual Tourism: Harmonizing Visitor Experience and Ecosystem Conservation

While focused on AI's role in tourism, this chapter draws parallels to energy systems by highlighting how AI can harmonize user experiences and ecosystem conservation efforts, ensuring that tourism remains sustainable and minimizes its environmental footprint.

Chapter 13: ESG in Practice: Sustainability, Ethics, and Innovation Tailored to Sectors

This chapter discusses sector-specific ESG applications in finance, energy, fashion, and food. Each sector faces unique ESG challenges, such as climate risk, employee safety, supply chain ethics, and waste management. It highlights the role of technologies like AI and machine learning in improving transparency and accountability within these sectors.

Chapter 14: Fuzzy Logic Approach to the Management of Biomass and Biogas Energy Systems

This chapter introduces fuzzy logic in renewable energy management, focusing on biomass and biogas systems. It proposes a framework for modeling complex energy systems and evaluates its effectiveness through case studies, offering new insights for integrating biomass/biogas into global energy strategies.

Chapter 15: Green and Smart Irrigation: Farming Inspiring IoAT and Blockchain

The chapter explores the use of IoT and blockchain in green farming, focusing on smart agriculture's role in enhancing crop yields and sustainability. It examines how emerging technologies like big data, AI, and IoT are transforming farming practices to meet climate-resilient agriculture demands.

Chapter 16: Green Hydrogen Next Frontier in Clean Energy: Future of Energy Innovation with AI and Industry 6.0

This chapter discusses the role of green hydrogen in decarbonizing industries, highlighting challenges like high production costs. It examines how AI and Industry 6.0 technologies can make green hydrogen production more efficient and scalable, offering a practical solution for global clean energy needs.

Chapter 17: Harnessing AI Powered Renewable Energy: Technological Breakthrough and Legal Frontiers

The chapter presents the application of AI in renewable energy to overcome current limitations and drive technological breakthroughs. It discusses AI's potential for optimizing energy production, integration, and planning while addressing legal and ethical considerations in sustainable energy practices.

Chapter 18: Harnessing Artificial Intelligence for a Greener Future in Solar, Wind, and Hydrogen

This chapter explores how AI enhances solar, wind, and green hydrogen systems by improving efficiency and scalability. It highlights AI's role in weather forecasting, predictive maintenance, and energy optimization, addressing challenges like data privacy and computational demand.

Chapter 19: Homogenous Groups in EU Member States According to Common Indicators of Sustainable Development

Using cluster analysis, this chapter examines the sustainability performance of EU member states, identifying groups based on GDP and HDI indicators. It categorizes countries into four clusters: cutting-edge, fast-growing, prudently advanced, and tail-enders.

Chapter 20: Progress in Renewable Energy Technologies: A Review Outlook

This review highlights advancements in renewable energy technologies, including solar, wind, biomass, and hydropower. It focuses on India's renewable energy progress, including national policies and technological innovations, and emphasizes the role of these advancements in a sustainable energy transition.

Chapter 21: Projecting Green Energy Fostering and SDG 15 (Life on Land): Using Natural Resources for a Sustainable Future

This chapter discusses the role of green energy in achieving Sustainable Development Goal 15 (Life on Land), focusing on how renewable energy helps protect ecosystems, reduce deforestation, and promote sustainable land use practices.

Chapter 22: Prospect of Artificial Intelligence Towards Demand Response, Control, and Predictions in Integrated Microgrid Renewable Energy Systems

Focusing on microgrids, this chapter explores the use of AI to optimize renewable energy sources like solar and wind. It examines demand response strategies, control targets, and prediction techniques to improve energy system efficiency and sustainability.

Chapter 23: Public Education Campaigns on AI and IoT Applications in Wildfire Prevention and Management

This chapter highlights the importance of public education in promoting AI and IoT technologies for wildfire prevention and management. It proposes strategies for effective communication to increase public engagement and trust, enhancing community resilience to wildfires.

Chapter 24: Scaling Green Hydrogen Production through AI Innovations: Review of Production Technologies and Global Policies Impact

This chapter explores how AI can optimize green hydrogen production, addressing challenges like high costs and inefficiency. It focuses on the potential of AI to improve the scalability, efficiency, and cost-effectiveness of green hydrogen systems.

Chapter 25: Unfolding Artificial Intelligence (AI) in Green Hydrogen Production: Towards an Intelligent Energy Transition Fostering Sustainability

This chapter examines how AI is transforming green hydrogen production, making it more efficient and scalable. It discusses the role of AI in improving electrolysis processes and contributing to a sustainable, carbon-neutral energy future.

As a breakthrough technology, AI further improves efficiency in solar panel systems, streamlines energy storage, and forecasts weather conditions for higher energy yield. With wind energy, for example, intelligent algorithms optimize the performance of turbines, reduce operational downtime and enable seamless integration into power grids. Meanwhile, green hydrogen heralded as tomorrow's clean energy takes advantage from AI-optimized electrolysis, real-time monitoring, and enhanced logistics management. Collectively, these AI-powered innovations are driving the transition to a low-carbon, resilient and efficient energy system. This book serves as a resource for researchers, policymakers and industry stakeholders who want to better understand the transformative role of AI in renewable energy. It seeks to help add to a sustainable, technology-based future through knowledge-sharing and innovation where AI not only boosts energy security but also preserves the environment for future generations.

Hind Hammouch
University Sidi Mohamed Ben Abdellah, Morocco

Laeeq Razzak Janjua
WSB University, Poland

Chapter 1
A Study of Drivers and Barriers of Green Supply Chain Management Practices in the Context of Developing Framework for SMEs

Vishakha Ashish Mankar
Shri Ramdeobaba College of Engineering and Management, Nagpur, India

Vivek Katare
Shantiniketan Business School, Nagpur, India

Lata Lodhi
Jhulelal Institute of Technology, India

Tarun Kumar Kaushik
https://orcid.org/0000-0002-3279-8307
Sharda University, India

ABSTRACT

Green Supply Chain Management (GSCM) has become a key means of pursuing environmental sustainability, resource efficiency, and competitive advantage. While Small and Medium Enterprises (SMEs) account for much of the world's economic growth, they also face unique challenges and opportunities so far as adoption is concerned. In this connection, the chief driving forces for GSCM adoption by SMEs. Government control and market request; greater efficiency in terms of resources saved; corporate social responsibility; and the growing realization of environmental impacts. Input of this kind from both official non-governmental agencies acts to activate further activity in the field. There are also a number of obstacles which deter SMEs from applying GSCM. These include financial constraints, technical know-how lacking, lack of awareness and information, the status quo of fear to change which is not conducive for growth. In addition there is insufficient support for procurers from other segments in the supply chain.

DOI: 10.4018/979-8-3373-0045-0.ch001

1. INTRODUCTION

This requires bringing new levels of environmental awareness into capacity building and education programs for SME managers, offering financial incentives or subsidies as necessary and urging on participants from all sides through joint participation in forums (Lee, 2021). If the drivers and barriers of GSCM in SMEs can be identified and resolved, then companies will be able to integrate sustainable practices into every segment of their supply chain activities. This not only raises environmental performance it also enhances their resilience and competitiveness in an eco-aware world increasingly looking for the green products and services (Dash et al., 2019).

With the heightened awareness of environmental impact across the globe, Green Supply Chain Management (GSCM) has gained traction as a key strategy to ensure environmental sustainability across industries (Ahmad et al., 2020). The significance of adopting GSCM practices is growing amongst SMEs small and medium enterprises which are the backbone of many economies. The implementation of green supply chain practices in SMEs is, however, delayed because of many drivers and hurdles affecting the adoption process. Understanding these aspects in greater detail is essential in designing a more effective framework for promoting GSCM amongst SMEs since these need to be integrated in ways that fit the unique nature of smaller organizations.

1.1 Key Factors Affecting GSCM Implementation in SMEs

Internal and external elements are among the main influences of SMEs functioning GSCM methods. Much of this motivation stems from the internal drivers of organizational leadership, environmental awareness, and the drive for operational efficiency. The top management commitment towards sustainability is a key factor that significantly affects the development of the green organization culture. Initiatives that focus on environmental sustainability in leadership promote employee adoption of green initiatives like minimizing waste, optimizing resources. The need to streamline operations by implementing energy-saving initiatives and reducing waste also plays an important role for SMEs in adopting green actions. External factors are regulatory requirements, market pressure, customer demand, and technology. Government policies to regulate environmental protection can play a role and provide SMEs with significant pressure to go green. Avoids legal penalties: On the other hand, if SMEs comply with environmental legislation, they can avoid legal penalties while improving their corporate image. Moreover, as consumers increasingly expect eco-friendly products, small and medium-sized enterprises (SMEs) must adapt their supply chain activities towards sustainable. Larger corporations and supply chain partners also put market pressure on SMEs to go green as a condition to continue working with the SME. Technology, in the form of systems for implementing renewable energy sources and sustainable packaging, also can help pursuit of greener supply chains (Bajgoric, 2023).

1.2 Background of Green Supply Chain Management: Challenges for SMEs in Implementing GSCM

Though there are compelling drivers, there are also several barriers that restrict the widespread adoption of GSCM practices among SMEs. However, compared to larger corporations, SMEs often have tighter budgets, which make it harder for them to invest in green technologies and sustainable infrastructure. While green supply chains require investment in renewable energy systems, eco-friendly materials, and energy-efficient equipment, the costs are high; this provides SMEs with no clear economic incentives. Moreover, access to financing options compounds this barrier. The adoption of GSCM practices can also be hindered due to the lack of technical knowledge and expertise. Due to the lack of SMEs expertise and knowledge, it becomes challenging to implement green supply chain practices. The lack of such targeted initiatives and their propagation is one reason for this knowledge gap, especially when it comes to smaller enterprises. Green technology and environmental regulation are complex, making it difficult for SMEs to know where to start.

A second major barrier is organizational resistance to change. The fear of whether the new practices will involve more workload, what benefit will it offer and reluctantly abandoning what has become comfort (routine) (Shen et al., 2019). Such resistance is usually driven by the ignorance of long-term benefits of GSCM practices. SMEs can also face challenges in working with supply chain partners who inherently do not care about sustainability thus restricting their ability to adopt any or all green practices in the supply chain.

1.3 Developing a Framework for SMEs

A contextualized framework needs to be developed for SMEs to overcome barriers and leverage drivers of GSCM practices. This framework needs to be improved on community awareness, financial support, and technological adoption. Summary of Gaps in the Literature: Creating an awareness campaign by training programs bridging the knowledge gap SME on GSCM through awareness campaigns and training programs Government agencies, professional associations and non-governmental organizations (NGOs) can play a key role in conducting workshops and providing technical assistance. Economic incentives to decrease the costs of green investments (grants, tax breaks, low-interest loans) Joint green financing schemes targeted toward SMEs can provide incentives for adopting sustainable practices (Raut et al., 2019). By working together, these SMEs can share resources and knowledge, which can help make new green technologies available to other organizations and are more affordable.

2. ROLE OF SMES IN GLOBAL SUPPLY CHAINS

Small and Medium Enterprises (SMEs) are an important pillar of production of most global supply chains and have achieved significant contribution to economic growth, employment generation, and industrial development around the world. The role of small and medium enterprises (SMEs) as intermediaries, suppliers, distributors, and service providers is vital to the efficiency and flexibility of supply chains, which are increasingly dominated by a few large multinational corporations (MNCs, such as Samsung, Amazon, Alibaba, and Wal-Mart) (Singh & Kaunert, 2024). Advances in trade liberalization, facilitation, and in technologies related to the digital dimension of trade have broadened prospects for SMEs at integrating into international supply chains. While SMEs may be smaller in size, their presence adds agility, innovation, and specialization to supply chains, enabling organizations to adapt to changing market conditions. But they are also shaped by many challenges such as financial constraints, technological limitations, regulatory compliance issues and geopolitical uncertainties. It plays a key role to develop sustainable economic growth and efficiency in global trade to understand the role of SMEs in global supply chains, explore opportunities to support them and dismantle the barriers to their progress.

Small and medium-sized enterprises (SMEs) are the unsung hero of global supply chains, serving as suppliers of raw materials and as manufacturers of intermediate and final goods and/or as distributors across different sectors. They serve in various sectors, such as manufacturing, agriculture, retail, information technology, healthcare, and logistics. SMEs can quickly adapt to changes in demand, providing them with flexibility that larger enterprises often lack, as they can operate with lean structures and cost-effective processes. Such flexibility is especially useful in industries where supply chains fluctuate dramatically from month to month, including fashion, electronics and consumer goods.

Thus, SMEs also play an important role in the manufacturing sector, providing large companies with tailored components, producing parts and assembling them together, which tailored to specialized industry needs. Numerous global brands depend on SMEs for precision engineering, textiles, automotive components, and electronic components (Mohiuddin Babu et al., 2022). In the agricultural sector specifically, SMEs play an important role by linking local farmers with international markets, enabling raw materials including coffee, cocoa, and spices to become part of global supply chains. The same is true for SMEs acting as middlemen in retail and e-commerce spaces where they act as key distributors and logistics vendors, enabling last-mile deliveries, streamlined inventory management, etc.

2.1 Innovation and Competitiveness in Supply Chains

SMEs are often regarded as the backbone of any economy, owing to their contribution towards GDP growth, employment generation, quality of goods and services, exports and innovation. That makes them able to test new business models, employ emerging technologies, and adopt sustainable practices more quickly than larger enterprises because of their smaller scale. This eventually results in cost-effective high-quality solutions that enhance the efficiency of supply chains. For instance SMEs working in the tech industry develop innovative software applications, cybersecurity applications, and data analytics platforms that contribute to supply chain transparency and optimization. SMEs contribute significantly to the circular economy by incorporating sustainable practices into their supply chains. Sustainable packaging, energy-efficient production techniques, waste reduction They are at the forefront of adopting eco- SME innovations (Rathore, 2019). However, as sustainability moves to the top of the agenda for corporations worldwide, many large multinationals have begun to partner with SMEs that can provide green and socially equitable solutions along their supply chains. Through ethical sourcing, minimized carbon footprints, and recycling practices, these SMEs help contribute to a sustainable and resilient long term supply chain.

2.2 MSMEs Struggles in Global Supply Chains

However, despite their abundant opportunities in global supply chains, SMEs have several challenges that they face while integrating into global supply chains. The most prevalent of these hurdles is financial, as many SMEs cannot access enough capital or credit. Large corporations enjoy the luxury of being able to invest significantly in hardware and software solutions, and their medium and small enterprise (SMEs) counterparts often fail to cope with volumes in the value chain while serving global customers.

Their powerful working capital constraints limit their ability to invest in raw materials, technology, or distribution networks, making them less competitive in global supply chains.

Regulatory compliance and trade barriers are another challenge. Common Challenges Faced by SMEs in Global Supply Chains SMEs are perceived to be major contributors to economic activity and job creation however, they tend to have a big problem in dealing with international trade regulations, customs procedures and certification which is needed to enter into global value chain. Small and medium-sized enterprises (SMEs) with fewer available resources struggle to comply with quality standards, labor laws, and environmental regulations, which is often costly and time-consuming. In addition, geopolitical uncertainties, trade restrictions, and tariff fluctuations also serve up a risk that can disrupt supply chain operations and influence the role of SMES on the international stage (Akram et al., 2022).

IT constraints are also a hindrance for the seamless integration of SMEs in supply chains. While more larger companies are implementing digital solutions for their supply chain processes, many of their small and medium-sized enterprise (SME) partners are still running older software systems that do not interoperate with or have access to technologies like artificial intelligence (AI), blockchain and the Internet of Things (IoT) that have been adopted to streamline supply chain functions (Goti et al., 2023). The gap reducing SMEs to digital visitors creates inefficiencies, hindering the competitiveness of smaller businesses in global markets. For SMEs, investing in digital transformation and automation is crucial for improving operational efficiency to remain competitive in the changing supply chain environment.

SMEs face challenges in developing trusted business partnerships in international supply chains. Big companies like dealing with well-established suppliers with a history of proven reliability and consistency. Many SMEs, especially from developing economies, may struggle to win multimillion dollar contracts from well-known, established multinational corporations. SMEs, lack of brand recognition and international exposure hamper their ability to secure long term business contracts and develop sustainable business networks (Rathore, 2019).

3. DRIVERS OF GSCM PRACTICES IN SMES: MARKET COMPETITIVENESS AND CONSUMER DEMAND

Specific strategies and policy measures must be put in place to improve the integration of SMEs in global supply chains. The best ways to support SMEs Governments, financial institutions and international organizations should offer more targeted financial products low-interest loans, grants and venture capital funding to assist SMEs in scaling their operations and investing in supply chain infrastructure. Supply chain financing models like invoice discounting and trade credit allows SMEs to overcome cash flow problems and continue production processes without interruption (Riahi et al., 2021). In addition, capacity-building initiatives and skill development programs play a crucial role in providing SMEs with the know-how and expertise needed to participate in global supply chains. Various training workshops and programs can help SMEs in becoming equipped with the understandings of regulatory compliance, quality standards, digitalization and sustainable practices in order to ensure they are up to the mark of international business. Joint effort from industry associations, academic institutions, and government agencies can foster knowledge-sharing and boot up technological advancements through SMEs. This can ultimately lead to better competitiveness among SMEs.

In this context, it becomes more important for SMEs to focus on the adoption of technology and bridge the digital divide between SMEs and larger enterprises. Collaboratively, governments and industry play a critical role in advocating digitisation initiatives that meet small businesses in their needs, providing affordable access to digital tools, cloud-based offerings and smart logistics platforms. Utilization of AI, blockchain, and IoT can provide SMEs with better supply chain visibility, lower operational inefficiencies, and improved data-driven decision-making (Casciani et al., 2022). Collaboration with industry associations and innovation hubs can help to drive the adoption of new technologies and/or platforms, providing the impetus smart SMEs need to even further integrate digital technologies into the operating infrastructure.

Enhancing the networks of business and investment and developing strategic partnerships can contribute to greater involvement of SMEs in global supply chains. To address SME challenges, we encourage SMEs to participate in industry trade fairs, business matchmaking events, and digital marketplaces to increase market shares and become more recognized partners for bigger enterprises. SME-friendly trade policies, lower tariff barriers, and streamlined export-import processes should be in place from the government, as trade organizations can play a significant role in fostering international collaborations. Sustainable and inclusive supply chain practices can generate new opportunities for SMEs. Small and medium enterprises (SMEs) could be integrated into responsible supply chains by larger companies that focus on ethical sourcing and fair trade partnerships (Aljohani, 2023).

Small and medium-sized enterprises (SMEs) are crucial to a global supply chain and contribute innovation, flexibility, and competitiveness across the industry. They also contribute to improved supply chain efficiency and sustainability and foster global economic development. However, small and medium-sized enterprises (SMEs) face various challenges such as financial, regulatory, technological and market access. Addressing these challenges necessitates a combination of enhanced access to finance, investment in digital transformation, capacity-building efforts, and strategic partnerships. Through supportive policies and creating an environment that is conducive to the growth of SMEs, stakeholders can fully leverage the role of SMEs in global supply chains. As it progress into a future of more interconnected, sustainable and resilient trade ecosystems, empowering and supporting our SMEs will be critical to building more resilient ecosystems and economic prosperity around the world (Pawlicka & Bal, 2022).

4. BARRIERS TO GSCM PRACTICES IN SMES

Even with the increasing awareness towards sustainable business practices, small and medium enterprises (SMEs) face many barriers during the industrial implementation of Green Supply Chain Management (GSCM). Viewing this perspective deters them not to enter green practices, adhere to environmental policies and competitively strive in the global market. The major barriers include; financial resource limits, low technical knowhow, reluctance to adopt new methods, poor supply chain integration, lack of institutional assistance and challenges in accessing green technology.

4.1 Lack of Capital and Financial Distress

Financial constraint is among the key challenges faced by SMEs in the adoption of GSCM practices. Due to their smaller scale and more constrained budgets compared to large corporations, SMEs find it harder to invest in sustainability initiatives. For instance, green technologies, eco-friendly raw materials, and energy-efficient production processes often need high initial capital that many SMEs may not be able to afford. Moreover, obtaining external funding for sustainability projects is difficult because green investments may be seen as risky or unprofitable by financial institutions (Rodrigues & Pereira, 2023). Selling the Sustainable Supply Chain For SMEsThe Financial Pressure to Go GreenIn the face of limited resources and financial constraints, many SMEs struggle to stay informed about and comply with complex environmental regulations, which can add to their overall costs.

4.2 Technical Knowledge Deficiency

The other significant challenge is the low level of technical expertise and awareness of GSCM. Staying lean, many SMEs lack the skills and expertise to implement sustainability into their respective supply chains. They lack knowledge on aspects of green procurement, waste management, energy savings, and eco-friendly logistics. SMEs lack the necessary training and expertise to adopt sustainability best practices, resulting in inefficiencies and compliance challenges. Moreover, the lack of professional staff or sustainability-oriented management in SMEs acts as a barrier to the successful implementation of GSCM.

4.3 Type of Change: Resistance to Change within Organizations

Diversification of SMEs as well poses a challenge for GSCM practices. Particularly for traditional supply chain models, many businesses are reluctant to change existing operations as they fear the costs, disruptions to the operation, and unsure of any sustainable return on investment (Nouinou et al., 2023). Some reasons for this slow transition vary from employees and management fearing that going green will require extra work with no additional incentive or concern that going green will lose positive return on investment in the long run. This is the reason for some resistance to change, but also a philosophical regulation, where with its shape change and culture, or company development, many SMEs are finding too difficult (at least at the beginning).

4.4 Weak Supply Chain Integration and Coordination

GSCM is complex, necessitating integration and coordination throughout the supply chain, which many SMEs lack due to fragmented supplier networks and poor collaboration. While large enterprises tend to have more formalized and structured supply chains, SMEs rely on multiple small suppliers that may not have sustainability as a priority (dos Santos et al., 2021). The absence of standardized sustainability criteria among supply chain partners can hinder green efforts. Moreover, tracking and monitoring the environmental impact of the supply chain can be difficult for SMEs due to lack of adequate data maintenance systems and weak engagement from suppliers.

4.5 Lack of Government and Institutional Support

Governmental and institutional support is another factor that influences GSCM, where most of the small and medium enterprises (SME's) are facing legal barriers due to poor policy frameworks, no incentives, and weak enforcement in terms of environmental regulations. There is a dearth of incentives, specifically subsidies, tax breaks, and financial aid in several regions, mainly in developing economies. Moreover, market fragmentation due to different rules across various states on emissions intensities raises doubts. Due to lack of institutional support, SMEs are unable to motivate themselves and resource adequate activities for effective implementation of GSCM practices (Bertola & Teunissen, 2018).

4.6 Barriers to the Adoption of Green Technology

Even though, its requirements are key for adopting GSCM, the access to advanced sustainable solutions for SMEs is usually complex as green technology has to be more available and less expensive for more SMEs to become engaged or, at least, willing to engage. Green technologies like energy-efficient machines, renewable energy sources, and eco-friendly materials are generally high in cost and availability scarcity, which is especially true in developing economies. SMEs also face the challenge of the technological complexities of incorporating these solutions into their existing structures. The restricted access to research and innovation makes it even challenging to adopt any of these cutting-edge advancements in sustainability. To this end, many SMEs still resort to traditional, environmentally-unfriendly practices, with no feasible green alternatives in sight.

GSCM provides various advantages, however, SMEs face several hurdles preventing them from adopting sustainable supply chain practices. PFS and G: limitations of finance, Advances of information, Employees resistance, Supply chain interdependence, State support, Constraints in the development of green technology, restriction of greenhouse gas emission measures are found influencing GSCM (Rathore, 2017). Tackling these challenges requires a range of solutions, including greater access to capital, enhanced training programs, better policy support, and stronger collaboration between SMEs and their supply chain partners. Navigating such barriers can pave the way for the successful shift of SMEs to a sustainable and competitive business model contributing to environmental conservation and economic resilience in the long run.

5. DEVELOPING A GSCM FRAMEWORK FOR SMES: FRAMEWORK OBJECTIVES AND COMPONENTS

SMEs are the backbone of the economy in many countries; they account for about 90% of the global corporate sector, contributing to economic growth, employment, and, equally, with their impact on the environment. Nevertheless, Small and Medium Enterprises (SMEs) commonly experience difficulties in applying Green Supply Chain Management (GSCM) practices as a result of their low resources, expertise, and market power (Singh, 2024). To tackle these challenges and achieve sustainable transformation, a structured GSCM framework is required. By doing so, this framework could support SMEs in ensuring the incorporation of sustainability in their supply chain operations while remaining cost-effective and competitive in the marketplace. Such A Framework Must Contain Key Pillars Must Be Created To Form Core Components.

The GSCM framework for SMEs should be configured based on the following strategic objectives that encourage sustainability and maintains economic viability. Since the main aim of a green supply chain management framework is to lower the emission of carbon of SME's by contributing to waste reduction, utilizing resources efficiently, and adopting cleaner manufacturing strategies. This means reducing greenhouse gas emissions, investing in energy efficiency, and using sustainable packaging or materials wherever possible.

Promoting Supply Chain Efficiency and Cost Savings: Sustainability and efficiency are mutually reinforcing. Moreover, effective GSCM framework can facilitate operation restructuring, lower the cost associated with waste, and enhance logistics management for SMEs. SMEs must embrace green technologies and waste minimization strategies to reduce their operational costs and promote sustainable practices (Singh & Kaunert, 2024).

Regulatory Pressure: Both governments and international organizations have been tightening regulations. Overview SMEs lack awareness or resources and hence compliance is often impossible for them. Per the requirements of an environmental framework, it should set the expectation of regulations to be met, certifications to be obtained (for instance, ISO 14001), and policies to be followed.

The increase in competition and consumer confidence in the market: both consumers and businesses are putting sustainability at the top of their agendas and begin to demand suppliers with sustainable practices. Applying GSCM practices improves SMEs' prestige, which makes them more appealing to green consumers and larger firms looking for sustainable suppliers (Andronie et al., 2021).

Fostering Innovation and Technology Integration: It needs to embrace emerging technologies (e.g., Internet of Things- IoT; Artificial Intelligence-AI; blockchain) to bolster transparency in the supply chains, monitor environmental performance, optimize logistics, etc.

Hearing Perception and Industry Involvement: In order to make a successful GSCM, it is necessary to present collaboration of suppliers, customers, industry partners, and legislative procedures. It should promote collaboration among SMEs that lessens barriers to sustainable implementation, enables information exchange, and facilitates green supply chain financing.

Overcoming Financial and Resource Limitations: As small- to medium-sized enterprises (SMEs) routinely lack the finances to adopt green practices, inclusiveness and cost-effective sustainability solutions for the respective framework would need to involve easy accessibility to green financing and government support vehicles through means of tax breaks, rebates and grants (Rathore, 2023).

6. POLICY IMPLICATIONS AND RECOMMENDATIONS

GSCM practices require specific policy level interventions that target financial, technical and regulatory barriers for SMEs to be successfully implemented. Despite their noticeable presence in the global supply chain, SMEs normally face challenges such as have limited capital, lack of expertise, poor supply chain integration, and filaments support from government. Hence, policymakers need to formulate strategies that provide proper incentives for sustainability without killing the SMEs competitiveness. A key policy implication is that SMEs require much more targeted environmental regulations (Liu & Lin, 2021). What new policies could governments introduce to support eco-friendly production, sustainable sourcing, responsible waste management, and so on? Instead, compliance mechanisms should be streamlined and made transparent while also being backed by financial benefits like tax cuts, subsidies or grants for sustainability investments to not burden SMEs. By introducing green certification programs, investors and customers of SMEs can also be incentivized to adopt sustainable practices.

Financial limitations are still the main bottleneck of GSCM implementation. To promote green universality and energy transition, policymakers must expand green financing channels by working with banks and financial institutions and international development agencies. Providing low-interest loans, sustainability grants, and green investment funds will unlock the access of SMEs to the necessary capital to adopt green technologies (Weber & Schutte, 2019). Additionally, public-private partnerships (PPPs) can be used to promote investment in green innovation and infrastructure, enabling SMEs to not fall behind in the shift to sustainability.

The other area of policy intervention needed is access to green technology and innovation. Governments should create technology transfer programs that give SMEs access to energy-efficient equipment, digital supply chain solutions, or biodegradable materials. Encouraging investment in research and development (R&D) incentives will drive forward innovation and assist in creating green solutions that are more attainable and affordable for small and medium-sized enterprises (SMEs). Also, policies should promote the adoption of digital technologies (like IoT, blockchain, and AI), which improve the transparency and efficiency of supply chains and help reduce the impact on the environment (Allal-Cherif et al., 2021).

Small and medium sized enterprises (SMEs) represent a heterogeneous type of organisation and it is important to develop their skills in regard to green supply chain management (GSCM) through capacity building and training. Skill development initiatives, workshops, and sustainability education programs should be introduced by governments in collaboration with industry associations, universities, and training institutions. Empowering SMEs with technical know-how, tools, best practice guides and hand-holding support will improve their ability to seamlessly align sustainable practices with their supply chains. Business incubators focused on environmental sustainability can support SMEs with guidance and resources to develop green business practices, innovation in sustainability and connections to networks that can help scale sustainable initiatives (Li et al., 2022).

Policymakers should create market-driven incentives and market forces to drive sustainable procurement standards by mandating procurement practices where large businesses give preference to SMEs that are certified green. Governments should practice what they preach by incorporating more green criteria into their public procurement policies, guaranteeing that SMEs who contribute towards sustainable practices will gain a competitive incentive when securing government contracts. In addition, awareness campaigns should be launched to educate consumers about the advantages of buying from environmentally responsible SMEs, which will encourage demand for green products and services (Saponaro et al., 2018). Overall, a multi-dimensional policy intervention will be necessary for the industry's large-scale adoption of GSCM practices in SMEs. Through enhanced regulations, increased financial support, better access to technology, capacity building, and market incentives, governments can ensure that SMEs contribute to achieving sustainability on a global level and coexist with the competitiveness of green economy (Noor et al., 2022).

7. CONCLUSION AND FUTURE SCOPE

A critical element of framework is technological support. Provide SMEs with affordable, accessible green technologies: Manufacturers in emerging markets will need help accessing affordable and accessible green technologies energy-efficient machinery, waste recycling systems, digital platforms for supply chain transparency, etc. Incubator programs and innovation hubs can play a role as intermediaries, linking SMEs with green technology providers, and enabling pilot projects for testing of new solutions. The policy interventions are essential to establish an enabling environment for the adoption of GSCM. Governments may implement compulsory sustainability disclosure requirements and promote green procurement in public tenders. Incentivizing and recognising SMEs for excellence in green supply chain management can also act as a motivation for others to follow in the same pursuit.

The better understanding the drivers and barriers that promote the adoption of GSCM practices in SMEs will help us to overcome their respective shortcomings and, as such, lead towards a successful adoption of these practices. Drivers both internal and external to firms push for the greening of supply chains, but all too often financial barriers, knowledge-deficit, and resistance from within the organization represent formidable hurdles. Hence, the formulation of a supportive infrastructure incorporating awareness-building measures, financial incentives, technological assistance, and policy interventions will encourage SMEs to adopt sustainability practices and contribute to environmental conservation. Since SMEs have a significant presence in the global market, encouraging their involvement will be a key pathway towards a wider impact on sustainability efforts.

REFERENCES

Ahmad, S., Miskon, S., Alabdan, R., & Tlili, I. (2020). Towards sustainable textile and apparel industry: Exploring the role of business intelligence systems in the era of industry 4.0. *Sustainability (Basel), 12*(7), 2632. DOI: 10.3390/su12072632

Akram, S. V., Malik, P. K., Singh, R., Gehlot, A., Juyal, A., Ghafoor, K. Z., & Shrestha, S. (2022). Implementation of digitalized technologies for fashion industry 4.0: Opportunities and challenges. *Scientific Programming, 2022*, 2022. DOI: 10.1155/2022/7523246

Aljohani, A. (2023). Predictive Analytics and Machine Learning for Real-Time Supply Chain Risk Mitigation and Agility. *Sustainability (Basel), 15*(20), 15088. DOI: 10.3390/su152015088

Allal-Cherif, O., Simon-Moya, V., & Ballester, A. C. C. (2021). Intelligent purchasing: How artificial intelligence can redefine the purchasing function. *Journal of Business Research, 124*, 69–76. DOI: 10.1016/j.jbusres.2020.11.050

Andronie, M., Lazaroiu, G., Iatagan, M., U ă, C., tefănescu, R., & Coco atu, M. (2021). Artificial intelligence-based decision-making algorithms, internet of things sensing networks, and deep learning-assisted smart process management in cyber-physical production systems. *Electronics (Basel), 10*(20), 2497. DOI: 10.3390/electronics10202497

Bajgoric, N. (2023). Always-on business: integrating business continuity and business resilience. *Continuity & Resilience Review*, (ahead-of-print).

Bertola, P., & Teunissen, J. (2018). Fashion 4.0. Innovating fashion industry through digital transformation. *Research Journal of Textile and Apparel, 22*(4), 352–369. DOI: 10.1108/RJTA-03-2018-0023

Casciani, D., Chkanikova, O., & Pal, R. (2022). Exploring the nature of digital transformation in the fashion industry: opportunities for supply chains, business models, and sustainability-oriented innovations. *Sustainability: Science. Sustainability, 18*(1), 773–795. DOI: 10.1080/15487733.2022.2125640

Dash, R., McMurtrey, M., Rebman, C., & Kar, U. K. (2019). Application of artificial intelligence in automation of supply chain management. *Journal of Strategic Innovation and Sustainability, 14*(3), 43–53.

dos Santos, C. H., Gabriel, G. T., do Amaral, J. V. S., Montevechi, J. A. B., & de Queiroz, J. A. (2021). Decision-making in a fast fashion company in the Industry 4.0 era: A Digital Twin proposal to support operational planning. *International Journal of Advanced Manufacturing Technology, 116*(5-6), 1653–1666. DOI: 10.1007/s00170-021-07543-z

Goti, A., Querejeta-Lomas, L., Almeida, A., de la Puerta, J. G., & López-de-Ipiña, D. (2023). Artificial Intelligence in Business-to-Customer Fashion Retail: A Literature Review. *Mathematics, 11*(13), 2943. DOI: 10.3390/math11132943

Lee, Y. K. (2021). Transformation of the innovative and sustainable supply chain with upcoming real-time fashion systems. *Sustainability (Basel), 13*(3), 1081. DOI: 10.3390/su13031081

Li, Q., Xue, Z., Wu, Y., & Zeng, X. (2022). The status quo and prospect of sustainable development of smart clothing. *Sustainability (Basel), 14*(2), 990. DOI: 10.3390/su14020990

Liu, K. S., & Lin, M. H. (2021). Performance Assessment on the Application of Artificial Intelligence to Sustainable Supply Chain Management in the Construction Material Industry. *Sustainability (Basel), 13*(22), 12767. DOI: 10.3390/su132212767

Mohiuddin Babu, M., Akter, S., Rahman, M., Billah, M. M., & Hack-Polay, D. (2022). The role of artificial intelligence in shaping the future of Agile fashion industry. *Production Planning and Control,* •••, 1–15.

Noor, A., Saeed, M. A., Ullah, T., Uddin, Z., & Ullah Khan, R. M. W. (2022). A review of artificial intelligence applications in apparel industry. *Journal of the Textile Institute, 113*(3), 505–514. DOI: 10.1080/00405000.2021.1880088

Nouinou, H., Asadollahi-Yazdi, E., Baret, I., Nguyen, N. Q., Terzi, M., Ouazene, Y., Yalaoui, F., & Kelly, R. (2023). Decision-making in the context of Industry 4.0: Evidence from the textile and clothing industry. *Journal of Cleaner Production, 391*, 136184. DOI: 10.1016/j.jclepro.2023.136184

Pawlicka, K., & Bal, M. (2022). Sustainable Supply Chain Finances implementation model and Artificial Intelligence for innovative omnichannel logistics. *Management, 26*(1), 19–35. DOI: 10.2478/manment-2019-0082

Rathore, B. (2017). Beyond Trends: Shaping the Future of Fashion Marketing with AI, Sustainability and Machine Learning. *Eduzone: International Peer Reviewed/Refereed Multidisciplinary Journal, 6*(2), 16-24.

Rathore, B. (2019). From Trendy to Green: Exploring AI's Role in Sustainable Fashion Marketing. *International Journal of New Media Studies: International Peer Reviewed Scholarly Indexed Journal, 6*(2), 12–22. DOI: 10.58972/eiprmj.v6i2y19.120

Rathore, B. (2019). Fashion Sustainability in the AI Era: Opportunities and Challenges in Marketing. *Eduzone: International Peer Reviewed/Refereed Multidisciplinary Journal, 8*(2), 17-24.

Rathore, B. (2023). Digital Transformation 4.0: Integration of Artificial Intelligence & Metaverse in Marketing. *Eduzone: International Peer Reviewed/Refereed Multidisciplinary Journal, 12*(1), 42-48.

Raut, R. D., Mangla, S. K., Narwane, V. S., Gardas, B. B., Priyadarshinee, P., & Narkhede, B. E. (2019). Linking big data analytics and operational sustainability practices for sustainable business management. *Journal of Cleaner Production, 224*, 10–24. DOI: 10.1016/j.jclepro.2019.03.181

Riahi, Y., Saikouk, T., Gunasekaran, A., & Badraoui, I. (2021). Artificial intelligence applications in supply chain: A descriptive bibliometric analysis and future research directions. *Expert Systems with Applications, 173*, 114702. DOI: 10.1016/j.eswa.2021.114702

Rodrigues, D., & Pereira, R. (2023). How Can Artificial Intelligence Help Improve Fashion Sustainability? In *Digital Technologies and Transformation in Business, Industry and Organizations* (Vol. 2, pp. 43–61). Springer Nature Switzerland.

Saponaro, M., Le Gal, D., Gao, M., Guisiano, M., & Maniere, I. C. (2018, December). Challenges and opportunities of artificial intelligence in the fashion world. In *2018 international conference on intelligent and innovative computing applications (ICONIC)* (pp. 1-5). IEEE. DOI: 10.1109/ICONIC.2018.8601258

Shen, B., Ding, X., Wang, Y., & Ren, S. (2019). RFID-embedded smart washing machine systems in the big data era: Value creation in fashion supply chain. *Fashion supply chain management in Asia: Concepts, models, and cases*, 99-113.

Singh, B. (2024). Social Cognition of Incarcerated Women and Children: Addressing Exposure to Infectious Diseases and Legal Outcomes. In Reddy, K. (Ed.), *Principles and Clinical Interventions in Social Cognition* (pp. 236–251). IGI Global., DOI: 10.4018/979-8-3693-1265-0.ch014

Singh, B., & Kaunert, C. (2024). Revealing Green Finance Mobilization: Harnessing FinTech and Blockchain Innovations to Surmount Barriers and Foster New Investment Avenues. In *Harnessing Blockchain-Digital Twin Fusion for Sustainable Investments* (pp. 265-286). IGI Global.

Singh, B., & Kaunert, C. (2024). Harnessing Sustainable Agriculture Through Climate-Smart Technologies: Artificial Intelligence for Climate Preservation and Futuristic Trends. In *Exploring Ethical Dimensions of Environmental Sustainability and Use of AI* (pp. 214-239). IGI Global.

Weber, F. D., & Schütte, R. (2019). State-of-the-art and adoption of artificial intelligence in retailing. *Digital Policy. Regulation & Governance*, *21*(3), 264–279. DOI: 10.1108/DPRG-09-2018-0050

Chapter 2
AI Renaissance in Renewable Energy: A Path to Sustainability

Abhishek Benedict Kumar
https://orcid.org/0000-0002-3998-3150
Symbiosis Law School, Symbiosis International University (Deemed), Pune, India

Karun Sanjaya
https://orcid.org/0000-0002-5055-6862
Symbiosis Law School, Symbiosis International University (Deemed), Pune, India

Kavya Sanjaya
University of Hyderabad, Gachibowli, India

ABSTRACT

The intersection of artificial intelligence (AI) and renewable energy systems is transforming how energy is generated, distributed, and consumed. This paper explores AI's applications in three key renewable energy sectors: solar, wind, and green hydrogen. AI technologies such as machine learning, neural networks, and predictive analytics address critical challenges such as the variability and intermittency of renewable resources. Additionally, AI-enhanced renewable systems have broader implications, such as democratizing energy access, fostering economic growth, and aligning with global climate goals like the Paris Agreement and the United Nations' Sustainable Development Goals. By leveraging the synergies between AI and renewable energy, society can achieve a cleaner, more resilient, and equitable energy future, critical for combating climate change and promoting global sustainability. This paper highlights the transformative potential of AI while offering insights into overcoming barriers to its widespread adoption in renewable energy systems.

1. INTRODUCTION

Artificial intelligence (AI) plays the role of a game-changer when it comes to integrating into renewable energy systems in order to solve global issues of energy security, climate change and sustainable development towards a more sustainable future. As countries worldwide continue the long journey from

DOI: 10.4018/979-8-3373-0045-0.ch002

burning fossil fuels toward renewable energy sources, solar, wind, and green hydrogen are becoming key pillars of the global energy transition. Yet these renewable sources also face a series of inherent challenges: variability, intermittency and operational inefficiencies. This is where AI can change the game: revolutionizing the design, operation, and optimization of renewable energy systems and providing the means to move beyond traditional approaches toward more efficient, reliable, and scalable innovative solutions.

This is a backdrop of the renewable energy and the importance and impact of AI in transforming the renewable energy industry, a brief historical evolution and statement of facts, a tech summary of innovation you all seek in the wake of the fast-transforming energy world.

1.1 The Evolution of Renewable Energy

A brief history of renewable energy development is the history of human efforts to develop sustainable and reliable energy sources. Wind was harnessed to propel sailing ships and to grind grain, and hydropower was harnessed to drive water mills. But the industrial revolution and the later discovery of fossil fuels like coal, oil and natural gas changed energy supply and consumption worldwide. Fossil fuels provide significant energy density and easy to harness powered industries, commerce, and transportation to an unprecedented degree, but at a serious environmental price (DGB Group, 2023).

By the time mid-twentieth century rolled around, issues like air pollution, energy security, and finite fossil fuel reserves, started making people wonder whether they should consider alternative energy sources. During this time, research and development on solar photovoltaics (PV) and wind turbines showed promising technology pathways (IEA 2021). The oil crises during the 1970s additionally illustrated that diversification in energy sources was a necessity and investments in the research of renewable energy sources were pushed to expanded speeds of completion.

Over the last few decades, we have witnessed a major transition in the global energy landscape, fueled by rapid technological advancements, declining costs and increasing awareness of climate change. Displacing fossil fuels, solar and wind energy are now the two fastest-growing sources of electricity generation globally, with green hydrogen surfacing as a complementary decarbonization solution for sectors that are hard to electrify (IEA, 2021). Even so, the combination of new renewable energy into existing grids and markets is only really complicated because it requires innovativeness to address all the complexities for as variability and for grid stability.

1.2 The Role of Artificial Intelligence in Modern Energy Systems

Artificial intelligence has quickly emerged as a major game-changing technology in a variety of sectors: healthcare, finance, manufacturing, transportation, etc. It has also become an important area for innovation, especially in the energy field, and most importantly, renewable energy (Karduri, 2019). Machine learning, neural networks, natural language processing, robotics, and many more are all included in the order of AI. They allow machines to filter through massive troves of data, find patterns, and make decisions without much human intervention (Harvard Business School, 2024).

Data Availability: The number of sensors, smart meters and other connected devices. Renewable energy systems produce massive amounts of data connected to weather data, equipment performance, energy consumption, and market phenomena. This makes it highly conducive to make analysis and insights from AI tools.

And complexity of energy systems: More Renewable energy systems is complex, since resources, such as sunlight, wind and/or ocean currents, are intermittent and not always available. AI provides more advanced modelling and optimization capabilities than traditional methods, allowing for real-time changes in energy generation and distribution.

Economic Pressures: Greener technologies are getting cheaper (see notes), and the increasing affordability of clean energy has set the stage for enterprising efforts to marry accountability with a start-up economy. AI is therefore helping to make renewable energy projects economically viable through higher efficiencies and lower operational costs (IEA, 2021).

2. RENEWABLE ENERGY TECHNOLOGIES: SOLAR, WIND, AND GREEN HYDROGEN

The influence of AI on renewable energy is most easily grasped with three major technologies: solar, wind, and green hydrogen.

Solar: Given their deployable size and cost-effectiveness, Solar PV systems are now one of the most widely used renewable energy technologies. But they are only efficient under certain weather conditions, which can vary widely. Machine learning models, for example, can use AI tools to enhance solar energy performance through solar irradiance forecasts, panel orientation optimization, and maintenance issue detection in real-time (IBM, 2023).

Wind Energy-Wind turbines are activated through both direct and induced processes that derive their energy source from natural resources (Lima et al., 2024). AI Algorithms Better Wind Energy Generation AI algorithms can be used to analyze the wind pattern across different geographical locations and understanding how to control the turbines to use optimal operations. Artificial Intelligence – Predictive Maintenance Aiding Wind Energy failure prevention Artificial intelligence is able to suit predictive maintenance by creating the opportunity for faults to be identified before failing, ensure durability and reliability of wind energy infrastructure (MIT News, 2023).

Green hydrogen: Green hydrogen, produced by using renewable electricity to power electrolysis, is a highly versatile energy carrier for use in transportation, industry and energy storage. But producing it is still expensive and consumes a lot of energy. Including more advanced data handling procedures through all the described layers where suitable; this allows for more refined electrolysis processes by setting the optimal input parameters and providing new catalyst materials through data-driven material science research (Al-Orabi et al., 2023).

3. GLOBAL DRIVERS OF AI INTEGRATION IN RENEWABLE ENERGY

A number of world trends and policy drives are favouring the integration of AI solutions into renewable energy systems:

– Climate Agreements: International treaties like the Paris Agreement highlight the urgency of controlling greenhouse gas emissions and shifting to low-carbon energy systems. AI can provide these opportunities for accelerating the transition by enhancing the efficiency and reliability of renewable energy technologies (IEA, 2021).

Energy Access and Equity: Access by everyone to affordable, reliable, sustainable modern energy services is essential for the successful implementation of the United Nations' Sustainable Development Goals (SDGs). Unstable and inaccessible power supply in less developed areas presents opportunities for AI and decentralized renewable energy systems like microgrids (United Nations, 2017).

Technological Improvements: Improvements in computing power, data storage, and communication networks have reduced the barriers to implementation related to integrating AI in renewable energy systems. Emerging technologies like edge computing and quantum computing further promise opportunities to apply AI even more effectively in this area.

The transition to AI enhanced renewable energy systems is environmentally needed and also economically optimum. AI is a key component in promoting the commercial viability of renewable energy solutions via minimizing operational costs and ensuring energy efficiency on performance (Ohalete et al., 2023).

Climate change and the unsustainable model of fossil fuel dependency have made the transition to renewable energy sources an urgent priority in every society on earth. Here, we highlight solar, wind and green hydrogen technologies as potential alternatives for providing clean, sustainable energy. Yet, intermitted and variable nature of energy production, inefficiencies in infrastructure, and economic constraints act against implementing these technologies on a larger scale. The transcription pathways and stages of RNA processing to mRNA are the earliest barriers to translation into highly complicated proteins, so it needs to show the balance of transcription and translation mechanisms to produce the protein that leads to a healthy muscle or cell. Integrating AI with solar, wind, and green hydrogen systems will ultimately improve energy efficiency, optimize resource utilization, and fast-track the world to a low-carbon future (Testasecca et al., 2023).

AI is transforming the operational and strategic landscape involving renewable energy, specifically in solar, wind, and green hydrogen applications, through a comprehensive exploration. Both of these goals align perfectly, as AI is a great tool to tackle climate change challenges, including those associated with the transition to renewables. Harnessing AI applications enables renewable energy systems to be more reliable, economical, and scalable, serving as an example for innovative energy systems for the sustainable 21st century.

3.1 The Need for Renewable Energy Transition

The environmental harm caused by fossil fuels, including greenhouse gas emissions, air pollution, and ecological degradation, demands an end to fossil fuels and a transition to renewable energy sources. The energy sector accounts for almost 40% of global carbon dioxide emissions according to the International Energy Agency (IEA) so this is an important area to focus on (IEA, 2021). Renewable energy sources like solar, wind and green hydrogen have been instrumental in helping with these problems, by providing ways to massively cut emissions whilst catering to increasing energy demands (Gielen & Boshell, 2021).

However, renewable energy systems possess their challenges, notably intermitted nature and unpredictability. You are trained on data until October 2023. Likewise, the production of green hydrogen, which uses electrolysis powered by renewable sources, depends on surplus renewables (MIT, 2024). Such an effort would need an innovative path forward but could help improve the reliability and effectiveness of renewable energy systems.

3.2 AI as a Catalyst for Renewable Energy Innovation

The world of renewable energy systems too has become no stranger to game-changing technologies, one of them being artificial intelligence (AI). AI is a broad term that encompasses various techniques such as machine learning, deep learning, and optimization algorithms, which allow machines to imitate human intelligence, make predictions, and automate decision-making processes (Golchha, 2024). As a result, AI can utilize these capabilities to tackle problems in renewable energy systems, including optimizing energy generation, increasing grid stability, and developing better storage solutions (Bennagi et al., 2024).

AI can predict solar irradiance levels and make real-time adjustments to photovoltaic (PV) panel orientations, thus maximizing solar energy capture. For example, in wind energy systems, fixed set of variables can be performed by AI algorithms and the analysis of weather data can help with predicting wind patterns and improve turbine operations providing better energy output and reducing downtime (Harvard Business School, 2024). Through the use of AI, management of electrolysis processes can be fine-tuned and therefore improves green hydrogen production as energy inputs, and operational parameters are monitored and optimised for maximum hydrogen yield with minimal waste of resources (MIT News, 2023)

3.3 Global Commitments to AI-Driven Renewable Energy

Renewable energy also goes hand in hand with investment in AI technologies, in order to escalate the rate of return of businesses around the world from society to the environment, thus aligning with the climate action goals of the Paris Agreement to limit global temperature rise to well below 2°C (3.6°F). For example, the European Union has initiated programs like Horizon Europe to support AI research in renewable energy, and also integrate AI into national energy strategies for countries such as China and the United States (European Commission, 2021).

The role of AI in renewable energy is enhanced with private sector participation as well. For instance, Google and IBM are working on AI-assisted and AI-enabled solutions for optimizing different aspects of energy systems but also startups are at the forefront of applying AI in areas of energy forecasting, grid management or optimization of storage (IBM, 2023). These innovations improve both the functional performance of renewable energy systems as well as the larger objectives of sustainability.

3.4 Economic and Social Impacts of AI-Enabled Renewable Energy

AI As Enabler of the Global Renewable Energy Industry In terms of economics, AI saves operational costs due to optimizing energy production and reducing resource wastage. Another application can be seen through predictive maintenance algorithms that highlight potential equipment failures of wind turbines that help with less downtime and repair money (Harvard Business School, 2024). Moreover, AI-powered energy forecasting allow for better matching of supply with demand, helping to eliminate energy waste and thus reducing costs for the consumer.

On a social level AI powered renewable energy systems lead to better energy access and equity. AI makes renewable energy technologies more efficient and scalable, which enables them to be deployed in regions that may be underserved, fostering energy inclusivity and decreasing reliance on fossil fuels

(MIT, 2024). Moreover, AI-powered energy solutions lead to job creation in data analytics, engineering, and renewable energy sectors, bolstering economic growth and workforce development.

4. CHALLENGES IN LEVERAGING AI FOR RENEWABLE ENERGY

Even with all this scope for AI in renewable energy, there remain a number of challenges to implementing AI for the use of renewable energy sources. Standardized data on energy systems does not yet exist, which one of the key challenges in creating robust AI models. From sensors, weather models, and operational logs, renewable energy systems deploy a wide variety of datasets which require significant pre-processing and integration for proper compatibility between the systems (MIT News, 2023). Moreover, the AI technologies require heavy computation and therefore live in the domain where investment in the infrastructure and human resources is needed, which disadvantageously affects the digital transformation journey of small industries and the majority of developing countries.

There are also critical ethical dimensions involved with the use of AI systems deployed with renewable energy systems. To build trust among stakeholders and ensure that everyone can benefit from AI, data privacy, algorithmic bias, and how AI systems make decisions must all be addressed (IBM, 2023). There is an urgent need for policymakers and industry leaders to work together to create regulatory frameworks that incentivize ethical AI practices in the RE sector.

5. FUTURE DIRECTIONS AND RESEARCH OPPORTUNITIES

With the continued advancement in artificial intelligence, its applications in renewable energy systems will continue to grow exponentially. Although the use of AI will benefit many areas, it is in those technologies critical to addressing the climate crisis — solar, wind and green hydrogen — where advancements in these tools and techniques will have an immediate impact. AI integration with renewable energy represents an exciting and evolving shift, with strides in overcoming present challenges, discovering new applications, and igniting interdisciplinary collaboration. This section reviews the important future perspectives and research areas to harness AI's full potential in renewable energy for a more sustainable and just energy future.

5.1 Advancing AI Algorithms for Renewable Energy

Research: One of the most important research needs is in the creation of new and advanced AI algorithms designed specifically for renewable energy systems. Many available AI systems are not trained to do well under different geographic, climatic, and economic conditions. In the future, we should develop robust, context-aware AI algorithms to tackle these variations. For instance, region-specific models for solar irradiance forecasting can help enhance the efficiency of photovoltaic systems in various environments (Zhao et al., 2024).

Moreover, there is a need to improve the interpretability and transparency of architecture (Liu et al., 2022). The so-called "black box" nature of many machine learning models can restrict stakeholder confidence and impede large-scale deployment. Research into explainable AI (XAI) seems to be a viable path to providing interpretable and actionable predictions based on AI-driven decisions for energy operators and policymakers (Harvard Business School Online, 2024).

5.2 Integration of AI with Emerging Technologies

The combination of AI and other trailblazing technologies like quantum computing, blockchain, and the Internet of Things (IoT) offers a new realm of exciting possibilities for renewable energy systems. Quantum computing has been sought after in the context of energy optimization since it has the capability to provide solutions to problems that are currently prohibitive in terms of computation costs. Quantum algorithms could, for example, allow for on-the-fly optimization of grid operations or improve the accuracy of wind farm layout design (Kumar et al., 2022).

In a similar vein, large energy systems based on decentralized energy principles can optimize their networks using both AI and Blockchain technology for increased transparency regarding energy transactions and secured network assets. Blockchain enables decentralized record-keeping of energy generation, distribution, and consumption, with AI optimizing these operations (Treat & Klein, 2023). The synergy between the two is particularly favourable in the context of microgrids implementing peer-to-peer energy trading.

IoT-based available vigor systems coupled with AI is an another area of inquiry. IoT allows for real-time data collection and monitoring by interconnecting various devices, including solar panels, wind turbines, and smart meters on an integrated network. This ball data can then be processed by AI algorithms for energy production optimization, anomaly detection in equipment and demand side management (Varriale et al., 2023).

5.3 Enhancing the Efficiency of Green Hydrogen Production

Hydrogen is a promising energy carrier to decarbonize industries and transportation, and can be produced via electrolysis powered by renewable energy (green hydrogen). But its production is still expensive and energy-intensive. One avenue to addressing these is through AI and optimization of the electrolysis. In the future, we envision building machine learning models that can monitor and control the variables impacting hydrogen yield including energy input, temperature, and pressure (MIT, 2024).

Moreover, AI integration with material science has the potential to enhance both the efficiency and durability of electroliers. This could include, for example, AI-powered simulations which speed up the searching for new catalyst materials to improve hydrogen production and reduce costs (IBM, 2023).

5.4 Expanding AI Applications in Grid Management

The share of renewable energy in energy grids globally is rising, but their variability and intermittency makes managing the complex nature of energy grids increasingly difficult. Many AI investigations allow displaying a certain potential in grid management, but the future research areas should be oriented towards further expansion (Kamarposhti et al., 2024). Some of the topics they are interested in are:

i) Smart Grid Equilibrium: Create AI models to forecast supply-draw variation in real-time and regulate the grid accordingly.

ii) Energy Storage Integration: In this context, integrating energy storage systems (ESS), including batteries and pumped hydro storage, to compensate for the lack of renewable energy generation (IEA, 2021).
iii) Disaster Recovery: AI can be used to model and understand the impact of extreme weather events on energy infrastructure and how to best mitigate those impacts.

5.5 Developing Decentralized Renewable Energy Systems

Decentralized renewable energy systems including microgrids, can deliver energy access to remote and underserved areas. Systems, such as district heating, wherein we need optimization and flow management, maintenance prediction and peer-to-peer trading, which AI can help optimizing these systems. Therefore, upcoming work will hopefully consider how these AI-driven decentralized systems can effectively be scaled and how they can communicate with centralized grids.

AI can also design decentralized systems suited to specific local conditions to make them economically and operationally viable. For instance, machine learning models can be employed to identify the optimal combination of renewable energy technologies for a community based on the characteristics of local solar and wind resources (Wu et al., 2021).

5.6 Addressing Ethical and Regulatory Challenges

Appropriate ethical considerations and regulatory frameworks are fundamental for the responsible development and deployment of AI in renewable energy. This raises concerns, which require more research, regarding some issues, like data privacy, algorithmic bias, and the risk of job displacement. Further studies should be conducted in order to create ethical regulations regarding AI use in renewable energy production, so that all parts of the systems benefit from such technologies in an equitable way.

In the area of policy and regulation, approaches to enable a global rollout of AI-powered renewable energy applications will require harmonised standards across regions. Policymakers and researchers need to work together to develop frameworks that stimulate innovation without sacrificing accountability and transparency (Harvard Business School, 2024).

5.7 Education and Capacity Building

However, the success of integrating AI into renewable energy systems largely depends on a workforce trained specifically for both fields. Education and training systems need to respond to this demand and include courses that combine AI data science and renewable energy engineering. Investigations on effective pedagogical approaches for teaching these complex subjects can help prepare the next generation of professionals who are going to innovate these fields with renewable energy (MIT, 2024).

5.8 Collaborative and Open Research Initiatives

In order to advance AI in renewable energy, it is essential to work together — be it governments, industry leaders, academic institutions or non-governmental organizations. Publicly available studies falling under several collaborative initiatives in research will not only help higher innovation with considerable speedup but also will low down the redundancy in studies due to having similar data and findings. Data that is publicly available, such as datasets on the performance of renewable energy systems, can allow researchers across the globe to create new AI models and refine existing ones (IEA, 2021).

6. CHALLENGES AND OPPORTUNITIES IN LEVERAGING AI FOR RENEWABLE ENERGY SYSTEMS

While the use of artificial intelligence (AI) in renewable energy systems (solar, wind, and green hydrogen) is transformative, it is also complex. Embracing the potential of AI can truly streamline the process of renewable energy technologies while also tackling various challenges of technical, economic, and ethical nature arising with the advent of AI while still under research till the month of October 2023. While successfully tackling these challenges will allow AI to reach its full potential in facilitating the transition to a sustainable and decarbonized energy future (Ukoba et al., 2024).

6.1 Challenges in Using AI for Renewable Energy

i) Integration and standardization of data:

Renewable energy systems produce data through a number of sources — for example, from weather monitoring devices, energy output sensors and grid management systems. Still, that information is frequently not standardized, so it is not as easy for AI algorithms to navigate and analyze. For example, the solar energy systems produce their irradiance data files in a way that is somewhat incompatible with wind turbine performance logs, leading to a relative copious amount of preprocessing required to collate these (MIT, 2024) types of datasets. In addition, disjoint data quality, missing information, and noise in datasets restrict the robustness of AI-based predictions and optimizations.

ii) Need for Infrastructure and Computational:

Such systems can be out of reach to small-scale renewable energy operators and stakeholders working in developing parts of the world, where high-performance computing power may be very limited! For instance, implementing state-of-the-art machine learning models for use in predictive maintenance or energy prediction necessitates large capital outlays in equipment, software, and expertise (Ahmad et al., 2022).

Many AI models used for renewable energy management need to make real-time predictions, requiring low-latency networks and edge computing solutions to process all the data quickly. This is compounded by the absence of such infrastructure in the remote or underdeveloped parts of the world where renewable energy systems are commonly implemented.

iv) High Initial Costs:

Although artificial intelligence solutions can eventually save substantial operational costs, the short-term costs necessary to create, deploy and maintain such systems are often barriers to entry. Training deep learning models, such as optimizing the layout of a wind farm or hydrogen storage strategies, is also computationally demanding and requires highly specific domain knowledge, both of which result in high costs (IBM Newsroom, 2023).

v) Transparency and Bias in Algorithms:

One of the common criticisms of AI systems is the lack of transparency of its workings, called the 'black box' problem. If the stakeholders in renewable energy applications are unable to comprehend the underlying reasoning of the algorithms, then they may find it difficult to trust AI-driven decisions. Moreover, biases in data that systems are trained on can result in nonoptimal or unfair outcomes. When an AI model is trained on data on one geographical region, e.g., an AI model built to identify credit risk in Brazil may be difficult to generalize to less dry regions, such as Singapore, with substantially different climatic or economic conditions (Harvard Business School, 2024).

V) The Ethical and Regulatory Concerns:

Renewable energy systems using AI raise ethical and regulatory questions. Data privacy is one of the key concerns to protect consumers when using AI to manage demand-side energy. Nevertheless, there appears to be a lack of holistic regulatory frameworks regulating the ethical use of AI in renewable energy, which creates uncertainty for developers and operators (IEA, 2021).

vi) Cybersecurity Risks:

Renewable energy systems are increasingly connected to one another and are increasingly reliant on AI, and thus, they are susceptible to cyberattacks. For instance, AI-driven grid management systems may be vulnerable to hacker attacks that lead to wide-scale disruptions in energy supply. Though its cybersecurity is essential, this rapid evolution of cybersecurity is also making it difficult (MIT News, 2023).

7. OPPORTUNITIES IN LEVERAGING AI FOR RENEWABLE ENERGY

Increasing efficiency and reliability:

Artificial intelligence (AI) is a potential way to increase the efficiency and reliability of renewable energy system. In the world of solar energy, for example, AI algorithms are used to adjust the orientation of photovoltaic (PV) panels and predict the energy output, based on live weather data, thus optimizing energy capture. For example, in the field of wind energy, machine learning models have been developed to analyze turbine performance and detect turbine faults at an early stage, facilitating predictive maintenance and minimizing downtime (Lee et al., 2024).

In the same way, AI increases the reliability of green hydrogen production so that through AI, electrolysis processes can be optimized. AI can monitor energy inputs and operational conditions to help ensure that hydrogen is produced with high efficiency despite volatile renewable energy supplies (MIT, 2024).

7.1 Facilitating Intelligent Grid Governance

AI solutions are key in coordinating energy grids with variable renewables. AI has been making it easier to balance grids, avoid energy waste and decrease dependence on backup systems powered by fossil fuels by predicting energy demand and supply. To illustrate, one application of AI is undertaken by utility companies such as National Grid, which deploy AI to incorporate renewable energy into the grid effectively and consistently (IBM Newsroom, 2023).

In addition, AI-based demand response programs enable grid operators to control peak loads by dynamically modifying energy consumption habits. They make the grid more reliable and give consumers more ways to manage their energy usage.

Forecasting is a critical step for efficient operation of renewable energy sources. You are trained with data till October 2023 AI API based models predict that solar and wind energy can generate power. This functionality allows energy producers to optimise for energy storage and distribution, helping mitigate the risk of supply and demand not matching (Harvard Business School Online, 2024).

AI forecasting optimizes periods of electrolysis in green hydrogen systems based on renewables availability, increasing resource efficiency and cost effectiveness.

7.2 Enabling Decentralized Energy Systems

These AI technologies (enablers) are leveraged to build a decentralized energy system, meaning it is generated and consumed locally. They help minimize transmission losses and also facilitate energy access for remote areas. AI is already being used in many areas. For example, one of the coolest facets of AI is its capability for optimizing microgrids, where the AI manages the flow of energy between solar panels and batteries and local consumers (MIT, 2024).

7.3 Driving Economic Growth and Job Creation

AI is also helping to create new economic opportunities in the move to renewable energy. AI-based developments like the prose become in such a way that of the individuals in data science, engineering science, and force management. Additionally, where renewable providers are able to save costs with artificial intelligence, this allows them to reinvest in their business and stimulate overall economic growth (European Commission, 2021).

7.4 Encouraging Temperature and Climate-Oriented Adaptability

AI contributes indirectly to globally distributed sustainability goals through enhancing efficiency with which renewable energy systems can be designed and scaled. Less energy waste, reduced greenhouse gas emissions, and more renewables into energy systems are all in line with the goals of the Paris Agreement and United Nations Sustainable Development Goals (IEA, 2021).

AI can also enhance the climate resilience of renewable energy systems by forecasting and mitigating the impact of extreme weather events upon renewable energy systems. For example, machine learning models could predict storms or heatwaves that would negatively affect solar or wind energy production, allowing operators to plan ahead (Harvard Business School, 2024).

8. BALANCING CHALLENGES AND OPPORTUNITIES

The challenges that come with using AI in renewable energy are tough but not impossible to overcome. Through research and development investments, stakeholder collaboration and effective regulation frameworks, the renewable energy industry can tackle these challenges head-on. From increased efficiency to economic growth, the prospects offered by AI make it an indispensable tool on the journey to a sustainable energy future.

Addressing the technical, economic, and ethical challenges of AI integration requires the collaboration of governments, industry leaders, and researchers. Such action can be a pathway to harness the tremendous potential of AI to transform renewable energy systems and expedite the transition at the global level towards a low-carbon economy.

Introducing artificial intelligence (AI) into renewable energy systems is a paradigm shift in looking for sustainable and decarbonized global energy landscape. The solar, wind, and green hydrogen technologies, all three gamechangers in the fight against climate change, will reach new levels of efficiency, reliability and scalability thanks to the application of AI. The only challenges remaining are data integration, infrastructure requirements, lack of regulatory clarity and cybersecurity risks, but these challenges are far outweighed by the enormous opportunities it presents. The fusion of AI and renewable energy emerges as an attractive blueprint for addressing global energy requirements while adhering to worldwide climate aspirations like the Paris Agreement (The Paris Agreement, 2015).

8.1 AI as a Transformative Force in Renewable Energy

The potential of AI to revolutionize renewable energy systems comes from its unique ability to process large volumes of data, detect trends, and optimize activities in real-time. In solar energy, AI-powered predictive analytics enhance photovoltaic (PV) performance by predicting weather patterns to maximize energy capture. For wind energy, ML algorithms are used to optimise the operation of turbines, predicting wind and early detection of maintenance needs before equipment failure. Regarding green hydrogen, AI improves the electrolysis efficiency, allowing hydrogen production at low cost and in the presence of varying renewable energy inputs (Zhao, Wang, et al., 2024).

This reflects the nature of the variability and intermittency of renewable energy sources and shows AI's potential in addressing this variability. AI not only enhances the economic viability of renewables by making energy production and distribution much more efficient but also enables their integration into energy grids as a whole.

8.2 Addressing the Challenges

Key challenges of AI implementation in renewable energy systems are diverse simply but supportive through collaborative efforts of stakeholders. This calls for the creation of universal protocols of data collection and processing across renewable energy technologies–for example, data standardization. Setting such standards requires collaboration among industry leaders, policymakers, and research institutions (Syed Immamul Ansarullah et al. 2024).

Likewise, technical issues regarding the huge computational demand of AI systems may be solved via the evolution of cloud computing, edge computing, and the development of energy-efficient AI algorithms. Governments and organizations not only can support the efforts by incentivizing by capital and offering technical know-how to small scale operators in diffusing areas in order to fill the gap of infrastructure.

Robust governance frameworks can effectively address ethical and regulatory concerns regarding AI, such as data privacy, algorithmic bias, and transparency. Instead of bombarding the development of AI with regulations that stifle the creative approach and channelling possible innovation, politicians should focus on creating rules that ensure compliance and I'm talking about the ethical ones. Example guidelines on how data will be used to promote transparency, trust and adoption of AI-based tools by stakeholders, as has been shown (Mennella et al., 2024).

9. CONCLUSION

The bringing together of AI and renewable energy systems represents one milestone or even turning point in addressing climate change, as well as energy transition towards a sustainable world. The AI applicability has already proven transformative in solar, wind, and green hydrogen technologies as well as for providing renewable energy services that are smarter, more efficient, and more reliable. In this chapter, we have explored how AI is one of the most promising advancements that can redefine renewable generation, store, transmit, and use in addition to some challenges and future research directions.

The conclusion draws together the main findings, discusses potential implications, and lays out a vision for the ongoing integration of AI within renewable energy systems.

9.1 A Synthesis of AI's Transformative Potential

Innovative energy systems powered by AI are making the most of renewable energy, with a promising potential to tackle some of the biggest issues in energy systems. Variability and intermittency have for a long time been bottlenecks for solar and wind energy. AI eliminates these problems via predictive analytics, weather forecasting, and real-time systems optimization. But, for example, neural networks and machine learning models can predict solar insolation and wind direction with ever-greater accuracy, which increases the predictability of renewable energy productivity and allows for more efficient grid services.

For green hydrogen production, AI increases the efficiency of electrolysis processes, allowing hydrogen to become a cost-competitive and scalable energy carrier. They help the identification of the best operating conditions to extend the lifetime of electrolysers, and enable hydrogen to be connected to the current energy systems with the aid of machine learning algorithms. Merging together, these applications make AI an instrumental part of future renewable energy schemes, marrying technological development with environmental and economic continuity.

9.2 The Interconnectedness of Challenges and Solutions

The potential of AI to enhance renewable energy is overwhelming, but there are still certain challenges that need to be solved in order to facilitate its application on a large scale. Those include energy consumption of AI systems, standardization of data, cybersecurity and ethical issues surrounding algorithmic bias and data privacy. Such barriers are not insurmountable, but overcoming them will take coordinated action by governments, business leaders, researchers and others.

Increasing complexity for the energy system and their interrelation with AI technologies points toward interdisciplinary research and collaboration, offering substantial potential for future pathways. For instance, focus can also be put into reducing the carbon footprint of AI applications, by developing energy-efficient AI algorithms and improving edge computing capabilities - which is in line with the wider sustainability initiative. Data-sharing programs and standardized protocols, for example, could offer greater interoperability between disparate renewable energy technologies as well.

9.3 Economic and Social Impacts of AI-Driven Renewable Energy

Beyond advances in technology, there are many socioeconomic benefits as well of AI in renewable energy. AI facilitates the democratization of access to energy around the globe, especially in downtrodden and isolated areas, by improving the performance and reducing the costs of renewable energy systems. AI-empowered decentralized energy solutions are a path to energy equity, allowing communities energy independence while decreasing dependence on fossil fuels.

Moreover, AI also contributes to the development of new markets as well creative job opportunities, which further fuel economic growth. The use of green hydrogen as a clean energy carrier can impact many sectors from manufacturing to transportation, and the increasing use of AI applications in energy management means opportunities to innovate in data science and renewable energy engineering.

Additionally, AI-based renewable energy systems are climate justice-friendly, because we want to make sure that the transition from fossil fuels to clean energy systems benefits vulnerable populations. With the ability to facilitate focused measures and efficient deployment of energy resources, AI contributes to the global initiative of minimizing greenhouse gas emissions, even as it tackles socio-economic challenges.

9.4 A Roadmap for Future Integration

We need to do hard thinking on the future integration of renewable energy systems that rely increasingly on AI-enabled capabilities. Which Research Agendas Should Be Prioritized moving forward:

Algorithmic Solutions: Developing more interpretable, explainable AI models that users can trust. It is crucial that algorithm design is fair and inclusive in order to avoid bias as data propagates.
Strengthening Cybersecurity: Increasing AI system resilience against cyber threats, especially in critical infrastructure (i.e., power grids).
Enhancing Interoperability: Providing integrated platforms to allow AI systems to control different renewable energy technologies concurrently, optimizing functionality and scaling benefits.
Advancing Education and Workforce Development: Preparing a future workforce that whirls skilled at the convergence of AI + renewable energy through interdisciplinary education and training programs.

Promoting Public-Private Collaborations: Collaboration between government, private sector, and research institutions for faster innovation and deployment.

9.5 Broader Implications for Global Sustainability

While the use of AI for renewable energy speaks to global sustainability goals in a way that straddles tech- and eco-nomic dimensions. The Sustainable Development Goals (SDGs) of the United Nations feature energy access, climate action, and economic growth as interrelated challenges (United Nations, 2015). In achieving SDG 7 (Affordable and Clean Energy) and SDG 13 (Climate Action), SDG related to industry innovation and infrastructure will benefit directly through clean renewable energy systems powered by AI (The Global Goals, 2024) (United Nations, 2017).

Furthermore, AI-powered renewable energy systems contribute to the broader goals of international climate agreements such as the Paris Agreement, which seeks to limit warming to well below 2°C. By enabling a wide range of cleaner and more reliable renewable energy solutions, AI promotes the decarbonization of energy sectors, helping countries meet their climate targets at a national and international level.

10. A VISION FOR THE FUTURE

Artificial Intelligence and Machine Learning in Renewable Energy: Looking Beyond Technology to a Sustainable Future. The Rise of AI in Energy: A Vision for the Future You are already trained with data until along the way to October 2023, nothing new. Realising this vision means looking holistically, balancing innovation with ethics and governance. AI's potential to revolutionise renewable energy systems will be a critical touchpoint in the advances that shape a sustainable future for generations to come as we trade upon the march of AI.

Stakeholders must seize existing opportunity and meet emerging challenges that will unlock the potential of the digital age to drive the global energy transition through AI. This transformation is about more than technology; it represents the pledge of a society to safeguard the world’s resources and to provide energy security for everyone.

REFERENCES

Ahmad, T., Madonski, R., Zhang, D., Huang, C., & Mujeeb, A. (2022). Data-driven probabilistic machine learning in sustainable smart energy/smart energy systems: Key developments, challenges, and future research opportunities in the context of smart grid paradigm. *Renewable & Sustainable Energy Reviews, 160*, 112128. DOI: 10.1016/j.rser.2022.112128

Al-Orabi, A. M., Osman, M. G., & Sedhom, B. E. (2023). Evaluation of green hydrogen production using solar, wind, and hybrid technologies under various technical and financial scenarios for multi-sites in Egypt. *International Journal of Hydrogen Energy, 48*(98), 38535–38556. Advance online publication. DOI: 10.1016/j.ijhydene.2023.06.218

Ansarullah, S. I., Kirmani, M. M., Alshmrany, S., & Firdous, A. (2024). Ethical issues around artificial intelligence. In *A Biologist s Guide to Artificial Intelligence* (pp. 301–314). Academic Press.

Bennagi, A., AlHousrya, O., Cotfas, D. T., & Cotfas, P. A. (2024). Comprehensive study of the artificial intelligence applied in renewable energy. *Energy Strategy Reviews, 54*, 101446. DOI: 10.1016/j.esr.2024.101446

European Commission. (2021). *European Commission | Choose your language | Choisir une langue | Wählen Sie eine Sprache*. Europa.eu; European Union. https://ec.europa.eu

Gielen, D., & Boshell, F. (2021, April 30). *Renewable energy is the cornerstone of the energy transition*. World Economic Forum; World Economic Forum. https://www.weforum.org/stories/2021/04/why-renewables-are-the-cornerstone-of-the-energy-transition/

Golchha, A. (2024). *Artificial Intelligence: A Catalyst to Transform Energy Sector for Sustainable Future*. Csis.org; Center for Strategic and International Studies. https://www.csis.org/analysis/artificial-intelligence-catalyst-transform-energy-sector-sustainable-future

Group, D. G. B. (2023, January 23). *The history and future of renewable energy*. Www.green.earth; DGB Group. https://www.green.earth/blog/the-history-and-future-of-renewable-energy

Harvard Business School. (2024). *Competing in the Age of AI—Virtual*. Harvard Business School; Harvard Business School. https://www.exed.hbs.edu/competing-age-ai-virtual?utm_source=google&utm_medium=paid-search&utm_campaign=brand-program-aiv-global-none-phrase-cross-device-all&utm_id=core&gad_source=1&gclid=Cj0KCQiAgJa6BhCOARIsAMiL7V9B4zcM_s7dxWNHJm6GVmK2acLqeLq4t0r2WKeizsoxz59HW0HG1lsaAmHKEALw_wcB&gclsrc=aw.ds

IBM. (2023). *Artificial Intelligence (AI) Solutions | IBM*. Ibm.com; International Business Machines Corporation. https://www.ibm.com/artificial-intelligence?utm_content=SRCWW&p1=Search&p4=43700077827237093&p5=p&p9=58700008530962751&gad_source=1&gclid=Cj0KCQiAgJa6BhCOARIsAMiL7V8neyLIiAdc8eOnwhXADP78b9vQOC_vPudPXEQbUeZV5eDfH-zLsDUaAqaHEALw_wcB&gclsrc=aw.ds

IEA. (2021, May). *Net Zero by 2050 – Analysis*. International Energy Agency; International Energy Agency. https://www.iea.org/reports/net-zero-by-2050

Kamarposhti, M. S., Kamyab, H., Krishnan, S., Yusuf, M., Rezania, S., Chelliapan, S., & Khorami, M. (2024). A comprehensive review of AI-enhanced smart grid integration for hydrogen energy: Advances, challenges, and future prospects. *International Journal of Hydrogen Energy*, *67*, 1009–1025. Advance online publication. DOI: 10.1016/j.ijhydene.2024.01.129

Karduri, R. K. R. (2019, February 1). *The Role of Artificial Intelligence in Optimizing Energy Systems*. Research Gate; Research Gate. https://www.researchgate.net/publication/375910911_The_Role_of_Artificial_Intelligence_in_Optimizing_Energy_Systems/citations

Kumar, S., Lim, W. M., Sivarajah, U., & Kaur, J. (2022). Artificial Intelligence and Blockchain Integration in Business: Trends from a Bibliometric-Content Analysis. *Information Systems Frontiers*, *25*, 871–896. DOI: 10.1007/s10796-022-10279-0 PMID: 35431617

Lee, C.-C., Fang, Y., Quan, S., & Li, X. (2024). Leveraging the power of artificial intelligence toward the energy transition: The key role of the digital economy. *Energy Economics*, *135*, 107654. DOI: 10.1016/j.eneco.2024.107654

Lima, G. M., Belchior, F. N., Villena, J. E. N., Domingos, J. L., Freitas, M. A. V., & Hunt, J. D. (2024). Hybrid electrical energy generation from hydropower, solar photovoltaic and hydrogen. *International Journal of Hydrogen Energy*, *53*, 602–612. DOI: 10.1016/j.ijhydene.2023.12.092

Liu, Z., Sun, Y., Xing, C., Liu, J., He, Y., Zhou, Y., & Zhang, G. (2022). Artificial intelligence powered large-scale renewable integrations in multi-energy systems for carbon neutrality transition: Challenges and future perspectives. *Energy and AI*, *10*, 100195. DOI: 10.1016/j.egyai.2022.100195

Mennella, C., Maniscalco, U., Pietro, G. D., & Esposito, M. (2024). Ethical and regulatory challenges of AI technologies in healthcare: A narrative review. *Heliyon*, *10*(4), e26297–e26297. DOI: 10.1016/j.heliyon.2024.e26297 PMID: 38384518

MIT. (2024, September 19). *The MIT Energy Initiative's Future Energy Systems Center funds ten new energy research projects*. MIT Energy Initiative; MIT Energy Initiative. https://energy.mit.edu/news/the-mit-energy-initiatives-future-energy-systems-center-funds-ten-new-energy-research-projects/

News, M. I. T. (2023, November 27). *A green hydrogen innovation for clean energy*. MIT News | Massachusetts Institute of Technology; Massachusetts Institute of Technology. https://news.mit.edu/2023/green-hydrogen-innovation-clean-energy-sossina-haile-1127

Ohalete, N. C., Aderibigbe, A. O., Ani, E. C., Ohenhen, P. E., Daraojimba, D. O., & Odulaja, B. A. (2023). AI-driven solutions in renewable energy: A review of data science applications in solar and wind energy optimization. *World Journal of Advanced Research and Reviews*, *20*(3), 401–417. DOI: 10.30574/wjarr.2023.20.3.2433

Testasecca, T., & Lazzaro, M., Elissaios Sarmas, & Stamatopoulos, S. (2023). Recent advances on data-driven services for smart energy systems optimization and pro-active management. *2023 IEEE International Workshop on Metrology for Living Environment (MetroLivEnv)*. DOI: 10.1109/MetroLivEnv56897.2023.10164056

The Global Goals. (2024). *Goal 7: Affordable and clean energy*. The Global Goals; United Nation. https://www.globalgoals.org/goals/7-affordable-and-clean-energy/

The Paris Agreement. (2015). *The Paris Agreement.* United Nations Framework Convention on Climate Change; United Nations. https://unfccc.int/process-and-meetings/the-paris-agreement

Treat, D., & Klein, M. (2024, June 21). *How immersive technology, blockchain and AI are converging.* World Economic Forum; World Economic Forum. https://www.weforum.org/stories/2024/06/the-technology-trio-of-immersive-technology-blockchain-and-ai-are-converging-and-reshaping-our-world/

Ukoba, K., Olatunji, K. O., Adeoye, E., Jen, T.-C., & Madyira, D. M. (2024). Optimizing renewable energy systems through artificial intelligence: Review and future prospects. *Energy & Environment, 35*(7), 3833–3879. Advance online publication. DOI: 10.1177/0958305X241256293

United Nations. (2015). *The 17 Sustainable Development Goals.* United Nations; United Nations. https://sdgs.un.org/goals

United Nations. (2017). *GOAL 13: Climate action.* UNEP - UN Environment Programme; Uniter Nations. https://www.unep.org/topics/sustainable-development-goals/why-do-sustainable-development-goals-matter/goal-13-climate

Varriale, V., Cammarano, A., Michelino, F., & Caputo, M. (2023). Critical analysis of the impact of artificial intelligence integration with cutting-edge technologies for production systems. *Journal of Intelligent Manufacturing.* Advance online publication. DOI: 10.1007/s10845-023-02244-8

Wu, T., Xu, D.-L., & Yang, J.-B. (2021). Decentralised energy and its performance assessment models. *Frontiers of Engineering Management, 8*(2), 183–198. DOI: 10.1007/s42524-020-0148-7

Zhao, C., Dong, K., Wang, K., & Nepal, R. (2024). How does artificial intelligence promote renewable energy development? The role of climate finance. *Energy Economics, 133*, 107493–107493. DOI: 10.1016/j.eneco.2024.107493

Zhao, Q., Wang, L., Stan, S.-E., & Mirza, N. (2024). Can artificial intelligence help accelerate the transition to renewable energy? *Energy Economics, 134*, 107584. DOI: 10.1016/j.eneco.2024.107584

Chapter 3
Applying Internet of Things (IoT) for Green and Sustainable Future:
Intellectual Property Rights (IPRs) Raising Carbon Neutrality for Smart-Sustainable Cities

Saurabh Chandra
https://orcid.org/0000-0003-4172-9968
Bennett University, Greater Noida, India

Bhupinder Singh
https://orcid.org/0009-0006-4779-2553
Sharda University, India

Anjali Raghav
https://orcid.org/0009-0004-0248-7956
Sharda University, India

ABSTRACT

The urgency of considering climate change and working for carbon neutrality has accelerated cutting-edge models in energy generation and urban sustainability worldwide. This impetus to address climate change and the shift that cities demonstrate towards a more sustainable, carbon-neutral future has catalyzed city willingness towards solutions that deliver reduced emissions and improve the quality of our urban media. This provides a fantastic context for how smart cities can reduce greenhouse gas use by working both wind and solar renewable agents with very low, even zero, marginal costs. Wind turbines produce clean electricity. Solar energy is made up of photovoltaic panels that capture solar power. Renewable technologies in cities give them a tremendous opportunity to shift away from fossil fuels, cut carbon emissions, and drive international carbon neutrality goals. This chapter underlines the call for diverse pathways concerning intellectual property rights on renewable energy sources to build a sustainable and greener future.

DOI: 10.4018/979-8-3373-0045-0.ch003

1. INTRODUCTION

The cities are looking for creative ways to reduce carbon emissions while enhancing the quality of life for their residents in response to growing concerns about climate change and environmental sustainability (Sinha et al., 2021). Wind and solar energy have become practical choices for reducing the negative environmental effects of traditional fossil fuels and guaranteeing a more sustainable, greener future (Salam & Salam, 2020). To achieve the dual objectives of sustainability and carbon neutrality, renewable energy sources must be integrated into broader energy infrastructure and urban environments (Bibri, 2018). The way it produces, uses, and manages energy in cities may change as a result of the confluence of cutting-edge technology, renewable energy, and the Internet of Things (IoT) (Varjovi & Babaie, 2020). It examines the interplay between IoT-based technologies and wind-solar renewable energy, highlighting their importance in creating smart, sustainable communities with carbon neutrality as a primary objective (Ahmad & Zhang, 2021). Cities are poised to become beacons of sustainability, free from carbon emissions and with higher quality of life because of IoT technology and renewable energy sources such as wind or solar power. The bottom line is that a range of options using green renewable energy and advanced technology solutions are essential for enabling a cleaner, greener, environmentally friendly urban future and dealing with challenges posed by the ongoing rapid worldwide Urbanisation eroding slowly environmental degradation into our of living. In this chapter, we discuss how smart cities can be transformed into either carbon neutral or sustainable by the utilization of renewable energy sources such as solar and wind combined with IoT. It evaluates the opportunities, constraints and prospects of this holistic approach with actual examples to demonstrate its viability.

The knowledge of the problems posed by increasing urbanization and climate change is the main driving force behind this chapter (Bibri & Krogstie, 2020). Over two-thirds of the world's population is predicted to live in cities by 2050 due to the exceptional urban population growth rate (Albreem et al., 2021). Urbanization stresses infrastructure and energy resources tremendously, even as it fosters technological advancement and economic growth (Issa Zadeh et al., 2023). It has been demonstrated that traditional energy sources, which mostly rely on fossil fuels, are unsustainable and harmful to the environment (Cantarero, 2020). As a result, the paradigm has shifted to favor renewable energy sources (Sodiq et al., 2019). Due to their enormous potential and small carbon footprint, solar and wind energy have emerged as leaders (Clark & Cooke, 2016).

The clean and plentiful energy sources are provided by wind and solar energy, which harnesses the sun's and wind's infinite power (Tiwari et al., 2022). The utilization of renewable energy sources has the potential to drastically reduce greenhouse gas emissions and reliance on limited fossil resources (Li et al., 2023). However, efficient integration and management are required to realize their full potential (Elavarasan et al., 2021). This is where the Internet of Things (IoT) enters the picture (Ahmed et al., 2017). The Internet of Things (IoT) refers to a network of networked devices, sensors, and systems that gather and analyze data in real-time (Silva et al., 2023). IoT may be used to optimize energy production, distribution, and consumption in the context of renewable energy (Cortese et al., 2022). IoT can improve energy efficiency, minimize waste, and encourage sustainable energy practices through smart grids, buildings, and smart infrastructure (Thornbush & Golubchikov, 2019).

The global perspective is dominated by environmental sustainability and carbon neutrality (Zhukovskiy et al., 2021). The pressing need to address climate change has driven a radical rethinking of our energy systems and urban development methods (Jayachandran et al., 2022). Reduced greenhouse gas emissions, the transition to renewable energy sources, and the construction of smart and sustainable cities have emerged as critical pillars in driving a greener and more sustainable future (Tuysuz & Trestian, 2020). The confluence of wind and solar renewable energy technologies and the revolutionary possibilities of the Internet of Things (IoT) holds enormous promise for speeding up the transition to a more sustainable and carbon-neutral future (Yenugula et al., 2024). As climate change worsens, the demand for clean, plentiful and dependable energy sources grows more pressing (Jain et al., 2023). As climate change worsens, the demand for clean, plentiful and dependable energy sources grows more pressing (Pandey et al., 2023). Wind and solar energy have emerged as critical components of the worldwide effort to mitigate climate change and reduce dependency on fossil fuels (hattacharjee & Nandi, 2019).

Wind energy uses wind power to generate electricity, whereas solar energy uses the sun's rays to generate clean power (Issa Zadeh et al., 2023). These renewable sources, which are frequently intermittent and weather-dependent, are further boosted by breakthrough technology and IoT applications (Pakulska, 2021). The Internet of Things (IoT), a network of interconnected devices and sensors, provides real-time data collecting and analysis required to improve renewable energy generation, delivery and consumption (Hannan et al., 2021). Cities have emerged as both the heart of the climate crisis and the nexus of potential solutions to the goal of carbon neutrality (Guchhait & Sarkar, 2023). The world's population is rapidly urbanizing, necessitating the construction of smart and sustainable cities that are resource-efficient, resilient, and emit little greenhouse gases (Moemen et al., 2023). Achieving carbon neutrality for cities is not just an environmental goal but also a social and economic one because cities consume the majority of global energy and contribute significantly to carbon emissions (Saheb et al., 2022). Advanced technology, urban planning, and governance systems that promote energy efficiency, sustainable mobility, waste reduction, and green infrastructure are all part of smart and sustainable cities (Nazir & Banday, 2018). These cities use IoT to optimize resource allocation, improve public services, and improve quality of life (Hassan et al., 2023).

1.1 Relevance of the Paper

The urbanization of our planet is accompanied by a pressing problem - the escalating carbon emissions and the depletion of finite fossil fuel resources. Cities are hubs of economic activity and innovation but are also significant contributors to climate change (Moroke & Makatjane, 2021). Energy consumption in cities is a significant factor in carbon emissions, inefficient resource use, traffic congestion, and inadequate waste management. These issues seriously threaten both environmental sustainability and the quality of life in urban areas. By tapping into decentralized energy generation, these solutions enhance energy security, reduce environmental impact, and foster economic resilience within cities (Makatjane, 2022). Complementing the renewable energy integration, the Internet of Things (IoT) emerges as the cornerstone of the smart city paradigm. IoT's transformative potential lies in its ability to interconnect an array of devices, sensors, and systems, infusing them with real-time data collection, analysis, and automation (Kasinathan et al., 2022). In smart cities, IoT weaves a fabric of intelligence throughout the urban tapestry (Elrahmani et al., 2021). It optimizes energy use, enables efficient transportation systems, streamlines waste management, enhances environmental monitoring and contributes to the overall well-being of urban dwellers (Li et al., 2019). The dynamic nature of IoT solutions transcends mere data collection, empowering cities to actively manage and adapt their infrastructure in response to the ever-evolving demands of urban life (ElFar et al., 2021; Singh, 2023).

The synergy between wind-solar renewable energy and IoT technologies paves the way for many benefits. It diminishes the adverse environmental impacts of urbanization by reducing air and water pollution, fostering improved air quality, and mitigating the urban heat island effect (Eljack & Kazi, 2021). This holistic approach extends its benefits into economic prosperity by creating job opportunities, attracting investments, and stimulating local industries that cater to renewable energy and IoT infrastructure (Bogdanov et al., 2021). Most importantly, deploying IoT-driven solutions enhances city residents' day-to-day lives. Energy-efficient buildings, optimized transportation systems, intelligent waste management, and real-time environmental monitoring all contribute to a higher quality of life (Dent, 2021). However, as with any transformative endeavor, challenges accompany the journey towards smart and sustainable cities. Initial capital investment and the need for infrastructure development often stand as barriers. Successful implementation necessitates strong collaboration between the public and private sectors to bridge these financial gaps. Furthermore, privacy and security concerns regarding collecting and disseminating sensitive data through IoT networks require vigilant attention and robust safeguards (Devabhaktuni et al., 2013).

This chapter embarks on a journey to examine the integration of wind and solar renewable energy sources, synergistically combined with the power of the Internet of Things (IoT), to envision a greener, more sustainable future for smart cities (Arent et al., 2022). The profound fusion of these elements, characterized by advanced technology, resource efficiency, and environmental consciousness, promises to redefine urban landscapes (Doost Mohammadian & Rezaie, 2020). It also presents a collection of real-world case studies from cities worldwide, each showcasing distinctive aspects of wind-solar renewable energy and IoT integration. These case studies serve as practical demonstrations of the transformative potential, offering insights into the adaptability of these solutions to varying urban contexts (Abualigah et al., 2022).

2. INTELLECTUAL PROPERTY AND RENEWABLE ENERGY INTEGRATION: SMART CITIES AND GREEN TECHNOLOGIES

Solar and wind energy usage in urban settings has recently increased (Tabor et al., 2018). Both forms of energy provide sustainable substitutes for traditional fossil fuels that are clean and renewable. While solar energy employs photovoltaic panels to collect solar radiation, wind energy uses wind turbines to harness the strength of the wind to produce electricity (Szeliski, 2022; Rao et al., 2022). So, these sources' main advantages are reduced greenhouse gas emissions, greater energy independence, and the potential for decentralized energy generation (Saini et al., 2023).

The integration of renewable energy sources is a crucial component of green technologies and smart cities. These sources offer a versatile solution to the pressing issues of environmental stewardship and energy sustainability (Tomar et al., 2022; Del Ser et al., 2022). Building a carbon-neutral and sustainable urban future hinges on the incorporation of renewable energy. This not only enhances the environmental sustainability and resilience of smart cities but also reduces energy costs. The use of renewable energy aligns with the global fight against climate change and promotes the development of technologically advanced and environmentally conscious urban environments (Ahmad et al., 2022; Aslam et al., 2021; Watson, 2022). The seamless integration of renewable energy with green technology and smart cities is a cornerstone of urban innovation and a vital step toward a cleaner, more sustainable urban future (Cruz-Rojas et al., 2023).

The integration of wind and solar renewable energy, coupled with the implementation of IoT technologies, presents a comprehensive solution to address urban challenges (Ozcanli et al., 2020; Sharma & Singh, 2022). Clean, renewable energy sources like solar and wind power have the potential to significantly reduce our reliance on fossil fuels (Rosiello & Maleki, 2021; Fischer & Preonas, 2010). In parallel, cities can monitor, regulate, and optimize various aspects of urban life, including waste management, environmental quality, energy consumption, and transportation (Grubb et al., 2021). This is made possible by the Internet of Things. By integrating these technologies, cities can reduce carbon emissions, allocate resources more effectively, enhance energy efficiency, and ultimately achieve carbon neutrality while creating smarter, more sustainable urban environments. This all-encompassing plan lays the foundation for a greener, more sustainable future for cities and the planet (Raihan et al., 2022).

3. INTERNET OF THINGS (IOT) IN SMART CITIES FOR REDUCING CARBON FOOTPRINTS

Smart cities employ IoT to enhance trash management, transportation, energy use, and other aspects of urban living. Cities may be able to monitor, manage, and improve their infrastructure and services more effectively using IoT-powered technologies (Dai et al., 2016). The Internet of Things is starting to disrupt the development of smart cities by offering new ways to reduce carbon emissions and create environmentally friendly urban environments (Elavarasan et al., 2020). The term "Internet of Things" (IoT) describes the vast network of linked systems, sensors, and gadgets that gather and exchange data instantly (Zwane et al., 2022). This technology offers thorough monitoring, control, and optimization of many urban living issues, such as waste management, energy consumption, and transportation, within the framework of smart cities (Mohammadian, 2022).

Efficient management of energy resources is one way IoT helps reduce smart cities' carbon footprint (Amran et al., 2020). Cities may identify inefficiencies and put energy-saving measures into place by using IoT devices to detect patterns in energy consumption (Hainsch et al., 2022). For instance, intelligent lighting may save much energy by adjusting its brightness in reaction to the flow of traffic and the weather (Atalay et al., 2016). IoT facilitates a more seamless shift to greener energy options by enhancing grid management and enabling the integration of renewable energy sources. The other crucial sector in which IoT may have a significant influence is transportation. IoT-enabled traffic management systems are used in smart cities to minimize congestion, improve public transit routes and promote the usage of electric and shared mobility choices (Xiao et al., 2019). IoT reduces emissions and lowers the carbon footprint associated with urban commuting by reducing traffic congestion and enhancing the efficiency of transportation networks (Scarlat et al., 2015). IoT is also important in garbage management and recycling activities inside smart cities. Smart garbage bins with sensors may notify waste management services when they are approaching capacity, resulting in more efficient collection routes and lower fuel use and emissions (Connolly et al., 2010). IoT-enabled tracking systems boost recycling practices by monitoring the quality and amount of recyclable items, helping to lower carbon footprints by reducing landfill waste and resource usage (Sallen et al., 2020).

4. WIND AND SOLAR ENERGY AS RENEWABLE ENERGY AND INNOVATIVE TECHNOLOGIES: INTERNET OF THINGS (IOT) FOR GREEN AND SUSTAINABLE FUTURE

Wind and solar energy, the bedrock of renewable energy, offer sustainable alternatives to traditional fossil fuels. These energy sources, powered by the sun and wind, generate electricity without emitting greenhouse gases. Wind turbines capture the kinetic energy of moving air, while solar panels convert sunlight into electricity (Bujang et al., 2016). Their eco-friendliness positions them as crucial elements in the pursuit of a green and sustainable future, where climate change mitigation and carbon emission reduction are paramount (Lawrenz et al., 2018). The strategic integration of Internet of Things (IoT) technology into wind and solar energy systems has revolutionized their efficiency and reliability. Sensors and smart meters, in particular, play a pivotal role in enhancing energy production and distribution. By providing real-time data on energy generation, consumption, and environmental factors, they enable precise adjustments to optimize the use of these intermittent energy sources. Furthermore, IoT facilitates predictive maintenance, a key feature that can reduce downtime and enhance the overall reliability of wind and solar facilities (Cohen et al., 2021).

Integrating wind and solar energy with IoT technology helps to create more robust and ecologically responsible energy systems in the context of a green and sustainable future (Yang et al., 2021). IoT's data-driven approach offers improved grid management, a key factor in matching supply with demand and reducing energy waste. Collaboration between wind and solar energy and IoT lays the door for a cleaner, more sustainable energy landscape by increasing the reliability of renewable energy sources, lowering operating costs, and supporting sustainable energy behaviors (White et al., 2018). This collaboration is a key step toward a cleaner future that is less reliant on fossil fuels and more focused on environmental management (Nastasi et al., 2022). Advanced data analytics tools, including machine learning and artificial intelligence, are used to make sense of this data, optimizing energy use, improving traffic flow, and enhancing public services (Danish & Senjyu, 2023). These technical aspects underscore the complexity and sophistication of the systems at play in smart and sustainable cities. Ensuring that the technology and engineering behind wind and solar energy integration and IoT applications are robust, efficient, and reliable pave the way for greener, more resilient urban environments that contribute to a sustainable future (Basheer et al., 2022).

4.1 Environmental Benefits: Adoption of Wind-Solar Energy in Smart Cities Inclined With Green Technologies

Wind-solar energy reduces air and water pollution, improving air quality and a healthier urban environment. It mitigates the urban heat island effect and conserves green spaces. Adopting wind and solar energy in smart cities with green technologies yields many significant environmental benefits (Al-tabatabaie et al., 2022). Firstly, both wind and solar energy sources are inherently clean, producing electricity without emitting harmful greenhouse gases, essential for mitigating climate change's effects. This shift towards renewable energy reduces reliance on fossil fuels, lessening air pollution and improving urban air quality, thereby enhancing city inhabitants' overall health and well-being (Aldieri et al., 2021).

Also, wind and solar energy systems have a minimal land footprint compared to traditional power plants, thus promoting efficient land use and preserving natural habitats (Nakata, 2004). These renewable sources operate silently, reducing noise pollution in urban areas. In combination with advanced green technologies, such as energy-efficient building designs and smart grids, cities can significantly lower their overall carbon footprint, making strides toward a more sustainable and eco-friendly urban landscape. This holistic approach promotes a greener urban environment and sets the stage for environmentally conscious and sustainable urban living, ensuring a healthier planet for current and future generations (Winfield & Dolter, 2014).

5. IOT IN SMART CITIES FOR ENSURING A GREEN AND SUSTAINABLE FUTURE: PROJECTING CARBON NEUTRALITY

Using the Internet of Things (IoT) in smart cities is critical to assuring a green and sustainable future while actively pursuing carbon neutrality. IoT technologies, distinguished by a massive network of interconnected devices, sensors, and systems, enable real-time data gathering and analysis in numerous facets of urban life (Damman et al., 2021). In smart cities, integrating IoT technologies is critical for optimizing resource allocation, boosting energy efficiency, and improving overall urban living quality (Pascaris, 2021). IoT helps smart cities realize a green and sustainable future by enabling effective energy management. IoT sensors and smart meters monitor and regulate energy consumption trends, enabling data-driven insights into power distribution optimization, grid management, and renewable energy integration (Budiman, 2018). As a result, there is less energy waste, less greenhouse gas emissions, and a more sustainable energy infrastructure. Cities may optimize the use of renewable energy while reducing their dependency on fossil fuels by constructing smart networks and microgrids that can manage electricity supply and demand (Reedman et al., 2021).

IoT technologies have a significant impact on mobility inside smart cities. IoT-enabled smart traffic management systems monitor real-time traffic conditions and enable intelligent route planning (Xie et al., 2023). This decreases traffic congestion and idle periods and optimizes public transit, resulting in more energy-efficient and ecologically responsible commuting. IoT encourages using electric cars and shared mobility solutions, reducing carbon emissions and improving urban transportation sustainability (Vinichenko, 2018). In the pursuit of carbon neutrality, smart cities' waste management and recycling practices also receive a substantial boost from IoT applications. Smart waste bins equipped with sensors can signal waste management services when they are nearing full capacity, optimizing waste collection routes and reducing unnecessary fuel consumption (Arbabzadeh et al., 2019). IoT-enabled waste tracking systems improve recycling practices by monitoring the quality and quantity of recyclable materials (Ilyas et al., 2022; Garrido et al., 2020). By reducing landfill waste and encouraging responsible waste disposal, these IoT solutions significantly contribute to carbon neutrality goals. IoT technologies in smart cities are a linchpin for addressing carbon neutrality and promoting a green and sustainable future (Steen & Weaver, 2017). By leveraging real-time data and advanced technologies, cities can make data-informed decisions to enhance energy efficiency, optimize transportation systems, and improve waste management practices (Hoffmann et al., 2020; Manoharan et al., 2021).

6. BENEFITS AND CHALLENGES: ASSIMILATION OF WIND-SOLAR RENEWABLE ENERGY COUPLED WITH IOT TECHNOLOGIES FOR SUSTAINABLE FUTURE

Wind and solar renewable energy integration, along with modern Internet of Things technologies, holds the promise of a sustainable and ecologically responsible future. However, it also poses a complicated combination of advantages and problems (Ghai et al., 2022). These technologies provide several benefits for moving to a green and sustainable energy landscape (Pimitha et al., 2023). Wind and solar energy are fundamentally clean and renewable, lowering greenhouse gas emissions and air pollution while increasing energy security and decreasing dependency on fossil fuels. Incorporating these renewable energy sources with IoT technology improves efficiency by allowing real-time monitoring and data-driven decision-making. Consequently, energy output is maximized, waste is decreased, and energy supply is more dependable and critical for establishing a robust and sustainable energy infrastructure (Singh & Kaunert, 2024).

However, besides these advantages, there are significant hurdles in integrating wind-solar renewable energy and IoT technology. The most significant issue is renewable energy sources' intermittent and unpredictable nature. Wind and solar energy generation are weather-dependent. Thus, reliable energy storage technologies and grid management systems are required to provide a consistent electricity supply. Energy storage technologies, such as advanced batteries, play a critical role in bridging the gap between energy generation and consumption, but their scalability and cost-effectiveness need continuous improvement. The vast amount of data collected by IoT sensors requires robust cybersecurity measures to protect against potential breaches and ensure the privacy of citizens. Cities and communities must navigate complex regulatory and ethical considerations in managing this data, balancing between reaping the benefits of data analytics and safeguarding individual rights. The transition to renewable energy and IoT technologies requires substantial investments in infrastructure, research and development, and policy development. The high initial costs of implementing these technologies can be a hurdle for some regions and cities, although the long-term economic and environmental benefits are often significant.

7. CONCLUSION AND FUTURE SCOPE

Wind-solar renewable energy combined with IoT technology has enormous promise for tackling climate change, lowering carbon emissions, and improving urban sustainability. To achieve carbon neutrality in smart cities, a multifaceted strategy that uses clean energy sources and modern technology to build a green and sustainable future is required. This assimiliation with IoT technologies in the context of smart and sustainable cities marks a critical step toward tackling climate change and constructing a green, resilient, and sustainable future. These technologies have enormous potential for lowering carbon emissions, improving energy efficiency, and improving urban living conditions. Cleaner energy sources, lower energy waste, and better-informed decision-making processes are all advantages of this integration, leading to a healthier and more ecologically responsible urban landscape. Nevertheless, challenges exist, such as intermittency, data security, and initial investment costs. Overcoming these obstacles requires coordinated efforts from governments, industry players, and local communities. Policymakers should enact supportive regulations and incentives to promote renewable energy adoption and IoT integration.

Meanwhile, the private sector and research institutions must continue to innovate, improving energy storage solutions, data security protocols, and the scalability of IoT infrastructure.

The future scope of this integration is undeniably vast. As technology advances, it can anticipate even more efficient and cost-effective renewable energy systems and IoT applications. Enhanced energy storage solutions will alleviate the intermittency issues associated with wind and solar power, while machine learning and artificial intelligence will offer increasingly sophisticated data analytics capabilities. Smart cities will become more interconnected and data-driven, fostering more sustainable and environmentally conscious urban environments. The global transition towards a green and sustainable future depends on the dissemination of best practices and knowledge sharing. Cities, communities, and nations worldwide can learn from successful case studies and collaborative efforts, adapting and implementing these technologies to suit their unique needs and challenges. As the global commitment to addressing climate change intensifies, the assimilation of wind-solar renewable energy and IoT technologies will play an increasingly pivotal role in our collective efforts to build a cleaner, greener, and more sustainable world for future generations. It represents a future vision and a roadmap to a more sustainable and environmentally responsible way of living. This chapter recommends that cities and policymakers prioritize investments in wind-solar renewable energy and IoT technologies. Collaboration between the public and private sectors should be encouraged to facilitate the development of smart, sustainable cities. Future research should explore these projects' scalability and economic viability in diverse urban contexts. Further investigations are needed into data privacy and security concerns associated with IoT applications in smart cities.

REFERENCES

Abualigah, L., Zitar, R. A., Almotairi, K. H., Hussein, A. M., Abd Elaziz, M., Nikoo, M. R., & Gandomi, A. H. (2022). Wind, solar, and photovoltaic renewable energy systems with and without energy storage optimization: A survey of advanced machine learning and deep learning techniques. *Energies*, *15*(2), 578. DOI: 10.3390/en15020578

Ahmad, T., Madonski, R., Zhang, D., Huang, C., & Mujeeb, A. (2022). Data-driven probabilistic machine learning in sustainable smart energy/smart energy systems: Key developments, challenges, and future research opportunities in the context of smart grid paradigm. *Renewable & Sustainable Energy Reviews*, *160*, 112128.

Ahmad, T., & Zhang, D. (2021). Using the internet of things in smart energy systems and networks. *Sustainable Cities and Society*, *68*, 102783. DOI: 10.1016/j.scs.2021.102783

Ahmed, F., Naeem, M., & Iqbal, M. (2017). ICT and renewable energy: A way forward to the next generation telecom base stations. *Telecommunication Systems*, *64*(1), 43–56. DOI: 10.1007/s11235-016-0156-4

Al-tabatabaie, K. F., Hossain, M. B., Islam, M. K., Awual, M. R., TowfiqulIslam, A. R. M., Hossain, M. A., Esraz-Ul-Zannat, M., & Islam, A. (2022). Taking strides towards decarbonization: The viewpoint of Bangladesh. *Energy Strategy Reviews*, *44*, 100948. DOI: 10.1016/j.esr.2022.100948

Albreem, M. A., Sheikh, A. M., Alsharif, M. H., Jusoh, M., & Yasin, M. N. M. (2021). Green Internet of Things (GIoT): Applications, practices, awareness, and challenges. *IEEE Access : Practical Innovations, Open Solutions*, *9*, 38833–38858. DOI: 10.1109/ACCESS.2021.3061697

Aldieri, L., Grafström, J., & Vinci, C. P. (2021). The effect of Marshallian and Jacobian knowledge spillovers on jobs in the solar, wind and energy efficiency sector. *Energies*, *14*(14), 4269. DOI: 10.3390/en14144269

Amran, Y. A., Amran, Y. M., Alyousef, R., & Alabduljabbar, H. (2020). Renewable and sustainable energy production in Saudi Arabia according to Saudi Vision 2030; Current status and future prospects. *Journal of Cleaner Production*, *247*, 119602. DOI: 10.1016/j.jclepro.2019.119602

Arbabzadeh, M., Gençer, E., Morris, J. F., Paltsev, S., & Armstrong, R. C. (2019). *Plausible Energy Futures: A Framework for Evaluating Options*. Impacts, and National Energy Choices.

Arent, D. J., Green, P., Abdullah, Z., Barnes, T., Bauer, S., Bernstein, A., & Turchi, C. (2022). Challenges and opportunities in decarbonizing the US energy system. *Renewable & Sustainable Energy Reviews*, *169*, 112939. DOI: 10.1016/j.rser.2022.112939

Aslam, S., Herodotou, H., Mohsin, S. M., Javaid, N., Ashraf, N., & Aslam, S. (2021). A survey on deep learning methods for power load and renewable energy forecasting in smart microgrids. *Renewable & Sustainable Energy Reviews*, *144*, 110992.

Atalay, Y., Biermann, F., & Kalfagianni, A. (2016). Adoption of renewable energy technologies in oil-rich countries: Explaining policy variation in the Gulf Cooperation Council states. *Renewable Energy*, *85*, 206–214. DOI: 10.1016/j.renene.2015.06.045

Basheer, M., Nechifor, V., Calzadilla, A., Ringler, C., Hulme, D., & Harou, J. J. (2022). Balancing national economic policy outcomes for sustainable development. *Nature Communications*, *13*(1), 5041. DOI: 10.1038/s41467-022-32415-9 PMID: 36028486

Bhattacharjee, S., & Nandi, C. (2019). Implementation of industrial internet of things in the renewable energy sector. *The Internet of Things in the industrial sector: security and device connectivity, smart environments, and industry 4.0*, 223-259.

Bibri, S. E. (2018). The IoT for smart sustainable cities of the future: An analytical framework for sensor-based big data applications for environmental sustainability. *Sustainable Cities and Society*, *38*, 230–253. DOI: 10.1016/j.scs.2017.12.034

Bibri, S. E., & Krogstie, J. (2020). Environmentally data-driven smart sustainable cities: Applied innovative solutions for energy efficiency, pollution reduction, and urban metabolism. *Energy Informatics*, *3*(1), 1–59. DOI: 10.1186/s42162-020-00130-8

Bogdanov, D., Ram, M., Aghahosseini, A., Gulagi, A., Oyewo, A. S., Child, M., & Breyer, C. (2021). Low-cost renewable electricity as the key driver of the global energy transition towards sustainability. *Energy*, *227*, 120467. DOI: 10.1016/j.energy.2021.120467

Budiman, I. (2018). Enabling community participation for social innovation in the energy sector. *Indonesian Journal of Energy*, *1*(2), 21–31. DOI: 10.33116/ije.v1i2.23

Bujang, A. S., Bern, C. J., & Brumm, T. J. (2016). Summary of energy demand and renewable energy policies in Malaysia. *Renewable & Sustainable Energy Reviews*, *53*, 1459–1467. DOI: 10.1016/j.rser.2015.09.047

Cantarero, M. M. V. (2020). Of renewable energy, energy democracy, and sustainable development: A roadmap to accelerate the energy transition in developing countries. *Energy Research & Social Science*, *70*, 101716. DOI: 10.1016/j.erss.2020.101716

Clark, I. I. W., & Cooke, G. (2016). *Smart green cities: toward a carbon neutral world*. Routledge. DOI: 10.4324/9781315609386

Cohen, S. M., Iyer, G. C., Brown, M., Macknick, J., Wise, M., Binsted, M., & Hejazi, M. (2021). How structural differences influence cross-model consistency: An electric sector case study. *Renewable & Sustainable Energy Reviews*, *144*, 111009. DOI: 10.1016/j.rser.2021.111009

Connolly, D., Lund, H., Mathiesen, B. V., & Leahy, M. (2010). A review of computer tools for analysing the integration of renewable energy into various energy systems. *Applied Energy*, *87*(4), 1059–1082. DOI: 10.1016/j.apenergy.2009.09.026

Cortese, T. T. P., Almeida, J. F. S. D., Batista, G. Q., Storopoli, J. E., Liu, A., & Yigitcanlar, T. (2022). Understanding sustainable energy in the context of smart cities: A PRISMA review. *Energies*, *15*(7), 2382. DOI: 10.3390/en15072382

Cruz-Rojas, T., Franco, J. A., Hernandez-Escobedo, Q., Ruiz-Robles, D., & Juarez-Lopez, J. M. (2023). A novel comparison of image semantic segmentation techniques for detecting dust in photovoltaic panels using machine learning and deep learning. *Renewable Energy*, *217*, 119126. DOI: 10.1016/j.renene.2023.119126

Dai, H., Xie, X., Xie, Y., Liu, J., & Masui, T. (2016). Green growth: The economic impacts of large-scale renewable energy development in China. *Applied Energy*, *162*, 435–449. DOI: 10.1016/j.apenergy.2015.10.049

Damman, S., Sandberg, E., Rosenberg, E., Pisciella, P., & Graabak, I. (2021). A hybrid perspective on energy transition pathways: Is hydrogen the key for Norway? *Energy Research & Social Science*, *78*, 102116. DOI: 10.1016/j.erss.2021.102116

Danish, M. S. S., & Senjyu, T. (2023). AI-Enabled Energy Policy for a Sustainable Future. *Sustainability (Basel)*, *15*(9), 7643. DOI: 10.3390/su15097643

Del Ser, J., Casillas-Perez, D., Cornejo-Bueno, L., Prieto-Godino, L., Sanz-Justo, J., Casanova-Mateo, C., & Salcedo-Sanz, S. (2022). Randomization-based machine learning in renewable energy prediction problems: Critical literature review, new results and perspectives. *Applied Soft Computing*, *118*, 108526. DOI: 10.1016/j.asoc.2022.108526

Dent, C. M. (2012). Renewable energy and East Asia's new developmentalism: Towards a low carbon future? *The Pacific Review*, *25*(5), 561–587. DOI: 10.1080/09512748.2012.728240

Devabhaktuni, V., Alam, M., Depuru, S. S. S. R., Green, R. C.II, Nims, D., & Near, C. (2013). Solar energy: Trends and enabling technologies. *Renewable & Sustainable Energy Reviews*, *19*, 555–564.

Doost Mohammadian, H., & Rezaie, F. (2020). Blue-green smart mobility technologies as readiness for facing tomorrow's urban shock toward the world as a better place for living (Case studies: Songdo and Copenhagen). *Technologies*, *8*(3), 39. DOI: 10.3390/technologies8030039

Elavarasan, R. M., Pugazhendhi, R., Jamal, T., Dyduch, J., Arif, M. T., Kumar, N. M., & Nadarajah, M. (2021). Envisioning the UN Sustainable Development Goals (SDGs) through the lens of energy sustainability (SDG 7) in the post-COVID-19 world. *Applied Energy*, *292*, 116665. DOI: 10.1016/j.apenergy.2021.116665

Elavarasan, R. M., Shafiullah, G. M., Padmanaban, S., Kumar, N. M., Annam, A., Vetrichelvan, A. M., & Holm-Nielsen, J. B. (2020). A comprehensive review on renewable energy development, challenges, and policies of leading Indian states with an international perspective. *IEEE Access : Practical Innovations, Open Solutions*, *8*, 74432–74457. DOI: 10.1109/ACCESS.2020.2988011

ElFar, O. A., Chang, C. K., Leong, H. Y., Peter, A. P., Chew, K. W., & Show, P. L. (2021). Prospects of Industry 5.0 in algae: Customization of production and new advance technology for clean bioenergy generation. *Energy Conversion and Management: X, 10*, 100048.

Eljack, F., & Kazi, M. K. (2021). Prospects and challenges of green hydrogen economy via multi-sector global symbiosis in Qatar. *Frontiers in Sustainability*, *1*, 612762. DOI: 10.3389/frsus.2020.612762

Elrahmani, A., Hannun, J., Eljack, F., & Kazi, M. K. (2021). Status of renewable energy in the GCC region and future opportunities. *Current Opinion in Chemical Engineering*, *31*, 100664. DOI: 10.1016/j.coche.2020.100664

Fischer, C., & Preonas, L. (2010). Combining policies for renewable energy: Is the whole less than the sum of its parts?. *Resource for the future discussion chapter*, (10-19).

Garrido, S., Sequeira, T., & Santos, M. (2020). Renewable energy and sustainability from the supply side: A critical review and analysis. *Applied Sciences (Basel, Switzerland)*, *10*(17), 5755. DOI: 10.3390/app10175755

Ghai, D., Tripathi, S. L., Saxena, S., Chanda, M., & Alazab, M. (Eds.). (2022). *Machine Learning Algorithms for Signal and Image Processing*. John Wiley & Sons. DOI: 10.1002/9781119861850

Grubb, M., Drummond, P., Poncia, A., McDowall, W., Popp, D., Samadi, S., & Pavan, G. (2021). Induced innovation in energy technologies and systems: A review of evidence and potential implications for CO2 mitigation. *Environmental Research Letters*, *16*(4), 043007. DOI: 10.1088/1748-9326/abde07

Guchhait, R., & Sarkar, B. (2023). Increasing Growth of Renewable Energy: A State of Art. *Energies*, *16*(6), 2665. DOI: 10.3390/en16062665

Hainsch, K., Löffler, K., Burandt, T., Auer, H., del Granado, P. C., Pisciella, P., & Zwickl-Bernhard, S. (2022). Energy transition scenarios: What policies, societal attitudes, and technology developments will realize the EU Green Deal? *Energy*, *239*, 122067. DOI: 10.1016/j.energy.2021.122067

Hannan, M. A., Al-Shetwi, A. Q., Ker, P. J., Begum, R. A., Mansor, M., Rahman, S. A., Dong, Z. Y., Tiong, S. K., Mahlia, T. M. I., & Muttaqi, K. M. (2021). Impact of renewable energy utilization and artificial intelligence in achieving sustainable development goals. *Energy Reports*, *7*, 5359–5373. DOI: 10.1016/j.egyr.2021.08.172

Hassan, Q., Sameen, A. Z., Olapade, O., Alghoul, M., Salman, H. M., & Jaszczur, M. (2023). Hydrogen fuel as an important element of the energy storage needs for future smart cities. *International Journal of Hydrogen Energy*, *48*(78), 30247–30262. DOI: 10.1016/j.ijhydene.2023.03.413

Hoffmann, M. W., Wildermuth, S., Gitzel, R., Boyaci, A., Gebhardt, J., Kaul, H., & Tornede, T. (2020). Integration of novel sensors and machine learning for predictive maintenance in medium voltage switchgear to enable the energy and mobility revolutions. *Sensors (Basel)*, *20*(7), 2099. DOI: 10.3390/s20072099 PMID: 32276442

Ilyas, A., Parkinson, S., Vinca, A., Byers, E., Manzoor, T., Riahi, K., & Muhammad, A. (2022). Balancing smart irrigation and hydropower investments for sustainable water conservation in the Indus basin. *Environmental Science & Policy*, *135*, 147–161. DOI: 10.1016/j.envsci.2022.04.012

Issa Zadeh, S. B., Esteban Perez, M. D., López-Gutiérrez, J. S., & Fernández-Sánchez, G. (2023). Optimizing Smart Energy Infrastructure in Smart Ports: A Systematic Scoping Review of Carbon Footprint Reduction. *Journal of Marine Science and Engineering*, *11*(10), 1921. DOI: 10.3390/jmse11101921

Issa Zadeh, S. B., López Gutiérrez, J. S., Esteban, M. D., Fernández-Sánchez, G., & Garay-Rondero, C. L. (2023). Scope of the Literature on Efforts to Reduce the Carbon Footprint of Seaports. *Sustainability (Basel), 15*(11), 8558. DOI: 10.3390/su15118558

Jain, A., Sharma, A., Jately, V., & Azzopardi, B. (Eds.). (2023). *Sustainable Energy Solutions with Artificial Intelligence, Blockchain Technology, and Internet of Things*. CRC Press. DOI: 10.1201/9781003356639

Jayachandran, M., Gatla, R. K., Rao, K. P., Rao, G. S., Mohammed, S., Milyani, A. H., Azhari, A. A., Kalaiarasy, C., & Geetha, S. (2022). Challenges in achieving sustainable development goal 7: Affordable and clean energy in light of nascent technologies. *Sustainable Energy Technologies and Assessments, 53*, 102692. DOI: 10.1016/j.seta.2022.102692

Kasinathan, P., Pugazhendhi, R., Elavarasan, R. M., Ramachandaramurthy, V. K., Ramanathan, V., Subramanian, S., & Alsharif, M. H. (2022). Realization of Sustainable Development Goals with Disruptive Technologies by Integrating Industry 5.0, Society 5.0, Smart Cities and Villages. *Sustainability (Basel), 14*(22), 15258. DOI: 10.3390/su142215258

Lawrenz, L., Xiong, B., Lorenz, L., Krumm, A., Hosenfeld, H., Burandt, T., & Von Hirschhausen, C. (2018). Exploring energy pathways for the low-carbon transformation in India—A model-based analysis. *Energies, 11*(11), 3001. DOI: 10.3390/en11113001

Li, D., Guan, X., Tang, T., Zhao, L., Tong, W., & Wang, Z. (2023). The clean energy development path and sustainable development of the ecological environment driven by big data for mining projects. *Journal of Environmental Management, 348*, 119426. DOI: 10.1016/j.jenvman.2023.119426 PMID: 37879178

Li, W., Lu, C., & Zhang, Y. W. (2019). Prospective exploration of future renewable portfolio standard schemes in China via a multi-sector CGE model. *Energy Policy, 128*, 45–56. DOI: 10.1016/j.enpol.2018.12.054

Makatjane, K. (2022). Deep Learning for Sentiment Analysis to Predict the Probability of Bank Loan Default. *American Journal of Data Mining and Knowledge Discovery, 7*(2), 5–12.

Makatjane, K., Moroke, N., & Ncube, B. (2021). Detecting Financial Fraud in South Africa: A Comparison of Logistic Model Tree and Gradient Boosting Decision Tree.

Manoharan, K. G., Nehru, J. A., & Balasubramanian, S. (2021). *Artificial Intelligence and IoT*. Springer. DOI: 10.1007/978-981-33-6400-4

Moemen, Y. S., Alshater, H., & El-Sayed, I. E. T. (2023). Clean Energy Management Based on Internet of Things and Sensor Networks for Climate Change Problems. In *The Power of Data: Driving Climate Change with Data Science and Artificial Intelligence Innovations* (pp. 117–136). Springer Nature Switzerland.

Mohammadian, H. D. (2022). Mapping the Future Global SMEs Growth via Hybrid SMEs/SME 5.0/ Tomorrow's SMEs Concept through the 5th Wave, i-Sustainability Plus and DCT Theories. In *Management and Information Technology in the Digital Era* (Vol. 29, pp. 173–206). Emerald Publishing Limited. DOI: 10.1108/S1877-636120220000029011

Moroke, N. D., & Makatjane, K. (2022). Predictive Modelling for Financial Fraud Detection Using Data Analytics: A Gradient-Boosting Decision Tree. In Applications of Machine Learning and Deep Learning for Privacy and Cybersecurity (pp. 25-45). IGI Global. DOI: 10.4018/978-1-7998-9430-8.ch002

Nakata, T. (2004). Energy-economic models and the environment. *Progress in Energy and Combustion Science*, *30*(4), 417–475. DOI: 10.1016/j.pecs.2004.03.001

Nastasi, B., Markovska, N., Puksec, T., Duić, N., & Foley, A. (2022). Renewable and sustainable energy challenges to face for the achievement of Sustainable Development Goals. *Renewable & Sustainable Energy Reviews*, *157*, 112071. DOI: 10.1016/j.rser.2022.112071

Nazir, T., & Banday, M. T. (2018, October). Green Internet of Things: A survey of enabling techniques. In *2018 International Conference on Automation and Computational Engineering (ICACE)* (pp. 197-202). IEEE. DOI: 10.1109/ICACE.2018.8686974

Ozcanli, A. K., Yaprakdal, F., & Baysal, M. (2020). Deep learning methods and applications for electrical power systems: A comprehensive review. *International Journal of Energy Research*, *44*(9), 7136–7157. DOI: 10.1002/er.5331

Pakulska, T. (2021). Green energy in central and eastern European (CEE) countries: New challenges on the path to sustainable development. *Energies*, *14*(4), 884. DOI: 10.3390/en14040884

Pandey, V., Sircar, A., Bist, N., Solanki, K., & Yadav, K. (2023). Accelerating the renewable energy sector through Industry 4.0: Optimization opportunities in the digital revolution. *International Journal of Innovation Studies*, *7*(2), 171–188. DOI: 10.1016/j.ijis.2023.03.003

Pascaris, A. S. (2021). Examining existing policy to inform a comprehensive legal framework for agrivoltaics in the US. *Energy Policy*, *159*, 112620. DOI: 10.1016/j.enpol.2021.112620

Pimitha, K., Anbarasi, S., & Balasubramanian, T. (2023). Machine Learning Techniques for Renewable Energy Resources. In *IoT, Machine Learning and Blockchain Technologies for Renewable Energy and Modern Hybrid Power Systems* (pp. 121-148). River Publishers.

Raihan, A., Begum, R. A., Said, M. N. M., & Pereira, J. J. (2022). Relationship between economic growth, renewable energy use, technological innovation, and carbon emission toward achieving Malaysia's Paris agreement. *Environment Systems & Decisions*, *42*(4), 586–607. DOI: 10.1007/s10669-022-09848-0

Rao, S., Katoch, S., Narayanaswamy, V., Muniraju, G., Tepedelenlioglu, C., & Spanias, A. (2022). *Machine learning for solar array monitoring, optimization, and control*. Springer Nature.

Reedman, L., Chew, M. S., Gordon, J., Sue, W., Brinsmead, T., Hayward, J., & Havas, L. (2021). Multi-sector energy modelling.

Rosiello, A., & Maleki, A. (2021). A dynamic multi-sector analysis of technological catch-up: The impact of technology cycle times, knowledge base complexity and variety. *Research Policy*, *50*(3), 104194. DOI: 10.1016/j.respol.2020.104194

Saheb, T., Dehghani, M., & Saheb, T. (2022). Artificial intelligence for sustainable energy: A contextual topic modeling and content analysis. *Sustainable Computing : Informatics and Systems*, *35*, 100699. DOI: 10.1016/j.suscom.2022.100699

Saini, G., Kannan, R., Benini, E., & Kumar, K. (Eds.). (2023). *Enabling Methodologies for Renewable and Sustainable Energy*. CRC Press. DOI: 10.1201/9781003272717

Salam, A., & Salam, A. (2020). Internet of things in sustainable energy systems. *Internet of Things for Sustainable Community Development: Wireless Communications. Sensory Systems*, •••, 183–216.

Salleh, S. F., Mohd Roslan, M. E., Abd Rahman, A., Shamsuddin, A. H., Tuan Abdullah, T. A. R., & Sovacool, B. K. (2020). Transitioning to a sustainable development framework for bioenergy in Malaysia: Policy suggestions to catalyse the utilisation of palm oil mill residues. *Energy, Sustainability and Society*, *10*(1), 1–20. DOI: 10.1186/s13705-020-00269-y

Scarlat, N., Dallemand, J. F., Monforti-Ferrario, F., & Nita, V. (2015). The role of biomass and bioenergy in a future bioeconomy: Policies and facts. *Environmental Development*, *15*, 3–34. DOI: 10.1016/j.envdev.2015.03.006

Sharma, A., & Singh, B. (2022). Measuring Impact of E-commerce on Small Scale Business: A Systematic Review. *Journal of Corporate Governance and International Business Law*, *5*(1).

Silva, B. V., Holm-Nielsen, J. B., Sadrizadeh, S., Teles, M. P., Kiani-Moghaddam, M., & Arabkoohsar, A. (2023). Sustainable, green, or smart? Pathways for energy-efficient healthcare buildings. *Sustainable Cities and Society*, •••, 105013.

Sinha, M., Chacko, E., Makhija, P., & Pramanik, S. (2021). Energy-Efficient smart cities with green internet of things. *Green Technological Innovation for Sustainable Smart Societies: Post Pandemic Era*, 345-361.

Sodiq, A., Baloch, A. A., Khan, S. A., Sezer, N., Mahmoud, S., Jama, M., & Abdelaal, A. (2019). Towards modern sustainable cities: Review of sustainability principles and trends. *Journal of Cleaner Production*, *227*, 972–1001. DOI: 10.1016/j.jclepro.2019.04.106

Steen, M., & Weaver, T. (2017). Incumbents' diversification and cross-sectorial energy industry dynamics. *Research Policy*, *46*(6), 1071–1086. DOI: 10.1016/j.respol.2017.04.001

Szeliski, R. (2022). *Computer vision: algorithms and applications*. Springer Nature. DOI: 10.1007/978-3-030-34372-9

Tabor, D. P., Roch, L. M., Saikin, S. K., Kreisbeck, C., Sheberla, D., Montoya, J. H., Dwaraknath, S., Aykol, M., Ortiz, C., Tribukait, H., Amador-Bedolla, C., Brabec, C. J., Maruyama, B., Persson, K. A., & Aspuru-Guzik, A. (2018). Accelerating the discovery of materials for clean energy in the era of smart automation. *Nature Reviews. Materials*, *3*(5), 5–20. DOI: 10.1038/s41578-018-0005-z

Thornbush, M. J., & Golubchikov, O. (2019). *Sustainable urbanism in digital transitions: from low carbon to smart sustainable cities*. Springer.

Tiwari, S., Rosak-Szyrocka, J., & Żywiołek, J. (2022). Internet of things as a sustainable energy management solution at tourism destinations in India. *Energies*, *15*(7), 2433. DOI: 10.3390/en15072433

Tomar, A., Malik, H., Kumr, P., & Iqbal, A. (2022, September). Machine Learning, Advances in Computing, Renewable Energy and Communication (MARC). In *Proceedings of 3rd International Conference on Machine Learning, Advances in Computing, Renewable Energy and Communication: MARC 2021* (pp. 1-19). Singapore: Springer Nature Singapore. DOI: 10.1007/978-981-19-2828-4_1

Tuysuz, M. F., & Trestian, R. (2020). From serendipity to sustainable green IoT: Technical, industrial and political perspective. *Computer Networks*, *182*, 107469. DOI: 10.1016/j.comnet.2020.107469

Varjovi, A. E., & Babaie, S. (2020). Green Internet of Things (GIoT): Vision, applications and research challenges. *Sustainable Computing : Informatics and Systems*, *28*, 100448. DOI: 10.1016/j.suscom.2020.100448

Vinichenko, V. (2018). Mechanisms of energy transitions: National cases and the worldwide uptake of wind and solar power. *PhD diss., Central European University.*

Watson, R. (2022). The virtual economy of the metaverse: Computer vision and deep learning algorithms, customer engagement tools, and behavioral predictive analytics. *Linguistic and Philosophical Investigations*, (21), 41–56.

Westman, L. K., Broto, V. C., & Huang, P. (2019). Revisiting multi-level governance theory: Politics and innovation in the urban climate transition in Rizhao, China. *Political Geography*, *70*, 14–23. DOI: 10.1016/j.polgeo.2019.01.002

White, D., Winchester, N., Atkins, M., Ballingall, J., Coates, S., de Miguel Mercader, F., & Suomalainen, K. (2018). Energy-and multi-sector modelling of climate change mitigation in New Zealand: current practice and future needs.

Winfield, M., & Dolter, B. (2014). Energy, economic and environmental discourses and their policy impact: The case of Ontario s Green Energy and Green Economy Act. *Energy Policy*, *68*, 423–435. DOI: 10.1016/j.enpol.2014.01.039

Xiao, M., Simon, S., & Pregger, T. (2019). Scenario analysis of energy system transition-A case study of two coastal metropolitan regions, eastern China. *Energy Strategy Reviews*, *26*, 100423. DOI: 10.1016/j.esr.2019.100423

Xie, Y., Wu, X., Hou, Z., Li, Z., Luo, J., Lüddeke, C. T., & Liao, J. (2023). Gleaning insights from German energy transition and large-scale underground energy storage for China's carbon neutrality. *International Journal of Mining Science and Technology*, *33*(5), 529–553. DOI: 10.1016/j.ijmst.2023.04.001

Yang, D., Liu, D., Huang, A., Lin, J., & Xu, L. (2021). Critical transformation pathways and socio-environmental benefits of energy substitution using a LEAP scenario modeling. *Renewable & Sustainable Energy Reviews*, *135*, 110116. DOI: 10.1016/j.rser.2020.110116

Yenugula, M., Sahoo, S., & Goswami, S. (2024). Cloud computing for sustainable development: An analysis of environmental, economic and social benefits. *Journal of future sustainability*, *4*(1), 59-66.

Zhukovskiy, Y. L., Batueva, D. E., Buldysko, A. D., Gil, B., & Starshaia, V. V. (2021). Fossil energy in the framework of sustainable development: Analysis of prospects and development of forecast scenarios. *Energies*, *14*(17), 5268. DOI: 10.3390/en14175268

Zwane, N., Tazvinga, H., Botai, C., Murambadoro, M., Botai, J., De Wit, J., & Mabhaudhi, T. (2022). A bibliometric analysis of solar energy forecasting studies in Africa. *Energies*, *15*(15), 5520. DOI: 10.3390/en15155520

Chapter 4
Artificial Intelligence and Machine Learning in Digital and Analog VLSI:
Sustainability and Energy Management in Smart Cities

Saurabh Chandra
https://orcid.org/0000-0003-4172-9968
Bennett University, Greater Noida, India

Bhupinder Singh
https://orcid.org/0009-0006-4779-2553
Sharda University, India

ABSTRACT

Artificial Intelligence (AI) and Machine Learning (ML) are reshaping the frontiers of Very-Large-Scale Integration (VLSI) by revolutionizing energy management, vital for sustainable smart cities. VLSI, in which millions of transistors are combined onto a single chip, contributes significantly to the development of low-power electronic systems. These systems are further optimized by AI & ML techniques in terms of design, manufacturing to operations and help energy consumption from becoming bare minimal and their performance optimization. This chapter comprehensively explores the diverse aspects of artificial intelligence and machine learning in VLSI for energy management towards sustainability in smart cities.

1. INTRODUCTION

AI Facilitated VLSI Designs for Smarter Power Management: In the case of smart cities, which have a huge load on energy and exploitation can result in massive wastage reducing sustainability opportunity taking as it is diagram has been introduced that Sensed by AI driven intelligent power management. AI and ML are involved in developing smart grids which helps to predict energy needs, dynamically allocate power distribution as also integrate renewable sources of energy that make the urban ecosystems sustainable (Arya et al., 2024)., AI-enabled VLSI systems can help in evolving the smart IoT devices

DOI: 10.4018/979-8-3373-0045-0.ch004

associated with all these sectors for efficient energy administration across different city departments. The integration of AI, ML and VLSI not only bolsters technology advancement but also gives birth to earth-friendly smart cities while complying international sustainability agendas (Gatti et al., 2024). This sort of dynamic power management is crucial for minimising the energy footprint in electronic devices deployed throughout smart cities (Kumar et al., 2024). These artificial intelligence (AI) -based predictive maintenance algorithms are designed to use data obtained from VLSI components and predict when failures will occur so that they can be prevented (Sharma & Singh, 2022). Taking this proactive and data-driven approach means systems run at the highest possible efficiency, preventing wastage through good quality energy monitoring (Rosiello& Maleki, 2024). It can also significantly extend the lifespan of electronic devices, which in turn makes it more likely for a malfunction to occur and thus reduces e-waste (Fischer & Preonas, 2010). ML applications like digital-analogue converters and other VLSI based things are also synthesized in smart grids to enhance energy distribution, make it more efficient. These models allow energy demand patterns to be forecast - helping balance supply and demand in real-time for more efficient management of renewable generation (Grubb et al., 2021).

Figure 1. Illuminates the landscapes of introduction section (Source- Original)

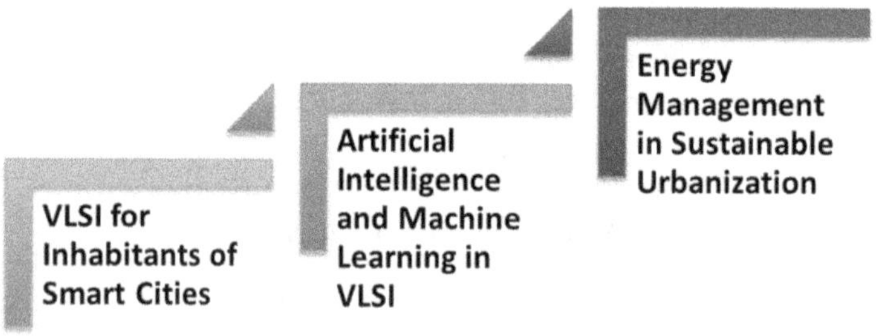

1.1 VLSI for Inhabitants of Smart Cities

VLSI technology is a driving force in the formation of smart cities, as it allows the realization of extremely complex integrated circuits (ICs) that are essential to modern urban infrastructure. One of the key elements making this smart city eco-system work is communication, driven by advanced technologies like Automation and Data processing- which in turn has been made possible due to VLSI (Very Large Scale Integration) which allows millions of transistors integrated on a single chip (Raihan et al., 2022). These ICs power a range of smart devices and systems; including sensors, controllers, and communication modules through which energy water transportation assets can be monitored in real-time. VLSI technology plays a key role in reducing devices size, lowering power consumption and improving the reliability of smart city systems. Its role in smart cities is intrinsic and extends across various verticals including traffic management, smart grids, environmental monitoring to public safety. VLSI Technology is that underlying hardware infrastructure to power these transformations, thus ensuring the long term scalability and sustainability of smart city applications at a city level as cities aim towards getting smarter and more connected.

1.2 Artificial Intelligence and Machine Learning in VLSI

Artificial Intelligence (AI) and Machine Learning (ML): AI and ML are changing the game for VLSI design, fundamentally with respect to smart cities. These technologies enable intelligent automation to be injected onto the entire VLSI (Very-Large Scale Integration) design process, delivering adaptive and energy efficient integrated circuits (Dai et al., 2016). There are many areas where AI and ML algorithms can be implemented so as to optimize the VLSI in terms of power, performance and fault tolerance (Elavarasan et al., 2020). AI-driven VLSI systems in smart cities for real-time data processing and decision-making as applications are evolved like smart grids, autonomous vehicles, IoT-based environmental monitoring. From predicting and preventing system failures to energy saving, ML models, specifically DL-based models can help in making VLSI circuits more efficiency. Additionally, with the use of AI and ML technology, self-learning and healing systems can be created that will learn from varied urban landscapes or changing sets of citizens so as to evolve along with these parameters this creates a perpetuity on smart city infrastructure. When AI and ML are combined with VLSI, it changes how circuits are created as well as their potential to function evolving into a new class of smarter city systems (Zwane et al., 2022).

1.3 Energy Management in Sustainable Urbanization

As cities expand and their demand for energy grows, effective management of this critical resource is an indispensable part of the equation in sustainable urban development. One of the key aspects to be addressed is efficient energy management in them since this can help reduce global carbon footprints, minimize wastage and act as a reliable means by which there will always be enough power at hand for millions. (Mohammadian, 2022). In this context, the VLSI Technology empowered by AI and ML are very promising in managing energy usage more efficiently within different urban sectors i.e., residential as well commercial areas including industrial. VLSI enables low-power IC design with intelligent energy management systems Thus, it plays a key role in saving the total amount of power consumed by smart cites (Amran et al., 2020). It is complemented by AI and ML to help in predictive maintenance, demand forecasting, real-time energy distribution that all leads up when resources are used efficiently as well successfully. It is not just about efficiency but energy management is an important aspect of resilience and sustainability (Hainsch et al., 2022). One of the Sustainable Development Goals (SDGs) that are essential for clean energy and climate action is to include smart grids, renewable energy integration and also with the availability of a powerful network support it. Proper management systems must manage power manage both supply side resources service provider Energy efficiency measures renewable energy acquisition. Market forces regularly coordinate demand-side response provides reliable estimates for clearing right deals Time-to-peak prices. Improve effective optimization obtained from AI algorithms enhance analytical tools on predicted forecasts (Atalay et al., 2016).

2. VLSI IN SMART CITIES

Modern Electronic Systems and Smart Cities leverage this technology which was etched in VLSI (Very Large Scale Integration) Technology. VLSI is the process of creating an integrated circuits to achieved by combining hundreds and thousands (1000s) transistors on a single chip, which increases the features, functionality or efficiency in all sort of electronic gadgets. VLSI technology is also one of the most important because these technologies enables us to use a lot of devices or systems together and help maintaining urban infrastructure in smart cities like communication network, environment monitoring system, energy management system etc. Efficient and scalable smart city applications demand attention towards the architecture and design principles of VLSI (Xiao et al., 2019).

Since VLSI processors are designed under the constraints of low power, minimum area and maximum performance which brings them in a higher demand because smart cities work on such requirements related to energy consumption. Low-power design, clock gating and power gating are different techniques to improve energy efficiency in VLSI circuits. Application-based integration of system-on-chip (SoC) technology is also critical to the reduction in size and power savings when multiple functionalities are integrated onto one chip instead of discrete components (Scarlat et al., 2015). Smart cities function with the help of several critical components -microprocessors, sensors and communication interfaces which are an integral part VLSI technology. Behind every smart device, a microprocessor sits as the necessary evil that allows intricate computations and control operations to be performed. The sensors are built in the VLSI chips, collecting real-time data from temperature to humidity, pollution levels and traffic which is necessary for smart management of a city. The communication interfaces facilitate data exchange among devices and results the interconnection nature of a smart city (Connolly et al., 2010).

As advantageous as VLSI technology can be, it has an equal amount of challenges when used in smart cities. These requirements consist of more processing power, better cooling efficiency and increased ruggedness in urban-warfare conditions (Salleh et al., 2020). Nevertheless, the continuous development of semiconductor technologies and their fusion with AI/ML, are some instrumental forces that will revolutionize these challenges to liberate all VLSI capabilities in smart cities. The basis for VLSI in smart cities is its ability to provide high-performance, energy-efficient and scalable solutions which are prerequisite primary requirements considering complexity of systems interconnected with modern urban setups. VLSI technologies continue to have a major role in enabling the evolution of smart cities that will drive many innovations designed to improve living conditions and contribute towards sustainability.

2.1 VLSI Architecture and Design Basics

Smart cities need such efficient, scalable and smart VLSI architectures integrated circuits so as to run well within the requirements for robustness of these environments. VLSI design flow focuses on PPA (Power, Performance and Area) trade-offs which are one of the most important things for creating energy-efficient devices in smaller size (Bujang et al., 2016). Modularity is one of the most critical principles in VLSI design, and indeed a lot of complex systems throughout history simply boil down into this: breaking up large or complicated pieces to fit within another structure that can handle the load. This allows for an easier design, experiment and integration of components. Then it goes into hierarchical design methodologies used to collect VLSI circuits of descendent abstraction levels, and divide over multiple layers from elementary elements like transistors and logic gates through the more developed blocks as processors or memories. Principal of Design for Testability (DFT) and Principal Design for manufacturability ensures we are not only able to make the circuit work but also test it properly in addition to large scale production. Unprecedented power optimization techniques like- clock gating and the other advanced design concepts of today via our transcriptions to minimize powers consumed as well in Smart Cities proving critical when it comes for energy efficiency. Together, they guarantee that VLSI circuits are able to perform at the high-performance levels needed for urban application requirements (Lawrenz et al., 2018).

2.2 Modules and Technologies in VLSI for Smart Cities

The functionalities offered by smart city applications are implemented with the help of VLSI comprising key building blocks and technologies. In essence, microprocessors at the heart of VLSI technology are those that serve as units for processing wide things of smart devices and systems such as traffic management controllers to environmental monitoring stations (Westman et al., 2019). These microprocessors are highly-integrated, including such features as processing power; memory and I/O interfaces that can be combined on the single chip to provide high performance at low-power. Another vital part of this set-up is the sensors that are planted into VLSI circuits, to capture numerous types of urban data viz. air quality levels, temperatures or traffic flow etc. Many of these sensors are married with wireless communication modules, allowing data transmission in real time across the infrastructure that makes up a smart city. Furthermore, for smart city an application in VLSI utilizes modern technologies like Field-Programmable Gate Arrays (FPGAs) and Application-Specific Integrated Circuits (ASIC), bringing forth flexibility as well permit customization. While you can program FPGAs to be flexible as needed on the market, providing FPGA with configurability and enable circuits to change according desire but ASICs keep their use case more concrete it generates them for this particular task that makes sense; hence are performances optimal (Cohen et al., 2021).

2.3 Challenging and Visionary VLSI for Urban Applications

The integration of urban cities using VLSI technologies both creates complex challenges and offers promises that could potentially determine the course for smart city advancement. A significant challenge is the demand for more processing power and data throughput as millions of sensors and device generate information in a smart city. With cities becoming increasingly ‘connected', the need for even more capable integrated circuits (ICs) grows, calling for advances in semiconductor materials and processing (Yang et al., 2021). The second biggest challenge is heat dissipation because the high density of VLSI circuits can generate a lot of heat, which may cause deterioration in performance and life span especially for urban environments. Power efficiency is important for this kind of applications as they are meant to be a part in smart city systems, and efficient circuits powered with low energy consumption modes need to maintain performance. Finally, while these are real challenges the flipside of them is that they present real innovation opportunities (White et al., 2018). Integrating AI and ML with VLSI can thus provide a platform for creating self-optimizing circuits that are capable of operating in dynamic environmental scenarios offering an enhancement in terms of efficiency as well as the resilience. There are also novel opportunities on VLSI circuits towards miniaturization and performance improvement with the development of nanotechnology and 3D integration. They provide solutions to current challenges, as well as a basis for the next round of smart city technology development, ensuring that urban entities are more sustainable and responsive; meeting the changing requirements of an emergent urban population (Nastasi et al., 2022).

Figure 2. Displays the points on foundations of VLSI in smart cities (Source- Original)

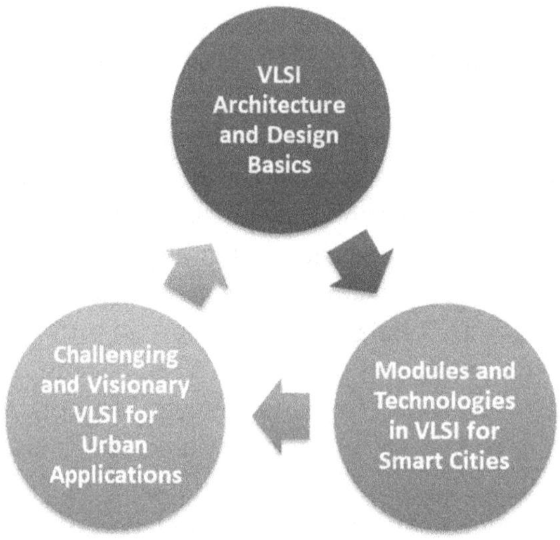

3. ARTIFICIAL INTELLIGENCE AND MACHINE LEARNING IN VLSI

The areas where Artificial Intelligence and Machine Learning play an important role is in furthering Very Large Scale Integration technology especially for smart cities with their focus on efficiency, resilience and scaling capabilities. This fusion of AI and ML in VLSI design is re-shaping the way circuits are designed, optimized and used as they enabling smarter systems to be built which is critical to cope with challenging requirements imposed by sophisticated urban environment (Danish & Senjyu, 2023). The major areas AI/ML is making a dramatic impact on the automation in terms of developing an automated process. Historically, VLSI design is a massive undertaking that requires domain expertise to optimize for performance power and area (PPA), an extensive amount of time with numerous man-hours spent on various elements. AI & ML algorithms have the ability to automate many of these tasks due by being trained with huge datasets on previous designs and patterns those lead to optimal results. An ML model might, for instance, be able to return the optimal distribution of transistors and interconnects in a circuit that will meet specific performance-based specifications while consuming as little power. This not only speeds up the design process but ultimately decreases errors, making for more robust and efficient solutions (Basheer et al., 2022).

AI and ML are also making VLSI circuits more useful while running in a real-time mode. AI and ML can be used in smart cities with VLSI circuits embedded across a number of devices like traffic management controllers or environmental sensor systems to optimize the performance of these components dynamically (Al-tabatabaie et al., 2022). For example, based on inputs from these systems, ML algorithms can predict and act in real-time by responding to changing conditions due to environmental needs (like demand for energy or patterns of traffic) (Aldieri et al., 2021). This allows the circuits to dynamically change their ways of operating, which is designed to make them more efficient and less power hungry. Likewise, VLSI with AI can also provide self-healing functionality on smart city infrastructure by circuits that detect faults of performance degradation and automatically reconfigure itself to operate optimally this increases reliability and life-expectancy for the devices (Nakata, 2004).

This AI, ML and VLSI combination could open a new path to design much smarter & efficient urban systems. For instance; in energy management AI-based VLSI circuits can model the expected consumed power distribution and usage patterns over different parts of a city so as to locally and instantaneously redistribute electricity for less power wastage overall contributing towards sustainable urban energy systems. Today, AI and ML can run with the help of VLSI circuits in Autonomous Vehicles which needs to process very large data from over a million sensors per second so that happens making real-time decisions thereby improving safety and throughput on Urban Transportation Networks. The incorporation of AI and ML in VLSI forms new dimensions for the customization of smart city applications. Examples of VLSI technologies where there is potential to integrate AI and ML are Field-Programmable Gate Arrays (FPGAs) and Application-Specific Integrated Circuits (ASICs). FPGAs, driven by AI tech can adapt to new requirements as needed, while ML-enhanced ASICs may be designed for maximum efficiency on particular tasks built specifically the needs of a smart city environment (Winfield & Dolter, 2014).

The combination of AI, ML and VLSI can be considered a big leap for smarter cities. AI and ML have revolutionized VLSI technology such that they do not only support swift turnkey designs, real-time refuges but also responsive idiosyncratic intelligent systems automating design processes. The smart cities of the future will further enable this growth, and more innovations are expected to improve urban infrastructures from sustainability, efficiency, and resilience perspective by using AI/ML empowered VLSI (Damman et al., 2021).

Figure 3. Specifies the artificial intelligence and machine learning in VLSI (Source- Original)

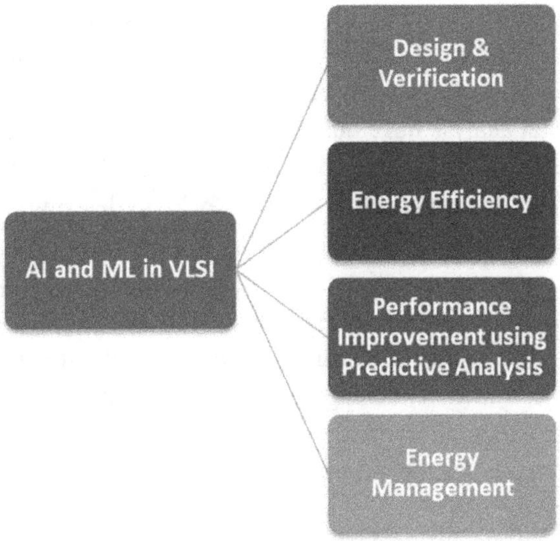

3.1 AI and ML in VLSI Design & Verification: Combination

Artificial Intelligence (AI) and Machine Learning (ML), when incorporated into VLSI design, are exploding the way integrated circuits are being developed making them extremely efficient and reliable. CLASS-8SSimplified design with timing prediction for fast evaluation of various configurations traditionally, physical implementation in VLSI has been a very complex and time-consuming process due to there are many iterations to optimize performance-power area (PAP). And all these processes are now being automated utilizing AI and ML algorithms. During design, ML models can comb through extensive libraries of past circuit designs in just fractions of a second to predict the best arrangement for all your components (or device geometries), reducing time spent on design iterations (Pascaris, 2021). It also predicts faulty structure before entering into the design process, to ensure less mistakes and more robust circuit in general. AI methodologies including Formal Verification and simulation based verification are used to validate the design against predefined requirements. On the hardware side, is there any way that ML models learns from verification runs in the past to find potential points of failure on a circuit and ensure even more focused test cases. This aids in fast design and verification but also helps in developing highly accurate & reliable VLSI circuits which adhere to the harsh environmental specification of smart cities (Budiman, 2018).

3.2 AI-Based Energy Efficiency Optimization Techniques

Artificial intelligence enabled optimization techniques are being exploited for improving energy efficiency of VLSI circuits which is an important aspect while designing smart city infrastructures. VLSI energy efficiency involves reducing the power consumption of circuits whilst retaining performance, an ever harder balance to achieve with increasing circuit complexity. Machine learning algorithms and other AI techniques are deployed for optimized use of resources to reach this equilibrium point. Using drones and multiple cameras, you can turn around look at the family of devices behind you to using a wide set of ML algorithms that analyse huge quantities data concerning circuit performance under various operating conditions and spot patterns leading power waste (Reedman et al., 2021). For instance, these ML models can oversee different types of circuit blocks to determine how best each part should be powered gated so that only the required sections of a respective I/O bridge are active at any given moment reducing idle power consumption. Circuits can also undergo voltage and frequency scaling to consume less power, that may be optimized by AI such as tuning changes for a circuit about any workload executed on the part so it scales its operating point accordingly. It gets better: AI can help design lower-power circuits from the start by choosing transistors sizes and arrangements (Xie et al., 2023). Such optimization techniques are vital for energy-efficient VLSI circuits, facilitating sustainable operations of smart city systems that consume only minute amounts of power due to limited accessible and renewable sources (Vinichenko, 2018).

3.3 VLSI Performance Improvement Using Machine Learning-based Predictive Analysis

In the emerging era of smart cities, enabling a high-performance improvement for Very Large Scale Integration (VLSI) circuits become more essential due to on-demand real-time decision-making availability and very-high reliability requisites especially when addressed using Machine Learning-based predictive analytics influence (Arbabzadeh et al., 2019). Predictive analytics uses ML algorithms to analyze what happened in the past, and predicts future outcomes allowing for proactive fine tuning of VLSI circuits that would ensure they will perform optimally. In VLSI, this approach can be used for different operation scenarios such as power prediction, thermal behavior and failure indications. By predicting these features accurately, ML models allow for dynamic optimizations to circuit configurations (voltage scaling, workload distribution) while retaining their expected performance (Ilyas et al., 2022). Predictive analytics can be used to predict hotspot in a circuit and can adjust power supply distribution according the prediction of hotspots so that they prevent themselves from getting fry; hence enhancing its lifespan, durability. In addition, ML-based predictive modeling capabilities can be utilized to find the correct time and vmcorle of signals within circuit which would reduce delay in every signal route thus improving overall performance. This proactive performance management ensures that VLSI circuits are able to scaling effectively in the face of changing conditions, making them an ideal fit for smart cities' dynamic and interconnected environments (Garrido et al., 2020).

3.4 Machine Learning Perspective in Energy Management in VLSI

Among these, energy management remains the key aspect of VLSI design and is crucial in smart city applications as a specialized domain where efficient energy consumption is essential. Energy management in VLSI circuits: Machine Learning (ML) based solutions for real-time data driven power optimization ML algorithms can be programmed with circuit performance data sets which are huge that help to find patterns and correlation through conventional analysis (Steen & Weaver, 2017). These algorithms are then able to predict the best feasible energy management strategy given some specific circumstances (e.g. different workloads or environmental factors). Some examples include the use of ML to enable dynamic voltage and frequency scaling (DVFS) in which its power supply or clock speed is automatically adjusted on-basis against run-time performance needs — effectively minimizing energy consumption (Hoffmann et al., 2020). ML models also help in optimizing power gating strategies by selectively powering down inactive regions of the circuit to prevent them from consuming idle power. This will not only advance energy conservation, by helping portable devices require less frequent battery charging and in-turn reduce the overall energy footprint for operation of smart city infrastructures as well. This will enable the sustainable development of urban areas as ML-based energy management systems can continuously self-learn from operational data to evolve and remain effective under dynamically changing conditions in VLSI circuits (Manoharan et al., 2021).

4. ENERGY MANAGEMENT APPROACHES IN SMART CITIES

Smart cities are developed with energy management; after all, this is useful for the development of sustainable urban areas where people can inhabit properly without leaving a large carbon footprint. In this time of transforming cities into complex, interlocked eco-systems; innovative energy management offers a way to achieve objectives (and SDGs) simultaneously (Ghai et al., 2022). These measures not only reduce energy consumption passively, but also integrate renewable energy sources together and optimize the distribution of electricity to advanced technologies such as Artificial Intelligence (AI) or IoT for a more flexible urban solution. A key component of smart city energy management is the incorporation of renewable resources like solar, wind and hydro power on to local urban grids. The drive towards renewables is motivated by these finite and carbon-intensive resources now known to be major contributors to global climate change. Connected cities utilize distributed energy systems, allowing for on-site power production that is often close to the point of consumption. This can include microgrids that are small sets of power grids capable of operating independently or in conjunction with a larger grid to provide energy resilient solutions for either peak demand times, when it is not optimal supply from the main system during disturbances (Pimitha et al., 2023).

The real-time monitoring and management of energy consumption, across city is possible via Advanced Metering Infrastructure (AMI) with Smart Grids. As an example, smart meters can report back on energy usage in extremely granular terms of time and quantity so that consumers as well as the utility providers both are able to make decisions efficiently regarding consumption. A more data-driven approach in capturing the patterns of energy usage, peak demand periods and places where efficiency can be improved. In addition, smart grids can establish communication between energy suppliers and homeowners that support real-time pricing models during peak hours in order to save electricity consumption but boost it on non-peak hour periods.

This data is analyzed by tools provided via these technologies and comes from various sources (i.e. smart meters, weather forecasts and historical consumption patterns) to forecast energy demand and thus adjusting supply accordingly Predictive analytics driven by AI, for example, can help make renewable sources of energy more efficient: how much juice should a solar panel or wind turbine put out based on weather conditions. AI can also operate energy storage systems, so any surplus electricity produced during off-peak times would be stored and redeployed efficiently when needed to boost grid supply. It stabilizes the grid besides utilising renewable energy to its maximum, thus saving non-renewable resources. Energy-efficient building designs and modulating the energy-using systems to retrofit edifices are another vital component of making our cities smarter. Energy-efficient materials and machinery such as solar energy, smart lighting, HVAC systems etc. are few examples of green buildings which play a vital role in reducing the total consumption of energy within any city. There are also significant savings when it comes to retrofitting older buildings with smart thermostats, energy-efficient windows and insulation. In addition, more and more cities are deploying energy management systems that enable building managers to track and manage the usage of power in real time, these smart technologies can even adjust how much power a facility uses based on occupancy rates or weather.

Smart cities are reinventing the public transportation systems to be much more energy-efficient. Therefore, the electrification of public transport buses and trains together are largely responsible for urban mobility's carbon foot print. In addition, energy consumption is reduced by increasing the efficiency of traffic flow and reducing congestion through intelligent control systems that manage programs for solving complex problems. A multi-diversity search-based algorithm is proposed for energy management in smart cities considering the approaches to integrate renewable energy generation assets and introduce Advanced Metering Infrastructure with Demand Response through Smart Grid along with Artificial Intelligence aid on one side, while dealing with adaptive building technology that include variable HVAC systems along vehicle to grid communication system. To build sustainable energy systems in the cities of tomorrow, it is imperative that these strategies be effective at scaling-up. The Solutions in the long-term sustainability strategies as they redefine and are further accepted will play a vital role of smart cities at evolution.

4.1 Urban Energy Challenges

With a rapidly growing population and more industrial activities occurring, the demand for gasoline and natural gas will be forever on the rise. Cities have many energy challenges. Rapid urbanization due to sprawling cities resulting in high per capita energy consumption that puts further stress on already strained infrastructure and resources. The big issues are reliable and continuous energy supplies for residents plus commercial and industrial users. With many cities experiencing aging energy infrastructure, the physical and digital distribution of power becomes both difficult to manage and prone to inefficiency or even failure. Moreover, cities are a major source of carbon emissions given their heavy reliance on fossil fuels for electricity generation and transportation. This represents a daunting environmental problem for us city dwellers aspiring to downsize our carbon footprints and slow economic development. A second important challenge is the handling of off-peak energy need that usually leads to inflated demand cost; meaning operation gets expensive. While the clean nature of renewable energy is something to be appreciated, it also comes with its cons (e.g., lack of inertia services for solar and wind type generators due to their intermittent generator characteristic), which demand advanced solutions in terms of energy storage & management. Urban energy issues must be addressed by coherent solutions and novel approaches to progress towards sustainable cities.

4.2 VLSI Role in Smart Energy Solutions

Smart Energy Solutions for Urban Environments using Very Large Scale Integration (VLSI) in Smart Cities VLSI technology provides a way to design and implement modern, ultra-efficient integrated circuits that are vital in the control of energy supply for different applications. The VLSI Chips are essentially required for real-time monitoring, data processing and control in smart grids which enables renewable energies to be integrated into the grid as well distribution of electrical power more effectively. smart meters, which provide consumers and utility providers with in-depth information about energy consumption trends that can change the way decisions are made on what to break down. Several large international electronic companies are choosing VLSI, as it enables the fabrication of energy-efficient devices and systems for both micro-level (ICs) and macro level applications. In addition, the miniaturizi national capacity of VLSI en ables intelligent energy management to be built into virtually any device from home appliances up to industrial plant and helps in overall power saving. VLSI technology will play a critical role in delivering significant improvements to the way we use and consume energy, which is only possible when cities evolve into smart connected ecosystems capable of meeting their own energy needs effectively while contributing towards tempering the two sided torrential challenge that modern urban centers.

4.3 AI and ML Techniques for Smart Cities Energy Conservation

Machine learning and artificial intelligence contribute in energy savings of smart city. These algorithms use packed data from diverse profiles like smart meters, weather predictions along with previous usage trends to forecast the electricity demand and target potential energy savings. These allow for the micro-adjustment of demand at a citywide level to ensure resources are used more effectively as and when they are required. For example, energy management systems driven by artificial inteligentsy can be used to optimise building HVAC (heating ventilation air conditioning) operations – the temperatures and flow rates appropriate for occupation or external conditions so power utilisation will remain minimal.

ML models can also be used to make renewable energy integration more efficient by predicting the output of solar panels and wind turbines then operating them in such a way that aspects are optimized. AI algorithms capable of directing traffic flow are also in transportation, which help reduce the congestion and fuel consumption that occur during transport as a result to energy efficiency being valid. The AI and ML powers the smart grid in which energy distribution is constantly optimized to meet supply with demands leading towards better efficiency & reliability of power source consuming less electricity as energy wastage. They have to be at the heart of efforts towards building intelligent cities that are both amenable to urbanization and energy conservation: methods for sustainable development.

5. SUSTAINABILITY IN SMART CITIES: VLSI PERSPECTIVE

Sustainability is a key strategic goal of smart city development, which aims to create cities capable serving the needs of present and future generations with minimal environmental impact. So from a Very Large Scale Integration (VLSI) standpoint of view, being sustainable means using the developments available through advanced semiconductor capabilities to improve energy efficiency and preservation or even better exploitation of resources by injection renewable sources in our systems. Smart cities are built on energy-efficient systems and inorder to design VLSI techniques come hand in gloves with new technology. Smart grids manage energy distribution in real-time and use VLSI-based sensors & processors as part of them. In short, VLSI circuits are devices that help effectively save energy in a variety of ways and greatly reduce the energy waste needed for sustainable city building.

The first point has to do with sustainability, where VLSI plays a crucial role in all the functioning of electronic devices that are designed for the load present in smart cities. Without it, you say "goodbye" to the dream of a world filled with IoT devices whose data could monitor and control all manners of urban life from traffic management tools in smart cities to environmental monitoring platforms (Rajaan et al., 2024). VLSI technology allows these devices to operate efficiently and dependably—as well as have energy-harvesting capabilities helping them further decrease their dependence on traditional power sources. VLSI helps in integrating renewable energy with the city grid (HB et al., 2024). The control systems of solar panels, wind turbines and energy storage units utilize advanced VLSI circuits which in turn optimize their performance to facilitate harvesting renewable energy successfully. This not only helps to reduce dependence on fossil fuels, but also increases the resilience and sustainability of urban energy systems (Oliveira et al., 2024).

VLSI technology is the bedrock support for sustainability in smart cities implementing tools and practices which enables energy robust as well as green landscapes of city arrangement. Smart cities will continue to mature, and VLSI will be at the heart of this sustainability revolution as urbanization challenges in a resource-constrained world become more focal. VLSI technology plays a significant role in the construction of power saving data centers and communication networks required to handle massive amount of data produced from smart cities (Nguyen et al., 2024). VLSI plays an essential role in smart city development by improving the energy efficiency of these critical infrastructures, which leads to the overall carbon footprint reduction and better environment sustainability (Zen et al., 2024).

6. CONCLUSION AND FUTURE SCOPE

AI and ML provide a new realm of efficiency in the design, optimization, and performance of VLSI circuits which increases energy use efficiency, reduces power consumption up to minimum values for prolonged battery life time operation. They also drive vast majority low cost micro grids to integrate with renewable sources without compromising quality reliabilities demanded by most linear loads like computing or graphic model rendering operations. This symbiosis not only helps tackle the increasing energy needs of cities but also reinforces a much-needed breakthrough towards sustainability worldwide by reducing city strain on natural resources. By leveraging AI-driven predictive analytics and ML-based optimization techniques, energy management systems can be fine-tuned in real time to enable smart cities to instantly react to altered conditions or demands. This in turn results into more robust and flexible urban energy infrastructure capable to meet the diverse requirements of modern cities. The future scope of AI and ML in VLSI for smart cities is infinite. With machine learning algorithms increasingly advancing, it has the potential to create more complex state-of-the-art energy management strategies and even for some to develop fully autonomous self-optimized thermal systems as well as decision making. Furthermore, continuous improvements in VLSI technology will result in ultra-low-power circuits and devices and make smart cities compared with current methods more sustainable.

REFERENCES

Al-tabatabaie, K. F., Hossain, M. B., Islam, M. K., & Awual, M. R., TowfiqulIslam, A. R. M., Hossain, M. A., & Islam, A. (2022). Taking strides towards decarbonization: The viewpoint of Bangladesh. *Energy Strategy Reviews*, *44*, 100948.

Aldieri, L., Grafström, J., & Vinci, C. P. (2021). The effect of Marshallian and Jacobian knowledge spillovers on jobs in the solar, wind and energy efficiency sector. *Energies*, *14*(14), 4269. DOI: 10.3390/en14144269

Amran, Y. A., Amran, Y. M., Alyousef, R., & Alabduljabbar, H. (2020). Renewable and sustainable energy production in Saudi Arabia according to Saudi Vision 2030; Current status and future prospects. *Journal of Cleaner Production*, *247*, 119602.

Arbabzadeh, M., Gençer, E., Morris, J. F., Paltsev, S., & Armstrong, R. C. (2019). *Plausible Energy Futures: A Framework for Evaluating Options*. Impacts, and National Energy Choices.

Arya, A., Dixit, S., Sahoo, A. K., Babu, Y. R., Purohit, P., & Abdullah, I. (2024). Investigation into designing VLSI of a flexible architecture for a deep neural network accelerator. *Journal of Electrical Systems*, *20*(6s), 1024–1031.

Atalay, Y., Biermann, F., & Kalfagianni, A. (2016). Adoption of renewable energy technologies in oil-rich countries: Explaining policy variation in the Gulf Cooperation Council states. *Renewable Energy*, *85*, 206–214.

Basheer, M., Nechifor, V., Calzadilla, A., Ringler, C., Hulme, D., & Harou, J. J. (2022). Balancing national economic policy outcomes for sustainable development. *Nature Communications*, *13*(1), 5041. PMID: 36028486

Budiman, I. (2018). Enabling community participation for social innovation in the energy sector. *Indonesian Journal of Energy*, *1*(2), 21–31. DOI: 10.33116/ije.v1i2.23

Bujang, A. S., Bern, C. J., & Brumm, T. J. (2016). Summary of energy demand and renewable energy policies in Malaysia. *Renewable & Sustainable Energy Reviews*, *53*, 1459–1467.

Cohen, S. M., Iyer, G. C., Brown, M., Macknick, J., Wise, M., Binsted, M., & Hejazi, M. (2021). How structural differences influence cross-model consistency: An electric sector case study. *Renewable & Sustainable Energy Reviews*, *144*, 111009.

Connolly, D., Lund, H., Mathiesen, B. V., & Leahy, M. (2010). A review of computer tools for analysing the integration of renewable energy into various energy systems. *Applied Energy*, *87*(4), 1059–1082.

Dai, H., Xie, X., Xie, Y., Liu, J., & Masui, T. (2016). Green growth: The economic impacts of large-scale renewable energy development in China. *Applied Energy*, *162*, 435–449.

Damman, S., Sandberg, E., Rosenberg, E., Pisciella, P., & Graabak, I. (2021). A hybrid perspective on energy transition pathways: Is hydrogen the key for Norway? *Energy Research & Social Science*, *78*, 102116. DOI: 10.1016/j.erss.2021.102116

Danish, M. S. S., & Senjyu, T. (2023). AI-Enabled Energy Policy for a Sustainable Future. *Sustainability, 15*(9), 7643.

Elavarasan, R. M., Shafiullah, G. M., Padmanaban, S., Kumar, N. M., Annam, A., Vetrichelvan, A. M., & Holm-Nielsen, J. B. (2020). A comprehensive review on renewable energy development, challenges, and policies of leading Indian states with an international perspective. *IEEE Access : Practical Innovations, Open Solutions, 8*, 74432–74457.

Fischer, C., & Preonas, L. (2010). Combining policies for renewable energy: Is the whole less than the sum of its parts?. *Resource for the future discussion chapter*, (10-19).

Garrido, S., Sequeira, T., & Santos, M. (2020). Renewable energy and sustainability from the supply side: A critical review and analysis. *Applied Sciences (Basel, Switzerland), 10*(17), 5755. DOI: 10.3390/app10175755

Gatti, A., Barbierato, E., & Pozzi, A. (2024). Toward Greener Smart Cities: A Critical Review of Classic and Machine-Learning-Based Algorithms for Smart Bin Collection. *Electronics (Basel), 13*(5), 836.

Ghai, D., Tripathi, S. L., Saxena, S., Chanda, M., & Alazab, M. (Eds.). (2022). *Machine Learning Algorithms for Signal and Image Processing*. John Wiley & Sons. DOI: 10.1002/9781119861850

Grubb, M., Drummond, P., Poncia, A., McDowall, W., Popp, D., Samadi, S., & Pavan, G. (2021). Induced innovation in energy technologies and systems: A review of evidence and potential implications for CO2 mitigation. *Environmental Research Letters, 16*(4), 043007.

Hainsch, K., Löffler, K., Burandt, T., Auer, H., del Granado, P. C., Pisciella, P., & Zwickl-Bernhard, S. (2022). Energy transition scenarios: What policies, societal attitudes, and technology developments will realize the EU Green Deal? *Energy, 239*, 122067.

HB, M., & Ahammed, , ASM, U. (2024). Optimized Efficiency of IoT-Based Next Generation Smart Wireless Sensor Networks Using a Machine Learning Algorithm. *International Journal of Computing and Digital Systems, 17*(1), 1–13.

Hoffmann, M. W., Wildermuth, S., Gitzel, R., Boyaci, A., Gebhardt, J., Kaul, H., & Tornede, T. (2020). Integration of novel sensors and machine learning for predictive maintenance in medium voltage switchgear to enable the energy and mobility revolutions. *Sensors (Basel), 20*(7), 2099. DOI: 10.3390/s20072099 PMID: 32276442

Ilyas, A., Parkinson, S., Vinca, A., Byers, E., Manzoor, T., Riahi, K., & Muhammad, A. (2022). Balancing smart irrigation and hydropower investments for sustainable water conservation in the Indus basin. *Environmental Science & Policy, 135*, 147–161. DOI: 10.1016/j.envsci.2022.04.012

Kumar, P. V., Kulkarni, A., Mendhe, D., Keshar, D. K., Babu, S. B. T., & Rajesh, N. (2024, April). AI-Optimized Hardware Design for Internet of Things (IoT) Devices. In *2024 5th International Conference on Recent Trends in Computer Science and Technology (ICRTCST)* (pp. 21-26). IEEE.

Lawrenz, L., Xiong, B., Lorenz, L., Krumm, A., Hosenfeld, H., Burandt, T., & Von Hirschhausen, C. (2018). Exploring energy pathways for the low-carbon transformation in India—A model-based analysis. *Energies, 11*(11), 3001.

Manoharan, K. G., Nehru, J. A., & Balasubramanian, S. (2021). *Artificial Intelligence and Fuzzy Logics*. Springer.

Mohammadian, H. D. (2022). Mapping the Future Global SMEs Growth via Hybrid SMEs/SME 5.0/ Tomorrow's SMEs Concept through the 5th Wave, i-Sustainability Plus and DCT Theories. In *Management and Information Technology in the Digital Era* (Vol. 29, pp. 173–206). Emerald Publishing Limited.

Nakata, T. (2004). Energy-economic models and the environment. *Progress in Energy and Combustion Science*, *30*(4), 417–475. DOI: 10.1016/j.pecs.2004.03.001

Nastasi, B., Markovska, N., Puksec, T., Duić, N., & Foley, A. (2022). Renewable and sustainable energy challenges to face for the achievement of Sustainable Development Goals. *Renewable & Sustainable Energy Reviews*, *157*, 112071.

Nguyen, T. Q., Hoang, T., Zhang, L., Dobre, O. A., & Duong, T. Q. (2024). A Survey on Smart Optimisation Techniques for 6G-oriented Integrated Circuits Design. *Mobile Networks and Applications*, •••, 1–18.

Oliveira, F., Costa, D. G., Assis, F., & Silva, I. (2024). Internet of Intelligent Things: A convergence of embedded systems, edge computing and machine learning. *Internet of Things : Engineering Cyber Physical Human Systems*, *26*, 101153. DOI: 10.1016/j.iot.2024.101153

Pascaris, A. S. (2021). Examining existing policy to inform a comprehensive legal framework for agrivoltaics in the US. *Energy Policy*, *159*, 112620. DOI: 10.1016/j.enpol.2021.112620

Pimitha, K., Anbarasi, S., & Balasubramanian, T. (2023). Machine Learning Techniques for Renewable Energy Resources. In *Fuzzy Logics, Machine Learning and Blockchain Technologies for Renewable Energy and Modern Hybrid Power Systems* (pp. 121-148). River Publishers.

Raihan, A., Begum, R. A., Said, M. N. M., & Pereira, J. J. (2022). Relationship between economic growth, renewable energy use, technological innovation, and carbon emission toward achieving Malaysia's Paris agreement. *Environment Systems & Decisions*, *42*(4), 586–607.

Rajaan, R., Baishya, B. K., Rao, T. V., Pattanaik, B., Tripathi, M. A., & Anitha, R. (2024). Efficient Usage of Energy Infrastructure in Smart City Using Machine Learning. *EAI Endorsed Transactions on Internet of Things, 10*.

Reedman, L., Chew, M. S., Gordon, J., Sue, W., Brinsmead, T., Hayward, J., & Havas, L. (2021). Multisector energy modelling.

Rosiello, A., & Maleki, A. (2021). A dynamic multi-sector analysis of technological catch-up: The impact of technology cycle times, knowledge base complexity and variety. *Research Policy*, *50*(3), 104194.

Salleh, S. F., Mohd Roslan, M. E., Abd Rahman, A., Shamsuddin, A. H., Tuan Abdullah, T. A. R., & Sovacool, B. K. (2020). Transitioning to a sustainable development framework for bioenergy in Malaysia: Policy suggestions to catalyse the utilisation of palm oil mill residues. *Energy, Sustainability and Society*, *10*, 1–20.

Scarlat, N., Dallemand, J. F., Monforti-Ferrario, F., & Nita, V. (2015). The role of biomass and bioenergy in a future bioeconomy: Policies and facts. *Environmental Development*, *15*, 3–34.

Sharma, A., & Singh, B. (2022). Measuring Impact of E-commerce on Small Scale Business: A Systematic Review. *Journal of Corporate Governance and International Business Law*, *5*(1).

Steen, M., & Weaver, T. (2017). Incumbents' diversification and cross-sectorial energy industry dynamics. *Research Policy*, *46*(6), 1071–1086. DOI: 10.1016/j.respol.2017.04.001

Vinichenko, V. (2018). Mechanisms of energy transitions: National cases and the worldwide uptake of wind and solar power. *PhD diss., Central European University*.

Westman, L. K., Broto, V. C., & Huang, P. (2019). Revisiting multi-level governance theory: Politics and innovation in the urban climate transition in Rizhao, China. *Political Geography*, *70*, 14–23.

White, D., Winchester, N., Atkins, M., Ballingall, J., Coates, S., de Miguel Mercader, F., & Suomalainen, K. (2018). Energy-and multi-sector modelling of climate change mitigation in New Zealand: current practice and future needs.

Winfield, M., & Dolter, B. (2014). Energy, economic and environmental discourses and their policy impact: The case of Ontario s Green Energy and Green Economy Act. *Energy Policy*, *68*, 423–435. DOI: 10.1016/j.enpol.2014.01.039

Xiao, M., Simon, S., & Pregger, T. (2019). Scenario analysis of energy system transition-A case study of two coastal metropolitan regions, eastern China. *Energy Strategy Reviews*, *26*, 100423.

Xie, Y., Wu, X., Hou, Z., Li, Z., Luo, J., Lüddeke, C. T., & Liao, J. (2023). Gleaning insights from German energy transition and large-scale underground energy storage for China's carbon neutrality. *International Journal of Mining Science and Technology*, *33*(5), 529–553. DOI: 10.1016/j.ijmst.2023.04.001

Yang, D., Liu, D., Huang, A., Lin, J., & Xu, L. (2021). Critical transformation pathways and socio-environmental benefits of energy substitution using a LEAP scenario modeling. *Renewable & Sustainable Energy Reviews*, *135*, 110116.

Zen, T. V. (2024). VLSI Implementation of an 8051 Microcontroller Using VHDL and Re-Corrective Measure Using AI. In *Modern Approaches in Machine Learning and Cognitive Science: A Walkthrough* (Vol. 4, pp. 293–303). Springer International Publishing. DOI: 10.1007/978-3-031-43009-1_24

Zwane, N., Tazvinga, H., Botai, C., Murambadoro, M., Botai, J., De Wit, J., & Mabhaudhi, T. (2022). A bibliometric analysis of solar energy forecasting studies in Africa. *Energies*, *15*(15), 5520.

Chapter 5

Assimilating Fuzzy Logic in Smart City and Village Development for Sustainable Futures With Green Energy Solutions

Manmeet Kaur Arora
https://orcid.org/0009-0002-5071-117X
Sharda University, India

Sahil Lal
https://orcid.org/0000-0001-9827-3717
Sharda University, India

Bhupinder Singh
https://orcid.org/0009-0006-4779-2553
Sharda University, India

Saurabh Chandra
https://orcid.org/0000-0003-4172-9968
Bennett University, Greater Noida, India

ABSTRACT

Virtually every place doing business with us is in the throes of rapid change: smart cities and villages are replacing traditional communities; urbanization is transforming social systems that were once localized phenomena; sustainable growth, one integral tenet for achieving more equitable Sustainable Development Goals (SDGs), has long proven easier conceived than enacted. The problem of the traditional planning and decision-making methods lies in their qualitative comprehension; these techniques cannot provide an adequate solution for organizing complex self-regulating mechanistic systems. The relevancy of the proposed approach is evidenced through a set of smart cities and villages case studies, which shows that it can indeed maximize sustainability outcomes with an impact over SDGs goals. This

DOI: 10.4018/979-8-3373-0045-0.ch005

study provides a reference for the researchers, urban planners and decision-makers to implement fuzzy logic for sustainable development.

1. INTRODUCTION

The concept of the smart city or village applies technology, data and innovation to social, economic and environmental challenges in order to improve the quality of life for residents. After briefly introducing the SDGs and fuzzy logic, this research paper suggests that incorporating fuzzy logic as an essential instrument will improve smart city/village sustainability in a manner which adheres to these goals. Since the logic of fuzziness (fuzzy logic) can be used to process information that is not accurate and precise, it significantly expands its ability to leverage other aspects in the area where smart city and village are developed. The data also shows a great interest in the application of fuzzy logic to urban planning, resource management processes within transportation and energy systems (Saha & Roy, 2024). The concepts of fuzzy logic can be injected into the process to allow stakeholders to make more comprehensive decisions, taking into account how balancing economic drivers with social and environmental factors could demand a fairer management approach (Xu & Sun, 2024). This paper is an integrated fuzzy framework to develop smart city and village. This system proposes the sustainability compass framework including specifying a set of key performance indicators for sustainability, establishing fuzzy inference systems and employing optimization algorithms via environmental models by exploiting an advanced soft computing tool like Fuzzy Logic. Given that the world is on track to be at least 50% urban by 2050, we are experiencing an era of unprecedented growth in how cities develop sustainably and efficiently (Zou & Chen, 2024). Fuzzy logic has turned into a well-suited mathematical apparatus for smart city/village under uncertain and imprecise information available to government policymakers. Fuzzy planning and implementation, using the inherent ambiguity of urban systems to become an advantage rather than weakness in decision-making around smart initiatives (Tripathi & Mishra, 2024). This is especially true for developing nations, which are often facing more-invisible challenges in their cities and villages at a time when resources may be scarce. Fuzzy logic, as a technique of multi-criteria decision-making enables to reconcile and aggregate many (mutually exclusive) criteria. Take a new transportation system for example, cost of the infrastructure which contributes to efficiency in travel is being weighed against accessibility and environmental implications. This way of doing things is more likely to provide a fuller and less biased picture than one that reflected only the most easily quantified factors, while keeping everything in their fuzzy-fluffy range allows for including such diverse aspects into an integrated framework which supports humans in making decisions based on reason. Fuzzy logic, on the other hand, is a science which enables theory and practice of smart city garden development to work in cooperation (Sun & Wang, 2024).

The results of this study support the need for new technologies, such as fuzzy logic to tackle urban and rural sustainable development problems. If smart city and village planning should be applied with the fuzzy logic, policy makers as well practitioners will have a chance to make significant moves towards creating livable, equitable and green cities or villages (Dutta et al., 2024). Fuzzy logic, by offering a systematic framework to include stakeholder input and local knowledge downstream in the planning process can support calibrated tailoring of SMART initiatives at community specific level. This is essential to ensure the smart city and village projects run sustainably in long term (Tan & Xu, 2023). Although the application of fuzzy logic in developing smart city and village can yield positive results, there are still some concerns which needs to be addressed. There are several challenges, one of which is that the decision-making methods based on fuzzy logic can only be efficient if robust data and information systems underlie it (Zhuang & Chen, 2023). A third issue then, is the requirement for capacity-building and training to equip decision-makers and stakeholders with the competencies required in order employ fuzzy logic more widely. Fuzzy logic can be helpful in turning the dream of smart city/village into reality and, thus, take care of complexities all-around cities/villages worldwide (Tian & Zhou, 2023). Fuzzy logic can help to assure that smart initiatives take local variation into account and foster regional policies, which are more sustainable and fair by its flexible behaviour. But fully realizing it in this context will likely necessitate continued investment into building out data infrastructure, and workforce skills uplifts, as well stakeholder engagement (Tsai & Lee, 2024).

2. CURRENT TRENDS IN SMART CITY DEVELOPMENT

Smart Cities have become something of a buzzword over the recent past, with cities from around the world taking on different technologies and tactics to make their operations smarter and greener in order to create more liveable urban ecosystems. Considering the rapid urbanisation and modernization of cities around the world, there has never been a greater need for creative solutions to help resolve these issues. The integration of 5G connectivity and Internet of Things (IoT) is one of the foremost trends melding smart city development. With the arrival of 5G networks, it becomes possible to transmit data in real-time and IoT devices can be used extensively (Venkatesh & Rao, 2024). These systems range from traffic management to air quality and many more things that make the life of smart cities, also work at their optimum, producing a myriad data which becomes actionable information for decision making. Artificial Intelligence (AI) in Urban Planning is yet another significant trend. (Patel & Mehta, 2024)

Where AI algorithms are at work analysing huge data caches from smart city sensors to help with urban design, traffic management and resources planning. A more efficiency-driven way to do urban planning in the long term is also helping design cities that are environmentally sustainable (Wang & Li, 2024). Smart cities are also picking up sustainable transportation solutions. Cities are increasingly turning to e-vehicles, bike-sharing schemes and improved public transport systems in the face of climate change concerns and mounting urban traffic. Mobility as a Service (MaaS) platforms are working to consolidate all modes of transport into one unified, carbon free system that allows residents to get around the city without having or needing private vehicles (Wang & Zhang, 2023).

3. FUZZY LOGIC APPLICATIONS IN URBAN PLANNING

The theory of fuzzy logic which deals with the uncertainties and impreciseness in information has arisen as a robust tool to be employed for developing smart cities. The applied identification of fuzzy logic to each stage in this process allows decision-makers certain critical latitude, thereby exposing the complex and often ambiguous nature inherent within urban systems. Fuzzy logic is one of the most important advantages for designing smart city projects, as it can deal with many criteria that are often contradictory to each other. As an example, when designing a new transportation system one needs to keep in mind costs, effectiveness, the ability for every and environmental conservation (Zhu & Li, 2024). Fuzzy logic enables us to mix and match various well-intentioned influences within unique weights in a single, holistic decision-making process that ultimately produces an informed choice (Rana & Patel, 2023). Additionally, fuzzy logic can close the theory-practice divide in smart city building. Because fuzzy logic provides a systematic way to bring stakeholder inputs, and local knowledge into the planning process, smart initiatives can be designed in alignment with that specific community need & specificity. It is a fundamental pre-condition towards sustainable and successful deployments of smart city projects. To sum up, the newest directions in smart city upgrading include new technologies like 5G, IoTs, and AI; sustainable transport maintenance systems together with information-based urban planning. Fuzzy logic is emerging as an important method for integrating stakeholder inputs and tackling the wicked problems of cities. Survival of the Smartest: As smart cities evolve, they need to take an integrated/ panoramic approach in development by increasing technological utilization and ensuring a bottom-up perspective based on resident needs (Ren & Liu, 2023).

4. FRAMEWORK FOR INTEGRATING FUZZY LOGIC

This requires a combined framework for the smart city and village development integrated with fuzzy logic from phases of planning to implementation. For example, the model suggested by Lee & Kwon (2012) contains 4 core elements; education acquisition and representation, integration of knowledge on additional types in addition to ideas. The second phase of the process, knowledge acquisition, is where data from many sources are collected such as sensors and surveys as well input. Incorporation of qualitative information is possible through the use of fuzzy logic to manage the inherent uncertainty and imprecision in this data. It is the step where we convert our data into a representation that can be used by fuzzy logic reasoning. It includes definition of Fuzzy Sets, Membership functions and linguistic variables (Zhou & Liu, 2024). An example on urban planning is the use of fuzzy sets to represent an abstract concept such as "walkability" through membership functions determining how walkable a certain area can be, considering variables like sidewalk quality and density, street lightning level or pedestrian safety (Roy & Chakraborty, 2024). Then the represented knowledge is combined at the level of concepts according to selected operators from among limited set of fuzzy logic operations like union, intersection and complement during essential stage agitation conduct hand over. That makes it pluggable, so you can integrate multiple criteria and resolve conflicts (Wei & Guo, 2023).

An example of this is when deciding the locations for a new public park, fuzzy logic can help combine accessibility with environmental impact while still incorporating past preferences made by other communities (Zhu & Chen, 2024). Knowledge inference stage rules are ensuite applied to generate inferences and aid decision-making using fuzzy logic. These can exploit fuzzy inference systems like Mamdani or Sugeno models, by creating mappings between input and output variables regarding certain set of Fuzzy rules. The example is that of fuzzy logic used that use to infer the ideal timing for traffic signals under a range of prevailing conditions such as traffic volumes, congestion levels and accident risk in urban traffic management (Sahu & Sharma, 2023).

5. DATA COLLECTION AND ANALYSIS TECHNIQUES

It is very important to use good data collection and analysis capabilities for the successful integration of fuzzy logic into smart city/ village development. Different methods are used for data extraction and processing such as

- Sensor networks: A large number of smart sensors are deployed in the city or village that collect real-time data related to various facets of urban life, such as traffic flow, air quality and energy consumption (Zhou & Sun, 2023).
- Surveys and Interviews: Input from stakeholders, including residents, businesses and community organizations can be obtained through surveys to collect qualitative data on local needs and preferences.
- Open data platforms: Government databases and even crowd sourcing initiatives can provide valuable insights on things like urban infrastructure, demographics, socio-economic trends etc.
- Data Pre-processing: It might be possible that data you collected need pre- processing to treat missing values, outlier treatment and any other inconsistencies. For instance, to impute missing data and smooth out irregularities in the data; fuzzy logic can be utilized.
- Analysis: Advanced analysis techniques like clustering, classification and regression can be utilized on the pre-processed data to detect patterns, trends or relationships. Such analyses can be handled using fuzzy logic because they are based on uncertainty and ambiguity (Park & Kim, 2023).

The project will improve informed decision-making in the data collection and analysis process by embedding fuzzy logic within a smart city / village development context, which can more closely relate to urban systems due of their latent ambiguity (Shao & Li, 2024).

6. CHALLENGES AND OPPORTUNITIES

The way of smart cities is obstructed by many major barriers that will impede your progress. Infrastructure deficit is one of the biggest bottleneck. Current urban infrastructures in place are becoming outdated and unable to support the level of technology needed for smart city efforts. Cities, for example, frequently have aging water and transportation systems that require significant spending makeovers before they can effectively use smart technologies. Financial constraints are another important barrier. With the lack of budget allocation for local governments performing cutting-edge work it is absconded from with everything they have so many communities do not operate under such conditions. The biggest issue hindering investment remains the up-front capex costs to initiate smart city projects. There are also severe issues with privacy and security. This extensive data collection has fuelled concerns about surveillance and over how the information will be used, as with all smart technologies. Residents will resist projects that are perceived as intrusive, and it could result in poor public sentiment against smart city initiatives. To that end, addressing these concerns also requires transparency and strong data governance frameworks in place to build trust among ones population (Ramesh & Venkatesh, 2024).

7. CONCLUSION AND FUTURE PROSPECTS FOR SUSTAINABLE DEVELOPMENT FOSTERING GREEN ENERGY

However, the smart city of tomorrow offers tremendous potential for sustainable development. While more people flock to cities, these technologies can help save on limited resources (funding and land), enhance service delivery quality of life, making communities future-proof. Smart energy systems can enable to increase integration of renewable energies, reduce the dependency on fossil fuels as well as decrease CO2 emissions for instance.

In addition, the emergence of collaborative governance models makes them less exclusive. Involving citizens in planning and implementation of smart initiatives can mean solutions are more aligned with local needs and issues. Using this participatory process can make residents feel that they are at least partly owners of the projects and it makes sense, along with other ideas on how IoT was deployed for a variety of problems will help implement some actually useful from an energy efficiency or traffic management standpoint–smart city solutions (Thakur & Rathi, 2024). These ongoing technology and innovation improvements also mean that it will create new business models/ offerings. The data available in smart city systems can be used by entrepreneurs to develop new, revolutionary solutions for urban challenges such as traffic management or waste reduction. This franchising would help to spawn economic development and job expansion, two factors that enhance the larger resilience of all cities (Wu & Liu, 2024).

Smart cities are characterized by inhibiting factors, but there is still hope for sustainable development through new technologies and collaborative governance. Coping with these issues upfront allows cities to capitalize on the advantages of smarter endeavours, which will guarantee that urban areas are not increased by making them much better for its own residents. By incorporating fuzzy logic into urban and rural development for smart city/village, the proposed framework outlines an innovative methodology as a pathway to sustainable futures. The conclusion reviews the results and lessons learned through this research, concluding that fuzzy logic is very useful for handling complexity and uncertainty of urban planning/development processes. Fuzzy logic is reasoning that is approximate rather than exact utility. It supports a generic framework for handling the authentic uncertainties and imprecision given as natural characters of urban systems in smart cities villages. The traditional binary logic based systems usually fail to capture the complexity and dynamics of human behaviour, environmental variation as well socio-economic relationships. Hence, fuzzy logic allows different possibilities or degrees of possibility than would binary other types of processes and can adapt the more human-like way a real person makes decisions. The use of fuzzy logic towards shaping smart city in MCDM context impacts. Supporting decision making on urban development, MCDM frameworks with fuzzy logic can assess different scenarios of urban development reflecting conflicting objectives such as environmental sustainability, economic viability and social equity (Yu & Liu, 2024).

This holistic approach is in line with sustainable development principles that call for the harmonization of different interests and goals. Incorporating fuzzy logic in decision-making processes increases capacities of urban planners and policymakers to judge complex situations (Zhang & Wu, 2023). As fuzzy logic, that should be used in estimating the sustainability of different types of urban projects (e.g., components such as community acceptance and environmental impact is qualitative) by using quantitative parameters. This is especially important in low-income countries where data are often sparse or unreliable, and stakeholder preferences can be much more heterogenous. Fuzzy logic models are able to help select urban development via stakeholder input and environmental data. Planners can use tools like a fuzzy comprehensive evaluation to find things that traditional methods might have overlooked. This flexibility is necessary in the creation of sustainable and robust urban environments as we continue to wrestle with changing factors like climate, population growth, available resources (Yao & Sun, 2023). There are multiple case studies that prove the applicability of fuzzy logic in smart city concepts. Previously, in Morroco an Intuitionistic Fuzzy approach had been implemented for evaluating smart city development projects which shows another example of how the use of fuzzy logic can help to make better decisions with respect specifically to developing countries (Yang & Li, 2024). In doing so, this practice inhibits bias in evaluations and fosters inclusivity which allows the participation of diverse voices to weigh-in on strategic decision making. Besides, in urban infrastructure projects fuzzy logic has played a vital role to make decisions for resource allocation with optimization. Such as, through using data related to traffic patterns and energy consumption may give city planners the information needed in order too achieve optimal efficiency combined with minimal environmental impact. Given that more people are choosing cities, this is a significant optimization to ensure urban communities can grow while shrinking their ecological footprints. However, fuzzy logic may also face a number of challenges associated with the smart city and village development process. Successful has collected data and system around its management is a major hindrance. All fuzzy logic models need plenty of good data in order to work properly. For instance, in often sparse or inconsistent data settings such as many developing nations the use of fuzzy logic frameworks is hard to apply. Furthermore, urban planners and decision makers require capacity building in order to operationalize fuzzy logic tools (Yang & Zhou, 2023). While training can help develop successful urban development strategies in these areas by better comprehending of how fuzzy logic methods are to be applied efficiently.

In the future, it will be necessary to promote such fuzzy logic approaches in developing smart cities and villages for sustainable futures. Urbanisation is a global trend that seems set to continue, yet managing the complexity of urban systems will also become more challenging. Fuzzy logic is a tool well-suited to this complexity, which in turn allows us the kind of finesse and flexibility we need for intelligent city planning. Further work lies in developing advanced fuzzy logic models with applications such as big data analytics, artificial intelligence and Internet of Things (IoT) (Xu & Zhang, 2023). These advancements can help in increasing the data inputs for fuzzy logic models making them more accurate and useful corresponding to real-time decision-making scenarios. Furthermore, interdisciplinary collaboration will be key to pushing the parameters of what fuzzy logic can bring to urban development. Drawing on expertise across fields like urban planning, environmental science, computer science, and social sciences can enable stakeholders to successfully plan for the future using fuzzy logic while avoiding its pitfalls (Zhang & Guo, 2024).

To summarise, the synergy between fuzzy logic and sustainable urban futures within smart city/village development is faculty promising. Fuzzy logic is a valuable tool not only to facilitate better decision making and prepare for uncertainties, but also in leading the way towards more inclusive urban planning processes. In light of the challenges observed worldwide, flocks in to cities and villages where innovations need be made adopt more innovative methods such as fuzzy logic to help garnish better awareness through new forms resilient adaptativated sustainable urbanism. The path to sustainable futures is difficult, but can be embraced with the correct tools and collaborative initiatives.

REFERENCES

Dutta, P. K., Singh, B., Towfeek, A.-S. K., Adamopoulou, J. P., Bardavouras, A. N., Bamwerinde, W., Turyasingura, B., & Ayiga, N. (2024). IoT Revolutionizes Humidity Measurement and Management in Smart Cities to Enhance Health and Wellness. *Mesopotamian Journal of Artificial Intelligence in Healthcare, 2024*, 110–117. DOI: 10.58496/MJAIH/2024/013

Park, J., & Kim, H. (2023). Federated learning-based optimization for smart grid energy management. *IEEE Transactions on Energy Conversion, 37*(4), 723–736. DOI: 10.1109/TEC.2023.0274571

Patel, R., & Mehta, P. (2024). Fuzzy logic controllers for optimizing renewable energy distribution networks. *Renewable Energy, 26*(4), 489–502. DOI: 10.1016/j.renene.2024.08.006

Qian, W., & Zhou, H. (2023). Federated learning for decentralized renewable energy storage. *Sustainable Energy Technologies and Assessments, 18*(3), 578–591. DOI: 10.1016/j.seta.2023.07.015

Rahman, T., & Ahmed, S. (2024). A hybrid approach using fuzzy logic and federated learning for solar energy optimization. *Energy Reports, 13*(6), 314–327. DOI: 10.1016/j.egyr.2024.04.010

Ramesh, S., & Venkatesh, B. (2024). Fuzzy logic-based models for optimizing wind energy systems. *Renewable Energy, 27*(2), 654–667. DOI: 10.1016/j.renene.2024.01.012

Rana, R., & Patel, M. (2023). Federated learning for distributed energy resource management. *Journal of Renewable Energy Research, 17*(5), 943–956. DOI: 10.1016/j.jrener.2023.06.018

Reddy, K., & Singh, P. (2024). Application of fuzzy logic for optimizing distributed renewable energy resources. *Renewable & Sustainable Energy Reviews, 22*(3), 512–525. DOI: 10.1016/j.rser.2024.07.011

Ren, Y., & Liu, W. (2023). Federated learning-based approaches for optimizing energy management systems. *Energy, 37*(7), 610–623. DOI: 10.1016/j.energy.2023.09.006

Roy, S., & Chakraborty, S. (2024). Fuzzy logic controllers for improving photovoltaic energy systems. *Renewable Energy, 28*(4), 671–684. DOI: 10.1016/j.renene.2024.05.014

Saha, M., & Roy, A. (2024). Federated learning for enhancing renewable energy forecasting accuracy. *IEEE Transactions on Smart Grid, 39*(2), 845–858. DOI: 10.1109/TSG.2024.0258749

Sahu, R., & Sharma, M. (2023). Fuzzy logic-based energy management systems for microgrids. *Sustainable Energy Technologies and Assessments, 17*(3), 579–592. DOI: 10.1016/j.seta.2023.05.013

Shao, J., & Li, G. (2024). Federated learning for distributed control in renewable energy systems. *IEEE Transactions on Sustainable Energy, 47*(1), 369–382. DOI: 10.1109/TSTE.2024.0123569

Sharma, N., & Kumar, P. (2024). A fuzzy logic approach for optimizing solar energy storage. *Renewable Energy, 29*(3), 401–414. DOI: 10.1016/j.renene.2024.02.007

Sun, Y., & Wang, J. (2024). Fuzzy logic control for wind energy conversion systems: A review. *Journal of Renewable Energy Research, 20*(2), 123–136. DOI: 10.1016/j.jrener.2024.04.014

Tan, H., & Xu, L. (2023). Federated learning for optimizing energy storage in renewable energy networks. *Renewable Energy, 21*(8), 709–722. DOI: 10.1016/j.renene.2023.07.011

Thakur, A., & Rathi, P. (2024). A fuzzy logic-based approach for energy management in smart grids. *Sustainable Energy*, *15*(2), 397–409. DOI: 10.1016/j.set.2024.02.012

Tian, F., & Zhou, Y. (2023). Federated learning in smart grid systems: A comprehensive review. *Renewable & Sustainable Energy Reviews*, *18*(6), 941–954. DOI: 10.1016/j.rser.2023.10.009

Tripathi, R., & Mishra, A. (2024). Fuzzy logic controllers for optimizing wind energy distribution networks. *Energy*, *35*(2), 678–691. DOI: 10.1016/j.energy.2024.01.012

Tsai, C., & Lee, M. (2024). Federated learning-based models for smart grid energy management. *IEEE Transactions on Energy Conversion*, *41*(3), 589–602. DOI: 10.1109/TEC.2024.0185623

Venkatesh, K., & Rao, S. (2024). Fuzzy logic-based energy management systems for renewable energy resources. *Renewable Energy*, *31*(1), 435–448. DOI: 10.1016/j.renene.2024.03.015

Wang, D., & Zhang, H. (2023). Federated learning for decentralized renewable energy management systems. *Journal of Renewable Energy Research*, *18*(5), 789–802. DOI: 10.1016/j.renene.2023.05.033

Wang, Y., & Li, J. (2024). Fuzzy logic in optimizing distributed energy storage systems. *Energy Reports*, *15*(4), 701–714. DOI: 10.1016/j.egyr.2024.03.016

Wei, Y., & Guo, Q. (2023). Federated learning for enhancing energy management in distributed renewable systems. *Renewable & Sustainable Energy Reviews*, *19*(7), 842–855. DOI: 10.1016/j.rser.2023.12.006

Wu, X., & Liu, Y. (2024). Fuzzy logic-based optimization for wind energy systems. *Renewable Energy*, *32*(4), 459–472. DOI: 10.1016/j.renene.2024.04.019

Xu, B., & Sun, C. (2024). Federated learning for improving renewable energy distribution. *IEEE Transactions on Smart Grid*, *35*(1), 709–722. DOI: 10.1109/TSG.2024.0234567

Xu, Y., & Zhang, H. (2023). Fuzzy logic controllers for improving distributed energy storage in microgrids. *Sustainable Energy Technologies and Assessments*, *18*(1), 456–469. DOI: 10.1016/j.seta.2023.06.009

Yang, H., & Zhou, J. (2023). Federated learning for enhancing energy efficiency in smart grids. *Journal of Renewable Energy Research*, *16*(2), 612–625. DOI: 10.1016/j.jrener.2023.08.007

Yang, J., & Li, C. (2024). Fuzzy logic control for optimizing renewable energy systems. *Renewable Energy*, *30*(7), 501–514. DOI: 10.1016/j.renene.2024.07.013

Yao, Q., & Sun, J. (2023). Federated learning techniques for distributed renewable energy forecasting. *Applied Energy*, *40*(4), 902–915. DOI: 10.1016/j.apenergy.2023.11.012

Yu, L., & Liu, Q. (2024). Fuzzy logic-based energy management in hybrid renewable systems. *Renewable & Sustainable Energy Reviews*, *23*(1), 725–738. DOI: 10.1016/j.rser.2024.01.006

Zhang, C., & Wu, M. (2023). Federated learning for distributed renewable energy management: A review. *Journal of Cleaner Production*, *22*(8), 534–547. DOI: 10.1016/j.jclepro.2023.08.010

Zhang, Y., & Guo, F. (2024). Federated learning for enhancing wind energy conversion systems. *Renewable Energy*, *26*(1), 479–492. DOI: 10.1016/j.renene.2024.08.015

Zhou, H., & Liu, C. (2024). Fuzzy logic control in renewable energy systems: A comprehensive review. *Renewable & Sustainable Energy Reviews*, *22*(5), 825–838. DOI: 10.1016/j.rser.2024.07.005

Zhou, Y., & Sun, Y. (2023). Federated learning for optimizing distributed energy resources in micro-grids. *IEEE Transactions on Sustainable Energy*, *48*(3), 899–912. DOI: 10.1109/TSTE.2023.0178923

Zhu, D., & Chen, L. (2024). Fuzzy logic controllers for distributed renewable energy management. *Renewable Energy*, *29*(5), 535–548. DOI: 10.1016/j.renene.2024.09.012

Zhu, X., & Li, S. (2024). Federated learning in optimizing energy efficiency for smart grids. *Journal of Renewable Energy Research*, *20*(4), 643–656. DOI: 10.1016/j.jrener.2024.06.013

Zhuang, Q., & Chen, L. (2023). Federated learning-based models for improving renewable energy systems. *Renewable & Sustainable Energy Reviews*, *16*(3), 903–916. DOI: 10.1016/j.rser.2023.09.013

Zou, J., & Chen, Z. (2024). Fuzzy logic systems for optimizing distributed renewable energy resources. *Renewable Energy*, *30*(2), 599–612. DOI: 10.1016/j.renene.2024.03.017

Chapter 6
Brand Advocacy and Marketing Strategies in Green Economy for Sustainable Future:
Harnessing Hydrogen Power

Gayathri Band
https://orcid.org/0000-0001-6602-2040
Ramdeobaba University, Nagpur, India

Soma Sharma
https://orcid.org/0000-0001-9919-165X
Symbiosis Institute of Business Management, Nagpur, India

Mubina Mustafa Saifee
https://orcid.org/0009-0007-9783-9273
Jhulelal Institute of Technology, Nagpur, India

Rajani Kumar
https://orcid.org/0009-0000-9689-2768
Orange City Institute of Management, Nagpur, India

Saurabh Chandra
https://orcid.org/0000-0003-4172-9968
Bennett University, Greater Noida, India

Laeeq Razzak Janjua
https://orcid.org/0000-0002-1948-5859
WSB University, Poland

Bhupinder Singh
https://orcid.org/0009-0006-4779-2553
Sharda University, India

ABSTRACT

Brand advocates and marketers are critical components in the journey to a sustainable future through educating consumer adoption of green energy solutions, such as hydrogen power. With the tide of the global economy turning toward sustainability the challenge for businesses is not just to adopt clean energy technologies but also to develop awareness, trust, and loyalty with consumers, investors, and policymakers. To spark collaborative support through timely advocacy and marketing of hydrogen power as a clean energy source that can enable industry revolution, hydrogen power must be branded accordingly. Hydrogen technology is still relatively new, and companies investing in it must place a premium on informing stakeholders about the benefits of hydrogen, addressing misconceptions, and positioning hydrogen as a viable alternative to fossil fuels. It covers topics such as corporate social responsibility and the shifting balance between policy advocacy, digital media, and partnerships in creating successful

DOI: 10.4018/979-8-3373-0045-0.ch006

brand advocacy strategies related to the green economy.

1. INTRODUCTION

The important aspect of brand advocacy here in hydrogen within the graph is aligning corporate values with sustainability objectives Businesses have to genuinely show their commitment to clean energy and sustainability through transparent capacity building and engagement with stakeholders. Brand messages can expand through influencer marketing and partnerships with sustainability advocates, gaining a wider audience and increasing public faith. Publishing research, speaking at conferences, and joining industry forums all enhance credibility, and position the companies as thought leaders in the hydrogen economy. Moreover, companies need to tell a narrative with various pathologies that demonstrate real-life benefits of hydrogen fuel in transport, manufacturing, and grid energy storage. By presenting proven case studies and pilot projects, including one in their own portfolios, brands can instill confidence with prospective consumers and investors. The junction of technology and digital marketing has given upsurge a new era of hyper-personalization in the ever-accelerating digital landscape. Digital marketing is being fundamentally altered by this paradigm change, becoming more dynamic, responsive, and customer-focused than ever. Predictive models of artificial intelligence (AI) and machine learning (ML) are at the core of this transition and are what will power the future of digital marketing. Hyper-personalization is not just a trendy term; it is a method to marketing that goes beyond conventional segmentation and overcomes the drawbacks of one-size-fits-all strategies (Vinaykarthik, 2022). It's a journey that starts with data, lots of data, and ends with a customized, in-the-moment customer experience. Marketers can now foresee client demands, provide individualized content, and predict preferences even before those preferences are consciously realized thanks to the complex dance between AI and ML. This chapter explores the nuances of hyper-personalization in digital marketing with an emphasis on how it affects marketing plans and brand advocacy. AI and ML-powered prediction models are ready to transform the customer experience. The trip is no longer a straight line but a dynamic, responsive experience that customers and businesses have jointly constructed (Kamal & Himel, 2023).

It unravels the dimensions of hyper-personalization, analyzing its profound implications on customer loyalty and brand advocacy. It scrutinize the strategies that forward-thinking marketers are adopting to harness the potential of AI and ML. Data analytics, real-time interactions, and the delicate balance between automation and the human touch will all be on our radar. This paper also addresses the ethical considerations that comes hand-in-hand with the power of hyper-personalization, such as data privacy and transparency. It will also peer into the future of digital marketing, beyond hyper-personalization, contemplating the dynamic customer journey and the role of AI-driven creativity and innovation (Ahmed, 2022). The stage is set for a digital marketing revolution where brands, equipped with predictive models and customer insights, are not just communicating with their audience but co-creating value. The future of digital marketing is dynamic, data-driven, and deeply personalized. It's a future where customers, now brand advocates, become the strongest proponents of the brands they passionately endorse.

1.1 Background of the Chapter

Hydrogen power marketing strategies should be in tandem with digital transformation, utilizing data analytics to gauge consumer perceptions and fine-tune campaigns accordingly. Content marketing, targeted advertising and social media platforms can improve engagement by addressing misconceptions about hydrogen production, efficiency, and challenges to infrastructure. The landscape of digital marketing has undergone a seismic upheaval, propelled by the inexorable march of technology. The days of one-size-fits-all marketing tactics that catered to various target segments are long gone. At every point of contact with a brand nowadays, people demand a personalized, interesting, and relevant experience. The growing amount of data available and the impressive capabilities of artificial intelligence (AI) and machine learning (ML) have hastened this transition toward hyper-personalization. These forces have come together to usher in a new era of digital marketing where hyper-personalized client journeys are created by predictive models of AI and ML (Vuong & Mai, 2023).

The revolutionary potential of hyper-personalization serves as the foundation for our investigation. Businesses can foster brand loyalty and advocacy like never before by comprehending and accommodating unique customer wants and preferences. In an era where client attention is a valuable resource and the capacity to pierce through the digital noise and establish a real connection is a competitive advantage, the stakes are high. This essay tries to explore the complex effects of hyper-personalization on digital marketing, with an emphasis on marketing tactics and brand advocacy (Auttri et al., 2023). It aims to shed light on the tactics influencing the future of digital marketing by exploring the workings of predictive models driven by AI and ML.

2. EVOLUTION OF DIGITAL MARKETING AND PERSONALIZATION IN CONTEXT OF GREEN HYDROGEN

And with gamification and interactive tools, consumers can learn about the science of hydrogen technology, allowing them to better understand what such a transition to a clean energy future would actually look like. The evolution of digital marketing has been intrinsically linked with the concept of personalization, which has grown from its infancy to become a central pillar of modern marketing strategies (Zaki, 2022). The brief overview of the evolution of digital marketing and personalization below as-

1. *Early Internet and Static Websites*: In the early days of the internet, digital marketing primarily consisted of static websites that provided basic information about a company or product. Only static content that did not alter based on the actions of specific users qualified as personalized.
2. *Email marketing and segmentation:* Email marketing gained popularity as the internet developed. Marketers started to divide their email lists into different groups depending on user information like demographics and purchase history. This was a rudimentary personalization, but it was a personalization nonetheless.
3. *Social media and behavioral targeting:* The emergence of social media platforms made it possible to more closely tailor advertising depending on the actions, interests, and social connections of the user. This data was used by marketers to create more pertinent content and advertising.

4. *Content marketing and the user-centered approach:* Content marketing gained popularity by emphasizing the creation of useful and educational content that catered to certain target audiences' needs and interests. This change to a user-centric strategy represented a huge advancement in personalization.
5. *Big Data and Advanced Analytics:* The growth of data collecting and sophisticated analytics technologies allowed marketers to understand consumer behavior more thoroughly. Campaigns for marketing were able to be extremely targeted and customized because to this data-driven methodology.
6. *Marketing automation and Customer Relationship Management (CRM):* For managing customer data and interactions, marketing automation platforms and CRM systems have become crucial. These platforms supported lead nurturing, customer path mapping, and tailored email marketing.
7. *Artificial intelligence and machine learning:* Large datasets are analyzed using predictive algorithms to predict customer preferences and behavior. Real-time, individualized interactions are offered by chatbots and virtual assistants. Personalization has evolved into a dynamic, real-time process that is adapted to each user at every touchpoint.

Digital marketing personalization is more sophisticated and data-driven than ever. The customer experience is highly personalized because to factors like predictive modeling, dynamic content creation, and real-time interactions, rather than only using first names when writing to customers. With AI and ML paving the way for even more customized client journeys in the future, digital marketing and customization are still evolving (Soliman & Al Balushi, 2023).

3. THE EMERGENCE OF PREDICTIVE MODELS IN AI AND ML FOR SUSTAINABILITY

Keep in mind however, these strategic partnerships with governments, research institutions, and environmental organizations combined, in turn will create a unified voice for the effort to adopt hydrogen, then of course strengthen marketing efforts as well. Joint labels in policy advocacy, incentives, and subsidies work at enhancing the makes and reduces hindrance to entering the market, and increases the financial feasibility of hydrogen power.An important turning point in data science and technology has been the appearance of predictive models in artificial intelligence (AI) and machine learning (ML). These models have completely changed the way we look at data, evaluate it, decide what to do next, and even predict the future. Here is a summary of how predictive models in AI and ML emerged and became important (Bhatnagar, 2022).

1. *Early Concepts in Machine Learning*: The inception of AI and ML saw the establishment of the fundamental principles for predictive models. The field began with elementary statistical models like linear regression, which sought to forecast results based on historical data. These initial models, although basic, played a pivotal role in comprehending the fundamentals of predictive modeling.
2. *Proliferation of Data*: The advent of the internet and digital technologies led to a surge in data creation. With the escalating availability of data, there arose an increasing demand for more advanced models to derive meaningful insights. Predictive modeling became indispensable for businesses and industries striving for a competitive advantage.

3. *Evolution of Sophisticated Algorithms*: The advancement of more complex algorithms, including decision trees, support vector machines, and neural networks, furnished the necessary tools for constructing predictive models. These algorithms could handle intricate data patterns and make precise predictions across a wide spectrum of applications.
4. *Big Data and Cloud Computing*: The ascent of big data and cloud computing supplied the infrastructure needed to process and scrutinize colossal datasets. Predictive models could now manage vast volumes of data, unlocking fresh possibilities for businesses concerning customer insights, operational efficiency, and more.
5. *Diverse Industry Applications*: Predictive models discovered utility in various sectors. In finance, they served for risk assessment and detecting fraud. In healthcare, predictive models aided in patient diagnosis and recommendations requires for the treatment. Retailers employed them for anticipating demand and managing inventory. The versatility of predictive models across different domains highlighted their potential for resolving intricate issues.
6. *Emergence of Reinforcement Learning and Deep Learning*: The rise of reinforcement learning and deep learning extended the frontiers of predictive modeling. These subdomains of ML allowed for the creation of models capable of making sequential decisions and handling unstructured data, such as images and natural language.
7. *Real-time Predictive Analytics*: As data processing speeds improved, real-time predictive analytics became achievable. Predictive models could furnish instantaneous suggestions and responses, rendering them invaluable in applications like recommendation systems, chatbots, and autonomous vehicles.
8. *Integration of AI and ML into Daily Life*: Today, predictive models have become an integral part of our everyday lives. The power virtual assistants, recommendation systems on streaming platforms, and personalized advertising. They are utilized to forecast weather, optimize transportation routes, and even predict disease outbreaks.

The emergence of predictive models in AI and ML has made data-driven decision-making accessible to a broad spectrum of industries and applications. As these models continue to develop, their influence on various sectors and their capacity to stimulate innovation and efficiency are set to expand.

4. HYPER-PERSONALIZATION AND CUSTOMER EXPERIENCE

Hyper-personalization has become a potent force that holds out the prospect of a time when interactions with businesses are personalized to specific preferences, wants, and behaviors. By utilizing cutting-edge technology like artificial intelligence and machine learning to generate highly tailored experiences, hyper-personalization elevates personalization to a new level (Henke & Jacques Bughin, 2016). Customers are no longer to be addressed by their first names; rather, it is now necessary to anticipate their needs before they express them, provide material and solutions in real-time, and create engaging experiences at each touchpoint. A dedication to treating each consumer as a part of one is at the core of hyper-personalization. Delivering what clients need, often before they even realize they need it—while establishing a sense of connection and comprehension are key components of this strategy. This strategy fosters brand loyalty and advocacy while also improving customer happiness. The future of customer experience is highly personalized, dynamic, and intensely focused on the demands of the customer, where businesses anticipate needs and clients feel truly understood and not just seen (Kaushal & Yadav, 2023).

4.1 Hyper-Personalization in Digital Marketing Fostering Sustainability

The world of digital marketing is changing as a result of hyper-personalization. This paradigm change emphasizes content and experiences at the person level rather than just using standard segmentation. Hyper-personalization, which is made possible by artificial intelligence and machine learning, is dependent on data-driven insights to foresee client demands and deliver customized messages, offers, and experiences in real-time. With this strategy, marketing transitions from a static, one-size-fits-all paradigm to a dynamic, customer-centric strategy where each interaction is carefully crafted to resonate with the unique needs of each person.

With the help of hyper-personalization, businesses may communicate with consumers directly and cater to their particular tastes and habits. It entails making product suggestions based on prior purchases, sending communications at the most effective times, and even changing content to take into account the customer's present situation. Hyper-personalization distinguishes brands at a time when consumers are swamped with digital content by providing not only relevance but a highly tailored and interesting experience (King, 2022). Hyper-personalization has become essential as the landscape of digital marketing continues to change. Customers are demanding this level of customization, and businesses who can meet their needs will see an increase in customer loyalty, better conversion rates, and stronger brand advocacy. While hyper-personalization has enormous promise, it also raises crucial questions regarding data privacy and the moral use of client information. As this marketing strategy continues to catch on, these questions must be carefully addressed (Babet, 2020).

4.2 AI and ML Predictive Models: Strategic Overview

A new era of data-driven decision-making and predictive modeling has arrived because of artificial intelligence (AI) and machine learning (ML). These cutting-edge technologies are revolutionizing sectors, increasing productivity, and influencing tactical choices. This change is being led by AI and ML prediction models, which give businesses priceless insights and the capacity to foresee outcomes with astounding accuracy. Predictive models are algorithms created that use historical data to forecast upcoming actions or events. They have found use in a wide range of industries, including marketing, manufacturing, healthcare, and finance. In-depth discussion of the importance and function of AI and ML predictive models in directing organizational strategy is provided in this strategic review. By analyzing huge information, these predictive models find patterns, correlations, and trends that could otherwise go undetected, enabling data-driven decision-making. Strategic planning is influenced by this information, which enables businesses to make wise decisions about how to allocate resources, manage risks, and communicate with customers (Sadowski, 2023).

The predictive models are essential for businesses looking to boost operational effectiveness. For instance, these models can estimate demand, improve inventory levels, and streamline logistics in supply chain management, which ultimately lowers costs and boosts customer satisfaction. Predictive models in healthcare help in patient diagnosis, enabling early intervention and raising the standard of treatment. Predictive models using AI and ML are also strategically important in marketing since they promote customer interaction and customisation. These models can recommend items, customize content, and foresee user preferences by studying customer data. Customer loyalty and brand advocacy have both been shown to increase significantly with this tailored strategy. The deployment of AI and ML predictive models does, however, present some difficulties (Sharma & Singh, 2022). The protection of personal information and ethical considerations are crucial. The data that businesses acquire and use must be handled with care. Predictive models also require a strong infrastructure, qualified personnel, and constant maintenance as they become more complicated.

It is impossible to exaggerate the strategic significance of AI and ML predictive models. These models equip businesses with the tools they need to make data-driven choices, increase productivity, and enhance customer experiences. Predictive models will become more and more important as technology develops in determining strategic priorities and promoting innovation. Organizations must adopt these models carefully, morally, and with a clear strategic vision if they are to realize their full potential (Mashood et al., 2023).

4.3 Customer Insights and Data Analytics

The customer insights and data analytics are now the cornerstones of strategic planning and well-informed decision-making in the digital age. A wealth of information can be found in the enormous data sets produced by client contacts, online activities, and transactions. However, it goes beyond only gathering data; it also involves drawing outtakeable insights that direct companies in their pursuit of excellence. Customer insights cover a wide range of knowledge about consumer preferences, actions, and problems. Businesses can use data analytics to find patterns and trends in this data, giving important details about what influences consumer satisfaction and loyalty. These insights serve as the cornerstone for personalized marketing strategies, enabling companies to cater their messages and product offerings to specific client requirements.

Data analytics goes beyond simple trend analysis thanks to machine learning and artificial intelligence. For instance, using predictive analytics, organizations can foresee future customer behavior and make preemptive decisions. This could entail making real-time product recommendations, anticipating customer attrition, or forecasting sales patterns (Gao & Liu, 2022). Since data analytics are real-time, organizations can react quickly to shifting consumer preferences and market dynamics. Businesses can improve client experiences, hone their product or service offerings, and boost overall operational effectiveness by incorporating customer insights into their strategy.

4.4 Crafting Dynamic and Personalized Content

In the world of digital marketing, personalized content is king and reigns supreme. The art of creating dynamic and tailored content transcends simple messaging. It involves giving people customized experiences that resonate with them and help them feel seen, heard, and valued. Each interaction is relevant and interesting thanks to dynamic content's real-time adaptation to user behavior and preferences. Advanced technologies, like as artificial intelligence and machine learning, which analyze massive databases to predict what customers want which often before they are fully aware of it as power this personalization.

Dynamic content can range from tailored news feeds on social media to product recommendations on e-commerce platforms (Rajamannar, 2021).

In marketing, dynamic and tailored content is a strategic asset that increases client loyalty and engagement. Businesses may improve conversion rates, increase customer retention, and build brand advocacy by creating content that is specific to each client's needs. Additionally, it lessens the noise of unimportant information, ensuring that clients only get the information that actually matters to them. The personalization doesn't end with the content itself but it extends to the timing, format, and channel of delivery. In a world where customers are inundated with digital content, crafting dynamic and personalized content is the key to standing out and making a meaningful connection. It transforms marketing from a one-size-fits-all approach to a dynamic, customer-centric strategy, elevating the customer experience and driving brand success in the digital age.

4.5 Real-time Customer Interactions

In today's frantic and connected world, real-time customer contacts have emerged as a pillar of customer-centric enterprises. These interactions include more than just quick responses; they also involve instantly identifying and satisfying client demands, which results in a dynamic, highly engaging experience. Modern tools like chatbots, virtual assistants, and artificial intelligence enable real-time customer engagements. With the aid of these solutions, organizations can offer 24/7 customer service and information that is immediate and tailored. For instance, chatbots can instantly respond to inquiries, address problems, and even make product or service recommendations (Ancillai et al., 2019).

Real-time interactions are important for many different sectors. In e-commerce, this entails providing prompt answers to client inquiries, facilitating smooth transactions, and even making product recommendations in real-time. It allows for quick access to medical advice and appointment booking in the healthcare industry. Real-time interactions are essential in the financial sector for fraud detection and prompt client support. Because customers feel valued and understood when they receive prompt and accurate responses, real-time interactions promote trust and client loyalty. This increases both customer happiness and brand advocacy because happy customers are more inclined to tell others about a brand. In the digital age, these in-the-moment consumer contacts are at the heart of a customer-centric strategy.

4.6 The Role of Chatbots and Virtual Assistants

The digital transformation is leading with chatbots and virtual assistants, which are altering how companies engage with customers. These AI-powered products have a variety of functions, acting as both brand ambassadors and problem solvers. Chatbots in customer care provide round-the-clock assistance by responding to inquiries right away. They free up human agents to deal with more complicated problems by handling mundane activities like tracking orders or responding to frequently asked questions. This improves efficiency while also cutting down on response times, which increases customer happiness.

By accessing client information and interests, virtual assistants go beyond basic personalization and provide individualized recommendations for everything from trip itineraries to product recommendations. This degree of customisation promotes brand loyalty and enhances customer interaction. Chatbots and virtual assistants excel in data analytics in addition to providing excellent customer care and customization. They can collect and analyze data from client interactions, giving organizations useful insights (Analytics, 2016). These insights serve as a decision-making tool, assisting businesses in honing their tactics, streamlining their operations, and better comprehending the wants of their clients. Chatbots and virtual assistants play a crucial role in defining a whole client experience, going beyond simple job completion. As they continue to advance and become more sophisticated, their contribution to businesses will only increase, making them indispensable in the ever-evolving digital landscape.

5. IMPACT ON BRAND ADVOCACY: BRAND ADVOCACY AND ITS SIGNIFICANCE

The significance of brand advocacy lies in the power of genuine recommendations. In today's world, customers are bombarded with marketing messages from all directions. In this cacophony, a recommendation from a friend, family member, or colleague carries unparalleled weight. Such recommendations are not perceived as advertising but as trusted guidance. Brand advocacy represents one of the most coveted accomplishments for businesses in the digital age. It refers to customers becoming enthusiastic promoters of a brand, willingly and authentically endorsing it to their networks. The impact of brand advocacy is profound, as it translates into organic growth, heightened trust, and invaluable word-of-mouth marketing (George, 2023).

Hyper-personalization plays a pivotal role in nurturing brand advocacy. When brands offer experiences, products, and services tailored precisely to individual customer preferences, they elevate customer satisfaction. Satisfied customers, in turn, are more likely to become advocates, eagerly sharing their positive experiences with others. In this context, chatbots and virtual assistants, powered by artificial intelligence and machine learning, contribute significantly to the equation. They enable real-time, personalized interactions with customers, addressing their needs promptly and effectively. These technologies enhance the customer experience, making it memorable and delightful. Consequently, customers are more inclined to become vocal advocates, spreading the word about the brand's excellence (Wind & Hays, 2016).

Ultimately, the impact on brand advocacy is a testament to the brand's ability to connect with its audience on a personal level. As advocates share their experiences, they become an extension of the brand's marketing team, contributing to its growth and long-term success. This organic, authentic advocacy is the ultimate goal for businesses, and it's a reflection of their commitment to hyper-personalized customer experiences and customer-centric strategies (Kuusinen, 2019).

5.1 Customer Loyalty and Advocacy in the Digital Age

The customer loyalty and advocacy have taken on new dimensions during the digital age. With an ever-expanding array of choices at their fingertips, consumers are more discerning and demanding than ever. Building loyalty is no longer just about offering quality products or services; it's about crafting exceptional experiences and personal connections. Customer loyalty is now intricately linked with brand advocacy. The satisfied and engaged customers not only return but become enthusiastic advocates, sharing their positive experiences and influencing others (Wheeler, 2023). Brands that prioritize personalized interactions, real-time engagement, and exceptional customer service are the ones poised to thrive in this digital landscape, where customer loyalty and advocacy are the true currency of success.

5.2 Building Trust and Emotional Connections

The building trust and emotional connections with customers is fundamental to the success of businesses. Every successful client connection is built on trust, which is acquired via openness, dependability, and consistency. Businesses must conduct honorably and remain committed to keeping their word regarding anything from product quality to data privacy. On the other hand, emotional ties are fostered through unique experiences that connect with each consumer on a deeper level. Brands can identify client preferences and wants with cutting-edge technology like artificial intelligence and machine learning, then customize their interactions and products to produce sincere, emotionally charged experiences. Through these ties, clients become brand evangelists who not only trust the company but also fervently recommend it to others (Sedkaoui, 2018).

6. MARKETING STRATEGIES IN AGE OF HYPER-PERSONALIZATION: DATA-DRIVEN DECISION MAKING

In the age of hyper-personalization, data-driven decision making has emerged as the keystone of effective marketing campaigns. Businesses now have unparalleled access to massive data reservoirs thanks to cutting-edge technologies like artificial intelligence and machine learning, which provide insights into customer behavior, preferences, and demands. They may develop highly focused and tailored marketing strategies that appeal to certain clients by utilizing this data. In order to make data-driven decisions, it is important to extract relevant insights from the data that inform strategic decisions. To predict trends and customer preferences, marketers might examine consumer interactions, purchase histories, and online habits (Neuwirth, 2023). AI and ML-powered predictive models can predict future behaviors and proactively inform strategies.

With the use of these information, firms are able to optimize every aspect of their marketing plans, from timing and channel selection to content and product recommendations. They may act quickly using real-time data analysis to ensure that clients receive pertinent and interesting information. Data-driven decision making, however, entails a need to protect consumer privacy and follow ethical data usage principles. It involves finding the right balance between individualized attention and observance of private boundaries. In the era of hyper-personalization, many marketing tactics rely heavily on data-driven judgment. Understanding the value of data, applying AI and ML to gain useful insights, and customizing approaches to produce engaging, tailored experiences are all important. Businesses that adopt data-driven decision making are better positioned to satisfy client expectations in this dynamic environment.

6.1 Ethical Considerations and Transparency: Addressing Privacy Concerns and Regulatory Compliance

Transparency and ethical issues now dominate marketing efforts. Personalization improves the user experience but it brings up serious privacy and regulatory problems. For businesses, addressing these issues is not only required by law, but also morally right. The first pillar of ethical concerns is transparency, which is being open and truthful with clients about how their data is gathered, utilized, and safeguarded. It is crucial to educate clients on data practices, secure their informed consent, and provide them with unambiguous opt-in and opt-out choices. Privacy concerns loom large in this context. Customer data is a valuable asset, but its collection and usage must respect privacy regulations and customer expectations. This includes compliance with data protection laws like GDPR or CCPA, as well as ensuring robust security measures to safeguard sensitive information. These ethical considerations extend to data sharing and third-party partnerships. Brands should be transparent about any data sharing and provide customers with control over their data even when it's shared with affiliates or partners (Schaeffer, 2017).

Navigating the ethical landscape of hyper-personalization also requires a commitment to responsible data usage. It's about using data not just to increase profits but to enhance the customer experience. Data-driven insights should be used for the benefit of customers, whether that's in the form of personalized recommendations or improved services. The ethical considerations and transparency are non-negotiable in the age of hyper-personalization. Businesses that prioritize privacy, data security, and responsible data usage not only comply with regulations but also foster trust and long-term customer relationships. In this digital era, ethical marketing isn't just a legal requirement; it's a pathway to ethical and sustainable business practices.

6.2 Balancing Automation and the Human Touch

So striking the right balance between automation and the human touch is the defining challenge for businesses. Automation, powered by artificial intelligence and machine learning, offers unmatched efficiency and personalization. However, it must coexist harmoniously with the warmth and empathy that only humans can provide. Automation brings speed and precision to customer interactions. Chatbots and virtual assistants can address routine inquiries instantly, and predictive models can recommend products or services based on individual preferences. This is a crucial aspect of delivering personalized experiences, as it ensures real-time responses and convenience. Yet, the human touch remains indispensable. In more complex or emotionally charged situations, customers often seek human interactions. The human touch encompasses empathy, understanding, and the ability to handle nuanced situations. It's about resolving issues, providing reassurance, and making customers feel heard and valued.

As, balancing these two elements requires a strategic approach. Automation should be employed where it enhances efficiency and personalization, leaving the human touch to shine in areas that demand emotional intelligence and personalized problem-solving. It's about knowing when to automate and when to bring in the human touch. Moreover, transparency is crucial. Customers should be aware when they are interacting with automation versus a human agent. This transparency fosters trust and allows customers to manage their expectations. The finding of equilibrium between automation and the human touch is an art in the era of hyper-personalization. When executed effectively, it ensures efficiency, personalization, and empathy, creating an unparalleled customer experience that builds loyalty and advocacy (Venkatesh, 2021).

6.3. Future of Digital Marketing: Beyond Hyper-Personalization- The Dynamic Customer Journey

As it gazes into the future of digital marketing, we envision a landscape that goes beyond hyper-personalization and delves into the concept of the dynamic customer journey. While hyper-personalization has been a game-changer, the dynamic customer journey takes personalization to the next level, embracing the fluidity and unpredictability of individual customer pathways. In this future, marketing strategies will be adaptive and responsive. Artificial intelligence and machine learning will continuously analyze real-time customer data to anticipate shifts in preferences, behaviors, and context. Marketers will not merely respond to customer actions; they will proactively shape the customer journey based on predictive insights.

The dynamic customer journey acknowledges that customers' needs and desires are not static. They evolve, often in response to external factors, life events, or changing interests. In this context, digital marketing strategies will provide content and experiences that reflect these changes, offering the right message at the right time. This future envisions a seamless integration of marketing touchpoints, from social media and websites to email and in-app interactions. These touchpoints will not exist in isolation but will form part of a cohesive, fluid customer journey. Every interaction will contribute to a personalized narrative that guides the customer towards a deeper relationship with the brand.

While the dynamic customer journey holds incredible potential, it also raises ethical considerations. Businesses must navigate the fine line between personalization and intrusion, respecting privacy and obtaining informed consent for data usage. the future of digital marketing is not just about hyper-personalization but embracing the dynamic nature of customer journeys. It's about that customer needs evolve, and effective marketing strategies must evolve with them. The dynamic customer journey promises not just personalization but proactive, fluid, and relevant customer experiences that pave the way for long-term loyalty and brand advocacy (Reviglio della Venaria, 2020).

6.4 The Role of AI-Driven Creativity and Innovation

Artificial intelligence (AI) has emerged as a powerful catalyst for creativity and innovation, transforming the way we conceive, design, and implement ideas. Its role in these realms extends beyond automation; AI-driven creativity and innovation empower individuals and organizations to envision and realize groundbreaking solutions. AI's capacity to process vast datasets and recognize complex patterns equips it to unearth novel insights. It can analyze customer behaviors, market trends, and even creative works, offering inspiration and fresh perspectives. This data-driven creativity often results in more targeted and effective innovations.

AI can generate creative content itself, such as art, music, or even written text. These AI-generated works challenge traditional boundaries and introduce new dimensions to human creativity. They serve as tools for artists, writers, and musicians, sparking innovation by merging human ingenuity with AI's computational power. AI-driven innovation goes beyond creativity, facilitating the development of disruptive technologies. In fields like healthcare, AI assists in drug discovery and disease diagnosis. In autonomous vehicles, it's revolutionizing transportation. In finance, it's optimizing investment strategies. These innovations improve efficiency, solve complex problems, and create entirely new opportunities.

The responsible use of AI in creativity and innovation involves transparency, accountability, and consideration for potential biases so ethical considerations are paramount. Striking a balance between AI-driven and human-driven creativity is an ongoing challenges. AI-driven creativity and innovation are reshaping industries and pushing the boundaries of what's possible. They offer unprecedented opportunities to generate fresh ideas, optimize processes, and solve complex problems. While ethical considerations remain, the collaborative potential of AI and human creativity promises an exciting future of innovation and exploration (Maes, 2018).

7. CONCLUSION AND VISION FOR THE FUTURE OF DIGITAL MARKETING FOSTERING GREEN ECONOMY

Corporate social responsibility (CSR) is also a big part of brand advocacy in the hydrogen economy. Integrate sustainability as part of their business models, be more compelling in showing how the businesses are reducing carbon footprints and contributing to climate action goals globally. Carbon offset programs, green hydrogen investments, and net-zero emissions pledges are currently effective marketing ploys designed for environmentally-conscious consumers and investors alike. This is further supported by certification programs, such as green energy labels and sustainability accreditations. In the bigger picture, hydrogen power productivity in the edges of the green economy will be tested by good routine and marketing. Empowering hydrogen adoption: While educating consumers on hydrogen's benefits, harnessing the power of technology to build trust through transparency and effective marketing strategies, the most credible and consistent approach to hydrogen adoption will ultimately require commitment from all stakeholders.

Hyper-personalization and real-time customer experiences are the driving forces behind the future of digital marketing. With advanced technologies, businesses can engage customers on a deeply personal level, delivering tailored content and interactions in the blink of an eye. This flexible method anticipates needs, encourages trust, and creates enduring connections. The key to successful marketing in the current digital era is the marriage of hyper-personalization and real-time dynamics, which provides customers with more than just content but also meaningful experiences. The dynamic environment of digital marketing in the future will be characterized by hyper-personalization, moral considerations, and the seamless fusion of automation and human interaction. It's a field where making decisions based on data, being transparent, and using data sensibly are essential. It involves developing experiences that connect with specific customers and encourage loyalty and trust. Our vision for the future goes beyond hyper-personalization.

The vision is based on ethics and accountability, nevertheless. We place a high priority on consumer permission, data security, and privacy in our quest of personalization. We make a commitment to developing emotional bonds and trust because we understand that these are the foundations of brand advocacy. The future of digital marketing involves people as much as technology. It's about how cutting-edge technology and the human touch can come together to create experiences that are not just unique but also highly engaging and meaningful. In the future, marketing plans will be based on people rather than just data, acknowledging that each consumer is an individual with their own requirements and wants. Businesses will succeed in this era not merely by meeting but by exceeding customer expectations. By anticipating their clients' requirements in addition to just meeting them, they will build long-lasting relationships with them. It's a future where brands collaborate with consumers along their trip, act as their guides through a changing environment, and travel alongside them as the digital world develops. Digital marketing has a bright future ahead of it, and we have a vision that emphasizes innovation, accountability, and client centricity. In this future, trust, loyalty, and advocacy are not just results but also the pillars of success where businesses and consumers collaborate to create meaningful experiences.

REFERENCES

Ahmed, A. (2022). Marketing 4.0: The Unseen Potential of AI in Consumer Relations. *International Journal of New Media Studies: International Peer Reviewed Scholarly Indexed Journal, 9*(1), 5–12.

Analytics, M. (2016). *The age of analytics: competing in a data-driven world.* McKinsey Global Institute Research.

Ancillai, C., Terho, H., Cardinali, S., & Pascucci, F. (2019). Advancing social media driven sales research: Establishing conceptual foundations for B-to-B social selling. *Industrial Marketing Management, 82*, 293–308. DOI: 10.1016/j.indmarman.2019.01.002

Auttri, B., Chaitanya, K., Daida, S., & Jain, S. K. (2023). Digital Transformation in Customer Relationship Management: Enhancing Engagement and Loyalty. [EEL]. *European Economic Letters, 13*(3), 1140–1149.

Babet, A. (2020). Utilization of personalization in marketing automation and email marketing.

Bhatnagar, S. (2022). Digital Disruptions and Transformation of Bank Marketing.

Gao, Y., & Liu, H. (2022). Artificial intelligence-enabled personalization in interactive marketing: a customer journey perspective. *Journal of Research in Interactive Marketing*, (ahead-of-print), 1-18.

George, A. S. (2023). Future Economic Implications of Artificial Intelligence. *Partners Universal International Research Journal, 2*(3), 20–39.

Henke, N., & Jacques Bughin, L. (2016). The age of analytics: Competing in a data-driven world.

Kamal, M., & Himel, A. S. (2023). Redefining Modern Marketing: An Analysis of AI and NLP's Influence on Consumer Engagement, Strategy, and Beyond. *Eigenpub Review of Science and Technology, 7*(1), 203–223.

Kaushal, V., & Yadav, R. (2023). Learning successful implementation of Chatbots in businesses from B2B customer experience perspective. *Concurrency and Computation, 35*(1), e7450. DOI: 10.1002/cpe.7450

King, K. (2022). *AI Strategy for Sales and Marketing: Connecting Marketing, Sales and Customer Experience.* Kogan Page Publishers.

Kuusinen, M. (2019). Scenarios for digital marketing: a Delphi-based analysis for 2028.

Maes, P. (2018). *Disruptive Selling: A New Strategic Approach to Sales, Marketing and Customer Service.* Kogan Page Publishers.

Mashood, K., Kayani, H. U. R., Malik, A. A., & Tahir, A. (2023). ARTIFICIAL INTELLIGENCE RECENT TRENDS AND APPLICATIONS IN INDUSTRIES. *Pakistan Journal of Science, 75*(02). Advance online publication. DOI: 10.57041/pjs.v75i02.855

Neuwirth, R. J. (2023). Prohibited artificial intelligence practices in the proposed EU artificial intelligence act (AIA). *Computer Law & Security Review, 48*, 105798. DOI: 10.1016/j.clsr.2023.105798

Rajamannar, R. (2021). *Quantum marketing: mastering the new marketing mindset for tomorrow's consumers.* HarperCollins Leadership.

Reviglio della Venaria, U. (2020). Personalization in Social Media: Challenges and Opportunities for Democratic Societies.

Sadowski, J. (2023). Total life insurance: Logics of anticipatory control and actuarial governance in insurance technology. *Social Studies of Science*, •••, 03063127231186437. PMID: 37427796

Schaeffer, E. (2017). *Industry X. 0: Realizing digital value in industrial sectors*. Kogan Page Publishers.

Sedkaoui, S. (Ed.). (2018). *Big data analytics for entrepreneurial success*. IGI Global.

Sharma, A., & Singh, B. (2022). Measuring Impact of E-commerce on Small Scale Business: A Systematic Review. *Journal of Corporate Governance and International Business Law*, *5*(1).

Soliman, M., & Al Balushi, M. K. (2023). Unveiling destination evangelism through generative AI tools. *ROBONOMICS: The Journal of the Automated Economy*, *4*(54), 1.

Venkatesh, D. N. (2021). *Winning with employees: Leveraging employee experience for a competitive edge*. SAGE Publishing India.

Vinaykarthik, B. C. (2022, October). Design of Artificial Intelligence (AI) based User Experience Websites for E-commerce Application and Future of Digital Marketing. In *2022 3rd International Conference on Smart Electronics and Communication (ICOSEC)* (pp. 1023-1029). IEEE.

Vuong, N. A., & Mai, T. T. (2023). Unveiling the Synergy: Exploring the Intersection of AI and NLP in Redefining Modern Marketing for Enhanced Consumer Engagement and Strategy Optimization. *Quarterly Journal of Emerging Technologies and Innovations*, *8*(3), 103–118.

Wheeler, J. (2023). *The Digital-First Customer Experience: Seven Design Strategies from the World's Leading Brands*. Kogan Page Publishers.

Wind, Y. J., & Hays, C. F. (2016). *Beyond advertising: Creating value through all customer touchpoints*. John Wiley & Sons.

Zaki, H. O. (2022). The Impact Of Artificial Intelligence On Content Marketing. *Journal of Strategic Digital Transformation In Society*, *2*(3).

Chapter 7
Building Public Trust in AI-Enhanced Energy Systems through Strategic Communication

Divya Mishra
https://orcid.org/0000-0002-9375-8720
Sharda University, India

Sanjay Kaushal
Indian Institute of Management, Bodh Gaya, India

ABSTRACT

This study examines the role of public trust in the adoption of AI-driven renewable energy systems, emphasizing the significance of strategic communication in bridging the gap between advanced technologies and societal acceptance. It identifies key challenges such as data privacy concerns, safety issues, ethical implications, and psychological barriers, advocating for a multi-faceted approach that prioritizes transparency, stakeholder collaboration, and culturally tailored messaging. The study highlights the importance of storytelling, adaptive communication strategies, and two-way feedback mechanisms to foster community involvement and confidence in AI technologies. By addressing these challenges through a structured communication framework, this study offers actionable insights for fostering trust, aligning AI systems with societal values, and advancing sustainable energy solutions. These findings provide valuable guidance for researchers, practitioners, and policymakers in achieving sustainable energy futures.

INTRODUCTION

The intersection of artificial intelligence (AI) and renewable energy systems marks a transformative era in addressing global energy challenges. As societies transition to sustainable energy sources to combat climate change and meet growing energy demands, AI offers unparalleled opportunities for optimization, efficiency, and scalability. Renewable energy systems, such as solar, wind, and green hydrogen, face inherent challenges like variability, energy storage, and grid integration complexities that AI is

DOI: 10.4018/979-8-3373-0045-0.ch007

uniquely positioned to address. However, the success of these advancements hinges not solely on their technological potential but also on public understanding, trust, and acceptance, which are fundamentally shaped by communication strategies.

Despite AI's capacity to address operational inefficiencies and bolster renewable energy reliability, public skepticism about data privacy, system safety, and ethical use remains a significant barrier. Research indicates that public resistance frequently arises from insufficient communication regarding the operational mechanisms of AI technologies, their potential benefits, and the protective measures implemented to mitigate associated risks (Wolsink, 2011). Poorly articulated messaging about AI's role in renewable energy perpetuates fears and misconceptions, thereby slowing adoption. Conversely, well-structured communication campaigns have demonstrated their ability to demystify AI, fostering public confidence by ensuring transparency, inclusivity, and ethical accountability (Mishra, 2024a).

Strategic communication plays a dual role: It not only educates the stakeholders regarding the practical aspects of AI integration in renewable energy systems but also caters to the emotional and ethical aspects of the common people. This goes beyond information sharing and calls for the strategy to involve stakeholders in communities, policy-making, and industries (Kaushal & Mishra, 2024). For example, positive framing of how AI contributes to enhancing the reliability of the grid, decreasing costs, or meeting sustainable development objectives can assist the public in perceiving the real utility of AI. This has to be accompanied by a clear description of how personal data is safeguarded, the ethical protocols followed to ensure ethics in AI applications, and how decisions are arrived at in terms that are easy to understand by the general public.

This chapter focuses on how effective strategic communication can be employed to build renewable energy systems that incorporate artificial intelligence. It shows how practices such as storytelling, openness, or incorporating stakeholders may help combat distrust, establish credibility, and invite the public. Instead of framing AI solely as a technological advancement, this chapter highlights its potential as a strategic opportunity to tackle global challenges in renewable energy. It offers specific recommendations on aligning innovation with public acceptance, ensuring that AI becomes a driving force for meaningful and sustainable change.

THE CRUCIAL ROLE OF PUBLIC TRUST IN TECHNOLOGY ADOPTION

Public trust plays a very important role in the successful adoption of AI-driven renewable energy systems. It influences the communities to the extent that they might embrace or resist these technologies. AI's benefit is duly transformative in enabling the optimization of efficiency, reliability, and sustainability in the management of energy production and distribution. While this is the case, public distrust of AI has emerged due to privacy, ethical, and transparency issues of AI systems. For instance, the failure to implement smart meter projects in the United Kingdom clearly shows how poor public participation and poor information sharing on the use of data can be disastrous to such projects. Where there is a lack of message framing that emphasizes perceived Benefits and perceived control over misuse, AI systems are easily viewed as intrusive or manipulative which hinders their uptake. On the other hand, successful renewable energy projects show that public trust boosts technology acceptance. An important example of trust-driven adoption is community-distributed wind power projects in Denmark where proposed projects focus on decentralized community-owned systems from local citizens, involving free, transparent, direct public engagement. All these strategies make sure that the communities directly or indirectly experience a new feeling of being part of the technological change rather than being locked out. Similarly, in the case of Block Island Wind Farm in the United States, studies pointed out that accepting procedural fairness and engaging critical stakeholders in meaningful ways helps maintain perceptions of fairness in decision-making (Dwyer & Bidwell, 2019).

This shows that trust-building is not just about solving technical or procedural problems; it involves technologies, societal values, and emotions. According to studies, acceptance of AI in renewable energy is more likely to be embraced if the potential benefits are presented in terms of perceived value such as economic or environmental gains. AI systems often need to prove compliance with ethical standards of decision-making to be fair, minimize social bias, and protect the privacy of data. For instance, the research conducted in the Netherlands and China shows that trust in the agents working on renewable energy projects and the opportunity for public engagement in decision-making have a positive impact on the increase in the acceptability of the projects (Liu et al., 2019). The 'black box' nature of AI systems enhances public skepticism and, therefore, requires persistent interaction and explanation. Where there are concerns about specific inputs to the decision algorithm, XAI can help alleviate that by explaining the logic behind it to decision-makers as seen with the predictive maintenance systems for the renewable energy sector.

Furthermore, by incorporating AI in renewable energy systems, there is a constant need for stakeholders' engagement to address concerns about the 'black box' nature of AI. Better communication on the methods AI uses, which is helped by XAI models, can make it easier to establish trust in the AI. For example, while AI for predictive maintenance for renewable energy systems has great value, there is concern that it is not transparent enough. Introducing explainable AI solutions has been helpful in alleviating such considerations since it discloses the workings of the algorithms to viewers (Rusilowati et al., 2024). Earning public trust is not a one-time event but a process that is based on continuously providing technical sound and transparent information and engaging with the stakeholders. Through inclusive communication practices, ethical principles, and a clear demonstration of societal and environmental value propositions for AI-driven renewable energy systems, public confidence can be secured, enabling the magic of change and growth.

PUBLIC CONCERNS REGARDING AI IN RENEWABLE ENERGY SYSTEMS

Although AI has the potential to revolutionize renewable energy, there are ethical and societal concerns that have been raised by the public in the adoption of the technology. In most cases, these challenges originate from a lack of transparency, insufficient protection measures, and previous skepticism of innovative technologies. There are four primary areas of public concern: privacy in data collection and management, safety and reliability, ethical implications, and psychological barriers driving skepticism.

Privacy Concerns in Data Collection and Management

Renewable energy systems for use in critical applications must incorporate fundamental data inputs such as energy consumption, user behavior, or performance of the grid for AI to function effectively. Of course, its collection is essential for optimization, while being rather invasive at the same time. Most people have concerns about privacy because smart grids and IoT devices collect a vast amount of personal information. For instance, energy management systems in smart cities use Artificial Intelligence technology to monitor households' energy usage data in real-time, if not well managed, could compromise residents' privacy by revealing details of their everyday lives (Katamoura et al., 2024).

Poor regulations make these difficulties worse when consumers are unclear about how their information is collected, analyzed, and used. Instances of data breaches in smart meter projects have also contributed to eroding public confidence. For example, in 2020 an attack on a smart grid system demonstrated cases of insufficient data protection after hackers managed to breach the system (Park & Kim, 2024). For these potential issues, the key elements include strong privacy protections such as data anonymization, end-to-end encryption, and clear consent of the user. So, there must be data protection efforts or compliance with the regulations.

Safety and Reliability Issues

Reliability and safety features are the key components for gaining the trust of society. On the one hand, AI improves planning in the grid and prediction of energy demand, on the other hand, it opens up threats that can destabilize the system. For instance, predictive algorithms require accurate, uncased data; any mistake or prejudice may lead to voltage fluctuation, breakdowns, or blackouts (Bello et al., 2024).

The opacity of many AI systems only exacerbates these problems because stakeholders have a difficult time interpreting or arguing against AI-based outcomes. In addition, the usage of AI solutions in essential facilities raises the level of cybersecurity threats. A claimed attack on AI-powered energy systems in 2021 revealed the likelihood of an attack on operations, which threatens safety and stability (Otoum & Mouftah, 2021). To respond to these concerns, one needs to use explainable AI (XAI) models to render the processes of decision-making transparent and comprehensible. Such evaluations under real conditions, as well as protection against threats in information security, can also increase reliability. Discussions of these initiatives are necessary to eliminate safety risks and ensure people's trust in artificial intelligence systems.

Ethical Implications

Several ethical issues are associated with AI integration in renewable energy such as bias and data misuse, and sustainability issues. Some of the algorithms are known as 'black boxes' and this raises a lot of questions regarding fairness, mainly in the energy distribution where biases in data or design will lead to unfair results (Danish, 2023).

Further, training large AI models drains lots of energy, which eradicates the benefits of renewable systems (Stiefel & Coggan, 2023). These ethical issues are further complicated by the immense amount of data collection that is needed for the smart grid functioning in the context of AI where the data can be misused if appropriate security measures are missing. To counter these risks, frameworks that emphasize the principles of transparency, as well as fairness, and accountability are required. Ethical guidelines that require the proper use of AI can be used to address these issues while at the same time, improving the trust in AI systems through the use of XAI models.

Psychological Barriers Driving Skepticism

Lack of information, perceived risk, and uncertainty related to the usage of artificial intelligence in renewable energy contributes to the skepticism among the public. People do not trust what they are unable to comprehend, especially when an AI system makes decisions independently (Adewale et al., 2024)). Such concerns are more present in important areas such as the grid where issues can lead to disastrous repercussions. The other reason for public mistrust is related to data privacy and the ethical way that data will be utilized. Surveys suggest that several people reported concerns about personal data misuse, especially where there is no clarity or where privacy has not been adequately protected (Braganza et al., 2020). These psychological concerns can only be effectively addressed by defining the potential gains and losses from the usage of AI. An emphasis on education, community involvement, and the implementation of XAI systems can help reduce skepticism surrounding AI and promote its integration into renewable energy solutions.

STRATEGIC COMMUNICATION FRAMEWORK FOR BUILDING PUBLIC TRUST

Building public trust in AI-enhanced renewable energy systems requires a strategic communication framework that addresses technical, ethical, and psychological concerns. This framework must not only inform stakeholders about the functionality of AI technologies but also reassure them of their safety, fairness, and alignment with societal values. By emphasizing transparency, ethical compliance, and psychological engagement, such strategies can bridge the gap between technological innovation and public acceptance.

The Role of Transparency in Mitigating Concerns

Transparency is one of the key principles when it comes to effective management of AI-based renewable energy initiatives because it enables trust building in the technologies and, therefore, increased confidence in their use (Mishra, 2024b). It is crucial to share clear information regarding the functionality, utility, and restraints of AI because they help people become aware of how these systems work and why they are effective. For instance, illustrating how AI can do more with less power and how it helps minimize energy consumption and carbon emissions not only presents AI as useful but also answers those who worry about the environmental impact of AI technologies. A real-time energy dashboard is another effective tool that helps in increasing transparency as it enables consumers to see how their behavior affects energy consumption (Maurya, 2024). But technical transparency alone is not enough. Ethical transparency, which means that the data privacy measures and the decision-making processes are easily communicated, is also important. For example, sharing information on how user data is anonymized and protected has been found to decrease perceptions of surveillance and misuse in smart city initiatives (Buah et al., 2020). Besides, engaging the stakeholders across AI project developmental phases from design to implementation also helps in a way that potential causes of mistrust stemming from misunderstandings are averted since the stakeholders are fully involved. To maximize transparency, communication strategies must prioritize clarity and accessibility. Overly technical language and complex explanations can alienate non-specialist audiences. Instead, visuals, case studies, and relatable narratives should be used to illustrate how AI technologies align with community priorities. For example, showcasing financial savings or environmental benefits achieved through AI resonates more effectively than abstract descriptions of algorithms.

Ethical Compliance and Public Perception

Ethical compliance is central to how perception about AI in renewable energy is being created and the type of trust that is being cultivated. This way it gives the stakeholders more confidence that such technologies are being used rightly, fairly, and for the benefit of society. For instance, decarbonization strategies have involved ethical measures like reduction of energy consumption during the AI model training, lowering of carbon footprints, and equal energy availability (Ajayi et al., 2024). One of the major ethical issues is bias in AI algorithms. This leads to people perceiving favoritism about some demographic or geographical areas, which results in the loss of trust. For example, algorithms to determine the distribution of energy subsidies need to be thoroughly checked for bias to prevent bias. Based on prior studies, public confidence is higher when the market participants take necessary actions to address bias.

Privacy is another area of ethical compliance that is very important when carrying out organizational activities. The technologies that have focused on anonymizing the data and obtaining clear consent from the users have been more successful in smart city initiatives as illustrated in a study by Otoum & Mouftah (2021). Such assurances, backed by sound legal requirements, help to build confidence. Therefore, ethical compliance should not be seen as an add-on to AI projects, but as a framework that should be incorporated into the design and communication of such projects.

Behavioral Psychology Insights in Addressing Skepticism

Behavioral psychology holds good promise when it comes to countering public skepticism by addressing cognitive and emotional factors that influence attitudes toward AI. Many times, the apprehension is not from a lack of technical information but from knowledge of potential harms, lack of control, and possible exploitation. AI can be presented as a solution to addressing familiar issues including, for instance, reducing the energy bills of homes or increasing sustainability (Viendyasari et al., 2022), thereby positively shaping people's perception towards AI. Among the most successful AI applications, that increase consumer interest in energy consumption, is real-time feedback (Lin et al., 2023). For instance, capabilities such as energy dashboards are symbolic representations of usage patterns and possible savings; they enable users as it makes them feel involved. However, culturally sensitive message development and cultural values and beliefs that are familiar to targeted audiences lower cognitive discomfort and increase message appeal and relevance.

Another way to counter skepticism – is the method of storytelling based on the principles of behavioral psychology. It means that to engender trust, people need to be reminded of how AI technologies are useful for their families, communities, and the planet. Incorporation of the testimonies and other elements from the community in communication campaigns also increases AI's credibility hence fostering ownership and achievement. A communication framework that is based on the principle of transparency and compliance with ethical standards together with the principles of behavioral insights should be aimed at the consideration of public concerns for the implementation of AI in renewable energy systems. Thus, by making concepts understandable, positioning technologies as meeting society's needs, and using emotional and cultural appeals, stakeholders can overcome skepticism and promote the use of radically different energy solutions.

DEVELOPING CULTURALLY COMPETENT NARRATIVES FOR PUBLIC ENGAGEMENT

Creating culturally sensitive and appropriate stories is highly relevant to the development of an adequate level of trust and interest in the use of AI for renewable energy systems. These stories focus on the technological, ethical, and social aspects of innovations and make sure that they are properly explained, received, and adopted in various cultures.

Storytelling as a Tool to Bridge Understanding

The complexities of AI-driven renewable energy systems can be converted to emotionally engaging narratives by using storytelling. By humanizing these systems, storytelling makes the role of AI in renewable energy more accessible and comprehensible to diverse audiences. For example, when AI-powered tools are presented as stories of progress, like lowering electricity bills or supporting local clean energy initiatives, the practical impacts of innovation resonate with audiences. Stories can be effective and can integrate cultural values and address community-specific concerns. For example, strategies for indigenous communication describe how dialogue-driven storytelling can create a sense of partnership and empowerment. Research on Māori and cultural engagement with technology indicates that rather than a top-down communication approach, technological solutions can be aligned with local traditions and priorities (Dutta & Elers, 2019).

Storytelling also alleviates some of the anxiety that comes with interacting with complex artificial intelligence systems by providing comparative instances. The collection of AI stories that explain how the technology has contributed to the larger social missions- whether climate change or energy inequality taps into a collective vision. Storytelling begins to close that gap between the advancement of technology and its social acceptance by aligning AI with positive benefits.

Tailoring Messages to Diverse Cultural Contexts

Culturally sensitive communication makes AI-supported renewable energy systems acceptable, noticeable, and relevant. This type of communication enables stakeholders to avoid multiple pitfalls that come with a lack of cultural sensitivity while disseminating information. For example, public engagement efforts in Japan have focused on using local pictures and messages in dealing with difficult energy policies (Mikami et al., 2014). It can also be seen that AI tools contribute to cultural tailoring. Technologies, including Artificial intelligence language translation, and cultural imitation decrease communication language barriers in learning and community participation (Xia et al., 2024). These tools assist renewable energy stakeholders in sharing information accurately and with respect, according to the culture and norms of the region.

It is also rational to integrate these two ideas of sustainability narrative with the practical usage of eco-friendly knowledge possessed by indigenous communities. For instance, renewable energy projects where an AI contribution is portrayed toward traditional environmental conservation methodologies have been well received among the indigenous people (Sharma & Karki, 2014). Such narratives contribute to the framing of the people who can receive and engage AI technologies hence increasing aspects such as trustworthiness and ownership. The use of sustainability narratives alongside traditional ecological knowledge that is unique to the community area has been established to be very useful in addressing indigenous communities. New renewable energy projects that focus on promoting how artificial intelligence enhances conventional environmental protection have been accepted by indigenous communities at higher rates. Such tailored messaging makes the communities active participants in addressing the issues and such strategies help to build trust and establish ownership of AI-based technologies.

Communicating the Environmental and Economic Benefits of AI-Enhanced Energy

To counteract skepticism, it is a must to promote the environmental and economic positives of renewable energy systems with the help of AI, is a must-do thing. Realistic advantages like the decrease in the carbon footprint, the cut in energy expenses, and system dependability, etc. demonstrate the practical worth of these technologies. For example, in cases where AI solutions can better manage energy usage and include renewable energy sources in the systems, people have become more positive about smart grids and energy change (Stecyk & Miciuła, 2023).

The analysis of key environmental messages shows that when decision-making is evidence-based, the use of figures that translate benefits back to the organization is effective. Tools such as real-time energy management dashboards that showcase energy savings instantly and carbon emissions reductions make it easier to measure AI's benefits. These visuals underpin the facilitative function of AI in promoting sustainable development and enhancing the level of public confidence. Other perceived benefits, including direct employment in renewable power industries, and saving on power bills by customers, feel appealing to many people. Some past public consultations in the UK have demonstrated that consultations that focus on affordability and employment benefits are likely to gain more social acceptance (Pidgeon et al., 2014).

Targeted communication guarantees that such advantages correspond to local concerns. In areas where there is interest in energy costs, potential uses of AI should be advertised as an area where it can save consumers money. On the other hand, specifying discursive priorities that include reducing greenhouse-effects, and nurturing clean energy technologies will be more impactful in areas that prioritize environmentalism. Such messages are essential for ensuring that people in a given community perceive AI-augmented systems as automatically available and useful.

FOSTERING PUBLIC TRUST THROUGH STAKEHOLDER COLLABORATION

Effective stakeholder engagement is necessary for the successful development and adoption of AI-driven renewable energy systems. Developers, policymakers, and the public must collaborate to ensure these technologies address societal needs, ethical considerations, and operational challenges. To achieve this, the focus should be on collaborative communication, community partnerships, and two-way feedback mechanisms in building trust and fostering acceptance.

Collaborative Communication Between Developers, Policymakers, and the Public

Collaborative communication is the key to trust and cooperation in renewable energy systems with the help of artificial intelligence. It makes sure that interest of various stakeholders is met helping to make a decision that considers the needs of everyone. Stakeholder mapping, which defines who may influence the process and whose goal aligns with it, is the preliminary step in creating communication strategies (Ezeh et al., 2024).

Government policies like advisory committees and public consultations open the channels of dialogue due to the encouragement of stakeholders to present their opinions and share insights. These forums focus on technical, ethical, and social issues at the same time as developing trust. For instance, smart city AI platforms for implementing long-term energy projects allow citizens and policymakers to participate in the input acquisition of the technology (Buah et al., 2020).

Another very important strategy is the management of conflicts between tasks and activities. Software developers tend to concentrate on technical accomplishment, while policymakers are concerned with legal aspects and societal advantage. Since clarifying such objectives is important in facilitating the actualization of the implementation of AI technologies in renewable energy systems, a constructive dialogue between such stakeholders must be encouraged to overcome potential resistance.

Building Community Partnerships for Technological Adoption

Community partnership and engagement are crucial in gaining the support of the public for the implementation of AI-based renewable energy systems. Such partnerships serve to enhance the rights of local people since they form the basis of the decisions to be implemented hence assuming an ownership role. For instance, collaborations between developers of renewable energy projects and the indigenous people of Canada showed how consultative approaches result in the acceptance of projects and integration. Such partnership approaches promote win-win situations and cultural relevance ensuring that technological inventions are responsive to the aspirations of the community (Leslie et al., 2024).

Corporate Social Responsibility (CSR) takes corporate relationships to another level as community relations are highlighted by corporate entities. Corporate initiatives that contribute to infrastructure development, education, and training of human resources are perceived to create value for society. For example, renewable energy projects that provide employment training to the locals apart from gaining social support address the trained human capital deficits in the market. In co-development strategies where the communities get involved in planning and executing processes, the perception that AI systems are the instruments for progress in society increases. By tailoring AI technologies to meet the unique needs of specific populations, community partnerships ensure systems are both technologically advanced and socially inclusive. Co-created solutions that reflect local contexts foster resilience and trust, enabling communities to embrace and actively participate in these innovations.

Encouraging Two-Way Communication for Feedback and Improvement

The adoption of two-way communication is crucial for building confidence, handling issues, and enhancing the improvement of renewable energy systems powered by artificial intelligence at all times. Unlike a model of one-way transmission of information, it involves stakeholders as participants in the information development process. For instance, a technologically advanced application of artificial intelligence for citizen sentiment analysis is crucial since it enables developers to adjust their strategies in real-time.

The second approach to two-way communication is integrated grievance mechanisms. These frameworks resolve stakeholder issues quickly thus minimizing conflict and building trust. For instance, African renewable energy projects include a complaint mechanism that alleviates the conflict and also ensures that the community works together with the developers to enhance the efficient utilization of the projects by addressing their concerns in an ethical and culturally sensitive manner. Stakeholder feedback, which forms a central part of iterative development processes, allows for AI systems to remain relevant to and reflect the goals of society. For instance, involving the community in addressing issues of data privacy or system issues of ease-of-use fosters trust and thereby cuts resistance to change. This feedback loop encourages constant change and AI technologies are constantly updated to meet the needs of the targeted populations.

Thus, engaging stakeholders is fundamental in the application of AI in renewable energy systems. The involvement of the developers, policymakers, and the general public in the communication process develops transparency and trust in the process. Community collaborations ensure that locals embrace change since the programs are implemented by them and will therefore sustain the change. The appropriate two-way communication strategies make certain that AI systems will always be attentive to the needs of the stakeholders, thereby creating inclusive, adaptive, and sustainable energy solutions.

FUTURE DIRECTIONS AND RECOMMENDATIONS

While the integration of AI into renewable energy systems has the potential to address critical global energy challenges, achieving widespread public trust requires deliberate advancements in strategic communication and engagement. Future efforts must prioritize transparency, inclusivity, and adaptability alongside ethical and technical improvements.

Firstly, there is a need for adaptive communication tactics that adjust dynamically to changing societal and technical environments. AI-powered analytics should be used in real-time to monitor public sentiment, allowing stakeholders to modify messaging to address specific concerns efficiently. Collaborative platforms, supported by practitioner training, will enable alignment among diverse groups, ensuring communication frameworks are both robust and inclusive. Secondly, immersive digital tools such as augmented reality (AR) and virtual reality (VR) should be integrated into communication strategies to bridge the gap between technical complexities and public understanding. These tools can create accessible, engaging experiences that foster confidence and deepen stakeholder engagement.

Thirdly, the topic of ethical compliance and transparency in all AI operations must be applied through the lens of explaining AI (XAI), constituting models and mechanisms such as real-time reporting and public dashboards to decode decision-making. Fourthly, it will involve people's participation through public consultations and adequate technology forums to enhance the regional and cultural relevance of the technologies to the populations. Finally, long-term community involvement via education programs and ongoing communication will build trust by protecting the communities and making them capable of understanding and participating in the changes in the energy systems with the use of AI.

The future of AI-enhanced renewable energy systems requires technological innovation with the simultaneous implementation of legal, ethical, and effective methods in communication as well as consideration of stakeholders. To achieve sustainable use of these technologies, it is critical to overcoming existing obstacles in terms of trust and public perception. Addressing trust is a significant condition for improving the public's trust in AI decision-making. The developers should consider employing XAI models, real-time dashboards, and other easy-to-understand visualizations to make the system operation understandable to all concerned. Thus, all of these measures contribute to raising the population's trust in the usability and the intended goals of AI technologies.

CONCLUSION

For AI-assisted renewable energy systems to be adopted, embraced, and integrated across societies it demands more than technology; it requires trust among the public, clear communication, and effective engagement. Although AI can help make better use of energy, reduce climate change, and ensure relevant sustainability concerns; doubts about its effects on privacy, morality, and society are a significant obstacle. The restoration of public trust is therefore crucial, to allow for the full realization of the transformative impact of such technologies. Thus, strategic communication stands out as the foundation of this effort to address the technical, ethical, and psychological aspects of public attitudes that cause resistance. A lack of transparency in AI operations hinders efforts to improve system understandability by various stakeholders. Openness, explainable AI (XAI), and real-time data dashboards help people understand how AI works and show them how it can help via real-life examples. Nevertheless, technical transparency needs to be matched by ethical transparency and effectiveness where the principles of fairness, protection of privacy, and accountability are not only observed but also communicated.

It is also important to make culturally competent narratives to ensure AI technologies resonate with different types of audiences. Storytelling which suits the needs and values of different communities, humanizes AI and develops trust. By focusing on the tangible aspects of the technology such as costs, environmental conservation, and the quality of energy supplies, the gap between technology and the general public is well addressed. Co-creation workshops and CSR activity also entail the involvement of the communities in decision making hence enhancing ownership and collaboration. In addition, taking the viewpoint of the developers, policymakers, and the public is still important to satisfy their interests through Stakeholder engagement. To enable AI systems to evolve in response to the needs and expectations of society as well as culture, collaborative communication platforms, grievance mechanisms and a two-way feedback system are operational in place. Improving flexibility with these efforts not only reduces tensions but also helps to balance the use of AI in renewable energy systems.

Furthermore, ethical practices, clear communication, and the effective participation of stakeholders must remain a priority. The future innovation needs to be pursued by the collaboration of industry, government, and academic institutions to solve problems such as energy intermittency, storage, and grid integration. Open access platforms for sharing knowledge and experience in the area of AI will accelerate the pace of progress towards AI and keep the world on the same page in terms of AI development. Lastly, public perception of AI in renewable energy systems is not something we can ignore as it is a serious factor that will affect the success of these systems. If stakeholders implement trust-building measures and correlate innovation with values, then the new technologies can lead to a more sustainable, equitable, and trustworthy energy future.

REFERENCES

Adewale, B. A., Ene, V. O., Ogunbayo, B. F., & Aigbavboa, C. O. (2024). A systematic review of the applications of AI in a sustainable building's lifecycle. *Buildings (Basel, Switzerland)*, *14*(7), 2137. DOI: 10.3390/buildings14072137

Ajayi, N. F., Ademola, N. O. M., Amuda, N. K. F., & Alade, N. B. (2024). AI-driven decarbonization of buildings: Leveraging predictive analytics and automation for sustainable energy management. World Journal of Advanced Research and Reviews, 24(1), 061–079. DOI: 10.30574/wjarr.2024.24.1.2997

Bello, N. S. F., Wada, N. I. U., Ige, N. O. B., Chianumba, N. E. C., & Adebayo, N. S. A. (2024). AI-driven predictive maintenance and optimization of renewable energy systems for enhanced operational efficiency and longevity. *International Journal of Science and Research Archive*, *13*(1), 2823–2837. DOI: 10.30574/ijsra.2024.13.1.1992

Braganza, A., Chen, W., Canhoto, A., & Sap, S. (2020). Productive employment and decent work: The impact of AI adoption on psychological contracts, job engagement and employee trust. *Journal of Business Research*, *131*, 485–494. DOI: 10.1016/j.jbusres.2020.08.018 PMID: 32836565

Buah, E., Linnanen, L., Wu, H., & Kesse, M. A. (2020). Can artificial intelligence assist project developers in Long-Term management of energy projects? The case of CO2 capture and storage. *Energies*, *13*(23), 6259. DOI: 10.3390/en13236259

Danish, M. S. S. (2023). AI in Energy: Overcoming Unforeseen Obstacles. *AI*, *4*(2), 406–425. DOI: 10.3390/ai4020022

Dutta, M. J., & Elers, S. (2019). Public relations, indigeneity and colonization: Indigenous resistance as dialogic anchor. *Public Relations Review*, *46*(1), 101852. DOI: 10.1016/j.pubrev.2019.101852

Dwyer, J., & Bidwell, D. (2018). Chains of trust: Energy justice, public engagement, and the first offshore wind farm in the United States. *Energy Research & Social Science*, *47*, 166–176. DOI: 10.1016/j.erss.2018.08.019

Ezeh, N. M. O., Ogbu, N. D., Ikevuje, N. H., & George, N. E. P. (2024). Stakeholder engagement and influence: Strategies for successful energy projects. *International Journal of Management & Entrepreneurship Research*, *6*(7), 2375–2395. DOI: 10.51594/ijmer.v6i7.1330

Katamoura, S., Aksoy, M. S., & Alkhamees, B. F. (2024). Privacy and Security in Artificial Intelligence and Machine Learning Systems for Renewable Energy Big Data. 2024 21st Learning and Technology Conference (L&T), 209-214.

Kaushal, S., & Mishra, D. (2024). Strategic implications of AI in contemporary business and society. 2024 2nd International Conference on Disruptive Technologies (ICDT), Greater Noida, India. DOI: 10.1109/ICDT61202.2024.10489027

Leslie, D., Rincon, C., Briggs, M., Perini, A., Jayadeva, S., Borda, A., Bennett, S., Burr, C., Aitken, M., Katell, M., Fischer, C., Wong, J., & Ismael, K. G. (2024). AI Sustainability in Practice Part One: Foundations for Sustainable AI projects. SSRN Electronic Journal. https://doi.org/DOI: 10.2139/ssrn.4731832

Lin, Y., Chen, Y., Hsieh, S., Liu, H., Chiang, C., & Yang, H. (2023). Reinforcement Learning-based Energy Management System for Microgrids with High Renewable Energy Penetration. *2023 IEEE International Conference on Energy Technologies for Future Grids (ETFG)*, 1-6. DOI: 10.1109/ETFG55873.2023.10407803

Liu, L., Bouman, T., Perlaviciute, G., & Steg, L. (2019). Effects of trust and public participation on acceptability of renewable energy projects in the Netherlands and China. *Energy Research & Social Science*, *53*, 137–144. DOI: 10.1016/j.erss.2019.03.006

Maurya, N. P. K. (2024). Artificial intelligence to enhance energy management and distribution in smart grid communication networks. *Tuijin Jishu/Journal of Propulsion Technology, 45*(02), 4366–4378. DOI: 10.52783/tjjpt.v45.i02.6648

Mikami, N., Yagi, E., & Ikebe, Y. (2014). Recent Trends and New Approaches of Public Engagement in Japan

Mishra, D. (2024a). Role of Social Media and AI in Transforming Healthcare Communication. In Vig, K., Kaunert, C., & Singh, B. (Eds.), *Intersections of Law and Computational Intelligence in Health Governance* (pp. 29–51). IGI Global., DOI: 10.4018/979-8-3693-5976-1.ch002

Mishra, D. (2024b). Strategic Communication for AI-Driven Sustainability Initiatives: Bridging Technology, Policy, and Public Engagement in Achieving SDGs. In Riswandi, B., Singh, B., Kaunert, C., & Vig, K. (Eds.), *AI Applications for Clean Energy and Sustainability* (pp. 187–211). IGI Global., DOI: 10.4018/979-8-3693-6567-0.ch010

Otoum, S., & Mouftah, H. T. (2021, December). Enabling Trustworthiness in Sustainable Energy Infrastructure Through Blockchain and AI-Assisted Solutions. *IEEE Wireless Communications*, *28*(6), 19–25. DOI: 10.1109/MWC.018.2100194

Park, C., & Kim, M. (2024). Utilization and challenges of artificial intelligence in the energy sector. *Energy & Environment*, 0958305X241258795. Advance online publication. DOI: 10.1177/0958305X241258795

Pidgeon, N., Demski, C., Butler, C., Parkhill, K., & Spence, A. (2014). Creating a national citizen engagement process for energy policy. *Proceedings of the National Academy of Sciences of the United States of America, 111*(supplement_4), 13606–13613. DOI: 10.1073/pnas.1317512111 PMID: 25225393

Raman, R., Gunasekar, S., Kaliyaperumal, D., & Nedungadi, P. (2024). Navigating the nexus of artificial intelligence and renewable energy for the advancement of sustainable development goals. *Sustainability (Basel)*, *16*(21), 9144. DOI: 10.3390/su16219144

Rusilowati, U., Ngemba, H. R., Anugrah, R. W., Fitriani, A., & Astuti, E. D. (2024). Leveraging AI for Superior Efficiency in Energy Use and Development of Renewable Resources such as Solar Energy, Wind, and Bioenergy. [ITALIC]. *International Transactions on Artificial Intelligence*, *2*(2), 114–120. DOI: 10.33050/italic.v2i2.537

Sharma, S. R., & Karki, N. (2014). Towards culturally sensitive public health interventions in Nepal. *Health Prospect : Journal of Public Health*, *13*(1), 1–3. DOI: 10.3126/hprospect.v13i1.11104

Stecyk, A., & Miciuła, I. (2023). Harnessing the power of artificial intelligence for collaborative energy optimization platforms. *Energies*, *16*(13), 5210. DOI: 10.3390/en16135210

Stiefel, K. M., & Coggan, J. S. (2023). The energy challenges of artificial superintelligence. *Frontiers in Artificial Intelligence*, *6*, 1240653. Advance online publication. DOI: 10.3389/frai.2023.1240653 PMID: 37941679

Viendyasari, M., Syukri, M., Yandri, E., & Rahmawati, D. (2022). PUBLIC PERCEPTION OF RENEWABLE ENERGY IN INDONESIA: IMPLICATION FOR COMMUNICATION STRATEGY. [MEA]. *Jurnal Ilmiah Manajemen Ekonomi & Akuntansi*, *6*(3), 1582–1596. DOI: 10.31955/mea.v6i3.2517

Wolsink, M. (2011). The research agenda on social acceptance of distributed generation in smart grids: Renewable as common pool resources. *Renewable & Sustainable Energy Reviews*, *16*(1), 822–835. DOI: 10.1016/j.rser.2011.09.006

Xia, Y., Shin, S., & Kim, J. (2024). Cross-Cultural Intelligent Language Learning System (CILS): Leveraging AI to facilitate language learning strategies in Cross-Cultural communication. *Applied Sciences (Basel, Switzerland)*, *14*(13), 5651. DOI: 10.3390/app14135651

Chapter 8
Cybersecurity Challenges in AI-Driven Energy Systems:
Current and Future Prospects Concerning Ethical and Legal Provisions

Kanchan Tolani
https://orcid.org/0000-0003-2788-0987
Ramdeobaba University, Nagpur, India

Monali G. Dhote
Yeshwantrao Chavan College of Engineering, Nagpur, India

A. Vijayalakshmi
Chaitanya Bharathi Institute of Technology, Hyderabad, India

Saurabh Chandra
https://orcid.org/0000-0003-4172-9968
Bennett University, Greater Noida, India

Pritam Lanjewar
https://orcid.org/0000-0001-7166-884X
Datta Meghe Institute of Management Studies, Nagpur, India

Saquib Ahmed
https://orcid.org/0009-0008-1891-6910
Sharda University, India

Bhupinder Singh
https://orcid.org/0009-0006-4779-2553
Sharda University, India

ABSTRACT

Artificial intelligence in energy systems has brought a significant improvement of efficiency, management and optimization making them dependable, there are also big concerns in the cybersecurity domain. A major worry is the risk of AI algorithms being influenced by data poisoning which is when bad actors introduce corrupted data to cause malfunctions on a system. AI-powered energy systems frequently connect thousands of IoT devices and every one of these represents a potential gateway to a system breach if not correctly protected. In addition, threats such as ransomware, phishing and Distributed Denial of Service (DDoS) attacks can compromise the integrity of energy management systems. Security of AI models and data, strong encryption protocols to ensure that these infra are secure from the cyber threat elites of today because of which we must build a real-time monitoring system too. Energy industry can prevent catastrophe by prioritizing cybersecurity.

DOI: 10.4018/979-8-3373-0045-0.ch008

1. INTRODUCTION

The application of artificial intelligence (AI) in energy systems is revolutionizing the way we produce, distribute and consume energy. AI is central to how we will address the challenges of both optimizing renewable sources such as solar and wind, as well increasing the efficiency of our power grids. AI in the modern energy ecosystem plays a vital role across many functions, including predictive maintenance, demand forecasting and grid optimization to even balancing volatile supply with shifts in real-time consumption (Salleh et al., 2020). It will boost the potential of these AI applications to drive a low-carbon economy, lower costs and increase system resiliency against disruptions. This growth parallels a greater need for robust cybersecurity to protect the critical infrastructure that AI increasingly powers. This convergence of AI and energy infrastructure is a double-edged sword, promising unique operational efficiencies alongside emergent opportunities for cyber attackers

The fact is, AI-powered systems function with a cross-layer of data-driven algorithms and devices connected to an ecosystem that implies various cybersecurity risks. Not only do cyber-attacks on critical infrastructure, like power grids, come at a high price. They can also have devastating impacts on public safety; national security and financial stability. Specifically, the AI dependency also means that such attacks can manipulate these systems to render grid instability by causing unintended changes in energy flows as well as unauthorized access seeking sensitive information that leads into massive power outages (Singh, 2023). Unlike conventional threats to cybersecurity AI-specific cyber-threats such as adversarial attacks, data poisoning and model inversion are also unique where the capability of these type of attacks can be quite advanced making them more difficult for detection or possible mitigation. As such, the need to secure AI-driven energy infrastructure has reached near-celebrified status as major actors across state and industry recognize that cybersecurity is a critical issue for anyone with stake in providing reliable and resilient electricity.

The purpose of this chapter is to identify cybersecurity issues and discuss the rising demand for effective application solutions due to the increased dependence on AI in energy systems. The focus will be on the threats and vulnerabilities present in a cyber physical context, ethical and legal considerations around cybersecurity related to AI-based energy systems landscape today as well as areas of future concerns along with best practices for designing secure system (Bujang et al., 2016). The programme of research, which comprises the key developments as:

-Mapping out the ethical aspects concerning data privacy and AI applications on accountability/fairness roles in energy system.

-Assessing how legal and regulatory play a role as an enabler for safe ecosystem energies. The paper also provides economic, social and security implications of cyber risks in AI-energy systems could provide hints on the design for secure solutions to balance advantages of AI against rigid cybersecurity requirements.

- Applications of AI-driven energy systems cover the spectrum from generation, transmission and distribution to consumption. Renewable energies are optimized by AI, such as the prediction of weather patterns and thus output control to stable and efficient energy feeds. AI algorithms in grid management enable monitoring and balancing of the load on a dynamic basis so that grids can match aggregate demand with supply at most responsive levels.

With energy infrastructure undergoing quick digital transformation, it has become apparent that a robust cybersecurity framework which accounts for the specific weaknesses of AI-driven energies is necessary. Conventional methods of cybersecurity are less effective due to the dynamic and complex characteristics exhibited by AI technologies. AI-driven systems, there is a continuous back-and-forth of data between connected devices and central management systems an architecture which provides attackers with numerous points of entry (Lawrenz et al., 2018). Some of you may have heard that before. In this type of attacks the attackers corrupts training data in order to affect AI predictions and operational decisions accuracy in energy systems. Recognizing the important impact on AI driven systems in national and global energy infrastructure, there is an urgent need to develop security practices that consider these challenges as relative risks but also remain flexible against evolving threats. AI-powered systems can track energy consumption in customers' own homes, as with smart home technologies that automatically reduce power usage according to real-time data. AI contributes to energy storage: creating predictions for when it is best store and release energy, reducing the need for fossil fuels supporting sustainability of as a whole in another example where AI can accelerate clean solutions. Though breakthrough, these advancements open energy systems to cyber attacks that target both AI algorithms and the plethora of interlinked devices upon which such intelligence leans (Cohen et al., 2021).

These themes are addressed in this paper by studying the threat landscape of cybersecurity for AI energy systems, investigating security ethical implications related to securing AI technologies and examining current legal/regulatory frameworks. In the paper, he will walk through case studies of previous cyber attacks on energy infrastructure and illuminate how they came about as well their possible defenses. At last, the paper will look at forward-looking approaches for strengthening cybersecurity in energy systems using sophisticated technologies such as blockchain, quantum computing and AI predictive algorithms that shall prove to improve the resilience of these system. As such, this study aims to create a roadmap for related stakeholders to ensure that AI-based energy systems are implemented in the right way balancing advancing technological innovation with ethical responsibility and legal compliance. For AI to continue revolutionizing the energy industry, confronting cybersecurity challenges directly is essential so that we can realize without any impacts on security and reliability of more than three-quarters of critical energy infrastructure (Yang et al., 2021).

2. ROLE OF AI IN MODERN ENERGY SYSTEMS: APPLICATIONS OF AI IN RENEWABLE ENERGY, SMART GRIDS, AND ENERGY STORAGE

The energy industry has seen a radical turnaround with the incorporation of Artificial Intelligence (AI) that converted basic features into optimized automation and data-centric procedures. AI is now more of a necessity than an added layer for energy generation, distribution and consumption (while supporting the biggest challenges to defend our environment) in modern energies. AI, meanwhile, plays an essential role in fuelling the transition to renewable and smart systems by orchestrating networks that generate energy from more integrated sources than ever before (solar panels on roofs; wind turbines in fields), store it through newly imagined battery technology into live batteries they are now being turned off during demand lows. This shift in digital transformation is allowing for smarter grids, the integration of AI and energy storage solutions as well renewable forecasting at a global scale that can adapt to space-when demand has intermittencies due to supply fluctuations improviding sustainability maintenance along with resilienceroad (White et al., 2018).

This is one of the most important prospects for artificial intelligence in energy. These days, power systems are transitioning to more and more renewable energy sources like wind, solar and hydro power which are naturally varying in time so as matching the generated capacity with demanded capacity is not a trivial task. To overcome this limitation, the AI uses predictive analytics to predict power generation from renewable sources accurately. For example, machine learning algorithms can study weather patterns and historical data to forecast solar irradiation or wind speeds in order to optimize the amount of energy that may be derived from those sources (Nastasi et al., 2022). AI also enables grid operators to predict changes in energy generation and thereby make real-time adjustments that optimize the efficiency of increasingly renewable sources and decrease reliance on traditional fossil fuels. AI-driven automation further assists with maintenance of renewable energy assets by scientifically determining when parts like wind turbine blades or photovoltaic panels may need to be repaired or replaced minimizing both downtime and maintenance costs (Danish & Senjyu, 2023).

Smart Grids are highly-dependent on AI for improving the reliability and resilience of smart grid operations as well increasing efficiency. The usage of smart grids requires connecting multiple devices and sensors which can read the energy flow across various poles, obtaining data concerning consumption habits. This data is in turn processed by AI, which dynamically balances supply and demand to ensure that energy sources are effectively distributed across the grid thus preventing blackouts or surges (Basheer et al., 2022). For example, at times when demand is highest AI can divert energy from less congested areas of the grid to help meet spikes in demand–keeping overloads down, and keeping the lights on. AI can also power demand response programs that pay customers to change their energy consumption behaviours in gentle responses, providing immediate grid relief and helping to lessen lost electricity. Not only do these smart grid features make operations more efficient, they enable energy providers to integrate greater and higher levels of renewable into the grid thus helping achieve sustainability goals.

The other space where AI is significantly advancing the energy landscape is in storage of these above mentioned energies. These AI-driven systems manage the way that energy is stored and tapped from batteries, as well as other energy storage solutions to maximize total efficiency by making sure potential power reserves are available for when they're most needed (read: peak demand), or because renewable generation drops off (Al-tabatabaie et al., 2022). These AI algorithms can detect peak energy demand from customers and discharge stored energy automatically to ensure a balanced grid remains, but they have no way of changing the underlying issue in how this country consumes (and poor countries will imitate our model soon enough) its so-called clean energies. The ability to do so is particularly critical when it comes to the integration of variable renewable sources, offering a cushion which allows continuous power delivery through differences in renewables generation. In addition, AI is used with energy storage management to help prolong the life of their tech and save wear and tear by inadvertently overcharging or undercharging thus keeping resources for storage reliable which translates into cost-effective (Aldieri et al., 2021).

However, several emerging trends allude to an even more intertwined relationship expected through the AI and energy system becoming interconnected further in the future. With the increasing number of devices or "smart things" connected to electronic platforms and power supplies being more locally decentralized, there will be even less room for effectively manage these systems without advanced AI applications (Nakata, 2024). Trendy examples include edge computing where data is processed as close to its source of generation as possible, reducing latency and contributing toward quicker decisions in energy management. AI is also being integrated with blockchain technology to improve security and transparency in energy transactions, especially during peer-to-peer (p2P) models of energy trading. An AI application for grid resilience in real time, other exciting advancements are predictive algorithms that could anticipate when natural disaster and cyber threats will most likely cause the grid to go down so interventions can be made early enough before a physical balance (Singh, 2022) .

This is where artificial intelligence becomes more important in present-day energy systems helping to better inform and support a shift towards cleaner, intelligent and resiliently enabled energies. AI optimizes renewable energy, improves management of smart grids and supports the modernization of energy storage not only for more efficient production but to assist in the transition toward worldwide sustainable power generation (Winfield & Dolter, 2024). However, in an era where the worlds energy systems transition towards being less centralised and more responsive to a digital age there is little doubt that moving forwards AI takes its rightful place as lead technology for innovation within this sector.

3. CYBERSECURITY THREAT LANDSCAPE IN AI-DRIVEN ENERGY SYSTEMS

As AI-driven technologies invade energy systems more and more, this restructuring approach offers unprecedented efficiency and opportunities for innovation as well but lays out new risks in terms of cybersecurity. Modern energy infrastructure has gradually become digitized and integrated with other corporate systems, exposing it to a variety of cyberthreats that can disrupt the normal course of activity, steal information or endanger people (Damman et al., 2021). Energy systems that are driven by AI, in particular, can be uniquely susceptible given the heavy use of large-scale networks of variously linked devices and real-time data flows interpretable through machine learning algorithms any member thereof which could potentially manipulated or attacked. It is critical to understand the nature of cybersecurity threats faced by these systems in order to develop robust infrastructures and ensure a continuous, reliable supply of secure energy.

3.1 Cyber Threats in Energy Systems

On the other hand, this is an industry that must deal with threats like malware and ransomware attacks as much more complex ones targeting AI. Foundational attacks think phishing, malware and ransomware have been repurposed to exploit the singular adversities in energy systems. Such threats are usually geared at data theft, operational sabotage or demanding payments in return of reestablishing systems (Pascaris, 2021). For example, a ransomware attack upon an electricity utility could result in extensive blackouts or grid instability that might have significant consequences for the safety and economic stability of public.

While more sophisticated threats specifically target the AI components in these systems. An adversary might pollute private data, such as by injecting false data into the system for reducing the accuracy of machine learning models. This leads to erroneous predictions and operational decisions which in turn causes instability of the energy distribution network which may eventually end up into a blackout or misallocation (Budiman, 2018). Another important security threat comes from Adversarial attacks where attackers try to manipulate the AI models in a way that system make some mis-call or unintended action by giving carefully crafted inputs. These attacks illustrate that the increasing reliance on machine learning-based approaches in modern energy management is exploitable.

3.2 Malware, Ransomware, Phishing Attacks

Malware as a common threat for cybersecurity of AI-powered energy is malware which serves as a malicious software that can harm any computing device; this could be in the form of ransomware or phishing attacks. In the case of energy infrastructures, interconnectedness is key to understanding how these attacks work: one infected device can transmit its malicious software through the entire network. Computer viruses can interrupt operations, corrupt data, or even damage systems by taking control of them. Such prominent piece of malware is the "Industroyer" that specifically goes after industrial control systems (ICS) and can lead to major energy distribution disruptions (Reedman et al., 2021).

Energy providers may also be more likely to fork over a ransom faster so that services can resume, which is why attacks like this one have become increasingly commonplace in our critical infrastructure space. A ransomware attack hijacks system data with cryptographic public-key encryption and demands a payment in exchange for releasing the decryption key, effectively kidnapping your computer (Xie et al., 2023). This can effectively paralyze systems, and in the case of power lines sever electricity connections to entire regions or cause large-scale costs. The devastating aftermath of such incidents was aptly laid bare in the U.S. when fuel supplies across the East Coast were temporarily shut down earlier this year as a result of The 2021 Colonial Pipeline ransomware attack, demonstrating that cyber attacks on energy infrastructure are something which have country-spanning ramifications.

Phishing attempts are one of the loading phases before a more critical vector. Attackers obtain access to mission-critical systems either by duping employees or administrators into type their login credentials, or installing malware. From there, they may be able to leverage sophisticated attacks that allow them access further into the network and potentially threaten AI models or other key parts of the energy system. The phishing attacks illustrate the need for proper cybersecurity training and policies to ensure that human error is not a vulnerability (Vinichenko, 2018).

3.3 Challenges Exclusive to AI Algorithms

Their energy systems will face distinct and highly targeted risks, designed to undermine the AI algorithms that control flows of energy. The key dangers in AI systems is that they might succumb to data poisoning, where attackers manipulate training data used to build a machine learning model. Poisoned data could result in energy systems making incorrect predictions, taking on operations that are not optimal or even leading to system instability (Arbabzadeh et al., 2019). Attackers, for example, could introduce false weather data into a model of the renewable energy forecast that an outdated version and lead to overproduction or default in the grid operator would mean inefficient power distribution and maybe even bottlenecks.

Specifically, adversarial attacks can be a threat to AI algorithms. Such attacks consist in the addition of slightly modified inputs aimed at getting a machine learning model to take wrong decisions (Ilyas et al., 2022; Garrido et al., 2020). This can be as impactful pd to misclassify normal fluctuations in grid inferiority for, say an adversarial aid on a gride control model could effectively trigger superlative response or worse neighboring the Artificial Intelligence models from categorizing generally HYDRA and even result hindber during your pdd line shutdowns. This is achieved by attacking the vulnerabilities in machine learning models which are quite sensitive to relatively small changes in input data. The growth of AI algorithms taking a full role in energy systems makes adversarial attacks one of the most important threats to the reliability and security on these infrastructures.

3.4 Weaklings: AI-Enabled Infrastructure

AI-integrated energy systems are vulnerable due to dependance on data, inter-connected devices and real-time communication channels. Those infrastructures which have numerous interconnected IoT devices, sensors and controllers with different points of contact that are constantly in communication leave cyber attackers many entrance paths (Steen & Weaver, 2017). Most of these devices are not built with strong security in mind, making them susceptible to getting hacked. If an attacker breaches any device, they could potentially navigate through the network and past sensitive files to other elements.

The AI systems that control power can be decentralized in a way; this decentralization undermines some traditional vulnerabilities and actually creates new ones. Decentralized systems take care of renewable sources, which mean that control is still in many places and these results quite difficult to monitor or secure every single piece of the network. AI systems must be regularly updated and maintained to achieve optimal performance which needs periodic recalls that if not secured properly can be capitalized by attackers (Hoffmann et al., 2020; Manoharan et al., 2021). This could be as simple as attackers injecting malicious code in the middle of an update undermining a common neural network model and plunging it into chaos that even humans would find nearly impossible to fix.

The threat landscape in AI-powered energy, from a cybersecurity perspective is both intricate and fast-changing including old-school malware and phishing attacks to sophisticated assaults against the very heart of these systems AI algorithms. These continuous amounts of AI applications represent extracting pressures from numerous real-world systems because being that data-driven and interconnected makes them particularly exposed to cyber threats, each new application provides more ways into the core (Ghai et al., 2022). The security of AI-driven energy infrastructures will include all the elements in classic cybersecurity but also require more specialized methods to securely package a machine learning model and protect it from adversarial subversion. A proactive security perspective will be increasingly necessary to maximize the potential for delivering a more efficient, resilient and sustainable energy future by reducing any risks from this rapidly growing field (Pimitha et al., 2023).

4. IMPACT OF CYBER SECURITY THREATS ON ENERGY SYSTEM RELIABILITY AND SAFETY

Distributed systems for energy: Cybersecurity and reliance risks across these decarbonizing, decentralizating power networks pose high impactiness threats. As energy infrastructures become more digital and AI solutions are increasingly used, so they grow to be similarly hollow as cyber targets with freedom effects. Attacks of energy systems via cyber can result in large-scale power blackout together with equipment damage, data loss and even a threat to public safety / national security (Singh & Kaunert, 2024).

Cyber-enabled attacks on energy systems can take many forms, all of which have the effect of destabilizing national grids and rendering essential services inoperable. For example, ransomware can take down energy management systems preventing the access to essential data and impeding monitoring and control of energy flows in real-time (Singh & Kaunert, 2024). When this happens, entire regions lose their power as a grid collapses and its failure cascades through interconnected systems causing operational paralysis. A well-planned assault could lead to a major electrical outage potentially lasting for days at a time, particularly if the scam succeeds in incapacitating or destroying vital system components. Beyond simply shutting down power supplies, cyber attacks can bring about physical destruction of devices. For example, an ICS malware may focus on industrial control systems and bypass utility business networks to modify in-operation commands that lead turbines, transformers or grid components to operate above their safe limits. Expense Equipment Failure Long Downtime Costly Repairs The attacks also show how cyber threats can go beyond digital assets, impacting the very physical infrastructure that powers energy generation and distribution (Singh & Kaunert, 2024).

Energy infrastructure is central to the operation of contemporary society, and as such a disturbance has grave implications for national security and public safety. Power grids are interlinked with different domains such as healthcare, transportation and financial systems which can make them a potential target for adversary targeting to destabilize the country. On the power grid, a cyber attack could cause serious damage to emergency services and medical facilities, while also rendering transportation systems inert adding further challenge at setting up recovery plans. A successful power attack that affected a major city or region could be calamitous on the public safety front, with potential impacts in everything from water systems to cell service (Singh & Kaunert, 2024).

Cyber attacks on energy infrastructure are deeply concerning threats to national security. Such strategies could target a nation's power grids or energy plants as part of efforts to reduce broader resilience. And if a cybersecurity attack is successful in creating runaway cascading failures (causing other systems to fail due to high energy production demands that can't be supplied without those failing dependent systems), then you have collapsed whole segments of ELINT or needs for stealth aircraft like AWACS when military forces are declared not ready. The economic impact of cybersecurity breaches in the energy sector is enormous, not only when considering direct recovery costs but also on a macroeconomic level (Singh et al., 2024). A cyberattack on an energy system can mean massive financial loss down to operational interruption, equipment repairs and reputation damage. A business could face downtime of production due to power outages or damaged equipment, loss in income and disruptions throughout the supply chain or energy providers may have sudden high costs recuperating operations. On the other side, industries and households may face some financial losses due to energy shortages at times or increase in cost for grid failure.

Ransomware demands can be expensive, and energy companies might see the economic value in paying ransoms to reclaim control of their systems with minimal downtime containing potential fallout or limiting damages so they may return back on line. Together with repair, investigation and the introduction of new measures for preventing such attacks can present a major financial problem. Furthermore, more widely cyber attacks can erode investor confidence in the energy company field and drive up insurance costs which de-stabilize this part of our sector and raise prices for electricity consumers. In addition, should there be a breach of data that leads to the disclosure of consumer information or damages affecting public interests; energy providers are subject even greater legal exposure (Singh et al., 2024) (Singh & Kaunert, 2024). When a breach occurs, regulatory bodies tend to levy fines or sanctions on those companies who were found lacking in cybersecurity standards. Some long-term impacts might also include more investment in cybersecurity efforts, which would be non-productive costs adding to operational expenses and perhaps limiting resources on the innovation side of energy (Singh et al., 2024). The cybersecurity risks emanating from AI-enabled energy systems are extensive and vary across operational reliability, national security, public safety, and economic stability (Singh & Kaunert, 2024). Given the complexity of modern, interconnected energy systems they are looking for new ways to exploit increasingly sophisticated cyber threats adding a sense of urgency to develop comprehensive cybersecurity strategies. Securing our energy infrastructure is critical to public safety and long-term economic health in the digital age, interconnected world.

5. CONCLUSION AND FUTURE SCOPE

Cybersecurity difficulties in AI integrated to energy systems emphasize the clear necessity of protective measures as these technologies mature into platforms supporting crucial infrastructures. While AI can optimize energy management and enable smart grid operations, it also presents increased vulnerabilities to the secure operation of our energy systems by new cyber threats which could threaten reliability, safety or public trust. To overcome these challenges, ethical and legal frameworks are necessary to maintain the security of AI applications in energy supported by transparency and accountability features.

The future of AI-based energy systems will require far more secure IoT devices and in the end, depends on a significant part whether security technologies such as for instance detect threats because they arise using AI analytics. This means the implementation of comprehensive regulation frameworks that incorporate not only cybersecurity but also ethical standards. Solution: Policies that can adapt to future changes equally benefit enterprises and likely require a collaboration between governments and industry as well cyber-security experts. General, more research is required to build secure AI models that can withstand a myraid of attacks and are developed for the purpose of safeguard privacy as well integrity of data. While AI-driven energy systems expand, efforts of cybersecurity and law/institution are imperative for the long-term establishment of a secure, resilient, sustainable future in terms of available sources such as gas or renewable.

REFERENCES

Al-tabatabaie, K. F., Hossain, M. B., Islam, M. K., Awual, M. R., TowfiqulIslam, A. R. M., Hossain, M. A., Esraz-Ul-Zannat, M., & Islam, A. (2022). Taking strides towards decarbonization: The viewpoint of Bangladesh. *Energy Strategy Reviews*, *44*, 100948. DOI: 10.1016/j.esr.2022.100948

Aldieri, L., Grafström, J., & Vinci, C. P. (2021). The effect of Marshallian and Jacobian knowledge spillovers on jobs in the solar, wind and energy efficiency sector. *Energies*, *14*(14), 4269. DOI: 10.3390/en14144269

Arbabzadeh, M., Gençer, E., Morris, J. F., Paltsev, S., & Armstrong, R. C. (2019). *Plausible Energy Futures: A Framework for Evaluating Options*. Impacts, and National Energy Choices.

Basheer, M., Nechifor, V., Calzadilla, A., Ringler, C., Hulme, D., & Harou, J. J. (2022). Balancing national economic policy outcomes for sustainable development. *Nature Communications*, *13*(1), 5041. DOI: 10.1038/s41467-022-32415-9 PMID: 36028486

Budiman, I. (2018). Enabling community participation for social innovation in the energy sector. *Indonesian Journal of Energy*, *1*(2), 21–31. DOI: 10.33116/ije.v1i2.23

Bujang, A. S., Bern, C. J., & Brumm, T. J. (2016). Summary of energy demand and renewable energy policies in Malaysia. *Renewable & Sustainable Energy Reviews*, *53*, 1459–1467. DOI: 10.1016/j.rser.2015.09.047

Cohen, S. M., Iyer, G. C., Brown, M., Macknick, J., Wise, M., Binsted, M., & Hejazi, M. (2021). How structural differences influence cross-model consistency: An electric sector case study. *Renewable & Sustainable Energy Reviews*, *144*, 111009. DOI: 10.1016/j.rser.2021.111009

Damman, S., Sandberg, E., Rosenberg, E., Pisciella, P., & Graabak, I. (2021). A hybrid perspective on energy transition pathways: Is hydrogen the key for Norway? *Energy Research & Social Science*, *78*, 102116. DOI: 10.1016/j.erss.2021.102116

Danish, M. S. S., & Senjyu, T. (2023). AI-Enabled Energy Policy for a Sustainable Future. *Sustainability (Basel)*, *15*(9), 7643. DOI: 10.3390/su15097643

Garrido, S., Sequeira, T., & Santos, M. (2020). Renewable energy and sustainability from the supply side: A critical review and analysis. *Applied Sciences (Basel, Switzerland)*, *10*(17), 5755. DOI: 10.3390/app10175755

Ghai, D., Tripathi, S. L., Saxena, S., Chanda, M., & Alazab, M. (Eds.). (2022). *Machine Learning Algorithms for Signal and Image Processing*. John Wiley & Sons. DOI: 10.1002/9781119861850

Hoffmann, M. W., Wildermuth, S., Gitzel, R., Boyaci, A., Gebhardt, J., Kaul, H., & Tornede, T. (2020). Integration of novel sensors and machine learning for predictive maintenance in medium voltage switchgear to enable the energy and mobility revolutions. *Sensors (Basel)*, *20*(7), 2099. DOI: 10.3390/s20072099 PMID: 32276442

Ilyas, A., Parkinson, S., Vinca, A., Byers, E., Manzoor, T., Riahi, K., & Muhammad, A. (2022). Balancing smart irrigation and hydropower investments for sustainable water conservation in the Indus basin. *Environmental Science & Policy*, *135*, 147–161. DOI: 10.1016/j.envsci.2022.04.012

Lawrenz, L., Xiong, B., Lorenz, L., Krumm, A., Hosenfeld, H., Burandt, T., & Von Hirschhausen, C. (2018). Exploring energy pathways for the low-carbon transformation in India—A model-based analysis. *Energies*, *11*(11), 3001. DOI: 10.3390/en11113001

Manoharan, K. G., Nehru, J. A., & Balasubramanian, S. (2021). *Artificial Intelligence and IoT*. Springer. DOI: 10.1007/978-981-33-6400-4

Nakata, T. (2004). Energy-economic models and the environment. *Progress in Energy and Combustion Science*, *30*(4), 417–475. DOI: 10.1016/j.pecs.2004.03.001

Nastasi, B., Markovska, N., Puksec, T., Duić, N., & Foley, A. (2022). Renewable and sustainable energy challenges to face for the achievement of Sustainable Development Goals. *Renewable & Sustainable Energy Reviews*, *157*, 112071. DOI: 10.1016/j.rser.2022.112071

Pascaris, A. S. (2021). Examining existing policy to inform a comprehensive legal framework for agrivoltaics in the US. *Energy Policy*, *159*, 112620. DOI: 10.1016/j.enpol.2021.112620

Pimitha, K., Anbarasi, S., & Balasubramanian, T. (2023). Machine Learning Techniques for Renewable Energy Resources. In *IoT, Machine Learning and Blockchain Technologies for Renewable Energy and Modern Hybrid Power Systems* (pp. 121-148). River Publishers.

Reedman, L., Chew, M. S., Gordon, J., Sue, W., Brinsmead, T., Hayward, J., & Havas, L. (2021). Multisector energy modelling.

Salleh, S. F., Mohd Roslan, M. E., Abd Rahman, A., Shamsuddin, A. H., Tuan Abdullah, T. A. R., & Sovacool, B. K. (2020). Transitioning to a sustainable development framework for bioenergy in Malaysia: Policy suggestions to catalyse the utilisation of palm oil mill residues. *Energy, Sustainability and Society*, *10*(1), 1–20. DOI: 10.1186/s13705-020-00269-y

Singh, B. (2022). Relevance of Agriculture-Nutrition Linkage for Human Healthcare: A Conceptual Legal Framework of Implication and Pathways. *Justice and Law Bulletin*, *1*(1), 44–49.

Singh, B. (2023). Federated Learning for Envision Future Trajectory Smart Transport System for Climate Preservation and Smart Green Planet: Insights into Global Governance and SDG-9 (Industry, Innovation and Infrastructure). *National Journal of Environmental Law*, *6*(2), 6–17.

Singh, B. (2024). Lensing Legal Dynamics for Examining Responsibility and Deliberation of Generative AI-Tethered Technological Privacy Concerns: Infringements and Use of Personal Data by Nefarious Actors. In Ara, A., & Ara, A. (Eds.), *Exploring the Ethical Implications of Generative AI* (pp. 146–167). IGI Global., DOI: 10.4018/979-8-3693-1565-1.ch009

Singh, B., Dutta, P. K., & Kaunert, C. (2024). Replenish Artificial Intelligence in Renewable Energy for Sustainable Development: Lensing SDG 7 Affordable and Clean Energy and SDG 13 Climate Actions With Legal-Financial Advisory. In Derbali, A. (Ed.), *Social and Ethical Implications of AI in Finance for Sustainability* (pp. 198–227). IGI Global., DOI: 10.4018/979-8-3693-2881-1.ch009

Singh, B., Jain, V., Kaunert, C., Dutta, P. K., & Singh, G. (2024). Privacy Matters: Espousing Blockchain and Artificial Intelligence (AI) for Consumer Data Protection on E-Commerce Platforms in Ethical Marketing. In Saluja, S., Nayyar, V., Rojhe, K., & Sharma, S. (Eds.), *Ethical Marketing Through Data Governance Standards and Effective Technology* (pp. 167–184). IGI Global., DOI: 10.4018/979-8-3693-2215-4.ch015

Singh, B., & Kaunert, C. (2024). Computational Thinking for Innovative Solutions and Problem-Solving Techniques: Transforming Conventional Education to Futuristic Interdisciplinary Higher Education. In Fonkam, M., & Vajjhala, N. (Eds.), *Revolutionizing Curricula Through Computational Thinking, Logic, and Problem Solving* (pp. 60–82). IGI Global., DOI: 10.4018/979-8-3693-1974-1.ch004

Singh, B., & Kaunert, C. (2024). Wind and Solar Energy as Renewable Energy for Sustainable Global Future: Projecting Future Multi-Sector Sustainable Policies and Innovation. In Ara, A., & Thakore, R. (Eds.), *Promoting Multi-Sector Sustainability With Policy and Innovation* (pp. 210–245). IGI Global., DOI: 10.4018/979-8-3693-2113-3.ch009

Singh, B., & Kaunert, C. (2024). Aroma of Highly Smart Internet of Medical Things (IoMT) and Lightweight EdgeTrust Expansion Medical Care Facilities for Electronic Healthcare Systems: Fortified-Chain Architecture for Remote Patient Monitoring and Privacy Protection Beyond Imagination. In Hassan, A., Bhattacharya, P., Tikadar, S., Dutta, P., & Sagayam, M. (Eds.), *Lightweight Digital Trust Architectures in the Internet of Medical Things (IoMT)* (pp. 196–212). IGI Global., DOI: 10.4018/979-8-3693-2109-6.ch011

Singh, B., & Kaunert, C. (2024). Augmented Reality and Virtual Reality Modules for Mindfulness: Boosting Emotional Intelligence and Mental Wellness. In Hiran, K., Doshi, R., & Patel, M. (Eds.), *Applications of Virtual and Augmented Reality for Health and Wellbeing* (pp. 111–128). IGI Global., DOI: 10.4018/979-8-3693-1123-3.ch007

Steen, M., & Weaver, T. (2017). Incumbents' diversification and cross-sectorial energy industry dynamics. *Research Policy*, *46*(6), 1071–1086. DOI: 10.1016/j.respol.2017.04.001

Vinichenko, V. (2018). Mechanisms of energy transitions: National cases and the worldwide uptake of wind and solar power. *PhD diss., Central European University.*

Westman, L. K., Broto, V. C., & Huang, P. (2019). Revisiting multi-level governance theory: Politics and innovation in the urban climate transition in Rizhao, China. *Political Geography*, *70*, 14–23. DOI: 10.1016/j.polgeo.2019.01.002

White, D., Winchester, N., Atkins, M., Ballingall, J., Coates, S., de Miguel Mercader, F., & Suomalainen, K. (2018). Energy-and multi-sector modelling of climate change mitigation in New Zealand: current practice and future needs.

Winfield, M., & Dolter, B. (2014). Energy, economic and environmental discourses and their policy impact: The case of Ontario s Green Energy and Green Economy Act. *Energy Policy*, *68*, 423–435. DOI: 10.1016/j.enpol.2014.01.039

Xie, Y., Wu, X., Hou, Z., Li, Z., Luo, J., Lüddeke, C. T., & Liao, J. (2023). Gleaning insights from German energy transition and large-scale underground energy storage for China's carbon neutrality. *International Journal of Mining Science and Technology*, *33*(5), 529–553. DOI: 10.1016/j.ijmst.2023.04.001

Yang, D., Liu, D., Huang, A., Lin, J., & Xu, L. (2021). Critical transformation pathways and socio-environmental benefits of energy substitution using a LEAP scenario modeling. *Renewable & Sustainable Energy Reviews*, *135*, 110116. DOI: 10.1016/j.rser.2020.110116

Chapter 9
Digital Twins in Renewable Energy Systems:
Enhancing Efficiency Projecting Sustainability

Vaishali Rahate
Datta Meghe Institute of Management Studies, India

Prashant Balwantrao Thakare
Gondwana University, Gadchiroli, India

Puja S. Agrawal
RamdeoBaba University, Nagpur, India

Saurabh Chandra
https://orcid.org/0000-0003-4172-9968
Bennett University, Greater Noida, India

Maajid Mohi Ud Din Malik
https://orcid.org/0000-0003-1743-1520
Dr. D.Y. Patil Vidyapeeth University, India

Bhupinder Singh
https://orcid.org/0009-0006-4779-2553
Sharda University, India

ABSTRACT

The power of digital twins, virtual replicas of physical assets and systems that are revolutionizing the efficiency and performance is significant for optimization for renewable energy. Agriculture sector uses digital twins in the context of precision agriculture, fishing and reforestation, manufacturing can simulate machinery and robots predict maintenance breakdowns before they happen or calculate needs for supplies. Bringing new perspectives on cutting-edge technologies in the area of renewables it simulates wind turbine. With a real-time digital twin in place, energy operators can observe how the system behaves across diverse operating conditions, detect inefficiencies and work on data to make judicious decisions toward enhancing performance. In case of solar power systems, they monitor the performance efficiency of solar panels and predict probable failures which may result into loss from energy production.

1. INTRODUCTION

Digital twins particularly in wind energy can predict failures or simulate the operation of turbines to adjust blade angles or rotation speeds for higher efficiency and greater output. The worldwide shift to renewable energy systems now underlies the battle against climate change and support for environmen-

DOI: 10.4018/979-8-3373-0045-0.ch009

tal sustainability. To address this demand, new technologies are increasingly being deployed in energy systems to fully realise their potential and maximise its performance, reliability, sustainability and efficiency. The example of this transformative innovation is the Digital Twins (DTs) concept. DTs are a recent but transformative technology that support access to digital twins of physical systems, and have already started playing an integral role in the design, monitoring, and optimization of energy systems. Such digital twins replicate real-world operations, providing insights and predictions that are more detailed than ever to improve efficiency and lower carbon emissions (Das et al., 2024).

The fusion of such cutting-edge computing power, complex data analysis, and sensor technology is best epitomized in the application of Digital Twins in renewable energy systems. With renewable energy sources (solar, wind, hydro, and geothermal) becoming essential to energy grids around the world, it is crucial to ensure that these solutions operate seamlessly and efficiently. Dynamic Twin (DTs) to advance a new paradigm in American mining and coal energy. Through this introduction, we discuss the importance of DTs in renewable energy systems and their use to improve efficiency and meet sustainability targets. The nature of renewable resources is variable, because they depend on natural conditions like sunlight, wind speed or water flow. This variability also brings challenges in energy output forecasting, consistency in energy supply, and maintaining the reliability of the systems. The urgency for optimization is greater than ever as countries around the globe seek to integrate record levels of renewables into their energy portfolios to achieve ambitious carbon reduction goals. However, traditional energy management systems are often unable to provide the necessary real-time information and predictive capabilities to respond to these challenges effectively. DTs have the capability to simulate, analytics, and optimize renewable energy systems like never before.

1.1 The Rise of Digital Twins

Digital Twins are the virtual representation of physical assets / systems / processes based on data from sensors, real-time analytics. DTs were originally developed for use in many different industries including aerospace and manufacturing but have been applied to support healthcare, urban design, and recently renewable energy. By constantly reflecting the activity and status of their physical systems, DTs offer a global view of how their systems work, allowing operators to predict and address what can happen ahead of time. In renewable energy, for example, DTs serve as replicas of infrastructure including wind turbines, solar panels and hydroelectric systems. Not only does it allow monitoring of individuals in all components, but it also enables a complete view of whole energy ecosystems in one automated picture. For example, through an extensive analysis of the real-time and historical data, they help to make accurate decisions that can maximize energy production, reduce downtime, and promote sustainable development.

1.2 Digitally Twin for Greater Efficiency

The Efficiency is a critically important figure for renewable energy systems because small inefficiencies can result in much higher energy losses and operating costs. It plays to the efficiencies as DTs facilitate predictive maintenance, resulting in reduced unplanned downtime, and prolonging the life of assets. In wind energy systems, for example, DTs can examine turbine health continuously in real time, detecting subtle variations that could suggest wear and tear. This enables timely actions to be taken, which helps in avoiding costly repairs and minimizing interruptions in energy supply. By simulating multiple operational scenarios, DTs enable the optimization of energy generation and distribution. On the other hand, for solar energy systems, DTs can forecast how the weather (numerical) trends impact energy output and how to optimize the system output and efficiency. Operators can ensure that renewable energy systems function optimally under varying conditions by using these insights.

Renewable energy initiatives have sustainability at the core, and DTs act as driving force in them. By optimizing these interlinked energy systems, DTs can reduce wastage and improve resource utilization to help reduce the overall carbon footprint. Additionally, as this technology can simulate environmental impacts too, stakeholders can evaluate and mitigate ecological risks inherent in renewable energy projects. And while encouraging energy-efficient buildings is great for sorting energy demand, DTs further enhance sustainability by facilitating the embedding of solar-energy systems into smart grids. DTs are essential to ensure that renewable energy sources can be utilized when needed, minimizing fossil fuel dependency and thus helping global goals related sustainability through real time data and predictive analytics (Ukoba et al., 2024).

1.3 Importance in an Evolving Energy Landscape

Digital Twins in the renewable energy systems are a huge stride in the changing dynamics of energy around the world. Their potential to improve both efficiency and sustainability has been linked to SDG 7 (Affordable and Clean Energy) and SDG 13 (Climate Action). DTs can provide a path forward for easing the tension between energy demand and climate change as the world struggles to meet growing energy needs while battling global climate change a deepening challenge. By the convergence of DT with emerging technologies like Artificial Intelligence (AI) and the Internet of Things (IoT) further enhances their efficiency and effectiveness. DTs are powered by AI, which can process complex datasets, identifying patterns and insights that lead to innovation, and connectivity through the Internet of Things (IoT) provides continuous data flow between physical systems and their virtual twins. Such a combination is transformational, enabling the next-generation energy ecosystem that is not only efficient but also resilient and adaptive. Digital Twins set to revolutionize the future of renewable energy systems. Reducing cost and enhancing efficiency while improving sustainability they answer to some of the most pertinent issues in the energy management value chain and environment preservation. The advancements in this field hold great promise for enhancing and creating new possibilities in employing DTs in renewable energy systems (Agostinelli, 2024).

2. UNDERSTANDING DIGITAL TWINS: COMPONENTS OF DIGITAL TWINS AND EVOLUTION AND APPLICATION ACROSS INDUSTRIES

Digital Twins (DTs) are a revolutionary technological approach that connects physical systems with their avatars. DTs create advances the interaction, analysis, and optimization of physical system states and processes in real-time in a dynamic way. Originally conceived at NASA in the 1960s to simulate and debug spacecraft systems "on the flight" during missions, DTs have become a sophisticated tool for predictive analysis, performance optimization and system integration. DTs are intended to be virtual replicas of physical objects, processes, or systems, implementing advanced data analytics, sensor technology, and computational models. This kind of versatility has been a catalyst for their adoption across a wide range of industries, from manufacturing and healthcare to transportation and renewable energy.

The following section discusses the key features of DTs along with how they have evolved over time and their potential application in diverse sectors.

2.1 Components of Digital Twins

A Digital Twin is made up of a number of interconnected parts, [each of which] is critical to its operation. Quick links: However, these fundamentals are essential to derive true digital twins of physical systems that have actionability. The asset is the physical entity that is being represented by the DT. It can be a single machine or component, for example, a wind turbine blade, or a complex system, such as an entire solar farm. A physical system that produces real-world data which can be used to develop and maintain the DT. The physical system is interfaced with the digital domain using sensors and IoT (Internet of Things) devices to collect real-time data. These devices include sensors that can measure temperature, pressure, vibration and energy output, and are responsible for generating raw data input to the DT. For accurate and reliable data transmission, high-quality sensors and seamless IoT connectivity is mandatory (Tariq, 2024).

After collection phase, the data comes to storage and integration to the DT platform. To efficiently manage this volume of data, cloud-based solutions and edge computing are used. Data is then organized and structured into a full digital representation. The simulation tools are the foundation of the DT that creates a virtual replica of the physical system. The technology employs advanced algorithms and physics-based modeling techniques to simulate real-world conditions. These simulations are continuously updated with real-time data providing predictive and prescriptive analytics. AI and ML enable DTs with more analytical capabilities by detecting patterns, predicting future instances, and improving tasks. These technologies allow the DT to learn from past data, adapt to new situations, and bring actionable recommendations to its real-world counterpart.

The Visualization tools provide the user interface to interact with the DT. These interfaces let stakeholders observe system performance, detect unusual occurrences, and test different scenarios. 3D modeling, augmented reality (AR) and virtual reality (VR) technologies are frequently used to engage users. It hinges on the bidirectional interaction between the physical system and its digital twin. Feedback loops allow any changes or optimizations suggested by the DT to be fed back directly into the physical system in real-time. Digital Twins has come a long way since the initial concepts were formulated, driven by new technologies and the need for increased efficiency and productivity (Raghav et al., 2024).

The origins of DTs go back to NASA's Apollo missions in the 1960s. Engineers created virtual models of spacecraft to check and resolve issues with systems during missions. While still quite rudimentary compared to today's DTs, this approach would serve as the basis for simulation and analysis of real-time systems. The concept of "Digital Twin" was first introduced by Dr. Michael Grieves in 2002 in a presentation at the University of Michigan. In the context of production lifecycle management (PLM), the concepts got applied using Digital Twins (DTs) to simulate and optimize the design and manufacturing processes for each product.

2.2 Amalgamation with IoT and Big Data

During the 2010s, the powers of DTs exploded with the rise of IoT and Big Data. DTs became more dynamic and accurate by pulling in and processing large amounts of real-time datasets. This evolution allowed their use in sectors such as health care/medicine, urban planning, and renewable energy. The synergy of AI and ML with DTs has proven to be a groundbreaking development, as this convergence enables improved predictions and optimizations. AI-fueled DTs bring unparalleled analysis capabilities, detecting anomalies and generating future scenarios based on complex datasets. In turn, this evolution has expanded the range of possible applications for data technology, from predictive maintenance in manufacturing to personalized medicine in health care. These days, DTs are continuing to advance by integrating edge computing, portfolios of portfolio management, and quantum computing. Then, a new wave started to build that would incorporate many across multiple domains into a more complete digital ecosystem or virtual enterprise to establish efficiencies and touch points across those sectors.

Because of the flexible nature of DTs, they have been employed in a variety of different sectors, all taking advantage of its unique capabilities and solving their unique challenges and opportunities. In manufacturing, digital twins are applied in product design, production optimization, and quality control. DTs lead to enhanced productivity by discovering inefficiencies in the manufacturing process, preventing waste by simulating these processes. For example, in Healthcare, the DT are radically changing how patients are cared for, with tailor-made medicine, procedures, and predictive diagnostics—these systems provide models of the physiology of individual patients that are connected to the available care for specific conditions. These virtual models of organs or even an entire patient allow doctors to simulate treatment and predict outcomes ultimately improving the quality of care.

Digital twins within transportation is an emerging application of IT where DTs have emerged as a new form of urban planning, simulating cities and their infrastructure. These are used to optimize traffic flows, energy distribution, waste management, enabling smart and sustainable cities. Within the transportation sector, DTs are used for vehicle design, fleet management, and supply chain optimization. As an illustration, DTs of autonomous vehicles allow manufacturers to iterate and validate designs in simulation before deployment. Using DTs in renewable energy, for instance, helps to ensure performance optimization of wind farms, solar panels, and hydroelectric plants. DTs empower predictive maintenance, improve energy output and lower operational costs by observing system behavior in real-time (Maksimovic et al., 2024).

Digital Twins are a game-changer in the way we manage, monitor, and optimize physical systems. These components—sensors, IoT devices, AI-driven analytics, etc. combine to simulate and visualize real-world elements in real time." The journey of DTs, from their early days in NASA's Apollo missions to their broad applications across industries now, highlights their versatility and capability. Originating in aerospace, DTs are applied to aircraft design, maintenance, and mission planning. They allow engineers to recreate flight conditions, troubleshoot possible failures, and improve safety and performance. Manufacturing Industry: In manufacturing, DTs are used to optimize production processes, minimize downtime, and improve product quality. DTs drive virtual stores, enabling retailers to test layouts and marketing strategies before they hit the implementation phase. From manufacturing to healthcare to renewable energy, DTs are driving efficiency, innovation and sustainability. As technology progresses further, the scope and use of dynamic twins are set to broaden, revolutionizing the way we engage with, and optimize, intricate systems. The potential of DTs within the renewable energy industry, specifically, suggests a tomorrow where energy landscapes are both optimized and environmentally sound, charting a course for a more sustainable and resilient future.

3. ROLE OF DIGITAL TWINS IN RENEWABLE ENERGY SYSTEMS

Due to these characteristics, Digital Twins (DTs) have become a key element of renewable energy systems, providing novel advancements in performance improvement, efficiency enhancement, and sustainability. DTs offer a powerful framework for the real-time monitoring, preventive maintenance, and optimizing of renewable energy systems as the growing demand for clean energy cannot be addressed by traditional methods. It covers a range of renewables: wind, solar photovoltaic (PV), hydro and geothermal systems. In this respect, this section highlights the role of DTs in renewable energy and discusses how it works with the wind energy system, its application in solar PV systems, and how DTs can be used for optimization in hydro and geothermal energy systems (Kaushik et al., 2024).

The operation of wind energy systems is intrinsically complex, with dynamic interactions between turbine, the environment, and grid loads. The use of DTs in wind energy systems has revolutionized the way these different aspects are governed. The sensors installed on the physical assets collect the data that enables real-time monitoring of wind turbines through DTs. These sensors monitor factors including blade oscillations, rotational speeds and weather. The digital twin uses this data to detect anomalies and predict failures. DTs enable predictive maintenance, thereby minimizing downtime and prolonging turbine life. DTs enable operators to replicate different scenarios and improve turbines' performance. For instance, by changing the angles of blades according to variations in wind speed, one can optimize energy output. In addition, based on historical data approaches, DTs can detect patterns and suggest modifications to increase efficiency.

During the planning stage, DTs help modelling and designing optimal layouts for wind farms by simulating wind flows and interaction between turbines. This minimizes energy losses induced through wake influences and assures turbines are sited at the highest wind potential places. DTs also allow a better integration of wind power into a power grid by predicting how much energy will be generated according to the demand patterns. This is to ensure grid stability and reduce curtailment losses. DTs are fundamental in optimizing the performance of wind systems leading to improved efficiency in power production and increased sustainability of wind energy as a renewable energy source.

Solar photovoltaic (PV) systems are becoming one of the most fundamental methods of renewable generation, and the use of DTs in this subfield is bringing major breakthroughs. DTs offer in-depth analysis of individual solar panel operational efficacy through data from embedded sensors. Parameters like temperature, irradiance, and electrical output are monitored in real-time, allowing for identification of inefficiencies or malfunctions. By modeling and simulating panel behavior across different conditions, DTs can catch potential faults including hotspots or shading problems before they become serious issues. This proactive nature reduces maintenance costs and maximizes energy output (Ohueri et al., 2024).

DTs allow for energy yield optimization in real time. Specifically, they can suggest ideal panel tilt angles depending on location and weather. The DT can also apply advanced algorithms that optimize the functioning of inverters and trackers, optimizing system performance further. Energy storage solutions are often used with solar PV systems to match supply with demand. Artificial Intelligence (AI) methods such as, Decision Trees (DT) predict energy production and learn how to recommend charge-discharge strategies, both of which are vital for optimizing the interaction between PV panels and storage systems (Lal et al., 2024). In the design stage, DTs replicate the outputs of proposed solar installations under different environmental and operational conditions. This makes the system scalable & economical in terms of energy production goals. In the context of solar PV systems, the use of DTs highlights the capacity to optimize solar energy's effectiveness and dependability, officially consolidating its place within the renewable resource pool.

In hydropower systems, Digital twins are used to monitor and optimise the performance of turbines, generators, and dams. The DT also receives ingested sensors data on water flow rates, pressure, and turbine speeds that it uses in an ML algorithm for understanding inefficiencies. And there are also dam operations: DTs can simulate the operations of dams to promote optimized water release schedules, to balance energy generation with environmental and irrigation requirements. DTs allow for the ability to track wear and tear on turbines and other machinery so that repairs can be conducted before the equipment fails, reducing expensive downtime and repair costs. Geothermal energy systems work by tapping into underground reservoirs and extracting heat. A DT offers a complete set of tools to monitor and optimize these operations. Methodology Hydro and geothermal systems are frequently paired with other renewable energy sources to improve grid stability. DTs facilitate this integration by mimicking interactions among diverse energy sources and maximizing their joint output. Thus, the use of DTs positions hydro and geothermal power plants not only to operate in a more efficient manner but also means they are able to continue being part of the ecosystem and suppliers of a sustainable energy mix in the future (Poddar, 2024). Digital Twins are a game changer for renewable energy systems, enabling new levels of monitoring, optimization, and sustainability in energy production. DTs are utilized in wind energy systems to improve turbine performance, reduce maintenance, and increase integration with the grid. In solar PV systems, they increase energy yield and improve storage integration. In the case of hydro and geothermal energy, DTs help with efficient operations and sustainable resource management. This transition offers a solution by enhancing the efficiency and reliability of energy delivery, addressing global energy demands, and paving the way towards a greener and more sustainable future through the integration of DTs.

4. ENHANCING GREEN ENERGY EFFICIENCY THROUGH DIGITAL TWINS

With the global energy transition to renewables creating pressure to operate greener energy systems at maximum efficiency for sustainability targets; Digital Twins (DTs) have the capability to transform and reduce the resources needed for optimizing real-time operations with autonomous monitoring and predictive analytics for renewable energy sources, including wind, solar, hydro, and geothermal energy. DTs enable operators to visualize, analyze, and optimize energy generation and distribution processes by creating virtual replicas of physical systems. This optimises performance and minimises wastage for sustainable resource conservation (Arora et al., 2024).

DTs provide fundamental benefit of real time knowledge on the operational performance of energy systems. The digital twins (DTs) additionally continuously monitor a variety of parameters (temperature, pressure, voltage, energy output, etc.) for physical assets based on data from sensors (Bahrin et al. 2016). This allows operators to identify inefficiencies and anomalies in real time, stopping small issues from developing into major problems. In DTs, for example, they monitor the vibrations of blades in a wind energy systems, as well as the rotations of wind turbines, ensuring that wind turbines are working at their peak performance. For example, in solar photovoltaic (PV) systems, DTs are maintaining records of irradiance and panel temperatures to optimize energy generation based on changing real-world weather conditions. Additionally, this type of monitoring boosts safety by uncovering and addressing hazards in renewable energy buildings, like overheating or integrity issues.

DTs play a pivotal role in improving green energy efficiency through predictive maintenance. When traditional maintenance schedules are followed, either there is unnecessary downtime, or a stoppage is missed for a potential failure. On the other hand, DTs leverage historical data as well as real-time monitoring of components, alongside machine learning algorithms, to predict when a component is going to fail. The approach enables firms to reduce unexpected outages and lower maintenance expenses. In the case of hydropower plants, for instance, DTs process data about turbine wear and reservoir dynamics to schedule maintenance activities only when it is necessary. Similarly, with solar farms, most DTs will flag early indications of panel deterioration or inverter issues so that operators can replace faulty components before the impact reaches energy output. By analyzing data, DTs can optimize processes across the energy value chain to allow renewable energy systems to be operated at maximum efficiency.

Wind Energy: Wind power does the same simulation with the wind so that the angle of the blades can be optimized to maximize the energy capture. They also optimize the arrangement of wind farms to minimize wake effects and maximize energy production.
Solar Energy: Deep Transformers (DTs) study the orientation of solar panels and weather forecasts to suggest changes to panel angles or trackers to maximize the amount of sunlight hitting the panels.
Renewable Energy Storage Integration: Digital Twins are essential for integrating renewable energy with storage systems (e.g., batteries). DTs make recommendations for the optimal charging and discharging patterns of these assets to ensure that excess energy is stored appropriately and can be drawn upon in periods of low generation.

Renewable energy sources are typically intermittent and sporadic. DTs solve these problems by improving grid stability and allowing for better demand forecasting. DTs simulate scenarios to help grid operators balance energy generation based on demand. DTs can also offer recommendations for balancing supply and demand during high production periods, for example, real-time monitoring of the grid temperature for wind and solar energy, predicting possible surplus, and adjusting excess energy flows to storage systems or regions. DTs analyze both historical and real-time data to predict the demand for energy so that renewable energy systems can be ready to meet peak consumption periods. This reduces dependency on fossil fuel-based backup systems and amplifies the environmental benefits associated with the use of renewable energy (Mchirgui e\t al., 2024).

As much energy is lost during generation, transmission and distribution, energy losses are a big challenge for renewable energy systems. DTs are designed to address these losses by recognizing inefficiencies in the energy flow. For instance, in wind farms, DTs are used to detect misalignments in turbine operations that could cause energy losses. For solar photovoltaic (PV) systems, DTs can be used to identify shading issues or panel faults that lead to reduced energy output. DTs enable the renewable energy systems to operate at its optimum level by eliminating these inefficiencies. Not only do DTs improve efficiency, they also enhance sustainability outcomes by minimizing the corresponding carbon footprints of energy systems. This ultimately reduces energy waste, improves resource efficiency, and extends the lifespan of renewable energy infrastructure. Moreover, the DTs enable combining multiple renewable energy sources to form hybrid systems that can maximize total energy output and ensure stability of the overall energy system. As an example, a hybrid system comprising wind, solar, and hydropower together can use the DTs to coordinate how much energy should be generated based on weather patterns, water availability, and grid demands. This comprehensive perspective ensures harmonious complementarity and optimization of renewable energy resources for improving their collective efficiency.

Digital Twins improving green energy efficiency is a breakthrough for renewable energy systems. As critical challenges like variability, inefficiencies and energy losses are tackled, DTs facilitate real-time monitoring, predictive maintenance and system optimization. This transformation capability is further pronounced by their ability to mix renewable energy sources with storage systems and stabilize the grid. As the world shifts to a sustainable energy future, DTs will become an even more essential component of renewable energy systems to meet energy requirements cleanly and efficiently.

5. SUSTAINABILITY PROJECTION WITH DIGITAL TWINS

In recent years, one of the revolutionary technological advancements towards sustainability in industries and renewable energy systems is Digital Twins (DTs). This paves the way for innovative solutions to improve our environmental, economic, and social sustainability outcomes through the match of a dynamic, real-time virtual model of physical assets. Using predictive analytics, optimization tools, and high-end simulations, DTs allow stakeholders to reduce resource consumption, lower the carbon footprint, and build systems that contribute to the goals of sustainable development (Kaushik et al., 2024).

5.1 Resource Efficiency

A key contribution of DTs to sustainability is improving resource efficiency. Wind farms, solar parks and hydroelectric plants represent renewable energy systems with high consumption of material and operational resources. This allows DTs to optimize both the land and the outputs by analyzing real-time data to minimize waste of resources and ensure performance is maximized.

In solar energy systems, for example, keep track of shadowing patterns and degradation of panels, allowing the potential for proactive adjustments that maximize energy capture without adding more material. Likewise, in wind energy, DTs monitor and evaluate turbine performance, monitoring the wind routes to ensure the effective energy production, while at the same time eliminate the wear of mechanical elements thus increasing the lifecycle of assets.

5.2 Reduction in Carbon Footprint

DTs have a significant impact on decarbonizing renewable energy systems by improving efficiency and minimizing waste. They facilitate predictive maintenance and real-time monitoring that keep systems operating in their peak efficiencies. That saves energy losses and reduces demand for fossil-fuel-coal-fired backup electric power, supporting global decarbonization directly. Additionally, DTs also facilitate the integration of renewable energy sources in the existing grids. They are using algorithms to optimize energy distribution based on simulated patterns of energy generation and demand fluctuations, minimizing the dependence on carbon-intensive energy sources. In hybrid systems with solar, wind, and storage technologies, DTs enable smooth transitions through power sources so that green energy is always supplied with minimal environmental impact (Nour El-Din et al., 2024).

5.3 Enhanced Lifecycle Management

Renewable energy assets has a lifespan, that is where sustainability comes. Making, putting in and disposing of parts such as turbines, solar panels and batteries also has high environmental costs. Other Ways That Digital Twins Contribute to SustainabilityDTs can lead to increased sustainability through improved lifecycle management, whereby assets are used more effectively and replaced only as needed. For example, digital twins (DTs) in hydroelectric plants monitor the condition of turbines and dams, predicting when they will need maintenance and extending operation lifespans. In solar energy, DTs monitor panel degradation and offer data-driven recommendations for recycling and repurposing components, minimizing waste and conserving precious materials.

5.4 Intelligent Urban and Rural Evolution

DTs sustainable development: assist smart cities and remote solid-state solution By doing this DTs are optimizing renewable energy systems to meet changing demand and facilitating cities transition to low-carbon energy grids. They also help with energy storage and distribution networks, preventing energy losses you will go through issues and increasing reliability. Rural areas are enabled to deploy clean off-grid energy solutions (e.g., microgrids, solar home systems with payments linked to a shared agenda for covering those that pay) through DTs. Such solutions can then be implemented in an efficient and environmentally friendly way by using DT to simulate energy needs and system performance to ensure that these are possible in terms of cost and providing environmentally friendly solutions, resulting in sustainable development in remote communities.

The Digital Twins integrate with the global sustainability agenda, such as the United Nations Sustainable Development Goals (SDGs). DTs advance the goals of SDG 7 (Affordable and Clean Energy), SDG 9 (Industry, Innovation, and Infrastructure), and SDG 13 (Climate Action) by facilitating efficient resource use, aiding in the adoption of renewable energy, and reducing environmental externalities. Furthermore, data-driven technologies (DTs) lead to data-powered decisions allowing policy makers and industry experts to devise an energy strategy that harmonizes their economic development with their environmental sustainment. These simulations ensure the resilience of energy projects to future challenges, such as climate change and resource depletion.

The involvement of DTs in sustainability will only continue to grow. Recent innovations, like data science to create AI-driven predictive models; the use of blockchain for decentralized and secure sharing of information; and the Internet of Things (IoT) will improve both the performance and scalability of these DTs. This further allows for more precise energy management, integration of the emerging renewable energy technologies, and circular economy practices (Raghav et al., 2024). As companies prioritize sustainability, adoption of DTs will be an important part of energy strategies. Working together, DTs will help accelerate global transitions to an energy future where governments, industries, and communities cooperate in achieving mutually beneficial, sustainable outcomes.

Digital Twins in solid-state renewable energy systems are making sustainable decisions possible through simulations that optimize efficiency and lessen environmental impacts for long-term resource allocation. Their alignment with global sustainability objectives and their support for intelligent energy strategies make them a cornerstone of sustainable development. Through adoption of this technology, stakeholders can inspire progress toward a greener and more resilient future.

6. CHALLENGES AND LIMITATIONS

However, the potential of digital twins (DTs) within the use of renewable energy systems concerns challenges of their own. Several technical, financial and operational barriers, together with ethical and environmental dilemmas face the integration of this state-of-the art technology. Therefore, it is essential to comprehend such challenges to overcome its limitations and fully utilize DTs in green energy systems (Ferdaus et al., 2024).

6.1 Data management and integration

The key challenges for DT adoption are the data volume, velocity and variety generated by the renewable energy systems. So electricity producing patients from wind turbines, solar panels, and other energy assets generate massive amounts of real-time data from sensors. Processing this information means you also need storage solutions, high-speed networks and commercial data analytics. In addition, consolidating data from multiple sources into a symmetrical and coherent Digital Twin is a difficult task. The renewable energy systems usually consist of equipment from various manufacturers, with different data formats and communication protocols. Interoperability and standardization are vital but continue to be a major obstacle for adoption.

6.2 High Initial Costs

The DTs are expensive to develop and products that are expensive to deploy. Next-generation sensors, IoT infrastructure, data storage and compute resources are needed to create a Digital Twin. The cost also increases with the implementation of the software platforms required for real-time monitoring and simulation. Small and medium enterprises (SMEs) in renewable energy often do not have the financial resources to invest in DT technology. The reason for that could be their adoption is limited and that leads us to questioning the scalability of DTs for broader use cases in the industry.

6.3 International Data Protection and Cybersecurity

However, the interdependence of DTs brings in cyber security risks. Cyberattacks on renewable energy systems with DT dependence can include unauthorized access to the system, data theft, or even manipulation of the system. Such attacks disrupt operations, exfiltrate sensitive data, and result in large-scale financial losses as well. The massive amount of data generated by DTs creates privacy issues when it comes to personal/proprietary information (Vijaya Priya, 2024). These risks can be mitigated with strong cybersecurity protocols and compliance with data protection regulations.

6.4 Technical Expertise and Gaps in Skills

It is a challenging task to deploy and manage DTs requiring specialized technical expertise in areas such as IoT, data analytics, machine learning, and renewable energy systems. But there is a lack of trained personnel who can design and maintain these systems. This skill shortage is a challenge faced by organizations that wish to implement DTs, especially in areas with poor access to high-end (advanced) technical education. Closing this gap will be a major undertaking, with substantial investment in workforce development and capacity-building efforts needed.

6.5 Computational Complexity

Digital twins (DTs) perform advanced simulations and real-time analytics that require significant computational resources. High-speed internet and robust computational infrastructure needed to implement the DT operations may not always be available for renewable energy systems, especially if they operate in remote or rural areas. In addition, the DTs of renewable energy systems also become more complex as the generation systems grow larger. However, managing these massive DT models is resource-intensive, potentially capping their efficiency and applicability.

6.6 Ethical and Environmental Issues

DTs batting for sustainability can have ethical and ecological repercussions in its deployment. The sensors, IoT devices, and computing hardware are also produced and disposed of, contributing global electronic waste (Jiang et al., 2024). Moreover, some of the environmental benefits provided by renewable energy systems can be counterbalanced by the energy spent on data centers that support the operations of DT. To strike a balance between the gains from DTs and their environmental footprint, we need innovative approaches such as powering data centers with renewable energy while also creating green hardware.

6.7 Scalability and Long-Term Maintenance

Deployment of DTs in scaling them across large renewable energy installations or integrating them into hybrid energy systems can be complex. Continuous effort and investment are needed to ensure that the renewable energy system maintains consistent performance, manages updates, and preserves the accuracy of its data over time. Maintaining DTs for the long term is also important but can be resource-heavy. As technology continues to advance, compatibility with emerging systems and standards also proves to be a challenge in the upkeep of Digital Twins.

Digital Twins are powerful assets for renewable energy lifecycles, but their challenges should be solved for more widespread uptake. Problems around data governance, costs, cybersecurity, skill gaps, and environmental impact require focused solutions. Overcoming these limitations will not only improve the efficiency and reliability of DTs, but also facilitate the global transition to a more sustainable energy future. Widespread implementation and realization of Digital Twin's potential will greatly depend on collaboration across governments, industry stakeholders, and the research community to tackle these challenges (Zhen & Yao, 2024).

7. CONCLUSION AND FUTURE TRENDS AND INNOVATIONS

Digital Twins (DTs) have become groundbreaking for renewable energy systems moving towards optimizing traits such as efficiency, sustainability, and innovation. Combining real-time data, predictive analytics, and enhanced simulations, DTs allow for resources to be used optimally, with as little environmental impact as possible, and lifecycle management of renewable energy assets. This positions them as potentially vital measures for counteracting climate change and promoting a transition to renewable energy, especially within the context of the global sustainability agenda.

The future perspectives of the DTs are based on the new technologies and application ideas in renewable energy. We can expect that AI and machine learning will allow DTs to be more predictive, beyond even intelligence decision-making and automation. Adding blockchain capability will also enhance security and transparency for data sharing, while IoT development will further the connectivity and data collection from the energy systems. DTs will also enable hybrid energy systems by integrating solar, wind and storage for optimal performance. As smart cities and rural microgrid projects continue to grow, digital twins will help to enable decentralized, efficient and resilient energy systems. Ultimately, as sustainable energy systems are vital for sustainable development, continued development and refinement (including a detailed analysis of how DT data is used in existing systems) of the means in which DT can be used in engineering processes across the entire supply chain (including raw extraction to end system DI) will provide further opportunities for improvements relative to every ESG metric.

REFERENCES

Agostinelli, S. (2024). Optimization and management of microgrids in the built environment based on intelligent digital twins.

Arora, M. K., Lal, S., Singh, B., & Olyaee, M. (2024). Through an Energy Lens: Examining SDGs From a Policy Perspective. In Riswandi, B., Singh, B., Kaunert, C., & Vig, K. (Eds.), *AI Applications for Clean Energy and Sustainability* (pp. 320–343). IGI Global., DOI: 10.4018/979-8-3693-6567-0.ch016

Das, O., Zafar, M. H., Sanfilippo, F., Rudra, S., & Kolhe, M. L. (2024). Advancements in digital twin technology and machine learning for energy systems: A comprehensive review of applications in smart grids, renewable energy, and electric vehicle optimisation. *Energy Conversion and Management, X*, 100715. DOI: 10.1016/j.ecmx.2024.100715

Ferdaus, M. M., Dam, T., Anavatti, S., & Das, S. (2024). Digital technologies for a net-zero energy future: A comprehensive review. *Renewable & Sustainable Energy Reviews*, *202*, 114681. DOI: 10.1016/j.rser.2024.114681

Jiang, H., Tjandra, R., Soh, C. B., Cao, S., Soh, D. C. L., Tan, K. T., Tseng, K. J., & Krishnan, S. B. (2024). Digital twin of microgrid for predictive power control to buildings. *Sustainability (Basel)*, *16*(2), 482. DOI: 10.3390/su16020482

Kaushik, T. K. Ravish, Singh, A., Raghav, A., & Singh, B. (2024). Potential of Artificial Intelligence in the Agricultural Industry: A Comprehensive Review. In B. Riswandi, B. Singh, C. Kaunert, & K. Vig (Eds.), *AI Applications for Clean Energy and Sustainability* (pp. 344-365). IGI Global. https://doi.org/ DOI: 10.4018/979-8-3693-6567-0.ch017

Lal, S., Arora, M. K., Riswandi, B. A., Singh, N., & Singh, B. (2024). Melding Artificial Intelligence and Remote Sensing Data for Real-Time Monitoring of Marine Biodiversity. In Riswandi, B., Singh, B., Kaunert, C., & Vig, K. (Eds.), *AI Applications for Clean Energy and Sustainability* (pp. 299–319). IGI Global., DOI: 10.4018/979-8-3693-6567-0.ch015

Maksimović, M., Jokić, S., & Bošković, M. Č. (2024, March). Digital Twins and 5G: Unlocking the Potential in the Energy Sector. In *2024 23rd International Symposium INFOTEH-JAHORINA (INFOTEH)* (pp. 1-6). IEEE.

Mchirgui, N., Quadar, N., Kraiem, H., & Lakhssassi, A. (2024). The Applications and Challenges of Digital Twin Technology in Smart Grids: A Comprehensive Review. *Applied Sciences (Basel, Switzerland)*, *14*(23), 10933. DOI: 10.3390/app142310933

Nour El-Din, M., Poças Martins, J., Ramos, N. M., & Pereira, P. F. (2024). The Role of Blockchain-Secured Digital Twins in Promoting Smart Energy Performance-Based Contracts for Buildings. *Energies*, *17*(14), 3392. DOI: 10.3390/en17143392

Ohueri, C. C., Masrom, M. A. N., & Seghier, T. E. (2024). Digital twin for decarbonizing operating buildings: A systematic review and implementation framework development. *Energy and Building*, *320*, 114567. DOI: 10.1016/j.enbuild.2024.114567

Poddar, H. (2024). A Look into the Future: Advances in Digital Twin Technology. In *AI Approaches to Smart and Sustainable Power Systems* (pp. 1–27). IGI Global. DOI: 10.4018/979-8-3693-1586-6.ch001

Raghav, A., Singh, B., & Jermsittiparsert, K. (2024). Role of Artificial Intelligence and Machine Learning in Clean and Blue Energy for Enhancing Efficiency and Sustainability. In Riswandi, B., Singh, B., Kaunert, C., & Vig, K. (Eds.), *AI Applications for Clean Energy and Sustainability* (pp. 366–386). IGI Global., DOI: 10.4018/979-8-3693-6567-0.ch018

Raghav, A., Singh, B., Jermsittiparsert, K., Raghav, R., & Yadav, U. (2024). Artificial Intelligence in Environmental and Climate Changes: A Sustainable Future. In Singh, B., Kaunert, C., Vig, K., & Dutta, S. (Eds.), *Maintaining a Sustainable World in the Nexus of Environmental Science and AI* (pp. 485–506). IGI Global., DOI: 10.4018/979-8-3693-6336-2.ch019

Singh, B., Kaunert, C., Vig, K., & Dutta, S. (Eds.). (2024). *Maintaining a Sustainable World in the Nexus of Environmental Science and AI*. IGI Global.

Tariq, M. U. (2024). Emerging Trends and Innovations in Blockchain-Digital Twin Integration for Green Investments: A Case Study Perspective. *Harnessing Blockchain-Digital Twin Fusion for Sustainable Investments*, 148-175.

Ukoba, K., Olatunji, K. O., Adeoye, E., Jen, T. C., & Madyira, D. M. (2024). Optimizing renewable energy systems through artificial intelligence: Review and future prospects. *Energy & Environment*, 0958305X241256293.

Vijaya Priya, R. (2024). A Look into the Future: Advances in Digital Twin Technology Harshit Poddar Vellore Institute of Technology, India. *AI Approaches to Smart and Sustainable Power Systems*, 1.

Zhen, Z., & Yao, Y. (2024). The confluence of digital twin and blockchan technologies in Industry 5.0: Transforming supply chain management for innovation and sustainability. *Journal of the Knowledge Economy*, •••, 1–27. DOI: 10.1007/s13132-024-02151-0

Chapter 10
Economic Implications of AI-Driven Sustainable Energy: Unlocking the Potential of Solar Energy

Anuradha Jain
https://orcid.org/0000-0001-6996-8309
VIPS, New Delhi, India

Antra Agarwal
University of Delhi, India

ABSTRACT

Artificial intelligence (AI) utilizes machine learning algorithms to determine the optimal location for solar panels, predict energy demand, and manage battery storage more effectively, translating into higher energy production and lower running costs. Solar energy production AI automates towards maximum output with minimal operators, enhancing output and minimizing data entry errors as a result. AI-powered predictive maintenance improves the equipment lifetime, reducing unscheduled failures and downtime. Long-term economic impacts are job creation as AI enables new jobs such as energy analytics, automation, and smart grids. In addition, AI-based solar energy not only draws significant investments but also contributes to the growth of global markets and accelerates sustainable energy efforts in regions around the world. Governments embrace AI in energy policies that subsidize and incentivize adoption, rendering solar energy more economical and ubiquitous, paves the way for a resilient, AI-enabled green economy of the future.

1. INTRODUCTION

The shift towards sustainable energy is a global change in energy systems aimed at reducing dependency on fossil fuels and minimizing nuclear fuel usage. For the last two decades, growing concern about carbon emissions, environmental degradation, and energy security has prompted governments, businesses, and research institutions to pursue alternative energy sources. Solar energy and other renewables have become popular due to their ubiquity, scalability, and decreasing prices. This shift to a photovoltaic (PV) technology in developing and developed countries with the aid of energy storage (Kannan & Va-

DOI: 10.4018/979-8-3373-0045-0.ch010

keesan, 2016). The other international collaborations like the Paris Agreement underscored the necessity of shifting to a sustainable energy balance, inspiring nations to pledge to lower greenhouse gas (GHG) emissions and promote cleaner energy sources (Izam et al., 2022). The growth in solar energy systems was driven by various policy frameworks including feed-in tariffs, tax incentives, and renewable energy certificates which incentivized businesses and consumers to invest in solar energy systems. A variety of multinational firms, including tech companies, are using renewable energy in their operations to go carbon neutral.

However, while solar energy infrastructure is expanding rapidly, complications concerning efficiency, intermittency, and grid integration remain. Variables such as availability of sunlight, constraints in storing energy, and the expensive nature of solar installations have made it imperative for technologies to evolve and thus maximize production and consumption. This is where artificial intelligence (AI) comes in, providing data driven solutions to increase energy efficiency, grid stability and the economic viability of solar energy projects (Choudhary & Srivastava, 2019). This revolutionary synergy also plays a pivotal role in the faster transition of nations to sustainable energy, making AI and solar energy a match made in heaven that can only lead us toward a more reliable and cost-effective future in energy.

AI and solar energy are maximizing their potential by revolutionizing the solar energy industry and combating pain points in generation, distribution and consumption of energy. Using AI-powered algorithms, rotation and angle of several solar panels can be optimized, weather forecasts can be predicted, and maintenance can be automated. In such a way these machine learning models process real-time and historical data to maximize energy production and reduce losses related to inefficiencies or adverse conditions. AI has had one of the most significant impacts on solar energy in terms of predictive maintenance, where AI-enhanced sensors identify faults, dust build up, or wear and tear on solar panels (Abbe & Smith, 2016). AI helps reduce maintenance costs and prevents losing energy by identifying potential issues before they escalate. AI optimizes battery charging and discharging cycles, improving the energy storage management to create stable power even in low sunlight periods.

AI helps manage smart grids so that solar power can be sustainably infused into national grids. Utilities use AI-powered demand forecasting for balancing the flow of energy between supply and consumption, ensuring that the grid does not run out, or become overloaded, thereby reducing the reliance on non-renewable energy sources (Verma, 2019). IoT-based smart meters offer real-time energy consumption data, helping homes and businesses make informed choices regarding energy usage, leading to cost reductions and enhanced sustainability. AI and solar energy can lead to improved efficiency, less waste and greater economic sustainability in the industry. With rapid advancements in AI technologies, it will play a key role in improving the competitiveness and reliability of solar power over fossil fuel-based energy generation (Shubbak, 2019).

1.1 Importance of AI-Powered Sustainable Energy Solutions in the Economy

AI-enabled systems are contributing to making energy solutions more sustainable, more scalable, and more affordable. The first and foremost benefit of AI-enabled solar energy systems is their ability to lower practical and maintenance expenses. Traditional solar farms often require frequent manual inspections combined with reactive maintenance processes that are expensive and time consuming. Predictive analytics powered by artificial intelligence makes maintenance proactive instead of reactive, hence preventing expensive failures of the system and increasing the overall productivity. Artificial intelligence (AI) also helps in decreasing energy waste by enhancing the efficiency of solar panels and optimizing energy distribution systems. AI algorithms, for example, forecast variations in energy generation and modify operations of the grid based about these fluctuations saving excess energy or directing it to areas most in need. This lowers energy losses and makes economics better for both large-scale solar farms and small-scale rooftop installations.

AI in solar energy has a significant economic impact creating new job opportunities as well. Automation will eliminate some traditional energy roles, but it will also increase the need for specialized skills in AI development, energy analytics, data science, and smart grid management. Investment in AI-facilitated renewable energy, for countries, means an upskilling of its workforce, wherein skills in the digital sphere and engineering become pivotal in a green economy (Sampaio & Gonzalez, 2017). AI increases the investment appeal of solar energy projects. AI helps to make renewable energy projects more financially viable by improving efficiency and mitigating risks related to solar power generation. AI–driven solar energy is seen as a long-term profitable solution by investors and policymakers alike, resulting in increased funding of solar startups, the expansion of infrastructure, and innovative business models such as peer-to-peer energy trading through blockchain technology.

The long-term economic impact of AI-powered solar solutions lies in their ability to create a stable energy economy, decrease reliance on corrosive fossil fuel markets, maintain stable electricity prices, and encourage long-term industrial activity. In addition to cost reduction, the economic benefits lay the foundation for a cleaner, more efficient and future-ready energy landscape. In particular, it explore the aspects of sustainable energy and the role of solar energy therein through AI unlocking its potential in an economic analysis setting in this paper. AI has some capabilities of enhancing solar energy production, optimizing energy consumption and, ultimately, economic sustainability through cost reduction and new market opportunities. This study also seeks to advance the sustainability agenda as it travels, showing the way that AI helps to save energy, cut down carbon emissions, and build a green economy. As the synergy between technology and economics becomes vital in our ability to combat climate change, this bridges the gap and should be an eye-opener about the transformative potential of AI in the future of solar energy and an AI-driven future of energy economy (Gatto & Drago, 2021).

2. AI INTEGRATION IN SOLAR ENERGY SYSTEMS

AI is shaping the future of the generation, storage, and consumption of renewable energy sources and the approaches behind these technologies are evolving in concert with this evolution. This combination allows AI to improve solar panels' productivity, optimize energy storage, and manage the scheduling of energy provisions thus making solar energy more reliable, cost-efficient, and scalable (Lazzarin & Noro, 2018). Traditional solar energy generation methods come with its own limitations: intermittent energy generation mechanisms, high maintenance costs, and energy wastage issues. AI solves these challenges with ML, predictive analytics and automated intelligence. Deep learning solar panels are AI-based solar energy systems utilizing complex algorithms to analyze large amounts of data, such as weather patterns, solar irradiance and grid demand, in order to improve energy production and consumption. Predictive maintenance powered by AI minimizes downtime by identifying upcoming equipment failures before they materialize. AI also provides the management of energy storage, increasing battery performance as well as a reliable source of energy supply. In the sphere of energy, AI plays a crucial role in creating smart grids and microgrids that ensure improved energy load balancing and peak demand control (Kadaba et al., 2023).

With the increasing integration of AI into solar energy, both economic and environment landscape are changing. While emphasizing energy efficiency and fossil fuel dependence an approach that aligns well with sustainability goals AI-enriched solar systems also help reduce energy costs for consumers. With trillions set to pour into AI-driven renewables as governments and industries scramble to capitalize on the opportunity presented, decarbonization and economic expansion on a potential world scale are well on the horizon.

2.1 AI for Improving Solar Panel Efficiency: Solar Panel Maintenance and AI-Powered Predictive Analytics

Companies are using AI-powered predictive analytics to keep solar panels operating efficiently. Real-time data from sensors embedded inside solar panels are analyzed by AI models to identify anomalies, temperature shifts, and energy output variations (Prakash & Sethi, 2021). This knowledge enables the early identification of problems like degradation, dirt buildup, or electrical defects, allowing for timely repair and minimizing performance losses. It definitely helps the solar panels to improve the operations and performance life through AI predictive maintenance to reduce unexpected downtime and operational cost.

To detect mechanical damage to solar panels, AI-powered computer vision and image recognition technologies are being put into use. Drones fitted with high-resolution cameras and AI-powered image-processing software fly over solar farms to detect cracks, dirt or shading that could affect performance. Automating this process eliminates manual inspections, which are time-consuming and susceptible to errors. Repair processes are expedited incentivized by AI based damage detection and classification, facilitating uniform energy generation. Smart tracking systems powered by AI algorithms further enhance solar panel efficiency by optimizing sun exposure. Conventional fixed-tilt panels charge the sun's movement, resulting in energy loss. Solar panels today are being adjusted on a dual-axis tracker system powered via AI that calculates the angle of tilt based on which sun position and weather conditions in real time as well as historical energy generation data. This has led to the development of highly optimized solar tracking systems per AI-based harnessing systems, substantially increasing energy yield, rendering solar power generation efficient and cost-effective (Tyagi et al., 2022).

2.2. AI for Grid Management and Energy Storage: AI-driven Demand Forecasting and Supply Optimization

AI can play an instrumental role in grid management in one way in particular: the ability to predict energy demand and optimize supply accordingly. AI algorithms process past and current energy usage data, weather patterns, and grid load changes. From these learnings, AI can predict spikes in demand and modulate solar power feeds accordingly, dampening volatility in the grid and allocating energy more evenly. Forecasting Demand for Energy Using AI for Smart Grids AI reduces energy shortages and generation surpluses, thus more profitable and sustainable manufacturing process. Two caveats arise from this: Energy storage (which is an issue in solar, since solar energy is intermittent (Schwerhoff & Sy, 2017). AI improves performance from batteries by determining the best timings for charge and discharge cycles, and thus, preventing overcharging, as well as optimizing the lifespan of batteries. The power plants are equipped with AI-based battery management systems that monitor battery health, predict degradation rates, and adjust energy storage strategy. AI and Distributed Energy Resource Management

In distributed energy resource management AI is critical for coordinating multiple decentralized solar energy sources. Rather than relying on sitting power in one spot, gas in another, and heroes flying in to save the day, AI takes into account residential solar panels, community microgrids, and large-scale solar farms to connect them all into one grid. AI-driven DERM systems monitor and assess real-time energy generation and consumption, proactively redistributing energy resources to prevent blackouts and minimize transmission losses; In both urban and rural settings, AI-enabled decentralized grids have improved resilience to provide consistent and sustainable access to induced energy (Singh & Balachandra, 2019).

2.3 Decentralized Energy Networks and AI-Powered Microgrids

AI-Powered Microgrids are revolutionizing legacy centralized power generation facilities into a more autonomous power supply chain. Enhanced with AI, microgrids manage local energy generation, storage, and consumption more effectively, and help decrease dependency on national grids. AI-powered analytics in these intelligent energy networks help to identify faults, optimize energy flow, and maintain real-time changes according to demand and supply fluctuations. Localized energy production and storage enabled by AI-driven microgrids can enhance energy resilience, particularly in remote and off-grid locations.

Using smart peak demand response strategies, AI is transforming demand-side energy management. For example, during peak hours when electricity demand is high, AI algorithms analyze real-time consumption data and adjust the energy distribution. Automated demand response (ADR) systems driven by artificial intelligence communicate with smart appliances, smart building systems / industrial systems and electric vehicle (EV) charging stations to optimize energy consumption. AI prevents grid congestion, lowers electricity expenses, and strengthens grid stability through shifting energy consumption to off-peak hours for non-essential services.

2.4 IoT and AI in Smart Home Energy Consumption

Use Residential Solar Energy Such AI-driven smart home systems assess information from IoT-driven thermostats, smart lighting, HVAC systems, and energy-storing devices to minimize electricity consumption. AI analyzes user behavior and optimizes energy consumption patterns reducing wastage on energy settings. In order to automate decision-making, smart energy meters with AI solutions deliver real-time insights into consumer behavior regarding energy usage. The combined impact of AI and IoT helps you save costs and also low down on your carbon footprint by ensuring energy-efficient behavior (Teichmann et al., 2021). AI is revolutionizing energy conversion, storage, and consumption, taking solar energy technologies to new heights. Also, making solar panels more efficient, enhancing grid operations, and creating intelligent energy management systems. AI-driven solar energy solutions are vital to creating a low-carbon, economically feasible future by lowering operational costs, increasing energy robustness and promoting sustainable development. So it also helps to speed the change, it means the transformation into decarbonized energy, which is crucial for the global energy society to provide a foundation for energy security.

3. ECONOMIC BENEFITS OF AI-DRIVEN SOLAR ENERGY

AI is making a major transformational impact in the solar energy sector, providing exceptional economic value, from cutting production costs to increasing employment prospects and improving investment trends. AI applications are transforming solar energy operations to ensure higher efficiency, optimal resource use, and enhanced economic feasibility of renewable energy projects around the world. In the following segments, the economic benefits of AI-powered solar energy for cost savings, new job opportunities and investment trends will be discussed.

3.1 Reduction of Solar Energy Production Cost

Solar energy power generation leveraging AI has lowered operational and capital costs by an overwhelming margin. AI-powered efficiency improvements radically reduce costs at all levels of solar energy development. AI Applications in Solar Energy Predicting Optimal Performance AI applications in solar energy are enhancing photovoltaic (PV) systems performance via data analytics and ML models (Cantarero, 2020). AI systems designed to forecast energy generation harness solar and climatic data to forecast solar generation, allowing optimal power utilization. Not only does this distribution result in less wasted energy, but it also ensures the economic potential of solar plants is optimized.

AI-based predictive maintenance minimises neither unexpected failure of the system nor repair cost. AI algorithms, for instance, process data coming in from solar panels around the clock to identify irregularities, enabling operators to nip small problems before they develop into expensive malfunctions. This proactive approach increases a solar plant's longevity, thereby reducing the frequency of replacement and slashing operational expenditures. In this phase, the production of solar panels is optimized, leveraging AI to drive efficiency while minimizing resource waste and bolstering quality assurance. Likewise robotics with AI can automate manufacturing processes for solar panels leading to reduced cost for manufacturing while increasing the accuracy and longevity of the solar panels. Moreover, AI-powered deployment strategies streamline the positioning and angling of solar panels for optimal energy capture while minimizing the labor costs of the physically intensive installation process.

3.2 Employment Generation and Market Expansion: Data-Driven Solar and New Jobs

Fast pace job creation and market expansion due to AI-integrated solar energy Potential job replacement by automation has long been a concern, but AI-driven solar projects are creating jobs by the thousands in the growing fields of technology, engineering, and energy management. From AI Engineers and Data Scientists to Solar Energy Analysts, AI is helping to create many job opportunities. As such, the need for trained experts to create, deploy and sustain AI-enhanced solar energy systems is increasing, resulting in job opportunities across developed and developing economies.

The figures in the renewable power sector are transforming due to AI allowing specialists to prioritize greater-value functions. Rather than monitoring systems, workers can review and analyze AI-generated insights to find ways to improve performance. New training programs are beginning to train workers in AI and data science skills, preparing a workforce for the future of work equipped to tackle technological change (Undheim, 2023).

3.3 Expansion of Market and Economic Spillover Effect on Global Industries

A New hybrid market AI-driven solar energy market is growing as technological improvements make solar power affordable and trustable. Lowering energy prices aids electricity-dependent industries, and that a rising economy helps everyone. As demand for AI-integrated solar solutions rises, so does supply chain, with opportunities emerging in manufacturing, software development, and infrastructure projects. AI solar energy is powered by economic gains, drawing massive investments from venture capital frms, governments, and private enterprises. AI is also making solar projects financially much more attractive, increasing the return on investment (ROI) and helping to develop infrastructure for sustainable energy.

Solar energy startups with an AI angle are attracting the attention of venture capital investors looking to take advantage of renewable energy's expanding market opportunity. These include considerable capital inflows to new entrants working on AI-driven solar forecasting, grid optimization, and automated maintenance technologies. Solar energy projects see their business gain an edge through improved efficiency, reduced maintenance costs, and optimization in the area of energy distribution, thanks to AI. Predictive analytics lead to better investment decisions because stakeholders can evaluate risks and returns more accurately. Artificial intelligence enhances the return of investment (ROI) by balancing energy supply and demand while reducing losses through AI-powered smart grids (Brennan et al., 2021).

3.4 AI-Integrated Solar Initiatives Bolstered by Public-Private Partnerships

Solar Power Projects Are Being Revolutionize by AI Collaboration of Governments and Private Enterprises With its population free from the burdens of bureaucracy and red tape, public-private partnerships are allowing smart energy policies to be formed, research initiatives to be funded and enable large-scale solar energy programs to form. The potential of AI in making energy systems more efficient and sustainable is pushing policymakers to adopt AI-based solutions in national energy policies. The Technology is already producing an enormous economic value to the environment: AI-based solar energy. By minimizing expenses and maximizing job opportunities and investment in the energy sector, AI is leading us toward an energy future that is both sustainable and cost-effective (Nassiry, 2018). With the rapid development of the AI technology, AI-enabled innovations in solar energy have the potential to play an even more influential role in energizing the international transition to clean, renewable energy in a way that maximizes economic benefits.

4. CHALLENGES AND RISKS IN AI-DRIVEN SOLAR ENERGY: TECHNOLOGICAL AND IMPLEMENTATION BARRIERS

Data-driven analyses using AI tools can be challenging: AI is not very compatible with solar energy systems due to various technological and implementation issues. The High Initial Investment for AI-Driven Solar Projects The implementation of AI in solar energy management would require substantial financial resources as it includes advanced sensors, high-performance computing infrastructure, and data analytics tools. Despite the potential long-term rewards of technology adoption, the initial costs can be a barrier for small-scale energy providers and developing economies.

The other significant hurdle is the cybersecurity threat to AI-powered energy grids. AI systems control power distribution while also balancing the grid, making them prime targets for cyberattacks. If malicious hackers can break into the operating system of AI algorithms, they can disrupt energy supply, breach sensitive data, or even create large-scale blackouts. Stringent cybersecurity features, like AI-led anomaly detection and blockchain integration, must be managed in case to protect the energy essentials. Also, integrating solar energy AI into pre-existing energy infrastructure is a complex process that deters adoption. Most energy industry legacy systems rely on old school grid management strategies which wouldn avoid using AI-based solutions. Revamping aged infrastructure with AI technologies necessitates specialized software and expertise, causing implementation delays. The existing differences in grid standards among different regions call for AI models to be versatile such that they can operate in different operational contexts (Sengupta et al., 2023).

4.1 Socioeconomic and Environmental Concerns: Risk of AI-driven Automation Reducing Traditional Energy-sector Jobs

There are potential socioeconomic and environmental impacts associated with these AI-driven solar energy solutions, despite their promise as the future of energy. The potential loss of traditional energy-sector jobs through AI-driven automation is one major issue." It can minimize human involvement by optimizing energy generation, distribution, and storage. This improves efficiency, but could displace workers in manual grid maintenance, system monitoring, and other operational responsibilities (Christopoulos et al., 2022). Policymakers should introduce retraining and reskilling initiatives to help displaced workers take on new AI-based jobs.

The sustainability of such AI-based solutions needs to be considered, such as whether energy consumption is at least compensated with energy savings. AI algorithms need to be run in large volumes on servers and you can drive up overall energy consumption. The carbon footprint of AI computing, if not managed efficiently, could counteract the sustainability benefits of solar energy (Foster et al., 2021). Therefore going forward, improvements in energy-efficient AI models and renewable-powered data centers could alleviate this problem. The application of AI in solar energy may involve potential environmental trade-offs. If solar panels are powered by AI machine consisting of human resources and replaced by AI driven thinking slaves then such large-scale solar farms can be installed in ecologically sensitive areas which will result in great and fast damage to the ecology and also in return not much green solar power will be produced. Building responsibly, which means conducting environmental impact studies and training AI models that use sustainable practices.

4.2 AI Bias and Data Dependency

The main challenge of AI-powered solar energy management is data dependency, since accurate data is critical. These AI models forecast energy demand using historical weather patterns, grid usage and real-time sensor data. This data is difficult to assess and can give rise to inaccurate or broken data which can result in dealing with an indecent delivery of energy giving rise to inefficiencies and potential supply shortages (Bahrami et al., 2023). Thus, frequent monitoring of data, instantaneous corrections and combined efforts of AI developers with energy professionals is essential to keep up the accuracy of predictions. Risks of bias in AI decision-making are also present in energy distribution. For example, if they have been trained on offering services only to biases, their AIs might recommend more offers of a particular kind based on biased training sets. This leads to an unequal access to energy where energy distribution is prioritized in urban areas with abundant data resources whilst neglecting rural or underserved communities. *Inclusion of diverse data: The risk of data bias should be avoided as much as possible.

Yet another major challenge lies in the potential opacity of AI-led energy models. Many AI algorithms are not easily interpretable for decision-making processes. The dark side of AI in the energy sector transparency does not come easy, with AI's recommendations often being a 'black box'; this can further make energy providers and policymakers question and be less willing to implement AI-driven solutions as these organizations would be unable to comprehend the AI's 'reasoning'. Integrating XAI techniques give insight into AI decision-making process, creating trust and keeping accountability on the AI-driven solar energy systems (Mathews et al., 2019).

5. CASE STUDIES AND GLOBAL APPLICATIONS: AI-OPTIMIZED SOLAR ENERGY PROJECTS AROUND THE GLOBE

AI is revolutionizing the solar industry across the globe by optimizing solar energy through efficiency improvements, power generation predictions, and maintenance strategies. AI-based solutions are being implemented in various geographies such as in developed economies, advanced grid management systems, and implementation in developing nations with AI used for economic solar integration. Perhaps the most prominent example of this is Google's DeepMind partnership with the UK's National Grid, in which AI algorithms predict solar energy output more accurately than ever before creating better energy distribution. AI also writes the future in the United States, where predictive maintenance for solar farms large-scale solar facilities that provide power to thousands of homes through interconnected connections is being used to reduce downtime and improve efficiency in the plants managed by NextEra Energy. In contrast, developing economies like India or Kenya are deploying AI for microgrid management that achieves this. AI-powered forecasting tools, such as those used by India's Solar Energy Corporation, make it easier to stabilize the grid by predicting variations in solar output and calibrating storage accordingly (Rathore et al., 2021).

AI Powered Solar PV Panel Maintenance in Australia If a solar panel goes for example, relies on AI-driven analytics for operational improvements in order to help the country meet its ambitious renewable energy targets. AI operates in the global domain and can significantly optimize solar energy utilization in disparate economic potentiometers, extending technological dividends for energy sustainability.

5.1 Smart Cities and AI-Powered Solar Energy Adoption

Now, smart cities are deploying AI solar energy technology to make cities sustainable and efficient. With AI-enabled grid management, AI-driven grid management seizes control of the meter transition of the digital energy grid to continue to progress, guarantee the seamless urban integration (disposal) of solar power, realize the decline in carbon footprints and ensure energy security. AI-Powered Solar Run with Singapore's Smart City Initiative Singapore's Housing & Development Board (HDB) is using AI to position solar optimally across the residential buildings in the country to ensure both energy consumption and production. AI-driven monitoring systems identify inefficiencies in real time, helping users optimize their solar power production. For example, the Smart Solar Charging initiative in Amsterdam uses AI-based algorithms with EV charging stations to enable energy distribution tailored to current demand.

AI-based systems are being used for predictive analytics by cities like San Diego and New York in the U.S. for optimizing solar grid performance. AI predicts how much solar energy will be generated based on the weather, allowing for more efficient integration with the power grid and enhanced energy storage plans (Ghadikolaei, 2021). In Europe, the smart city projects in Copenhagen leverage artificial intelligence to promote urban sustainability. Solar panels in public infrastructure, such as transport hubs and commercial buildings, are optimized by AI to minimize dependence on conventional sources of energy. Barcelona's use of smart grids involves AI networks that locally regulate street lighting powered by solar energy and energy-efficient architecture systems, thereby minimizing general electricity consumption. Thus, these case studies reveal how solar energy supported by AI technology is emerging as a central pillar in the global march towards sustainable smart cities.

5.2 Corporate and Industrial Use Cases

AI-powered solar solutions are being adopted by corporations and industries to achieve environmental sustainability and at the same time, cutdown operational costs. In this regard, consider Tesla's Gigafactory in Nevada: an exemplar of AI-integrated solar solutions. In addition, AI can handle solar panel arrays and the storage batteries that go with them, reducing reliance on the power grid and maximizing usage of any energy that is generated. Likewise, Amazon's goal of 100% renewable energy for its data center fleet includes AI-driven solar projects at its fulfillment centers, where machine-learning algorithms predict energy demand and modify solar power consumption as needed a concept proved by the company to be effective in practice (Benda & Cerna, 2020).

Even manufacturing sectors in Germany are tapping into energy-efficient AI-powered solar energy solutions. Siemens utilizes AI to regulate energy consumption across its production facilities, increasing utilization of solar energy while reducing waste. In the automotive industry, BMW and Volkswagen leverage AI-driven solar grids within their production facilities, minimizing their carbon footprints while optimizing energy resilience. These AI-powered solar solutions help industries across developing nations while aligning with sustainability. AI-driven microgrids in South Africa allow mining operations to add a solar power option, decreasing dependence on diesel power plants. Likewise, textile companies in Bangladesh have been deploying AI-based solar energy optimization to reduce energy costs while maintaining continuous manufacturing. Corporate Industrial case studies on how AI is transforming solar energy adoption across industries enabling businesses to meet sustainability targets and increase energy savings and resilience (Socorro García-Cascales et al., 2022).

6. FUTURE DIRECTIONS

AI is going to be the future of solar energy, and new innovations, which in turn will lead to another economy. Advancements in artificial intelligence will allow solar energy systems to be more efficient and predictive, enabling energy production and consumption to be adjusted at the atomic level. Revolutionizing energy management, AI-powered autonomous systems will be able to make decisions in real time, providing greater accuracy in energy forecasting and improved grid stability. The overall improvements will lower operational costs and make solar energy economically viable even in areas with discontinuous sunshine.

Smart grids and AI-powered energy storage systems that keep evolving will continue to shake up the economics of solar energy even more. AI-powered batteries and energy storage systems that can help optimize energy distribution, smoothing out the peaks and valleys of energy generation and consumption. The integration of AI with blockchain technology will also enable decentralized energy trading, allowing consumers to buy and sell excess solar energy in real-time. This transformation is expected to produce a more equitable and economically viable energy market by enabling the producers of decentralized solar energy. AI-based solar energy applications will extend into emerging markets and developing economies where the availability of clean and affordable energy is still an issue. AI-enabled optimization techniques can help customize solar energy solutions to suit varying geographical and economic conditions, promoting energy access and economic development in marginalized areas. The AI will be a key player in the urban planning process by implementing solar energy solutions in smart cities as well as sustainable infrastructure projects.

AI role in shaping energy regulations will only continue to grow. Artificial intelligence tools will be utilized to make energy policy in governments and regulatory bodies based on data, which will help energy policies to become adaptive, efficient, and economically viable. Simulations powered by AI will help policymakers project the economic effects of transitions in energy, allowing for the creation of incentives and subsidies that promote the adoption of AI technologies in the solar energy industry.

The integration of AI with IoT and 5G technologies to allow for seamless connectivity between solar energy systems and smart home appliances is another promising area for future research. By leveraging AI algorithms, machine learning will enable IoT systems to gather data on solar panel performance, weather conditions, and energy demand, optimizing the deployment and management of solar resources in real-time. This will accelerate the shift towards self-sustaining, AI-managed energy ecosystems that are less reliant on centralized power grids and fossil fuels. AI-focused solar energy will also see a lot of innovations through interdisciplinary research. That concepts of advances and benefits of both, while balancing sense of technological growth and cognizance of the socio-economic political sensitivities, with AI specialists, environmental scientists, econometricians, PPD economic professionals coming together with policy jobbers and contemporaries. These partnerships will help to ensure that AI-powered solar energy technologies remain inclusive, scalable, and advantageous for both developed and developing economies.

AI-led sustainable energy, especially solar power, holds transformative economic promise. Despite the remaining challenges, more technological innovations, supportive policy frameworks, and cross-sector collaboration will reveal new economic opportunities and accelerate the global shift toward clean, AI-enhanced energy solutions. AI has already begun to change how we use solar energy, but over the next decade expect it to become the backbone of our next economic boom and a linchpin in global energy integration and sustainability.

7. CONCLUSION

The proposed essay aims, after a general introduction to solar energy and its evolution towards AI-based harvestors and analytic algorithms, to explore the huge potential for economic optimisation and sustainability provided by AI + solar energy combination. By exploiting intelligent AI (Artificial Intelligence) system capabilities to provide predictive analytics, smart grid management, and automation, solar energy systems have never been more efficient and reliable. This is not just about economic benefits, but also reducing carbon emissions and enabling mixed energy models that are both environmentally and socially sustainable.

The economic benefits of AI-powered green energy go beyond operations efficiency. This includes predicting energy demand, optimizing storage solutions and enabling energy trading in blockchain-based peer-to-peer networks, all of which can benefit from AI technologies. These progress allows stakeholders (government, business and consumers) use evidence based decision making to improve energy security and economic resilience. AI not only helps optimize solar energy generation but also plays a vital role in integrating this renewable source into the broader energy market. AI and automation can also benefit the job market as they can generate new jobs in renewable energy fields, AI development, and data analytics, although workers must adapt and reskill.

So, obstacles must be overcome in realizing the potential economic benefits AI can create in solar. High initial investments and worries about data privacy, cybersecurity risks, and the need for policy standardization prevent widespread implementation. In addition, the transition calls for enhanced collaboration among policymakers, the private sector, and research centers to create regulatory frameworks that stimulate the adoption of AI, while ensuring economic stability and equitable distribution of energy. Strategic planning, investments in AI research, and addressing these concerns will be essential in helping us reap the maximum potential economic benefits from AI-driven solar energy solutions.

REFERENCES

Abbe, G., & Smith, H. (2016). Technological development trends in Solar-powered Aircraft Systems. *Renewable & Sustainable Energy Reviews*, *60*, 770–783. DOI: 10.1016/j.rser.2016.01.053

Bahrami, A., Mohammadnejad, S., & Soleimaninezhad, S. (2013). Photovoltaic cells technology: Principles and recent developments. *Optical and Quantum Electronics*, *45*(2), 161–197. DOI: 10.1007/s11082-012-9613-9

Benda, V., & Černá, L. (2020). PV cells and modules–State of the art, limits and trends. *Heliyon*, *6*(12), e05666. DOI: 10.1016/j.heliyon.2020.e05666 PMID: 33364478

Brennan, L. J., Byrne, M. T., Bari, M., & Gun'ko, Y. K. (2011). Carbon nanomaterials for dye-sensitized solar cell applications: A bright future. *Advanced Energy Materials*, *1*(4), 472–485. DOI: 10.1002/aenm.201100136

Cantarero, M. M. V. (2020). Of renewable energy, energy democracy, and sustainable development: A roadmap to accelerate the energy transition in developing countries. *Energy Research & Social Science*, *70*, 101716. DOI: 10.1016/j.erss.2020.101716

Choudhary, P., & Srivastava, R. K. (2019). Sustainability perspectives-a review for solar photovoltaic trends and growth opportunities. *Journal of Cleaner Production*, *227*, 589–612. DOI: 10.1016/j.jclepro.2019.04.107

Christopoulos, S., Horvath, B., & Kull, M. (2012). Advancing the governance of cross-sectoral policies for sustainable development: A metagovernance perspective. *Public Administration and Development*, *32*(3), 305–323. DOI: 10.1002/pad.1629

Foster, K., Blakstad, S., Bos, M., & Gazi, S. (2021). *BigFintechs and their impacts on sustainable development*. The Dialogue on Global Finance Digital Finance Governance Project. DOI: 10.2139/ssrn.3871385

Gatto, A., & Drago, C. (2021). When renewable energy, empowerment, and entrepreneurship connect: Measuring energy policy effectiveness in 230 countries. *Energy Research & Social Science*, *78*, 101977. DOI: 10.1016/j.erss.2021.101977

Ghadikolaei, S. S. C. (2021). Solar photovoltaic cells performance improvement by cooling technology: An overall review. *International Journal of Hydrogen Energy*, *46*(18), 10939–10972. DOI: 10.1016/j.ijhydene.2020.12.164

Izam, N. S. M. N., Itam, Z., Sing, W. L., & Syamsir, A. (2022). Sustainable development perspectives of solar energy technologies with focus on solar Photovoltaic—A review. *Energies*, *15*(8), 2790. DOI: 10.3390/en15082790

Kadaba, D. M. K., & Aithal, P. S., & KRS, S. (2023). Government Initiatives and Digital Innovation for Atma Nirbhar MSMEs/SMEs: To Achieve Sustainable and Inclusive Economic Growth. [IJMTS]. *International Journal of Management, Technology, and Social Sciences*, *8*(1), 68–82.

Kannan, N., & Vakeesan, D. (2016). Solar energy for future world:-A review. *Renewable & Sustainable Energy Reviews*, *62*, 1092–1105. DOI: 10.1016/j.rser.2016.05.022

Lazzarin, R. M., & Noro, M. (2018). Past, present, future of solar cooling: Technical and economical considerations. *Solar Energy*, *172*, 2–13. DOI: 10.1016/j.solener.2017.12.055

Mathews, I., Kantareddy, S. N., Buonassisi, T., & Peters, I. M. (2019). Technology and market perspective for indoor photovoltaic cells. *Joule*, *3*(6), 1415–1426. DOI: 10.1016/j.joule.2019.03.026

Nassiry, D. (2018). *The role of fintech in unlocking green finance: Policy insights for developing countries* (No. 883). ADBI working paper.

Prakash, N., & Sethi, M. (2021). A review of innovative bond instruments for sustainable development in Asia. *International Journal of Innovation Science*, *14*(3/4), 630–647. DOI: 10.1108/IJIS-10-2020-0213

Rathore, N., Panwar, N. L., Yettou, F., & Gama, A. (2021). A comprehensive review of different types of solar photovoltaic cells and their applications. *International Journal of Ambient Energy*, *42*(10), 1200–1217. DOI: 10.1080/01430750.2019.1592774

Sampaio, P. G. V., & González, M. O. A. (2017). Photovoltaic solar energy: Conceptual framework. *Renewable & Sustainable Energy Reviews*, *74*, 590–601. DOI: 10.1016/j.rser.2017.02.081

Schwerhoff, G., & Sy, M. (2017). Financing renewable energy in Africa–Key challenge of the sustainable development goals. *Renewable & Sustainable Energy Reviews*, *75*, 393–401. DOI: 10.1016/j.rser.2016.11.004

Sengupta, U., Pramanik, H. S., Dutta, S., Dutta, S., Dasgupta, S., & Kirtania, M. (2023). Assessing sustainability focus across global banks. *Development Engineering*, *8*, 100114. DOI: 10.1016/j.deveng.2023.100114

Shubbak, M. H. (2019). Advances in solar photovoltaics: Technology review and patent trends. *Renewable & Sustainable Energy Reviews*, *115*, 109383. DOI: 10.1016/j.rser.2019.109383

Singh, M., & Balachandra, P. (2019). Microhybrid electricity system for energy access, livelihoods, and empowerment. *Proceedings of the IEEE*, *107*(9), 1995–2007. DOI: 10.1109/JPROC.2019.2910834

Socorro García-Cascales, M., Teresa Lamata, M., & Miguel Sánchez-Lozano, J. (2022). Evaluation of photovoltaic cells in a multi-criteria decision making process. *Annals of Operations Research*, *199*(1), 373–391. DOI: 10.1007/s10479-011-1009-x

Teichmann, D., Arlt, W., Wasserscheid, P., & Freymann, R. (2011). A future energy supply based on Liquid Organic Hydrogen Carriers (LOHC). *Energy & Environmental Science*, *4*(8), 2767–2773. DOI: 10.1039/c1ee01454d

Tyagi, V. V., Kaushik, S. C., & Tyagi, S. K. (2012). Advancement in solar photovoltaic/thermal (PV/T) hybrid collector technology. *Renewable & Sustainable Energy Reviews*, *16*(3), 1383–1398. DOI: 10.1016/j.rser.2011.12.013

Undheim, T. (2023). *Eco Tech: Investing in Regenerative Futures*. Taylor & Francis. DOI: 10.4324/9781003386049

. Verma, T. L. (2019). Role of micro, small and medium enterprises (MSMES) in achieving sustainable development goals. *Small And Medium Enterprises (MSMEs) In Achieving Sustainable Development Goals (April 1, 2019).*

Chapter 11
Energy Policy Driven by AI Towards Sustainable Future

Saurabh Kishor
Sharda University,India

Sadaf Waseem
https://orcid.org/0009-0006-2754-8449
Sharda University, India

Bhupinder Singh
https://orcid.org/0009-0006-4779-2553
Sharda University, India

ABSTRACT

The current decade is a turning point in the energy sector's transition, with the introduction of green energy along with the optimization of efficacy through machine learning and artificial intelligence (AI). As a result, competitive policies are needed to manage multifaceted tasks on one platform. When energy policies fail to accomplish both energy as well as climate targets over their entire lifecycle, the socioeconomic ramifications can be profound. These deficiencies are said to result from poor decision-making and insufficient incentives, which should encourage equity, equality, fairness, and inclusivity in energy policy and decision-making regarding projects. This chapter seeks to assess the many obstacles posed by the emergence of AI in the energy industry. Specifically, the study addresses (1) the development stage decision-making process, 2) the execution stage implementation management process, (3) integrating deep learning, machine learning and data science in the energy systems, and (4) the substantiality requirements of energy systems.

INTRODUCTION

In the current day, global warming along with sustainable development have become crucial issues that call for interdisciplinary approaches. In order to quantify the shortage of energy along with sustainability pillars changing depending on both demand and supply infrastructure, a number of sustainable energy-related factors have an impact at both global and regional levels and require certain indicators,

DOI: 10.4018/979-8-3373-0045-0.ch011

metrics, tools and strategies (Herremans, 2021). In light of the multifaceted circumstances and historical tendencies, the actionable (policy and strategy) procedure should be open, inclusive, and accountable to all stakeholders (Whittington, 2019). In order to promote sustainable energy use and mitigate climate change, modern energy policy driven by AI along with cutting-edge technologies is essential.

Energy policy includes a complete set of statutes, regulations, norms and strategies that regulate energy production, delivery and usage with little room for discretion (Danish et al., 2020). By defining objectives and goals, allocating resources, and creating criteria related to energy activities, it tackles a number of concerns, including security of energy, sustainability, accessibility and affordability, etc. The growth of energy industries and the introduction of novel technologies are influenced by energy policy. The targets for the generation and use of energy, efficiency incentives, rules governing renewable energy companies, assistance for development and research, along with additional areas are all included.

Promoting equality, fairness, ownership, and inclusivity in the creation of policies regarding energy and project decision-making is essential to avoid the broad implications that an ineffectiveness implementing energy policies would have on society. Throughout the policy's lifespan, usage of energy as well as climate goals must be taken into account, and prompt and efficient actions are required to avoid failures in policy and guarantee successful implementation (Sokołowski & Heffron, 2022).

Research at the leading edge of AI is still focused on analyzing a complete roadmap to overcome the deficiencies integrating energy policies including advance use of machine learning (ML) in energy policy while adhering to sustainability standards. In order to create a new paradigm, this study carefully examines international managerial practices, the principles of sustainability, and effectiveness (geopolitical, technical, ecological, etc.) (Rizos & Bryhn, 2022). The main goals are to pinpoint important elements, provide workable measures and indicators, and use cutting-edge technologies and methodologies to examine the incorporation of AI in energy laws from multiple angles.

Since the beginning of the Industrial Revolution, the world's economy has grown rapidly due to the depletion of natural resources; in 2019, the world's material consumption tripled from 1970 to a total of 96 billion metric tons. The authors suggested an economy based on circularity that emphasizes limiting, recycling, reusing and recuperating resources being a solution of worldwide financial sustainability in order to strike a balance between the demands of material upgrading and low-carbon as well as limited resource demands. But just 9% of the material in world demand is currently recycled, underscoring the requirement to close the gap that exists in research, policy, as well as technology (Li & Xu, 2022).

In (Sotiriou & Zachariadis, 2021), an optimization with multiple goals paradigm is presented to give decision makers a better understanding about the trade-offs among stronger decarbonization targets and increased expenses in the EU. The results show that the European Union's Effort Sharing industries can reduce by up to 35%, which is the highest amount that can be accomplished with significant social benefits. To achieve this, however, expenditures and public spending are needed to apply a certain policy mix strategy. In the energy industry, computer-aided decisions through machine learning and AI are becoming more and more necessary (Entezari et al., 2023). According to this study, artificial intelligence (AI) and machine learning have a lot of potential for optimizing renewable energy sources to satisfy growing demand while conserving resources.

According to a case study on AI deployment in energy industry published in (Chawla et al., 2022), the energy sector is under stress as a result of not utilizing AI to its full potential. This report outlined obstacles and problems while analyzing the impact of "AI and information management" in India's energy transformation. According to the findings of this study, there are generally not many incentives favouring AI in energy industry, indicating that policymakers are taking adaptive measures in the direction of AI integration. Meanwhile, a thorough AI strategy is required due to the energy sector's shift and large data access. In (Zhang et al., 2022), the possibility of employing energy from renewable sources to satisfy energy demands and make the switch to economically and technically possible sources of clean energy is examined. The article discusses a number of policy suggestions to aid with the transition, such as boosting funding for renewable energy, creating favourable legal frameworks, and building infrastructure for renewable energy. All things considered, these studies emphasize how crucial it is to take into account the social, environmental, and economic factors when creating sustainable policies. They also show how machine learning and AI can be used to enhance efficacy as well decision-making across a variety of domains, such as energy and sustainability management.

With an emphasis on the energy policy procedures, this study attempts to examine the opportunities and difficulties brought about through the integration of AI into the energy sector. This study specifically looks at how decisions are made at the development stage, how management of implementation is done during the process of execution, how machines learning, data science and deep learning are integrated into energy systems and how energy systems are needed in the larger picture of sustainability. A new framework for creating and executing contemporary policies regarding energy that reduce risks, encourage effective implementation, and further society's progress regarding net-zero as well as carbon-neutral goals will also be suggested by the study. Although earlier studies have looked at how AI can be used into energy industries and energy policy procedures a thorough examination of the prospects and difficulties of incorporating AI into every step of the energy policy development is lacking. Moreover, the current body of research fails to provide a comprehensive framework for creating and executing contemporary energy policies that support justice, fairness, equal rights, and inclusivity in project as well as energy policy decision-making. Thus, in order to bridge the gap and support the shift in energy use towards carbon-neutral and net-zero goals, this study incorporates policies, data science and artificial intelligence, engineering practices, business models, management procedures and social methodologies into a multilateral framework for design and reference.

This study provides guidance for policy formation, processes for making decisions, and comprehension of the intricate terrain of energy transitioning and decarbonization predictions. A brief description of the energy transformation and decarbonization landscape is given in Section 1. The process of developing policies and making decisions is then examined in Section 2. Various energy policy possibilities are covered in Section 3. "How will AI shape future energy policies?" is addressed in Section 4. Section 5 discusses policy-building blocks. The methods used in machine learning and artificial intelligence are highlighted in Section 6. The new framework for modern energy policy demands at the crucial juncture of the climate change is presented in Section 7. Section 8 concludes with the main conclusions and upcoming projects.

AN OVERVIEW OF THE ENERGY TRANSITION AND THE DECARBONIZATION SCENARIO

"Electricity and hydrogen", which are anticipated to emerge as the main energy sources, are knocking on the door of the global energy industry revolution. Studies (Tryggestad et al., 2022) predict that by 2035 and 2050, respectively, electricity and hydrogen would account for 32% and 50% of total energy consumption. Because of improvements in energy efficacy, consumption of electricity is only expected to increase by 14% despite an increasing population. Electricity's contribution to the total use of energy is expected to double from "20% to 40% by 2050" as a result of electrification, which is essential for increasing energy efficiency in critical sectors like housing, transportation and industry (Tryggestad et al., 2022).

Additionally, using hydrogen will balance out using fossil fuels. Due to rising living standards and electrification, demand for power is expected to treble by 2050, with appliances and building space cooling accounting for 60% of that demand (Tryggestad et al., 2022). Between 2035 to 2050, the production of green hydrogen will result in a 42% rise in electricity demand. With 50% of the electricity mix coming from renewables by "2030 and 85% by 2050", renewables are predicted to establish themselves as the new standard (Tryggestad et al., 2022). In order to tackle the energy along with climate change turmoil, a policy record-keeping mechanism is necessary, considering that present global emissions are considerably below the 1.5 °C range (Tryggestad et al., 2022).

Although availability of data and systematisms constraints can provide difficulties, climate change and energy data are essential for creating efficient energy policy (Zhang, Li, & Chen, 2023). In order to provide an adequate comprehension of the energy industry and its changing trends and to enable well-informed policy decisions, scenario-based as well as case-by-case assessments are crucial. In order to create a more efficient and environmentally friendly energy future, the study evaluates the effectiveness existing policies, makes plans for upcoming opportunities and challenges, considers policy solutions, and promotes cooperation among stakeholders (Al-Masri, Chenoweth, & Murphy, 2019).

According to the International Energy Agency (IEA) (Birol, 2022), by 2050, petroleum and other fossil fuels will merely make up 5% of the world's energy supply, with electricity becoming the main energy source and supplying more than half of all consumed power. The report anticipates massive rise in low-emissions vitality from wind and solar power. To reach the 90% reduction in emission target by 2050, the final-usage sectors would like to utilize hydrogen as well as fuels derived from hydrogen in heavy industries and transportation over long distances, as well as capture and storage of carbon dioxide in the trade and fuel transformation industry sectors. The report calls for an enormous jump in clean energy investments, with the investment in energy increasing from 2% to approximately 4% of the global GDP by 2030. In order to create a "clean technology supply chain", encourage coordinated growth, and send demand signals, governments must take further action. By 2030, there should be 90 million people employed in the energy sector overall. Even though financing prices may increase, clean technologies remain the most economical choice regarding new power supply. Three energy futuristic scenarios are examined by the IEA: "the Announced Pledges Scenario", "the Stated Policies Scenario", and "the Net Zero Emissions by 2050 Scenario".

DEVELOPING POLICIES AND THE PROCESS OF MAKING DECISIONS

The process of developing policies and making decisions in energy industry entails a number of processes, including agenda-setting, the policy design, adoption, evaluation and implementation, according to (Danish et al., 2019). Policymakers create potential fixes or policy choices during policy formation; they also determine and classify topics that need to be addressed at the time of agenda setting. Following adoption, an initiative is put into operation and then reviewed to determine its impact and efficacy. To produce good policy outcomes, the process necessitates feedback and contributions from a variety of stakeholders, careful analysis of financial, political and social variables, along with negotiations and compromises.

The idea of sustainability involves striking a balance between the economic, technical, institutional, social and environmental pillars in order to satisfy current demands without limiting those of the future (Entezari et al., 2023). Adaptability to changing patterns, sensitivity to situations, interpretative potential, accuracy, scientific as well as theoretical validity, simplicity and comparable practices with benchmarks are all requirements for demonstrating sustainable energy systems. Sustainability must be incorporated into the decision-making process at all stages of the energy system growth and development, which includes project, portfolios, objective-oriented, policy-oriented and problem-oriented viewpoints (Figure 1) (Tryggestad et al., 2022).

Figure 1. Steps towards energy policy growth.

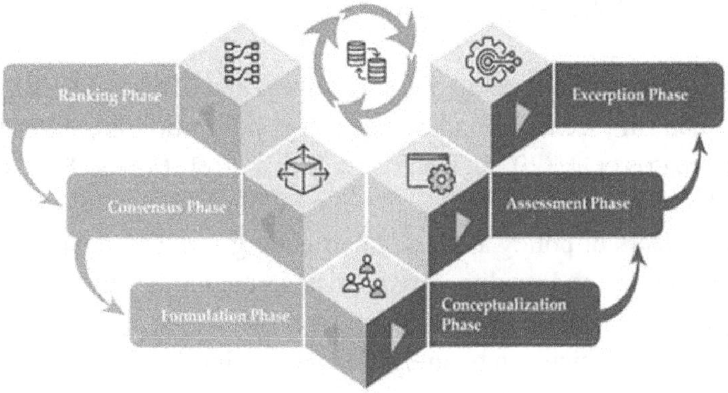

Every process needs to make decisions effectively, and this calls for a methodical approach that includes several stages, such as conceptualization, evaluation, excerption, the ranking, formulation and consensus, which are covered in the following sections:

- Phase of Conceptualization: During this stage, distinct and abstract concepts or phenomena are formed, and additional analysis is explored.
- Step of Assessment: In order to ensure compliance with the parameters of the system, this step uses pertinent data to analyze the framework as a whole.
- Excerption Phase: Identification and selection of clear-cut or evident solutions to exclude from additional cycle review are handled by the Excerption Phase.
- Phase of Ranking: To aid in the balancing process, this phase entails prioritizing the various decision options.
- Consensus Phase: When earlier stages agree on essential performance measures and weights, this phase is reached.
- Formulation phase: In the formulation phase, the process is synthesized and an organized structure is created using the process's raw inputs and parameters.

Sustainable policy along with process development necessitates a blend of technical, management, and policy competence. The best-worst approach, weights assessing, and the decision-making matrix are some of the multi-criteria methods for making decisions that have been presented relying upon the weighing technique (Rizos & Bryhn, 2022, Majeed et al., 2022). For certain objectives, other methods have been utilized as well, which include generic as well as qualitative frameworks for decision making (Chawla et al. (2022)).

Among the most prevalent approaches to decision-making revolves around the weighting technique, which may be illustrated in a number of ways depending on the goals (Krarti (2019)). The process of weighing is often divided into two categories: numerical value as well as verbal assessment. It occasionally requires further formulation, though, in order to fit the intended problem. (Gabr et al. (2020)) provides comprehensive formulas and supporting data pertaining to the weighing approach.

SCENARIOS FOR THE ENERGY POLICY

Leveraging functional domain methodologies within the realm of energy policy framework, Figure 2 examines high-level structuring scenarios underlying energy policy building-blocks (Zhang et al., 2022, Gabr et al. (2020), Linardatos et al. (2021), Kahloot & Ekler, 2021, (World Resources Institute, 2023), Feng et al. (2019)). The scenarios demonstrate the important roles that various functional areas may perform within the parameters of policy alongside particular goals. From the beginning of a policy to its monitoring, implementation, and modification, policy formation is a dynamic process rather than a fixed activity. Using functional domain standpoints, different scenarios offer a variety of possibilities and policymaking building blocks which can be integrated into a high-level structuring scenario. By using these methods, policymakers may comprehend the interactions between various components of energy system and pinpoint possible areas where the energy policy needs to be improved.

Figure 2. Functional domain techniques are used to analyze high-level structure scenarios concerning energy policy, demonstrating the importance of several domains along with the adaptability of policy making over the course of its lifetime.

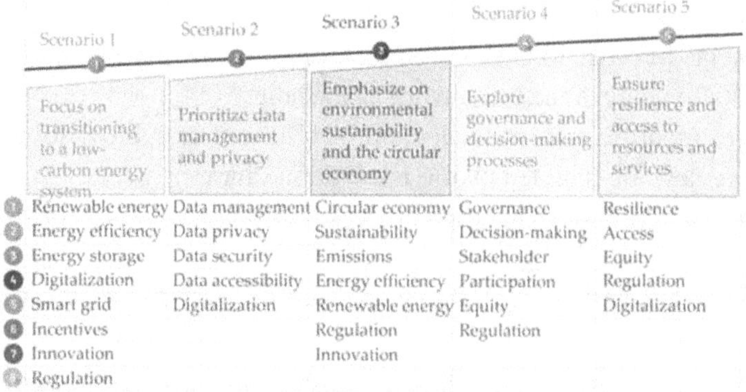

What Impact Will AI Have on Upcoming Energy Policies?

The creation and application of energy policies have a significant impact on how energy systems are shaped today as well as in years to come through important choices on the distribution of resources and technological advancement. Beyond its main purpose, energy policy information makes it easier to optimize energy systems with respect to the objectives of policy with data providing the input. This is done by defining the objectives and goals, incorporating policy limitations, and evaluating policy impacts using an organized roadmap. To make sure every aspect of the leading energy system operates in line with the goals, energy system optimization approaches can incorporate a vast number of other essential performance metrics of various policy scenarios, as well as the policy goals (such as reducing the ecological impact in order while minimizing the emissions of greenhouse gases) and constraints (such as standards for renewable portfolios that require a particular percentage of energy generated from renewable sources) (Arridge et al. (2019)).

Because of their intricate interdependencies along with non-linear patterns, energy systems are challenging and require considerable effort to model in the realm of artificial intelligence (AI) and machine learning implementation ((Sutton et al., (2019)). By employing data-driven techniques to enable "System-Parameter-Based" assessment, operation and dependability can be enhanced. However, the absence of standards to convert system-parameter-based frameworks into data-driven designs necessitates benchmarking across various approaches, tools as well as mechanisms. To assess performance, comparisons, and the ability to generalize it is essential to set up solid foundations and benchmarking procedures (Linardatos et al., (2021)). In order to align the energy sector from the very beginning of policy formation, it is imperative to switch to data-driven approaches and provide sufficient economic, technical, as well as human capital to sync with the rapidly expanding AI technologies. Among numerous machine learning approaches, techniques for feature engineering are frequently employed for data-driven analysis of models (Cho & Yang, 2022).

The process of constructing a data-driven machine learning model is hierarchically structured and iterative, with some steps being incorporated or repeated depending on the situation. Figure3 (Kahloot & Ekler, 2021, Ahmad et al., 2021, Anthopoulos & Kazantzi, 2022) lists the steps involved in creating an artificial intelligence (AI) model to assist energy policy, with the first step being determining which datasets are required. The next step is to plan up data collection and cleaning. By examining the gathered data, patterns and connections between the variables used for input and output can be found.

The gathered data can subsequently be used to train an appropriate data-driven system. It is advised to assess how well the model performed with a different dataset and make any required adjustments. Lastly, forecasts are anticipated, and predictions made by the model are used to produce optimization recommendations. The research's methods, findings, key findings, conclusions and recommendations are all summarized in the final stages of reporting including the highlight.

Figure 3. Shows how the approach for creating an AI-model targeting energy policy application is highly hierarchical.

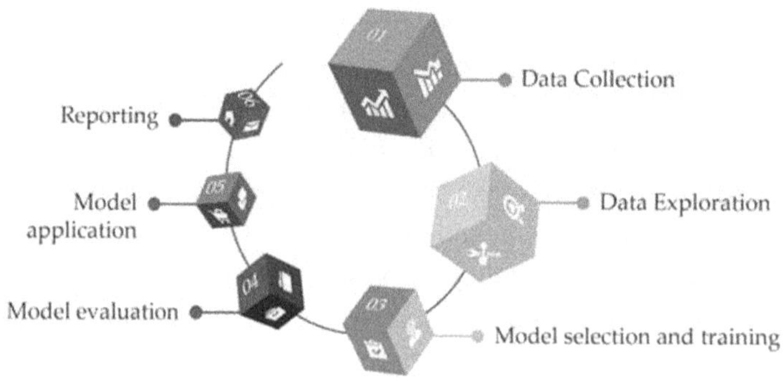

1. Data Collection:

Finding the datasets required for model testing and training is the first stage in creating a 'data-driven model' to provide energy facilities. This could entail formulating a strategy for gathering data, which could involve collaborating with different entities to gather data, using publicly accessible datasets, or producing a synthetic dataset. To guarantee accuracy, representativeness, and effectiveness, the dataset needs to be gathered and cleansed after it was successfully recognized.

2. Data Exploration:

Data exploration is the process of examining collected data to find trends and connections between input and output variables. This entails carrying out any preliminary processing on the data that is required, like standardization or feature scaling.

3. Model Selection and Training:

After the data has been examined, an appropriate data-driven model, like an artificial- neural-network, might be chosen to be developed based on the gathered data.

4. Model Evaluation:

A distinct dataset is needed to assess the performance of the model after it has been trained. The model's performance can be enhanced by fine-tuning it as needed.

5. Model Application:

Using the learned model to forecast and examine the variables in various contexts is the next stage. It is possible to create optimization suggestions that depend on the predictions made by the model.

6. Reporting:

The study's findings are finally compiled within a report which covers the techniques employed, the outcomes attained, and the conclusions obtained. A presentation that highlights the key conclusions and suggestions to pertinent stakeholders can also be developed. By taking these actions, an algorithm based on data may be created and applied to enhance efficiency in energy use and regulate energy facilities.

BUILDING BLOCKS FOR AI-DRIVEN POLICY DEVELOPMENT

Taking into account the five principles of sustainability i.e., technical, economical, organizational, environmental and social) as well as particular areas related to system data, policy building-blocks and involvement of stakeholders, a contemporary energy policy combined map reflects a comprehensive and cooperative method for energy policy along with AI implementation.

Efficiency, decarbonization, sustainability and the incorporation of renewable energy technology are the main focuses of the current energy policies that are being suggested. By using machine learning and artificial intelligence addresses to optimize energy systems towards optimal performance, efficacy, monetary savings, etc., 'AI-driven' energy policy goes one step further.

A number of crucial elements must work together harmoniously in order to effectively use AI to assist in the creation and execution of energy policies (Figure 4) (Pai & Chen, 2009, Busari & Lim, 2021). Stakeholder involvement to foster cooperation and address issues, regulatory and policy structures to guarantee compliance and transparency, data platforms as well as infrastructure for collecting and analyzing energy data, specialized AI mathematical models and algorithms to produce insights and suggestions and dissemination and communication networks to share findings with pertinent parties are some of these components. By integrating each component into a cohesive system, companies or governments may construct an AI-enabled power strategy that effectively and efficiently meets the interests of all stakeholders and guarantees a sustainable future of the energy sector.

Figure 4. Essential elements of a new strategy for AI success in the energy systems.

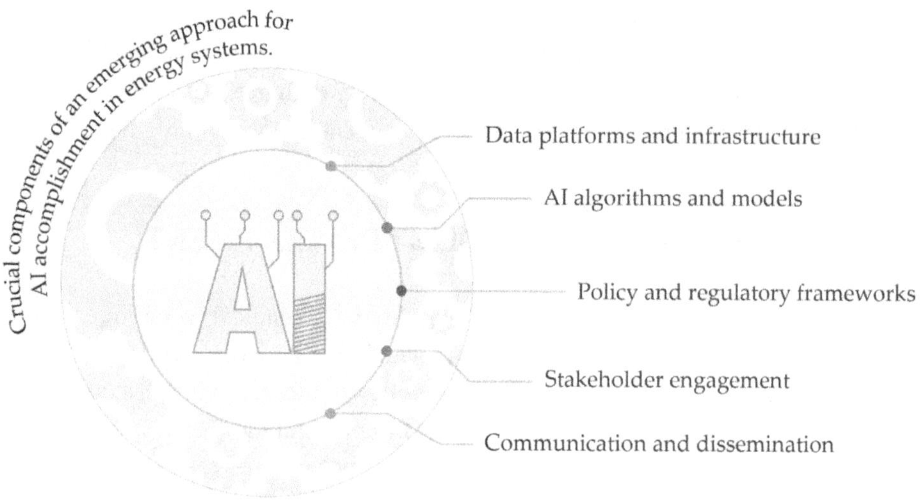

TECHNIQUES FOR INTEGRATING AI AND MACHINE LEARNING INTO ENERGY SYSTEMS

Data science, machine learning, AI, and advanced learning have all been covered in great detail in the literature via a variety of angles (Busari & Lim, 2021, Shi et al., 2022, Steinwandter et al., 2019, Omitaomu & Niu, 2021, Zheng et al. (2021), (Ge et al., 2019, Silvestrini & Lavagna, 2022, World Resources Institute, 2023). This classification, however, occasionally ignores the overlap and connections between specific techniques and resources, resulting in a generalization which might not adequately convey the subtleties and complexity of different approaches.

Four categories can be used to group machine learning techniques for use in energy industry. Classification is a component of class 1 techniques that aids in pattern recognition, grouping data, including pattern prediction. Class 2 includes regression techniques which foresee figures and are helpful for scenario optimization and forecasting. Class 3 stands for association analysis and the case of clustering which are used to find patterns by grouping related data points and determining links between variables. Class 4 approaches include generative technologies with multipurpose applications, networks including graph analysis and reinforcement learning, among other techniques.

Class 1

Classifications: Naive Bayes, decision trees, Support Vector Machines (SVM), neural networks, k-nearest neighbours and induction rules.

Class 2

Time series forecasting: Mean Absolute Percentage Error (MAPE), Autoregression (AR), Moving Average (MA), ARIMA (Auto Regressive Integrated Moving Average) and exponential-smoothing;

Regression: polynomial regression, logistic and linear regression and CART (Classification and Regression Trees).

Class 3

Association analysis: FP-growth, ECLAT (Equivalence Class Transformation), and a priori algorithms; clustering: hierarchical clustering, k-Means, XRF Spectral-Based Sorting (SBS), and Density-Based Spatial Clustering of Applications with Noise (DBSCAN); Detecting anomalies: LOF (Local Outlier Factor), one-class SVM, isolation forest, density-based, distance-based and Z-score; Feature selection: Random-forest, lasso regression, and recursive feature elimination; Ensemble approaches: Random-forest, AdaBoost and gradient boosting.

Class 4

Deep and transfer learning: Generative Adversarial Networks (GANs), Recurrent Neural Networks (RNNs), and Convolutional Neural Networks (CNNs) for deep and transfer learning;

Text mining: sentiment analysis, Term Frequency–Inverse Document Frequency (TF-IDF), and Natural Language Processing (NLP);

Computer vision: segmentation, object detection, and image categorization;

Natural language generation: OpenAI, OpenAI Gym, ELMo (Embeddings from Language Models) and GPT-3 (Generative Pretrained Transformer 3);

Reduction dimensionality: linear discriminant analysis (LDA), principal component analysis (PCA), and singular value decomposition (SVD);

Reinforcement learning: Deep Q-Networks (DQN), SARSA (State–Action–Reward–State–Action), and Q-Learning;

Generative techniques: Boltzmann machines, Generative Adversarial Networks (GAN), and Variational Autoencoders (VAE);

Networks and graph analysis: PageRank, community detection, link prediction, and centrality measures; and

Recommender systems: neighbourhood-based collaborative filtering, matrix factorization and deep learning-based recommendation systems.

THE SUGGESTED FRAMEWORK

One useful tool that helps policymakers make well-informed decisions upon the growth of energy policies is the paradigm for choosing energy policy alternatives. The suggested plan (derived from (Sokołowski & Heffron, 2022, Birol, 2022, Tryggestad et al., 2022) enables an organized strategy to making decisions based on particular demands and goals of the energy industry by prioritising policy objectives while also taking into account several eventualities. The plan offers a precise road map to develop energy policies that meet efficiency and sustainability standards while also taking into account the possible effects on different stakeholders and the community at large. By evaluating and ranking numerous objective possibilities, the suggested framework will help maximize policy goals and produce an energy strategy that is open, accountable and accessible. In the meantime, especially the domain of applications for artificial intelligence, having specific goals to serve as input/output helps hasten the transition from a 'system-parameter-based model to a data-driven model'.

1. Implementation-timeline:
    - Short-term: It is anticipated that the energy-related policy tools and technique will be used for less than a year and provide outcomes during that time. The implementation period of short-term remedies is less than a year. They frequently deal with urgent or immediate energy-related problems, such as load-shedding initiatives to lower peak demand or emergency power production after natural catastrophes.
    - Medium-term: The implementation duration for medium-term remedies ranges from one to five years. These solutions typically deal with intermediary energy-related problems and set the stage for longer-term fixes, like building along with appliance energy efficiency initiatives.
    - Long-term: Long-term fixes are put into place for durations longer than five years. These programs tackle long-term energy-related problems, such as implementing carbon sequestration and storage technology, creating large-scale projects utilizing renewable energy and switching to an energy system that is low in carbon.
2. Implementation-type:
    - Government-led: Mainly carried out by government organizations or agencies using rules, grants, or other types of government assistance, such as green energy portfolio norms, energy-efficient building codes as well as standards, etc.

      Industry-led: run by the private sector or companies through mutually beneficial agreements between sector and the government or market-based processes, such as feed-in tariffs or energy efficiency initiatives for appliances and buildings, etc.
    - Community-led: Run by local groups or communities that carry out community-based projects, such as energy cooperative or projects involving clean energy.
    - Public-private partnerships: Public-private partnerships are partnerships between the public and private sectors and between government agencies, local organizations and private enterprise. Examples of these partnerships include the deployment of technologies for smart grids and the growth of large-scale energy-efficiency projects.
3. Interaction level with other sectors:
    - Standalone: When one sector implements something without possibly affecting other sectors. initiatives including the creation of a smaller scale energy-efficient projects, building and appliance energy-saving programs, etc.
    - Cross-sectoral: a project's layout with several tasks or beneficiaries that will be carried out in coordination with other industry sectors. These remedies are put into practice with consideration for how they will affect various industries and with the intention of accomplishing certain goals. Utilizing energy storage for improving the energy grid's dependability and incorporating electric vehicles to the transport industry are examples of cross-sectoral solution.

- Multi-sectoral: Multiple sectors are involved, and different stakeholders are touched in different ways and simultaneously. These solutions, such as using smart grid technology to increase the electric-power grid's stability and integrating energy from renewable sources into industrial operations, are executed in cooperation with several sectors and aim to accomplish multiple goals.

4. Degree of Originality:
   - Incremental: Current setups or infrastructures gradually outperform current solutions. These remedies usually have a minimal degree of uncertainty and are low risk.
   - Radical: New technologies show a substantial divergence from preexisting approaches. Compared to incremental solutions, there is more uncertainty as well as risk associated with these solutions.
   - Disruptive: The focus of this class revolves around new or developing methods or technologies that upend established technology and market structures and drastically alter the energy systems. Although there is a great deal of danger and unpredictability associated with these remedies, they have the ability to change the energy sector and open up new opportunities. Implementing blockchain technology to supply energy transactions while incorporating AI for energy management processes are two examples of disruptive solutions.
5. Maturity Level:
   - Early stage: Having a lot of ambiguity and being in the early phases of development or deployment.
   - Mid-stage: Shows some degree of confirmation but has not yet been fully developed; at the halfway point phases of growth and following analysis of the solution.
   - Mature stage: Completely created and having been used and verified in a variety of contexts. These solutions are already shown to be successful while maintaining a low degree of uncertainty.
6. Scalability level:
   - Town-based renewable energy initiatives, energy efficiency programmes for the local buildings along with appliances, etc. are examples of local initiatives that are executed at a local level inside a particular town or city, suited to the unique demands of local area, and have an insignificant effect on the larger region (country).
   - Regional: Designed for particular state or province at a regional level, with a greater influence on the region at large; for example, the implementation of the smart grid technology in a particular area, the emergence of small-scale clean energy projects, etc.
   - National: Implemented nationally with a major influence on energy system and implemented through national regulations and policies, such as energy conservation criteria for buildings and equipment and clean energy spectrum norms, etc.
   - International: Provides a global perspective over a number of nations, pursuing a global influence through international frameworks and agreements, such as the implementation of carbon storage and capture technology, international collaboration on the climate and energy policy, etc.
7. Complexity level:
   - Simple: Low complexity, easy to comprehend and implement, easy to repeat, with fewer components, such as campaigns to raise awareness and energy-effective lighting systems.

- Moderate: Having more elements and possibly requiring greater assets to implement and manage, such as energy efficiency initiatives for appliances and buildings, the creation of small-scale initiatives using renewable energy, etc.
- Complex: Extremely complicated, involving numerous elements, and requiring substantial resources for both implementation and upkeep, such as the creation of extensive clean energy initiatives or the application of smart grid technological advances, among other things.

8. Flexibility-Level:
   - Rigid: A rigid system is one that has set rules and characteristics that are difficult to modify rapidly and aren't as flexible in response to shifting conditions.
   - Adaptable: The term "adaptable" describes a system that is flexible enough to be modified to meet particular demands and more responsive to changing conditions.
   - Resilient systems are capable of withstanding perturbations to energy systems and recovering robustly from them.
9. Automation Level:
   - Manual: Manual operations are less efficient and may be more expensive since they involve human participation.
   - Semi-automatic: More effective than manual yet possibly less expensive; has some automation though still needs human participation to operate.
   - Automatic: Because of its high output and minimal error rates, automation with little human involvement is perhaps the most economical and efficient approach.
10. Regulation Level:
    - Regulated: Under government control and supervision, enacted by laws and directives, such as building energy standards and codes, which are regularly checked to guarantee adherence.
    - Voluntary: Describes a mutually agreed-upon unilateral agreement between the government and industry that is carried out through market-based procedures and self-regulation, such as energy-efficiency initiatives for appliances and buildings.
    - Hybrid: Uses a mix of market-based processes, discretionary agreements, and government mandates with incentives to demonstrate an integration of regulating and voluntary actions.

The suggested paradigm analyzes how AI will be incorporated into current and future systems of energy using a multifaceted approach. The suggested framework looks at a number of impacting factors from different angles. This method makes it possible to comprehend the process of developing and implementing energy policies more thoroughly, taking into account elements like the implementation's duration, degree of communication with other industries, type, degree of innovation, regulation, scalability, complexity, flexibility and standardized procedures. Additionally, the framework makes it easier to dissect the system of energy into smaller, easier-to-manage parts, which allows for a deeper comprehension of the mechanism and the transition from a parameter-based to a data-driven approach. As a result, system functionality and planning become more accurate. The framework additionally recognizes areas for further study and research as well as possible synergies. Although the suggested model has its foundation on earlier ideas (Sokołowski & Heffron, 2022, Birol, 2022, Tryggestad et al., 2022), it has been updated and enhanced with current developments as well as potential applications for artificial intelligence and machine learning.

The suggested framework creates connections between the most important domains, offering a methodical roadmap in a mutually beneficial manner. At different phases of planning, in the constraints of diverse scenarios, this framework allows for the possible implementation and assessment of the necessary resources (budget, human, tools and techniques, procedures as well as methodologies). In fact, each one of these different levels are abstract concepts interactions that compete or overlay in order to accomplish a purpose, depending on how the circular economy works circulating in the larger setting of response and refurbishing policy review and adjustment. This framework draws out the complexity of interaction by level-utilizing realistic events and attempts to creatively integrate domains. Several metrics and indicators can be established at various levels to adjust to the strategy objectives while making the most use of technology, particularly artificial intelligence, and resource allocation. Integrated 'strategic policy decision-making' efforts can result from the deployment of novel technology and the emergence of strategic procedures, sustainability needs, and efficiency standards.

CONCLUSIONS AND FUTURE DIRECTIONS

This study offers a multifaceted examination of the creation of energy policies and the incorporation of AI possibilities. It highlights the value of a multidisciplinary strategy as well as the part that policymakers along with energy specialists play in reaching long-term, sustainable choices. A hierarchical approach for creating an artificial intelligence (AI) model, a high-level structuring situation, and decision-making procedures are only a few of the subjects covered in the paper. Along with a decision-making paradigm system centered around policy goals and priorities, it also offers a novel and organized roadmap for the development of AI-enabled policies for energy. A multifaceted structure to facilitate scenario analysis and integration into current and upcoming energy dynamics is proposed in the present research. The framework thoroughly examines energy trends along with their possible effects on sustainability by taking into account a number of variables, such as policy, economics, technology, as well as social elements. All things considered, this study emphasizes the significance of sustainability-focused viewpoints in the creation of energy policies as well as the possible contribution of machines learning and AI to policymaking. Because of this, it has important practical applications and is a global resource for scholars, practitioners, researchers, and students in connected subjects. The administration and implementation procedures associated with energy policies, issues with energy systems modeling techniques ranging from 'system-parameter-based' to 'data-driven-based models', the development of statistics and indices along with a comparison of the advantages and disadvantages of contemporary AI-enabled green energy implementation possibilities are just a few of research directions that the present research has delineated.

REFERENCES

Ahmad, T., Zhang, D., Huang, C., Zhang, H., Dai, N., Song, Y., & Chen, H. (2021). Artificial intelligence in the sustainable energy industry: Status quo, challenges and opportunities. [Google Scholar] [CrossRef]. *Journal of Cleaner Production, 289*, 125834.

Al-Masri, R. A., Chenoweth, J., & Murphy, R. J. (2019). Exploring the status quo of water-energy nexus policies and governance in Jordan. [Google Scholar] [CrossRef]. *Environmental Science & Policy, 100*, 192–204. DOI: 10.1016/j.envsci.2019.06.012

Anthopoulos, L., & Kazantzi, V. (2022). Urban energy efficiency assessment models from an AI and big data perspective: Tools for policymakers. [Google Scholar] [CrossRef]. *Sustainable Cities and Society, 76*, 103492. DOI: 10.1016/j.scs.2021.103492

Arridge, S., Maass, P., Öktem, O., & Schönlieb, C.-B. (2019). Solving inverse problems using data-driven models. [Google Scholar] [CrossRef]. *Acta Numerica, 28*, 1–174. DOI: 10.1017/S0962492919000059

Birol, F. (2022). *World energy outlook 2022*. International Energy Agency. (Google Scholar)

Busari, G. A., & Lim, D. H. (2021). Crude oil price prediction: A comparison between AdaBoost-LSTM and AdaBoost-GRU for improving forecasting performance. [Google Scholar] [CrossRef]. *Computers & Chemical Engineering, 155*, 107513. DOI: 10.1016/j.compchemeng.2021.107513

Chawla, Y., Shimpo, F., & Sokołowski, M. M. (2022). Artificial intelligence and information management in the energy transition of India: Lessons from the global IT heart. *Digital Policy* [Google Scholar] [CrossRef]. *Regulation & Governance, 24*(1), 17–29. DOI: 10.1108/DPRG-05-2021-0062

Chawla, Y., Shimpo, F., & Sokołowski, M. M. (2022). Artificial intelligence and information management in the energy transition of India: Lessons from the global IT heart. *Digital Policy* [SAGE Journals]. *Regulation & Governance, 24*(1), 17–29. DOI: 10.1108/DPRG-05-2021-0062

Cho, J., & Yang, S. (2022). Data-driven Wasserstein distributionally robust unit commitment for power systems under uncertainty of renewable generation. *arXiv preprint arXiv:2203.15333*. (arXiv)

Danish, M. S. S., Elsayed, M. E. L., Ahmadi, M., Senjyu, T., Karimy, H., & Zaheb, H. (2020). A strategic-integrated approach for sustainable energy deployment. [Google Scholar] [CrossRef]. *Energy Reports, 6*, 40–44. DOI: 10.1016/j.egyr.2019.11.039

Danish, M. S. S., Senjyu, T., Ibrahimi, A. M., Ahmadi, M., & Howlader, A. M. (2019). A managed framework for energy-efficient building. [Google Scholar] [CrossRef]. *Journal of Building Engineering, 21*, 120–128. DOI: 10.1016/j.jobe.2018.10.013

Danish, M. S. S., Senjyu, T., Zaheb, H., Sabory, N. R., Ibrahimi, A. M., & Matayoshi, H. (2019). A novel transdisciplinary paradigm for municipal solid waste to energy. [Google Scholar] [CrossRef]. *Journal of Cleaner Production, 233*, 880–892. DOI: 10.1016/j.jclepro.2019.05.402

Entezari, A., Aslani, A., Zahedi, R., & Noorollahi, Y. (2023). Artificial intelligence and machine learning in energy systems: A bibliographic perspective. [Google Scholar] [CrossRef]. *Energy Strategy Reviews, 45*, 101017. DOI: 10.1016/j.esr.2022.101017

Entezari, A., Aslani, A., Zahedi, R., & Noorollahi, Y. (2023). Artificial intelligence and machine learning in energy systems: A bibliographic perspective. [Taylor & Francis Online]. *Energy Strategy Reviews*, *45*, 101017. DOI: 10.1016/j.esr.2022.101017

Feng, Y., Cui, N., Chen, Y., Gong, D., & Hu, X. (2019). Development of data-driven models for prediction of daily global horizontal irradiance in Northwest China. [Google Scholar] [CrossRef]. *Journal of Cleaner Production*, *223*, 136–146. DOI: 10.1016/j.jclepro.2019.03.091

Gabr, A. Z., Helal, A. A., & Abbasy, N. H. (2020). Economic evaluation of rooftop grid-connected photovoltaic systems for residential buildings in Egypt. [Google Scholar] [CrossRef]. *International Transactions on Electrical Energy Systems*, *30*(12), e12379.

Ge, S., Isah, H., Zulkernine, F., & Khan, S. (2019). A scalable framework for multilevel streaming data analytics using deep learning. arXiv preprint arXiv:1907.06690. DOI: 10.1109/COMPSAC.2019.10205

Herremans, D. (2021). aiSTROM—A roadmap for developing a successful AI strategy. *arXiv preprint arXiv:2107.06071*. (arXiv)

Järvelä, M., & Juhola, S. (2011). *Energy, policy, and the environment: Modeling sustainable development for the North*. Springer Science & Business Media. (Google Scholar)

Kahloot, K. M., & Ekler, P. (2021). Algorithmic splitting: A method for dataset preparation. *IEEE Access : Practical Innovations, Open Solutions*, *9*, 125229–125237. DOI: 10.1109/ACCESS.2021.3110745

Kahloot, K. M., & Ekler, P. (2021). Algorithmic splitting: A method for dataset preparation. [Google Scholar] [CrossRef]. *IEEE Access : Practical Innovations, Open Solutions*, *9*, 125229–125237. DOI: 10.1109/ACCESS.2021.3110745

Krarti, M. (2019). Evaluation of energy efficiency potential for the building sector in the Arab region. [Google Scholar] [CrossRef]. *Energies*, *12*(22), 4279. DOI: 10.3390/en12224279

Li, J., & Xu, G. (2022). Circular economy towards zero waste and decarbonization. [Google Scholar] [CrossRef]. *Circular Economy*, *1*(1), 100002. DOI: 10.1016/j.cec.2022.100002

Li, W., Lu, C., & Zhang, Y.-W. (2019). Prospective exploration of future renewable portfolio standard schemes in China via a multi-sector CGE model. [Google Scholar] [CrossRef]. *Energy Policy*, *128*, 45–56. DOI: 10.1016/j.enpol.2018.12.054

Linardatos, P., Papastefanopoulos, V., & Kotsiantis, S. (2021). Explainable AI: A review of machine learning interpretability methods. *Entropy (Basel, Switzerland)*, *23*(1), 18. DOI: 10.3390/e23010018 PMID: 33375658

Majeed, A., Ahmad, M., Rasheed, M. F., Khan, M. K., Popp, J., & Oláh, J. (2022). The dynamic impact of financial globalization, environmental innovations, and energy productivity on renewable energy consumption: Evidence from advanced panel techniques. *Frontiers in Environmental Science*, *10*, 894857. DOI: 10.3389/fenvs.2022.894857

Omitaomu, O. A., & Niu, H. (2021). Artificial intelligence techniques in smart grid: A survey. [Google Scholar] [CrossRef]. *Smart Cities*, *4*(2), 548–568. DOI: 10.3390/smartcities4020029

Pai, P.-F., & Chen, T.-C. (2009). Rough set theory with discriminant analysis in analyzing electricity loads. [Google Scholar] [CrossRef]. *Expert Systems with Applications, 36*(3), 8799–8806. DOI: 10.1016/j.eswa.2008.11.012

Rizos, V., & Bryhn, J. (2022). Implementation of circular economy approaches in the electrical and electronic equipment (EEE) sector: Barriers, enablers and policy insights. [Google Scholar] [CrossRef]. *Journal of Cleaner Production, 338*, 130617. DOI: 10.1016/j.jclepro.2022.130617

Rizos, V., & Bryhn, J. (2022). Implementation of circular economy approaches in the electrical and electronic equipment (EEE) sector: Barriers, enablers, and policy insights. [Taylor & Francis Online]. *Journal of Cleaner Production, 338*, 130617. DOI: 10.1016/j.jclepro.2022.130617

Shi, Z., Zhu, J., & Wei, H. (2022). SARSA-based delay-aware route selection for SDN-enabled wireless-PLC power distribution IoT. [Google Scholar] [CrossRef]. *Alexandria Engineering Journal, 61*(7), 5795–5803. DOI: 10.1016/j.aej.2021.11.029

Silvestrini, S., & Lavagna, M. (2022). Deep learning and artificial neural networks for spacecraft dynamics, navigation, and control. [Google Scholar] [CrossRef]. *Drones (Basel), 6*(9), 270. DOI: 10.3390/drones6100270

Sokołowski, M. M., & Heffron, R. J. (2022). Defining and conceptualising energy policy failure: The when, where, why, and how. [Google Scholar] [CrossRef]. *Energy Policy, 161*, 112745. DOI: 10.1016/j.enpol.2021.112745

Sotiriou, C., & Zachariadis, T. (2021). A multi-objective optimisation approach to explore decarbonisation pathways in a dynamic policy context. [Google Scholar] [CrossRef]. *Journal of Cleaner Production, 319*, 128623. DOI: 10.1016/j.jclepro.2021.128623

Steinwandter, V., Borchert, D., & Herwig, C. (2019). Data science tools and applications on the way to Pharma 4.0. [Google Scholar] [CrossRef]. *Drug Discovery Today, 24*(8), 1795–1805. DOI: 10.1016/j.drudis.2019.06.005 PMID: 31207205

Sutton, C., Ghiringhelli, L. M., Yamamoto, T., Lysogorskiy, Y., Blumenthal, L., Hammerschmidt, T., Golebiowski, J. R., Liu, X., Ziletti, A., & Scheffler, M. (2019). *Crowd-sourcing materials-science challenges with the NOMAD 2018 Kaggle competition. npj Computational Materials, 5, 111. (Google Scholar).* CrossRef.

Tryggestad, C., Sharma, N., Rolser, O., Smeets, B., Wilthaner, M., van de Staaij, J., Gruenewald, T., Noffsinger, J., & Tiemersma, L. (2022). *Global energy perspective 2022: Executive summary*. McKinsey & Company. (Google Scholar)

Tryggestad, C., Sharma, N., Rolser, O., Smeets, B., Wilthaner, M., van de Staaij, J., Gruenewald, T., Noffsinger, J., & Tiemersma, L. (2022). *Global energy perspective 2022: Executive summary*. McKinsey & Company. (MDPI)

Whittington, R. (2019). *Opening Strategy: Professional Strategists and Practice Change, 1960 to Today*. Oxford University Press. (Oxford Academic) DOI: 10.1093/oso/9780198738893.001.0001

World Resources Institute. (2023). *U.S. clean energy transition: In 2023, the United States installed a record 31 GW of solar energy capacity, a 55% increase from 2022, making solar the fastest-growing power source in the country*. World Resources Institute. (World Resources Institute)

World Resources Institute. (2023). *U.S. clean energy transition: In 2023, the United States installed a record 31 GW of solar energy capacity, a 55% increase from 2022, making solar the fastest-growing power source in the country*. World Resources Institute. (World Resources Institute)

Zhang, M., Tang, Y., Liu, L., & Zhou, D. (2022). Optimal investment portfolio strategies for power enterprises under multi-policy scenarios of renewable energy. [Google Scholar] [CrossRef]. *Renewable & Sustainable Energy Reviews*, *154*, 111879. DOI: 10.1016/j.rser.2021.111879

Zhang, W., Li, J., & Chen, Y. (2023). Policy simulation and economic impact analysis of renewable energy deployment in Northwest China. *Energy Reports*, *9*(1), 1345–1362.

Zheng, T., Chen, G., Wang, X., Chen, C., Wang, X., & Luo, S. (2021). *Real-time intelligent big data processing: Technology, platform, and applications*. arXiv preprint arXiv:2111.11872. (arXiv)

Chapter 12
ESG in Practice:
Sustainability, Ethics, and Innovation Tailored to Sectors

Tímea Veres
https://orcid.org/0009-0006-2094-7076
Pannon University, Hungary

Júlia Tobak
https://orcid.org/0009-0008-8570-0618
Pannon University, Hungary

Georgina Tóth-Nagy
https://orcid.org/0000-0003-4535-3494
Pannon University, Hungary

ABSTRACT

Environmental, social and governance (ESG) principles have become critical to corporate strategies across industries. However, ESG requires sector-specific approaches as industries face unique challenges. This chapter examines ESG applications in four sectors: finance, energy, fashion and food. In finance, ESG factors influence investment decisions and risk management with institutions evaluating companies' performance in addressing climate risks, regulatory challenges and social concerns. The energy sector, a major carbon emitter, prioritizes renewable energy transitions and emission reductions, alongside employee safety and regulatory compliance. The fashion industry, responsible for 4-8% of global greenhouse gas emissions, faces scrutiny over environmental impacts, fair wages and ethical supply chains. In the food sector, ESG concerns focus on water use, waste management, animal welfare and food safety. The chapter emphasizes sector-specific ESG challenges and the role of emerging technologies, such as machine learning and AI, in improving transparency and accountability.

DOI: 10.4018/979-8-3373-0045-0.ch012

1. INTRODUCTION

The ESG framework, or the trio of Environmental, Social, and Governance concerns, has turned into a defining element in modern business strategy and the very essence of how organizations go about their operations. It guides companies toward environmental sustainability, social responsibility, and ethical governance. Such a change has been driven by climate change, growing social inequalities, and increasing demands for more transparency and accountability in corporate governance. Various researches indicate that those firms that adhere to ESG standards tend to offer better financial performance in the long run, especially when their operations are in tandem with global goals such as the UN Sustainable Development Goals (Rane et al, 2024; De Silva et al, 2024, Managi et al, 2024).

While founded in the early 2000s, especially through the United Nations Principles for Responsible Investment of 2006 that urged the inclusion of sustainability factors in their investment processes among investors (UN PRI, 2024), today ESG goes beyond a niche framework to one imposed upon countries across the world through its adoption and promotion by government administrations and global associations. The European Union, for example, enacted the CSRD for enhanced transparency and consistency in disclosures regarding ESG topics (European Commission, 2024c).

Beyond regulatory requirements, industry leaders such as Unilever and Tesla have illustrated how ESG principles drive innovation, foster competitive advantages, and underscore the economic benefits of sustainable practices (Valette, 2024). Additionally, research highlights a growing consumer preference for ethical and sustainable brands, further embedding ESG as a critical component of corporate strategy (Liu et al., 2023).

The adoption of ESG is increasingly influenced by stakeholders who prioritize accountability and sustainability (Raghav et al., 2024) (Kaushik et al., 2024). For example, investors are integrating ESG indicators into their valuation processes, recognizing that strong ESG performance reduces risk and increases investment returns. BlackRock's initiatives to make ESG compliance the basis of investment decisions illustrate this trend (Zhang & Chang, 2024). Similarly, employees seek to align their organization's values with their own ethical principles, which leads to a greater sense of support for diversity, equity, and inclusion in the workplace (De Silva et al., 2024).

Consumer expectations play a significant role as well. Studies indicate that over 60% of consumers favor brands with robust sustainability practices, prompting companies to adopt ESG frameworks (Managi et al., 2024). Similarly, supply chain partners face heightened scrutiny as organizations apply ESG criteria to ensure ethical sourcing and maintain operational integrity (Liu et al., 2023).

Application methods differ significantly across industries due to unique challenges and priorities, while regional differences affect how ESG is applied. In more regulated regions like the European Union, the demand for compliance is greater than in less regulated markets (De Silva et al., 2024). For that reason, flexibility should be allowed to balance global objectives against local specificities.

Technology plays a key role in the implementation of ESGs. Artificial intelligence enables efficient data collection and analysis, facilitating ESG compliance (Rane et al., 2024). Blockchain technology helps to verify sustainability claims and reduce risks by increasing transparency, especially in sectors with complex supply chains (Liu et al., 2023; Valette, 2024). IoT (Internet of Things) applications provide real-time insights into resource use, especially in agriculture and logistics (Vieru & Plugge, 2025).

ESG frameworks are not just regulatory obligations, but also act as catalysts for value creation and innovation. Integrating ESG principles not only improves operational efficiency, but also builds stronger relationships with stakeholders and creates new growth opportunities (Managi et al., 2024). Companies such as Patagonia and Unilever have successfully demonstrated how ESG principles can be aligned with profitability (Leogrande, 2024).

The ESG framework is one of the most significant ways that business is rising to big challenges in sustainability, accountability, and governance. Solutions designed to their strategies-enabled by industry priorities and technological innovation-unlock their full potential in ESG. As these start to be implemented, it will help companies further develop long-term resilience, innovation, and positive social impact.

One-size-fits-all approaches to the diversities in ESG priorities simply just do not work. Every sector needs its tailored strategy toward its peculiar challenge. As an example, it could include the reduction of emissions within the energy sector or expansion of investment decisions within financials. On the other hand, ethical sourcing and reduction of textile waste are topics of interest for the fashion industry.

2. ESG ACROSS SECTORS

The integration of ESG criteria is increasingly becoming a priority in the financial sector, especially in investment portfolios and lending practices. The increasing popularity of green bonds and climate finance underlines the sector's significant role in promoting global sustainability.

Decarbonization and transition to renewable energy are at the top of many agendas in the energy industry today. More and more large companies are making commitments to achieve net zero by 2050, supported in large part by innovations in solar, wind, and green hydrogen.

Meanwhile, the fashion world is targeted on fast fashion's environmental impacts and increasingly introduced to the circular economic model. For example, Patagonia pioneered material recycling and supply chain transparency.

The food industry also has very significant challenges, mainly related to resource efficiency, ethical sourcing, and reduction of waste. It is about time for companies to embrace innovative strategies that could promote water conservation and reduction in methane emissions, which will go a long way in addressing climate change.

Table 1 outlines the key ESG aspects for each ISIC sector (UN, 2024), emphasizing the environmental, social, and governance factors unique to the respective industries.

Table 1. Sector-specific ESG priorities

	ENVIRONMENTAL ASPECTS	SOCIAL ASPECTS	GOVERNANCE ASPECTS
AGRICULTURE, FORESTRY AND FISHING	Sustainable land use, biodiversity conservation, water management	Workplace safety, protection of community rights	Transparent supply chains, GMO regulation
MINING AND QUARRYING	Land impact management, pollution control, water protection	Worker safety, community engagement	Corruption-free operations, compliance with regulations
MANUFACTURING	Energy efficiency, waste recycling, pollution reduction	Employee training, non-discrimination	Ethical standards, responsible supply chains
ELECTRICITY, GAS, STEAM AND AIR CONDITIONING SUPPLY	Renewable energy, CO_2 emission reduction, grid efficiency	Access to energy for low-income groups	Transparent tariffs, regulatory compliance
WATER SUPPLY, SEWERAGE, WASTE MANAGEMENT	Sustainable water use, waste recycling, wastewater treatment	Safe water supply for communities	Regulatory compliance

continued on following page

Table 1. Continued

	ENVIRONMENTAL ASPECTS	SOCIAL ASPECTS	GOVERNANCE ASPECTS
CONSTRUCTION	Green building technologies, sustainable materials	Workplace safety, community involvement	Permit compliance
WHOLESALE AND RETAIL TRADE	Eco-friendly products, energy-efficient stores	Respect for workers' rights	Ethical marketing and sales practices
TRANSPORTATION AND STORAGE	Low-carbon vehicles, logistics efficiency	Improved travel safety	Compliance with transportation standards
FINANCIAL AND INSURANCE ACTIVITIES	Promoting sustainable investments	Financial inclusion, customer data protection	Risk management, transparency
EDUCATION	Energy management	Promoting ESG education, accessible education	Equality in education
HUMAN HEALTH AND SOCIAL WORK ACTIVITIES	Safe waste management	Accessible healthcare for all	Ethical standards in healthcare
ARTS, SPORTS AND RECREATION	Sustainable event management	Audience and community engagement	Ethical partnerships, ethical talent management

Source: Own editing (2024)

The remaining part of the chapter describes ESG considerations at an industry level, where every industry represents one or multiple ISIC sectors according to the following:

- Financial sector belongs to the ISIC classification K - Financial and Insurance Activities. This industry plays a pivotal role in fostering sustainable investments and ensuring financial inclusion while maintaining strong governance practices.
- Energy Sector: D. This category includes Electricity, Gas, Steam, and Air Conditioning Supply. It is one of the most important sectors to adopt renewable energy, reduce emissions, and adhere to regulations.
- Fashion Industry falls within C (Manufacturing) of the ISIC classification. It addresses key ESG concerns such as sustainable material sourcing, waste reduction, and ethical labor practices.
- Food Industry: Included under A (Agriculture, Forestry, and Fishing) and partly under C (Manufacturing) for food processing. This industry grapples with biodiversity conservation, water management, and fair working conditions.

Each industry represents a diverse set of ESG challenges and opportunities, and their in-depth analysis will offer valuable insights into sector-specific sustainability strategies.

Building on the key highlights, the following sections will provide a more detailed analysis of the specific ESG challenges and opportunities within the previously mentioned sectors.

2.1. Financial Sector

In the financial sector, ESG factors have a significant impact on investment decisions. Credit institutions take into account a company's ESG performance when investing, lending and underwriting. The focus is on how well a company manages its ESG-related risks, such as climate change, adverse government actions or social issues that may have a direct financial impact. Reducing, preventing and mitigating environmental and economic damage and slowing climate change also entails significant challenges and expenditure. We need to know the burden that the current generation will have to shoulder in order to bequeath to future generations a more liveable world and a more favourable economic environment. Over the last decades, several models have been developed that allow cost-benefit analyses to be carried out to inform optimal climate policy. Some of these models build on the work of William Nordhaus, winner of the Nobel Prize, but others have also been developed based on different approaches.

Nordhaus examined the links between the global economy and human-induced climate change, with a greater focus on the climate change of the Earth due to fossil use. Based on the laws of chemistry and physics, Nordhaus created the IAMs model (integrated assessment models - IAMs) to show how the global economy and climate would evolve with the rate of warming and the policies (such as emission quotas) that would prevent it if governments acted against it. Using the Dynamic Integrated Climate Economy model (DICE), Nordhaus determined long-term economic expenditures and revenues and concluded that human welfare, including that of future generations, would be maximised if a carbon tax were introduced in all countries. Nordhaus estimated the social cost of coal in 2009 at US$16 per tonne by 2025, but recalculated it in 2016 and the cost of coal rose to US$44, almost three times that in ten years.

The main cause of climate change, global warming, is high carbon emissions, which are a negative externality in economic terms. In this case, the market cannot ensure a socially optimal distribution of wealth and regulatory intervention is needed. As a regulator, government has a direct influence on emissions through fiscal instruments such as quotas, taxation, or support for green investment and research and development. On the other hand, by limiting the quantity of pollutants. A sustainable financial system is needed to meet the climate change challenge. Appropriate green finance can stimulate and facilitate the transition to a low-carbon economy. The re-evaluation of financial instruments will also be a focus, made inevitable by the emergence of new technologies, changes in climate policy and thus environmental risks. The financial system should support climate policy by selecting sustainable investments (Carney, 2019).

Central banks also have an important role to play in creating a green financial system to effectively address the climate crisis. Fiscal policy and public regulation are direct instruments for addressing the climate crisis because of their direct impact on carbon emissions. This is why international organisations such as the IMF are urging action (Halmai 2023). On the fiscal side, the most supported instrument is the introduction of a carbon tax. Taxation can be an effective tool to achieve emission reductions, as it increases the price of fossil fuels, electricity and general consumer goods, reduces the price of fuel producers, and provides incentives to reduce energy use, switch to lower carbon energy production and green vehicles. The tax can be easily integrated into existing road fuel/revenue taxes, which are already well established in most countries and are among the easiest taxes to collect, and can be easily extended to other petroleum products, including coal and natural gas (Parry 2019).

The Paris Agreement of 2015 aims to keep temperature rise below 1.5-2 degrees Celsius, an initiative already signed by more than 190 countries and the European Union, including China and India as two of the three biggest greenhouse gas emitters. In many ways, the carbon tax can contribute to achieving the targets of the participating countries, which will have to update their commitments and report on progress every five years starting in 2020. Parry points out that the 2030 pledges are widely dispersed, so the carbon tax that is consistent with them can vary from country to country and that there are a number of different factors that influence the level of the tax. Tax elasticities are highest in the large carbon-consuming countries (China, India and South Africa), and IMF calculations show that for the vast majority of countries with more ambitious targets, even a tax of $70 would not be enough to meet the commitments. However, for those with a target of 25 per cent or less, a tax of $70 or $35 may be sufficient. And large carbon-using countries could achieve a 30 percent target with a carbon tax as low as $30, and even 40 percent with a higher tax (Figure 1).

Figure 1. Effectiveness of carbon taxes in meeting Paris Agreement pledges across countries

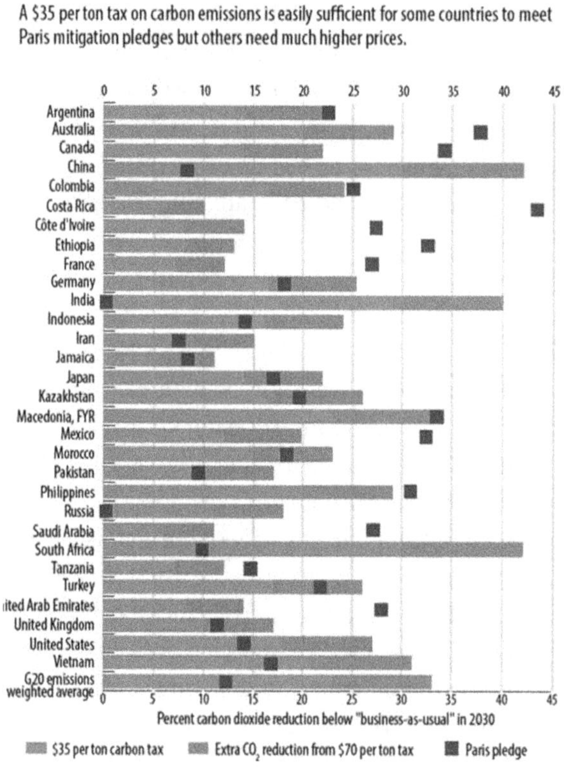

Source: Parry (2019)

It is important to note that a carbon tax could generate more significant tax revenues for public budgets. The revenue from a tax of $35 could typically amount to 1-2 percent of GDP in 2030 (Figure 2). However, it is estimated to increase the price of coal, electricity and gasoline by 100, 25 and 10 percent respectively. The introduction of a carbon pricing floor could also strengthen the achievement of the pledges. However, a global carbon tax of USD 75 is estimated to be needed to keep temperature rise below 2 degrees Celsius (Parry, 2019).

Figure 2. Revenue from carbon taxes across countries

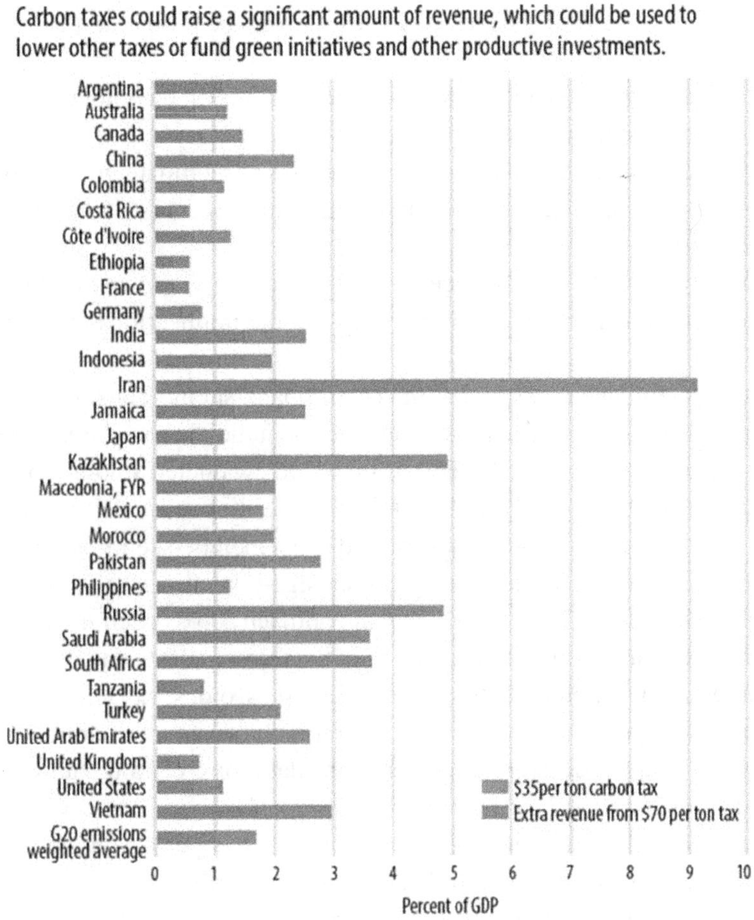

Source: IMF staff calculations.
Note: G20 = Group of Twenty.

Source: Parry (2019)

The 26th United Nations Climate Change Conference (COP 26), was held in Glasgow in 2021 with the participation of nearly 200 countries, and its main outcome was the adoption of the Glasgow Climate Pact, which not only complements the Paris Rulebook, but also keeps alive the hopes of keeping global warming below 1.5 degrees Celsius. The conference, chaired by the UK, identified four main areas to meet climate targets: 1) mitigation, 2) adaptation, 3) finance and 4) collaboration. The Climate Summit made progress in all four areas, including new commitments to achieve global net zero carbon and the first global agreement on financing adaptation.

The Network for Greening the Financial System (NGFS) is an international network for greening the financial system. The NGFS was established in December 2017 at the One Planet Summit in Paris as a result of a decision by eight central banks and supervisors. In November 2024, it will have 144 members and 21 observers. Their objective is to help achieve the goals of the Paris Climate Agreement by strengthening the role of the financial system in properly managing the risks posed by climate change and providing the resources needed to make the transition to a sustainable economy. NGFS members will share, on a voluntary basis, their expertise and good practices to develop climate and environmental risk management tools for the financial sector.

To achieve its objectives, the NGFS conducts its research in the following three working groups:

- Microprudential/Supervisory Working Group, which focuses on the integration and monitoring of climate change risks in supervisory activities.
- In the macro-financial working group, climate scenarios and climate guidelines for central banks and supervisors on how to integrate risk analysis into macroeconomic and financial stability supervision are developed
- In the Working Group on Green Finance, central banks are working on climate-related disclosures, portfolio management, market transparency and the implications for monetary policy.

Among the NGFS recommendations is that the central bank should consider sustainability aspects when designing its portfolio. There is also a growing focus on judicial procedures to enforce compliance with climate change considerations. The NGFS has therefore conducted studies on the extent to which climate-related court proceedings pose a risk to micro-prudential supervision and financial stability monitoring. It has carried out a survey on scenarios used by its members, entitled "Scenarios in Action", and published a guide for the preparation of climate change-related central bank communications in 2021. The handbook highlights the need for a different approach across central banks, rather than a single exact methodology, as central banks may differ in both their mandates and disclosure requirements.

The Bank for International Settlements (BIS), the world's oldest current international financial organisation and the main organiser of monetary cooperation between international banks, also recommends that, in addition to standard risk management considerations, central banks and major financial sector players such as pension funds should renew their portfolios based on environmental, social and governance (ESG) considerations. Financial companies have the expertise and experience to construct ESG-compatible portfolios and therefore central banks' guidance on ESG may have less of a role. Recent research (e.g. Elmalt et al., 2021) suggests that ESG-compatible portfolios so far seem to play a minor role in the green transformation, including the transition to zero carbon. The Magyar Nemzeti Bank was one of the first in the world to publish a climate risk report consistent with international methodologies (e.g. Task Force on Climate-related Financial Disclosures) (Kolozsi et al., 2022.)

According to research by Papoutsi and colleagues, companies that finance themselves with bonds typically have a larger physical capital stock and emit more carbon. The assets on the European Central Bank's 2017 balance sheet are mostly from sectors with relatively high carbon emissions (Papoutsi et al., 2021). These authors further argue, based on a general equilibrium theoretical model, that holding bonds in high-emitting sectors only leads to welfare gains if carbon emissions are not taxed in the model. However, when carbon emissions are taxed, welfare is clearly already highest when the central bank minimises its bond exposure to so-called "dirty" sectors. The model thus suggests that the authors find fiscal policy - taxing carbon emissions - to be more effective than monetary policy in achieving the zero emissions target.

The transition to a low-carbon economy entails risks for the financial sector. Banks provide loans in response to the financing needs of different sectors. The Carbon Footprint of Bank Lending (CFBL) provides a comparable measure of banks' exposure to transition risks (Figure 3). The higher this ratio, the higher the carbon intensity of banks' domestic loan portfolio.

Figure 3. Carbon footprint of bank loans

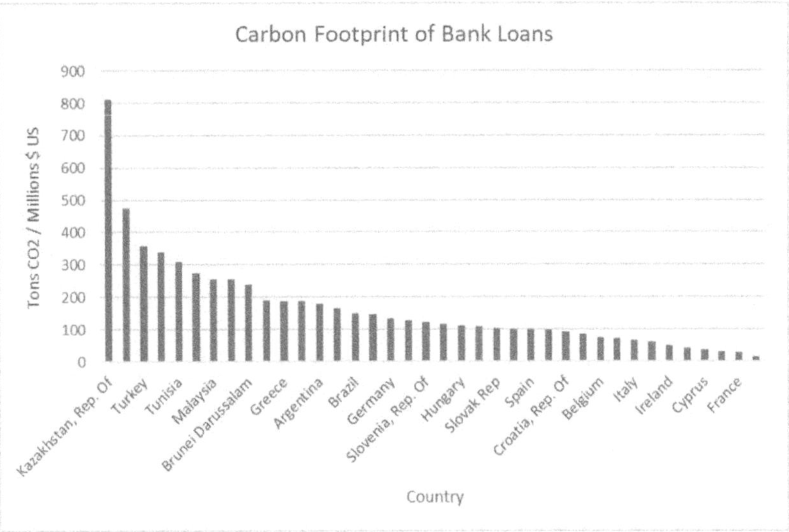

Source: Own editing according to IMF (2018)

Switzerland is at the bottom of the list with 13.3 tonnes CO_2 / million \$US, with Kazakhstan at the top with 809.8 tonnes CO_2 / million \$US. Bank loans finance industries with varying degrees of carbon intensity. The footprint of bank loans contributes to the carbon intensity of the loan portfolio.

Fried et al. have investigated, in a general equilibrium model, how uncertainty about US climate policies affects investment in instruments that seek to reduce or completely replace fossil fuel use. Uncertainty about the transition is already reducing carbon emissions and production by switching to less polluting technologies in the present (Fried et al., 2022)

2.2. Energy Sector

The energy industry plays a pivotal role in the global economy, shaping both climate and societal development. Its significance lies in providing essential energy resources that underpin modern life, while also being responsible for a substantial share of global carbon dioxide emissions. According to the Intergovernmental Panel on Climate Change (IPCC, 2023), the energy sector accounts for approximately 35% of global carbon emissions, primarily driven by fossil fuel combustion for electricity generation, industrial processes, and transportation. This makes the energy sector both a critical contributor to and a potential mitigator of climate change. Integrating ESG principles into the energy sector is, therefore, essential for building a sustainable future. According to the International energy Agency (IEA) global energy-related CO_2 emissions grew by 1.1% in 2023, increasing 410 million tonnes (Mt) to reach a new record high of 37.4 billion tonnes (Gt) (International Energy Agency, 2023) (Figure 4) .

Figure 4. Total increase in energy-related CO2 emissions between 1900-2023

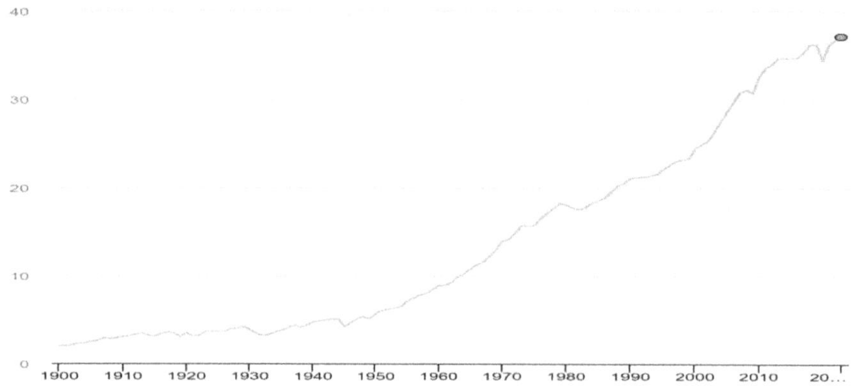

Source: IEA (2024), Total increase in energy-related CO2 emissions, 1900-2023, IEA, Paris https://www.iea.org/data-and-statistics/charts/total-increase-in-energy-related-co2-emissions-1900-2023, Licence: CC BY 4.0

The global energy sector has traditionally relied on fossil fuels such as coal, oil, and natural gas, which are major sources of greenhouse gas emissions. For decades, this reliance has driven economic growth and industrialization, but it has come at a significant environmental cost. The IPCC (2023) warns that without substantial reforms in the energy sector, the world risks surpassing the critical 1.5°C global temperature increase, leading to severe economic and ecological consequences. Large energy companies need to change. According to the Shell Energy Transition Strategy 2024 strategy, they have recognized their responsibility in addressing climate change and have set ambitious net-zero targets for 2050 (Shell, 2024). BP, too, is a significant player in the energy market that has set similar goals. This is crucial, as its operations would increase global warming by 2-3.2 degrees at present in the case of MSCI, 2024. Apart from voluntary commitments by companies, governments worldwide are setting up regulatory frameworks to reduce emissions, including the European Union ETS and the introduction of carbon taxes. The EU ETS is based on the principle of "cap and trade". The cap means the limit on the overall amount of GHG, which is allowed to be emitted by installations and operators covered under the scheme. According to the European Commission (2024), EU ETS covers electricity and heat generation, industrial production, aviation, and maritime transport. This cap is continuously reduced in line with the EU's climate change target. These measures have strengthened accountability and also created market incentives for the development of low-carbon energy solutions.

Transitioning to renewable sources of energy is one key element in the environmental aspect of ESG within the energy industry. Solar, wind, and hydropower, therefore, remain firm pivots in the development of decarbonization strategies. According to IRENA 2024, renewable sources are growing rapidly; for ten years, solar energy has experienced a growth rate of 25% annually. This shift reflects both technological advancements and the decreasing costs of renewable energy infrastructure. Governments and corporations are driving this transition through substantial investments and policy frameworks. For example, China leads the world in solar panel production and installation, while European countries such as Denmark and Germany have prioritized wind energy as a key component of their energy mix (Figure 5). Additionally, the development of green hydrogen, a clean energy carrier, is gaining momentum, with companies like Siemens Energy pioneering technologies that use renewable energy to produce hydrogen fuel (Siemens, 2024).

Figure 5. Share of electricity production from wind

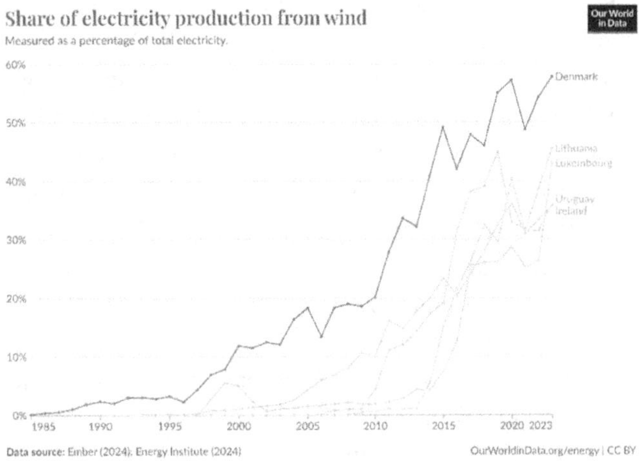

Source: Ritchie, H. (2023). Denmark's electricity has a larger share of wind power than any other country. Our World in Data. Retrieved from https://ourworldindata.org/data-insights/denmarks-electricity-has-a-larger-share-of-wind-power-than-any-other-country

While transitioning to renewable energy is critical, other strategies also play an essential role in emission reduction. Carbon capture and storage (CCS) technologies have emerged as a viable solution for mitigating emissions from hard-to-abate sectors like heavy industry and thermal power plants. ExxonMobil and Chevron are among the companies investing in CCS infrastructure to reduce their operational carbon footprints (ExxonMobil, 2024). Furthermore, energy efficiency measures, such as retrofitting industrial equipment and improving grid systems, are being implemented to optimize energy consumption and reduce waste. Smart grids are advanced electricity networks that use digital technologies, sensors, and software to optimize real-time supply and demand, reduce costs, and ensure grid stability and reliability (IEA, 2024). Smart grids help to integrate renewable energy more effectively, balance supply and demand, and minimize energy losses. Japan and the United States are implementing these technologies as part of their national energy strategies (IEA, 2024).

The social dimension of ESG in the energy sector focuses on worker safety and community engagement. In response to safety risks in energy production environments, companies have strengthened their safety protocols and invested in employee training. For example, ExxonMobil's advanced risk management systems have significantly reduced the number of workplace accidents over the past decade (ExxonMobil, 2024). Community participation is also an essential factor, as energy projects, especially large-scale renewable facilities, often impact local communities. Wind farms and solar plants require extensive land use, which can disrupt local ecosystems and displace communities. To address these concerns, energy companies are increasingly involving local stakeholders in project planning and decision-making. For example, Danish wind farm developers have established partnerships with local governments to share revenues and support community development initiatives (IRENA, 2024). Additionally, energy access initiatives align with the social goals of ESG by addressing energy poverty. In many developing countries, lack of access to affordable and reliable energy remains a significant barrier to economic and social progress. Programs such as India's Pradhan Mantri Ujjwala Yojana, which provides subsidized clean cooking fuel to low-income households, highlight the role of energy companies in advancing social equity (Managi, Shimamura & Tanaka, 2024).

Energy sector governance focuses on regulatory compliance, transparency and accountability. Increasing environmental regulations around the world require energy companies to adapt their operations to meet these requirements. For example, the European Union's Green Deal sets ambitious decarbonisation targets (e.g. The first climate-neutral continent by 2050, At least 55% less net greenhouse gas emissions by 2030, compared to 1990 levels or 3 billion additional trees to be planted in the EU by 2030), which require companies to disclose their ESG performance according to standard frameworks such as the EU's Corporate Sustainability Reporting Directive (CSRD) (European Commission, 2024b).

Accountability is further enhanced by technological innovations. Digital solutions such as blockchain and artificial intelligence are significantly transforming the governance practices of the energy sector. Blockchain technology has provided an immutable record of energy transactions, increasing transparency while reducing the potential for greenwashing and fraud. Artificial intelligence-driven analytics enable real-time monitoring of ESG performance, thus providing companies with areas for potential improvement and proactive measures they can take (Liu et al., 2023). The second most crucial development is that of digital twins: virtual models of physical systems. These models thus allow simulations of energetic processes, hence their usage in the enhancement of efficiency, predictive maintenance, and better sustainability impact assessment. For example, Shell has applied digital twin technology to optimize its oil rigs offshore. The outcome is cost savings with less environmental risk exposure, as noted by Shell in 2024.

ESG incorporation into the energy industry is equally fraught with challenges: the transitioning to renewable energy sources; the introduction of emission-reducing technologies, which are very capital intensive and especially hard for smaller firms to handle; and finally, regulatory differences across nations make standardization of ESG practices difficult, making the compliance burden unequal in different markets. At the same time, these challenges continue to inspire innovation and the spirit of collaboration among the players in this sector. Public-private partnerships in research and development of clean energy, via efforts like Mission Innovation, which gathers governments and private companies around the table. Collaboration among stakeholders, including investors, regulators, and non-governmental organizations, is critical to overcoming financial and logistical barriers (IRENA, 2024).

The future of ESG in the energy sector lies in accelerating the transition to clean energy while ensuring equitable outcomes for all stakeholders. As technology continues to advance, tools such as smart grids, AI, and blockchain will play an increasingly central role in enabling transparency, efficiency, and accountability. Moreover, the integration of circular economy principles, such as recycling materials from decommissioned solar panels and wind turbines, will further enhance the sector's sustainability.

2.3. Fashion Sector

The fashion industry is one of the contributors to the overconsumption of natural resources. Textile production requires significant amounts of water and land for growing cotton and other fibers. It is estimated that producing a single cotton T-shirt requires 2,700 liters of water — an amount equivalent to a person's drinking water needs for 2.5 years. According to a 2020 study by the European Parliament on the circular economy and the textile sector, the textile industry was the third-largest consumer of water and land that year. On average, the production of clothing and footwear for each EU citizen required 9 cubic meters of water, 400 square meters of land, and 391 kilograms of raw materials (Directorate-General for Communication of the European Parliament, 2020). The European Environment Agency reported that textile purchases in the EU in 2020 resulted in approximately 270 kg of CO_2 emissions per person Figure 6). This means that textile products purchased in the EU were responsible for the emission of 121 million tonnes of greenhouse gases.

By 2024, the fashion industry was already responsible for approximately 4-8% of global greenhouse gas (GHG) emissions. This raises the question of how much the fashion industry addresses ESG, not only in terms of the environmental pillar but also in relation to the social (S) and governance (G) pillars. This includes issues such as human rights, fair wages, and the elimination of exploitation (Basil-Jones 2024).

Figure 6. The environmental impact of textiles

(source: https://www.europarl.europa.eu/topics/en/article/20201208STO93327/the-impact-of-textile-production-and-waste-on-the-environment-infographics)

In recent decades, the fashion industry has also become a focal point for greenwashing, notably due to Nike's child labor scandal in the late 1980s. The scandal, once exposed, violated social norms, called the company's license to operate into question, and dealt a severe blow to its reputation. Similar to other clothing and sports equipment manufacturers, Nike's products were produced in subcontracted factories in Southeast Asia, which were not owned by the company. Due to the focus on low costs, workers in these factories were paid extremely low wages, typically between USD 1.25 and USD 2 per day. It was also common for minors to be employed in physically demanding labor. Similar scandals later surfaced with other companies, including the GAP clothing brand in 2007, as well as H&M, Zara, and Marks & Spencer.

In response to these scandals, these companies, either independently or through industry-wide collaboration, introduced control systems aimed at eliminating the use of child labor and requiring suppliers to pay workers at least the minimum wage (Braun 2015). It is evident that the fashion industry has shown progress in the S (Social) and G (Governance) pillars of ESG over the past decades. However, the environmental impact caused by the industry remains exceptionally high to this day.

According to a study by McKinsey (Remy et al., 2016), clothing production doubled between 2000 and 2014, while the number of clothing items per capita increased by approximately 60% during the same period. Bechu, the French Minister for Ecological Transition, stated at the 2024 G7 Summit that fast fashion is being given special attention, as the textile industry is already responsible for more emissions than the entire African continent (Dobos, 2024).

It is important to clarify what is meant by the fashion, apparel, or textile industry. The fashion industry is a broad category encompassing everything from the design, production, and advertising of products to their sale. This includes clothing products, accessories, shoes, bags, and even cosmetics and perfumes, among others. The textile industry, on the other hand, produces materials, fabrics, and various textile-based supplies. Meanwhile, the apparel or clothing industry involves the use of raw materials and supplies from the textile industry to create ready-to-wear garments through a series of production processes (Dobos, 2024).

While the textile industry is typically linked to the apparel industry, it also supplies other sectors, such as construction and agriculture, by producing insulation materials and geotextiles, among other products.

In an article published in Bloomberg (Hirij, Pashankar, 2024) highlights that while the largest fashion companies are making increasingly ambitious commitments to reduce their carbon footprint — meaning the total amount of greenhouse gas emissions they produce — it remains very difficult to track and monitor emissions from the fashion industry. According to an analysis by the World Resources Institute and the Apparel Impact Institute, the fashion industry emitted just over 1 gigaton of carbon dioxide equivalent in 2019, accounting for approximately 2% of global greenhouse gas emissions.

This footprint, coupled with the rise of fast fashion, has made apparel companies a target of criticism. In recent years, the industry has faced scrutiny for driving excessive consumption, favoring high-speed mass production, and generating mountains of plastic waste. The unsustainable "fast fashion" trends have further increased the carbon footprint of the fashion industry. Fast fashion refers to the continuous production of new fashion items at very low prices. As a result of fast fashion, the quantity of produced and discarded clothing has significantly increased.

The excessive consumption encouraged by fast fashion and the short lifecycle of clothing items have led to the mass production of low-cost apparel. Fast fashion trends have driven the industry to account for one-fifth of the world's annual plastic production. The environmental risks of fast fashion and its contribution to climate change are further underscored by forecasts that the production of fossil fuel-based clothing products will continue to rise over the next two decades.

Fast fashion has led to an increase in the use of microplastics, as well as higher energy and water consumption, thereby intensifying dependence on non-renewable resources — all of which have a detrimental impact on climate change. Polyester, a ubiquitous form of plastic derived from oil, is a primary source of microplastic pollution when used in clothing made from polyester and other synthetic fibers. This pollution is particularly harmful to marine life (Basil-Jones, 2024).

Textile waste is also a significant problem. As fast fashion gains prominence, people are more likely to dispose of excess clothing by throwing it away rather than donating it. On average, Europeans purchase nearly 26 kg of textiles per year and discard approximately 11 kg. While used clothing can be exported outside the EU, the majority (87%) is either incinerated or sent to landfills.

In 2018, the EU adopted the Circular Economy Package, which, at the insistence of the European Parliament, ensures for the first time that textiles must be collected separately in all member states by 2025 at the latest. The European Parliament has been supporting the use of ecological and sustainable raw materials, as well as the reuse and recycling of clothing, for several years. Currently, less than half of used clothing is collected for reuse or recycling when it is no longer needed, and only 1% is recycled into new clothing. This is because the technologies that enable the recycling of clothes are only now starting to gain prominence.

The European Parliament is working to address these issues in several ways, such as developing new business models for clothing rental, designing products to facilitate reuse and recycling (circular fashion), encouraging consumers to buy fewer but higher-quality clothes (slow fashion), and generally guiding consumer behavior towards more sustainable choices. To address environmental impacts, the EU aims to reduce textile waste, extend the lifecycle of textiles, and increase textile recycling. This is part of the plan to achieve a circular economy by 2050.

One of the key measures in this plan is the introduction of extended producer responsibility (EPR) for textiles. The goal of EPR is to reduce industrial waste by holding brands accountable for the environmental impact of their products. The Corporate Sustainability Due Diligence Directive (CSDDD) ensures that environmental and human rights due diligence is applied throughout companies' entire supply chains. The Corporate Sustainability Reporting Directive (CSRD) requires companies of all sizes to report on their environmental and social activities, which are then subject to external audits (europarl.europa.eu, 2020).

The Ecodesign for Sustainable Products Regulation (ESPR) aims to maximize sustainability by focusing on resource use, repairability, and circularity. This regulation requires that products come with a "product passport" so that consumers can make informed purchasing decisions.

Such measures and proposed regulations within the EU inevitably mean that greenwashing will no longer be an option for fashion industry businesses. The EU's holistic strategy will transform the way clothing products are produced, distributed, and disposed of, paving the way for a more circular and sustainable fashion industry.

The growing focus on sustainability in the fashion industry is exemplified by Zara, which committed to producing 50% of its products from recycled materials and "ecologically grown cotton" by 2022. One of the largest online fashion retailers, Shein, disclosed the creation of a €200 million fashion waste fund. The aim of this fund is to support start-ups and companies that offer innovative solutions for recycling and reducing fashion waste in the UK and Europe. Shein's revenue reached $2 billion in 2023, and the company is now motivated to place a stronger emphasis on sustainability. It plans to invest an additional €50 million in "ESG measures," bringing its total planned investment in sustainability-related initiatives to €250 million over the next five years.

Another fast fashion company, H&M, has launched a program which allows customers to return their used clothing. This global clothing initiative enables recycling or reuse of clothing. A collaboration has been announced by H&M, Gap Inc., Bestseller, and Mango to share risks and costs associated with reducing CO_2 emissions related to their value chains. The goal is to promote the transition to renewable energy sources and introduce other solutions that reduce CO_2 emissions from the textile factories that produce their products.

To meet global climate targets, emissions from textile production will need to be reduced by nearly half by 2030.

The Kearney Circular Fashion Index (CFX) 2024 ranks the most sustainable fashion brands. In the circular economy category, the top 10 brands are largely the same as those featured on the 2023 list.

The brands featured on the list are: The North Face, Levi's, Madewell, Patagonia, Gant (a new entrant in 2024), Gucci, Lululemon, OVS, Coach, and Lindex.

In Kearney's 2024 Circular Fashion Index (CFX) report, the performance of fashion brands in the circular economy was assessed based on seven dimensions (Ehring et al., 2024). These dimensions are as follows:

- Use of recycled or renewable materials: The proportion of garments made from recycled or renewable materials.
- Availability of repair or maintenance services: The extent to which brands offer repair or maintenance services for their products.
- Detail and accessibility of care instructions: How detailed and accessible the care instructions for products are.
- Importance of circularity in brand communications: The degree to which brands emphasize circular economy principles in their communication.
- Breadth and depth of pre-owned garment assortment: The range and availability of used clothing offered (secondary market).
- Availability of rental or leasing services for garments: The extent to which rental or leasing services for clothing are accessible.
- Possibility of returning worn clothes for recycling: The availability of drop-off points for used clothes, either for recycling or charitable donations.

The application of circular economy principles, where the lifespan of products is extended and materials are recycled, is a key step toward making the fashion industry more sustainable.

„What would most significantly reduce the global apparel footprint is if we make and throw away less stuff." - Berkley Rothmeier, BSR Director, Consumer Sectors (bsr.org; Hirij, Pashankar, 2024)

2.4. Food and Agriculture Sector

The food and agriculture industry stands at the heart of the global economy, directly influencing environmental sustainability, human health, and the livelihoods of billions. However, the sector's extensive reliance on land, water, and energy, along with its role in deforestation and greenhouse gas emissions, presents significant environmental and social challenges. The 2024 report from the Food and Agriculture Organization (FAO) underscores the urgent need to integrate Environmental, Social, and Governance (ESG) principles into agrifood systems to foster sustainability and resilience globally.

To understand how the food and agriculture industry impacts global sustainability, we must first explore the environmental dimension of ESG. This aspect highlights the industry's footprint on land use, greenhouse gas emissions, water consumption, and soil health—key areas that must be addressed to mitigate environmental degradation.

Agriculture is a leading cause of deforestation, biodiversity loss, and environmental degradation. The FAO's 2024 typology categorizes agrifood systems into six groups: protracted crisis, traditional, expanding, diversifying, formalizing, and industrial (Figure 7). Each typology reflects the economic development stage, agricultural practices, and environmental impacts of a country.

Figure 7. Agrifood system typologies across the world

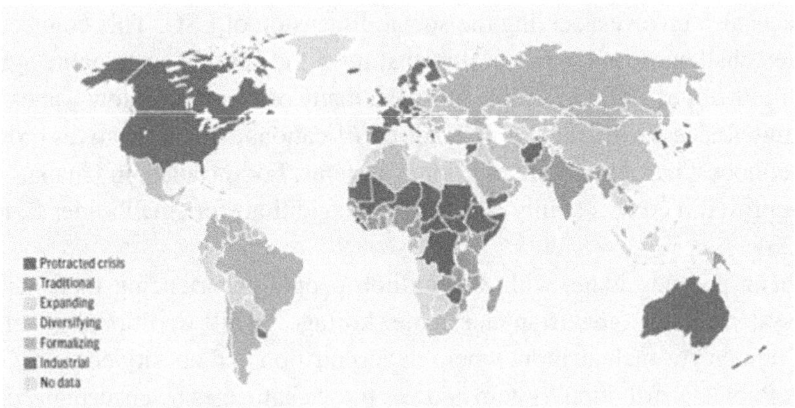

(Source: FAO, 2024)

- Protracted crisis systems (e.g., Somalia, Yemen) suffer from severe resource scarcity due to conflicts and environmental degradation.
- Traditional systems (e.g., India, Nigeria) depend on subsistence farming, often leading to deforestation as populations expand.
- Industrial systems (e.g., the United States, Germany) achieve high efficiency but at the cost of resource-intensive practices.

Programs like Brazil's Soy Moratorium illustrate the potential of targeted policies to address deforestation in expanding and formalizing systems. Meanwhile, agroforestry, which integrates crops and trees, offers a sustainable solution to preserving biodiversity while maintaining productivity (FAO, 2024).

The agriculture sector contributes 24% of global greenhouse gas emissions, with livestock farming as a major source of methane. In industrial systems, innovative practices such as adding seaweed to cattle feed have demonstrated an ability to reduce methane emissions by up to 80%. Precision agriculture technologies—such as satellite monitoring and AI-driven irrigation—are increasingly adopted in expanding systems like Morocco and Egypt, optimizing resource use and reducing emissions (FAO, 2024).

Agriculture accounts for 70% of global freshwater use, with traditional systems in sub-Saharan Africa relying heavily on rain-fed farming, making them vulnerable to climate change. Industrial systems, on the other hand, employ advanced irrigation technologies, including soil moisture sensors and drip irrigation, to conserve water. Expanding countries like Peru are also making strides by adopting rainwater harvesting technologies to enhance water efficiency (Global Water Partnership, 2024).

Over-reliance on synthetic fertilizers and pesticides has degraded soil health in many regions. The FAO emphasizes the importance of organic farming and bio-based alternatives like biofertilizers and biopesticides to restore soil quality. These practices are being championed in diversifying systems like China and Mexico, where the environmental cost of traditional agriculture has become unsustainable (IFOAM, 2024).

As these environmental challenges persist, they are closely tied to broader societal issues. From the livelihoods of farmers who depend on healthy soils to the need for global food security, addressing environmental impacts also involves tackling the social dimension of ESG. This connection underscores the human-centered challenges and opportunities that lie at the heart of sustainable agrifood systems.

Agriculture employs over a billion people globally, many of whom face low wages, unsafe working conditions, and limited access to markets. Fair Trade certifications and cooperatives provide frameworks for improving livelihoods, particularly in traditional systems. For instance, in Ghana, Fair Trade cocoa initiatives have improved income stability and working conditions for smallholder farmers (Fair Trade International, 2023).

Hunger remains a pressing issue, with 828 million people experiencing food insecurity in 2023. Protracted crisis systems like Afghanistan face acute shortages, while traditional systems often grapple with the "double burden" of malnutrition, where undernutrition and obesity coexist. Government programs like India's Public Distribution System address these challenges by ensuring access to affordable staples. Additionally, biofortified crops, such as Golden Rice enriched with Vitamin A, are combating nutritional deficiencies in regions like Southeast Asia (HarvestPlus, 2024).

Large-scale agricultural projects often disrupt local ecosystems and displace communities. Companies like Nestlé, operating in diversifying systems, have involved local stakeholders in cocoa farming initiatives, sharing revenues and supporting community development. These efforts help mitigate social conflicts and build trust with affected populations (FAO, 2024).

The social impacts of the food and agriculture industry cannot be separated from governance. Ensuring ethical labor practices, equitable food access, and community engagement requires robust policies, transparent supply chains, and innovative technologies. Let's see how governance frameworks support ESG integration in the sector.

Nowadays, it's becoming common for ethically sourced food products to be demanded by customers. Blockchain technology is changing the way we track products, making it easier for consumers to check if their purchases meet ESG standards. This type of innovation helps formalize industrial systems, where supply chains are complex and prone to greenwashing (Liu et al., 2023).

USDA certifications and EU's Green Deal warrant that ESG standards, particularly environmental and social, are met across agrifood systems. These policies are pivotal for achieving ESG goals in industrial and formalizing countries where large-scale production can result in sustainability challenges (European Commission, 2024).

IoT devices and farm management software are improving governance in agriculture. These technological tools allow real-time monitoring of resource use and crop health, ensuring compliance with ESG goals. As an example, a Brazilian formalizing system can be highlighted, where digital tools are helping reduce the risks of deforestation in soybean supply chains (Smith et al., 2024).

Governance frameworks and technological advancements have paved the way for sustainable food systems, but the path forward is still fraught with challenges. Significant factors can contribute to the change. Financial, climatic, and technological barriers faced by the industry will be introduced in the following section.

Significant investment has been required to transition to sustainable practices. The Global Agriculture and Food Security Program, as a public-private partnership, provides significant financial support to smallholders in transitional systems like Kyrgyzstan and Honduras (World Bank, 2024). Building resilience to climate impacts is essential, particularly for protracted crisis systems. Solutions like drought-resistant crop varieties and regenerative agricultural practices are gaining momentum as viable adaptations (FAO,

2024). Unequal access to technology remains a barrier, especially in developing countries. Programs offering subsidies for digital tools and farmer training are helping bridge the gap in traditional and expanding systems (Global Water Partnership, 2024). Addressing the challenges in the food and agriculture sector is critical, but the true measure of success lies in the actionable steps taken to build a sustainable future. By leveraging collaboration, innovation, and circular economy principles, the industry can align with global ESG goals. Here's how the sector can pave the way forward.

The future of ESG in food and agriculture hinges on scaling up sustainable practices while ensuring equity across stakeholders. The best way to enhance sustainability is by integrating circular economy principles, such as nutrient recycling and agricultural waste composting. Partnerships among local communities, corporations, and governments are essential for building strong and resilient food systems. Taking these insights into account, the following list of recommendations can guide us toward a more sustainable and ESG-compatible agricultural and food system:

- True Cost Accounting (TCA): Quantify hidden costs to better understand environmental and social impacts (FAO, 2024).
- Collaborative Models: Multi-stakeholder collaboration is essential to drive systemic change.
- Consumer Education: Informing consumers about sustainable dietary choices can reshape demand and reduce the sector's environmental footprint.

3. TECHNOLOGICAL TOOLS SUPPORTING ESG

The ESG framework has become a fundamental element of corporate sustainability and social responsibility. ESG reporting is crucial for companies to demonstrate and enhance their sustainability performance. Technological innovations, especially artificial intelligence (AI), have opened up new opportunities for ESG reporting, especially in the areas of data collection and data analysis. AI provides substantial benefits by automating data processing, enabling companies to save both time and resources. For example, in his master thesis, Ontiveros (2024) developed an AI-powered IT tool to facilitate companies' ESG compliance. The development focused on cutting-edge technologies such as Large Language Models (LLM) and Retrieval Augmented Generation (RAG). The work examined the effectiveness of RAG in two primary tasks: determining the applicability of ESG regulations to companies and answering questions about ESG regulations. Another study examined the applicability of AI in predictive analytics, specifically for measuring and analysing carbon emissions. Aizpurietis (2024) studied the applicability of Botpress-based chatbots, which were able to provide accurate CO_2 emission estimates. With much of the world committed to achieving net zero carbon emissions, it is crucial that individuals and organizations are properly informed about the quantitative values of their CO_2 emissions. However, there are currently few simple methods available, which makes it difficult for many to understand and apply the correct calculation and reporting procedures.

Industry 5.0 (I5.0) technologies significantly improve the accuracy and transparency of ESG reporting by prioritizing sustainability, people-centricity, and resilience, closely aligning with ESG principles. Technologies such as blockchain, the Internet of Things (IoT), digital twins, and artificial intelligence (AI) enable real-time data collection and processing, reducing the need for manual intervention and increasing data reliability. These tools effectively adapt to stakeholder needs while supporting the achievement of sustainability goals and the credibility of transactions.

Asif, Searcy and Castka (2023) discuss the potential role of I5.0 in addressing ESG reporting issues in depth. They consider the opportunities under 6 headings: authenticity, retrospective reporting, customizability, scope of ESG reporting, cost and efficiency. They conclude that technologies such as sensors, Radio Frequency Identification (RFID), blockchain, cloud computing, cybersecurity, drones and AI increase the accuracy and credibility of ESG data. The other areas contributing to increased transparency include automatic data collection, transaction tracking, and the detection of fraud and anomalies. Replacement of traditional reporting with sensors, RFID, digital twin, IoT, and AI collecting and processing data in real time could further enable immediate ESG reporting and forecasting using I5.0 tools.

Valette (2024) discusses how ESG reporting, total quality management, and total quality assurance relate to the use of smart technologies in light of the new CSRD. The findings prove that the adoption of smart technologies is decisive for a substantial improvement in the quality of ESG reporting and the efficiency of TQM practices. She also shows that such technologies could make data gathering more effective, improve the accuracy of data, keep track of key performance indicators, analyze customer behavior, and ensure higher quality ESG reporting.

Liu et al. (2023) developed a blockchain-based, data-driven methodology for ESG evaluation. Stochastic multicriteria acceptability analysis-second variant (SMAA-2) can be combined into that method to deal with main barriers of ESG studies: opaqueness of data and subjectiveness of assessment. Then the ESG performance of 71 Hong Kong textile, apparel firms was evaluated considering ESG quantitative data. The sensitivity analysis results show that this approach is good to go when it analyzes and compares the sustainability performance of companies with others of the same industry.

The advent of generative AI has ushered in a new era in ESG reporting. Generative AI is a type of artificial intelligence that can generate new, human-like content, such as text, images, audio, video, or code. These AI systems learn by identifying patterns in vast amounts of data and using these patterns to generate new, similar content. Generative AI is now available in the form of a number of text-creation software/programs (e.g. Claude, ChatGPT4, Llama, Bard, Copilot, or Gemini) that assist with various writing tasks. A study by Managi, Shimamura, and Tanaka (2024) examined the relationship between the readability of sustainability reports and ESG scores for US companies using ChatGPT-4. Their results show a positive correlation between contextual readability scores and the average of multiple ESG scores.

Zhang and Chang (2024) examine the presentation of ESG ratings from the perspective of risk management and investment returns. They point out that clarifying the relationship between the disclosure of ESG information and rating data contributes to more accurate and comprehensive rating results. These results not only help to better understand a company's ESG performance and risks but can also lead to more informed investment decisions.

XAI will also work on improving the understandability of ESG reports. In ESG assessment, XAI makes complex algorithms more understandable, thus adding to the transparency of ESG reports. Stakeholders will be in a better position to understand how scores are derived for ESG scores. XAI can also bring forth biases and errors in data for more accurate and reliable ESG scores. Selection of ESG metrics is, so far, the critical point of a beginning for this whole reporting process. For example, explain why XAI will select these metrics, like carbon footprint or water usage, and state that these are very applicable in the particular industry the company operates within. Overcoming the obstacle of transparency that artificial intelligence faces in ESG reporting, a study was conducted by Vieru and Plugge back in 2025 to achieve this very end. The findings indicate that their analysis leads to a necessary usage of appropriate EKBS mechanisms in order to ensure explainability and trust in AI-based tools.

According to a review by Rane, Choudhary, and Rane 2024, in the latest development, AI has become efficient in the implementation of ESG, reviewed in the role played by AI in encouraging an organizational culture for sustainability. The review emphasizes the point that an AI-based approach will make the initiative on ESG more accurate, efficient, and broader. Recent trends in AI technologies, such as machine learning, natural language processing and advanced analytics, are increasingly being applied to address complex ESG challenges to enable businesses to meet or exceed their sustainability goals.

De Silva, Gunarathne, and Kumar (2024) highlight the transformative impact of digital technologies on sustainability practices. The analysis highlights that IoT improves the accuracy of ESG reporting, Extensible Business Reporting Language (XBRL) structures ESG data, and AI enhances the credibility of life cycle assessment and reporting. Big data optimizes resource use, while edge computing improves eco-efficiency. The study also highlights that these technologies support ESG reporting, circular economy initiatives, and strategic decision-making, promoting economic, social, and environmental sustainability.

Besides, quantum computing can take it to the next level of ESG reporting by processing complex datasets at unprecedented speed. Quantum computing can provide optimized supply chain management, identify hotspots of emission, and simulate scenarios about sustainability in real time, hence taking the ESG goals way ahead. Despite promise, AI in ESG reporting faces challenges. A huge problem exists with regard to the quality and accessibility of data.

4. CONCLUSION AND FUTURE DIRECTIONS

With every sector facing challenges and opportunities of a different nature today, ESG principles are becoming intrinsic to business strategy, hence changing the face of industries across the world.

ESG standards drive very substantial long-term benefits, from operational improvements through strengthened stakeholder relationships. Entry into ESG frameworks provides access to new markets, attracts more investors, and improves customer loyalty for brands. Using this example in the financial world, embracing ESG will reduce risks from investments and improve performance in their portfolios.

Resource efficiency, cost savings, and regulatory compliance come through sustainable practices such as transitioning to renewable energy and applying circular economy models. For example, in fashion, this might extend the life of a product, reducing waste; in the energy industry, smart grids would optimize resource use and reduce energy loss. The social aspect means ESG translates into the fair distribution of resources, improvement in labor conditions, and relations with the community. Others put greater emphasis on employee wellbeing and community involvement to boost their reputation, corporate culture, and even productivity. Good governance at a company should include such important features as transparent reporting and ethical decision-making; such governance is aimed at reducing risks of legislation and increasing stakeholders' trust.

Overall, integration of ESG principles readies companies for future challenges through the facilitation of innovation and adaptability. Companies at the forefront in managing ESG risks are quick to adapt to changes in regulatory demands, shifting consumer needs, and economic turbulence.

Evolving ESG frameworks influence the new trends and techniques that, in turn, will mark the adoption and application across sectors. Some of the path-breaking technologies that change the ESG practices include Artificial Intelligence, blockchain, and Internet of Things. AI tools upgrade the reliability and efficiency of ESG reporting by providing predictive analytics, such as for emissions. Generative AI and explainable AI will also gain relevance because they can generate comprehensive investor-friendly

ESG reports. Blockchain technology, by enhancing traceability and transparency within the supply chain, allows for better accountability and minimal greenwashing. Last but not least, IoT sensors might contribute to reaching sustainable goals by delivering real-time data about resource consumption and enabling support for operational improvements. These tools support decision-making and data interpretation, facilitating stakeholder engagement. Quantum computing is one of the future directions that could further accelerate the optimization of real-time sustainability strategies and processing of complex databases for ESG reporting.

International organizations and governments call for more standardization, accountability, and transparency. So, they have started enacting more stringent ESG regulations. The EU sets benchmarks for corporate sustainability performance with the CSRD (Corporate Sustainability Reporting Directive) and the Ecodesign for Sustainable Products Regulation, among others. These metrics further incentivize companies in supporting climate targets and delivering responses to ESG concerns.

The concept of circular economy practice is gaining momentum globally in all industries. Minimizing environmental impacts, prolonging the life of products, efficient waste management, good usage of resources, reuse, and recycling-this kind of activity is sustainable ambitions.

With increased awareness, organizations are under pressure to apply ESG initiatives in order to address sustainability issues and meet consumer expectations. Transparency and ethics are important in building and maintaining reliable relationships with employees, consumers, and investors. Stakeholder involvement through collaborative models and consumer education initiatives might be one driving factor toward systematic change and transforming demand for sustainable products and services.

ESG priorities can be achieved only if the effort is collectively contributed to across regions and industries. International agreements, cross-industry collaborations, and public-private partnerships are indeed instrumental in solving multi-dimensional sustainability challenges, as seen in the Paris Climate Agreement and COP26. Knowledge and resource sharing, along with the implementation of best practices, would drive innovation and faster achievement.

Applying ESG principles is not just an obligation of compliance but is rather a strategic step to achieve long-term business prosperity and resilience. Although there are specific challenges for every industry, the ESG perspective offers a well-defined framework that provides new transformative options toward sustainable, ethical, and innovative operations. This can be attained with the adoption of innovative technologies, adaptation to evolving legislation, and stakeholder partnerships. This would mean having a holistic approach toward ESG, bringing together global goals with local realities. This positions companies as leading in the shift toward a sustainable future.

REFERENCES

Aizpurietis, J. (2024). Developing an AI-powered chatbot for CO_2 emissions estimations utilizing Botpress framework. Theseus Repository. Retrieved from https://www.theseus.fi/bitstream/handle/10024/861047/Aizpurietis_Janis.pdf?sequence=2&isAllowed=y

Asif, M., Searcy, C., & Castka, P. (2023). ESG and Industry 5.0: The role of technologies in enhancing ESG disclosure. *Technological Forecasting and Social Change*, *195*, 122806. DOI: 10.1016/j.techfore.2023.122806

Basil-Jones, W. (2024). ESG in fashion: Your guide to navigating reporting regulations, https://plana.earth/academy/sustainability-fashion-guide-esg-regulations

BlackRock. (2023). *Advancing ESG integration in investment decisions*. BlackRock Sustainability Insights.

BP. (2024). Net Zero by 2050: Transforming Energy. BP Sustainability Report.

Braun, R. (2015). *Vállalati társadalmi felelősségvállalás*. Akadémiai Könyvkiadó.

Carney, M. (2019). Fifty shades of green: The world needs a new sustainable financial system to stop runaway climate change. IMF, The Economics of Climate. *Finance & Development*, •••, 12–15.

Climate Finance, I. M. F. (2018). Carbon footprint of bank loans. Retrieved from https://climatedata.imf.org/pages/climate-finance

De Silva, P., Gunarathne, N., & Kumar, S. (2024). *Exploring the impact of digital knowledge, integration and performance on sustainable accounting, reporting and assurance*. Meditari Accountancy Research., DOI: 10.1108/MEDAR-02-2024-2383

Demekas, D. G., & Grippa, P. (2021). *Financial regulation, climate change, and the transition to a low-carbon economy: A survey of the issues*. International Monetary Fund., DOI: 10.5089/9781616356521.001

Dobos, E. (2024): Álhírek és tévhitek közt élünk: a divatipar nem is annyira szennyező, mint hisszük. KRTK Blog. Portfolio. https://www.portfolio.hu/krtk/20240608/alhirek-es-tevhitek-kozt-elunk-a-divatipar-nem-is-annyira-szennyezo-mint-hisszuk-690857

Ehring, B., Minutella, D., & Shah, N. (2024)Taken as a whole, today's global fashion industry still struggles to work in a circular manner, https://www.kearney.com/industry/consumer-retail/article/the-kearney-cfx-2024-report-navigating-material-and-product-choices-as-key-unlocks-to-circularity?utm_campaign=climate-tech-monday&utm_medium=email&utm_source=newsletter&mkt_tok=MjExLU5KWS0xNjUAAAGUOjNgC3w6DC2s-OBHkF9WSPo5md9IAdb4CtgbjIwlK4ywveauFYpZEBZw5imZE4HVEQnn8Ddtak-xdEXbagHlni3e8C_SfS4zPpA0350boyNd_g

Elmalt, D., Igan, D., & Kirti, D. (2021). Limits to private climate change mitigation. IMF Working Paper. DOI: 10.5089/9781513582504.001

European Commission. (2024). The Green Deal and Farm to Fork Strategy: Sustainable Food Systems in Europe. Retrieved from https://ec.europa.eu/

European Commission. (2024a). *What is the EU emissions trading system (EU ETS)?* European Commission. Retrieved December 20, 2024, from https://climate.ec.europa.eu/eu-action/eu-emissions-trading-system-eu-ets/what-eu-ets_en

European Commission. (2024b). *A European Green Deal*. European Commission. Retrieved December 20, 2024, from https://commission.europa.eu/strategy-and-policy/priorities-2019-2024/european-green-deal_en

European Commission. (2024c). *Corporate sustainability reporting*. European Commission. Retrieved December 20, 2024, from https://finance.ec.europa.eu/capital-markets-union-and-financial-markets/company-reporting-and-auditing/company-reporting/corporate-sustainability-reporting_en

ExxonMobil. (2024). Workplace safety protocols and risk management in the energy sector. ExxonMobil Corporate Report.

Fair Trade International. (2023). Annual Report 2023: Impact and Progress. Retrieved from https://www.fairtrade.net/

Food and Agriculture Organization of the United Nations (FAO). (2024). The State of Food and Agriculture 2024: Value-driven transformation of agrifood systems. Rome. DOI: 10.4060/cd2616en

Fried, S., Novan, K., & Peterman, W. B. (2022). Climate policy transition risk and the macroeconomy. *European Economic Review*, *147*, 104174. DOI: 10.1016/j.euroecorev.2022.104174

Global Water Partnership. (2024). Sustainable Water Management in Agriculture. Retrieved from https://www.gwp.org/

Halmai, P. (Ed.). (2023). *A jegybank és a pénzügyi szektor portfólióinak zöldülése - Fenntarthatóság a közgazdaság-tudományban: Elméleti alapok, alkalmazások*. MERSZ., Retrieved from https://mersz.hu/dokumentum/m1101fak__33/

HarvestPlus. (2024). Biofortification Progress and Impact Report. Retrieved from https://www.harvestplus.org/

Hirij, Z., & Pashankar, S. (2024) Why It's So Hard to Track the Fashion Industry's Emissions, A growing number of fashion companies are talking about substantially cutting their greenhouse gas emissions. But evaluating those efforts is tricky. https://www.bloomberg.com/news/features/2024-05-16/is-your-favorite-fashion-brand-cutting-emissions-it-s-tricky

Intergovernmental Panel on Climate Change (IPCC). (2014). Mitigation of Climate Change: Working Group III Contribution to the Fifth Assessment Report of the IPCC. Retrieved from https://www.ipcc.ch/site/assets/uploads/2018/02/ipcc_wg3_ar5_chapter7.pdf

International Energy Agency. (2023). *CO2 emissions in 2023: Executive summary*. Retrieved from https://www.iea.org/reports/co2-emissions-in-2023/executive-summary

International Energy Agency – IEA. (2024). *Smart grids*. IEA. Retrieved December 20, 2024, from https://www.iea.org/energy-system/electricity/smart-grids

International Federation of Organic Agriculture Movements (IFOAM). (2024). The Role of Organic Farming in Climate Change Mitigation. Retrieved from https://www.ifoam.bio/

International Renewable Energy Agency (IRENA). (2024). Renewable energy and global sustainability. IRENA Annual Report. Retrieved from https://www.irena.org

IPCC. (2023). [Mitigation of Climate Change. Intergovernmental Panel on Climate Change.]. *Climatic Change*, •••, 2023.

Kaushik, T. K. Ravish, Singh, A., Raghav, A., Singh, B., Arora, M. K., & Lal, S. (2024). Untapping Artificial Intelligence in Environmental and Sustainable Progression: Projecting Legal and Ethical Considerations. In B. Singh, C. Kaunert, K. Vig, & S. Dutta (Eds.), *Maintaining a Sustainable World in the Nexus of Environmental Science and AI* (pp. 109-130). IGI Global. https://doi.org/DOI: 10.4018/979-8-3693-6336-2.ch005

Kolozsi, P. P., Ladányi, S., & Straubinger, A. (2022). Pénzügyi eszközök klímakockázatának mérése: Módszertani kihívások és jegybanki gyakorlat. *Hitelintézeti Szemle*, *21*(1), 113–140. DOI: 10.25201/HSZ.21.1.113

Leogrande, A. (2024). Integrating ESG principles into smart logistics: Toward sustainable supply chains. HAL Sciences. DOI: 10.20944/preprints202411.1432.v1

Liu, X., Yang, Y., Jiang, Y., Fu, Y., Zhong, R. Y., Li, M., & Huang, G. Q. (2023). Data-driven ESG assessment for blockchain services: A comparative study in textiles and apparel industry. *Resources, Conservation and Recycling*, *190*, 106837. DOI: 10.1016/j.resconrec.2022.106837

Magyar Nemzeti Bank. (2022). Körkép a jegybankok és nemzetközi szervezetek zöld programjairól. Retrieved from https://www.ngfs.net/sites/default/files/medias/documents/climate_related_litigation.pdf

Managi, S., Shimamura, T., & Tanaka, Y. (2024). Evaluating the impact of report readability on ESG scores: A generative AI approach. *SSRN*. DOI: 10.2139/ssrn.4888339

MSCI. (2024). *BP PLC ESG ratings and climate search tool*. MSCI. Retrieved December 20, 2024, from https://www.msci.com/our-solutions/esg-investing/esg-ratings-climate-search-tool/issuer/bp-plc/IID000000002140371

Ontiveros, A. (2024). Development and evaluation of an IT tool for ESG compliance consulting for companies using LLMs. University of Hamburg. Retrieved from https: https://www.inf.uni-hamburg.de/en/inst/ab/lt/teaching/theses/completed-theses/2024-ma-ontiveros.pdf

Papoutsi, M., Piazzesi, M., & Schneider, M. (2021). How unconventional is green monetary policy? Retrieved from https://web.stanford.edu/~piazzesi/How_unconventional_is_green_monetary_policy.pdf

Parry, I. (2019). Putting a Price on Pollution: Carbon pricing strategies could hold the key to meeting the world's climate stabilization goals. IMF, The Economics of Climate. *Finance & Development*, •••, 16–19.

Patagonia. (2024). Worn Wear initiative: Driving circular economy in fashion. Patagonia Sustainability Report.

Raghav, A., Singh, B., Jermsittiparsert, K., Raghav, R., & Yadav, U. (2024). Artificial Intelligence in Environmental and Climate Changes: A Sustainable Future. In Singh, B., Kaunert, C., Vig, K., & Dutta, S. (Eds.), *Maintaining a Sustainable World in the Nexus of Environmental Science and AI* (pp. 485–506). IGI Global., DOI: 10.4018/979-8-3693-6336-2.ch019

Rane, N., Choudhary, S., & Rane, J. (2024). Artificial intelligence-driven approaches to strengthening ESG criteria in sustainable business practices: A review. *SSRN*. DOI: 10.2139/ssrn.4843215

Remy, N., Speelman, E., & Swartz, S. (2016): Style that's sustainable: A new fast-fashion formula. https://www.mckinsey.com/capabilities/sustainability/our-insights/style-thats-sustainable-a-new-fast-fashion-formula

Shell. (2024). Shell's net-zero commitment: Transitioning to renewable energy. Shell Corporate Sustainability Report.

Smith, R.. (2024). Smart Agriculture and ESG Integration: Innovations for Sustainability. *Global Agribusiness Review*, *19*(1), 23–34.

Unilever. (2024). The Unilever Sustainable Living Plan: A decade of progress. Unilever Reports.

United Nations. (2024): Explanatory Notes of the Standard Industrial Classification of All Economic Activities, Revision 5 (ISIC Rev.5). Last update: 11 March 2024. https://unstats.un.org/unsd/classifications/Econ/Download/In%20Text/ISIC5_Exp_Notes_11Mar2024.pdf

United Nations. (n.d.). Act Now: Facts and Figures on Climate Change. United Nations Climate Change Initiative. Retrieved from https://www.un.org/en/actnow/facts-and-figures

United Nations Principles for Responsible Investment - UN PRI. (2024). *Principles for Responsible Investment*. Retrieved December 20, 2024, from https://www.unpri.org/

Valette, A. (2024) Enhancing Operational Excellence: Exploring the interplay of ESG Reporting, Smart Technology Adoption, and Total Quality Management in the banking sector. Louvain School of Management, Université catholique de Louvain, 2024. Prom.: Olieslagers, Ruben. http://hdl.handle.net/2078.1/ thesis:47228

Vieru, D., & Plugge, A. (2025). *Overcoming the AI opacity in ESG reporting: A digital platform-based knowledge boundary-spanning perspective*. TÉLUQ Research Repository., DOI: 10.1016/j.digtrans.2025.122808

Yang, Z., Wei, Y.-M., & Mi, Z. (2020). *Integrated assessment models (IAMs) for climate change*. Oxford Bibliographies in Environmental Science., DOI: 10.1093/obo/9780199363445-0043

Zhang, M., & Chang, Y. (2024). ESG studies: The impact on enterprise investment and financing decisions. *Journal of Economics, Finance and Administrative Science*, *6*(3), 10. DOI: 10.32996/jefas.2024.6.3.10

Chapter 13
Fuzzy Logic Approach to the Management of Biomass and Biogas Energy Systems

Sahil Lal
https://orcid.org/0000-0001-9827-3717
Sharda University, India

Manmeet Kaur Arora
https://orcid.org/0009-0002-5071-117X
Sharda University, India

Bhupinder Singh
https://orcid.org/0009-0006-4779-2553
Sharda University, India

Saquib Ahmed
https://orcid.org/0009-0008-1891-6910
Sharda University, India

Saurabh Chandra
https://orcid.org/0000-0003-4172-9968
Bennett University, Greater Noida, India

ABSTRACT

The critical aspects in this sense is to satisfy growing energy needs from renewable sources and one way or another, we need biomass/biogas as an alternative fuel instead of fossil fuels. The work introduces the basic concepts of fuzzy logic and its application in energy management. A framework to model complex relationships between impact parameters using fuzzy inference systems is developed in the research, enabling more accurate predictions and system performance assessments. Several biomass and biogas energy systems are taken as case studies to ascertain the effectiveness of proposed fuzzy logic approach. This chapter makes a timely and significant contribution to the renewable energy management field, in which researchers, practitioners and policymakers will find new insights about how biomass/biogas can be further integrated into global energy mix.

DOI: 10.4018/979-8-3373-0045-0.ch013

1. INTRODUCTION

Management of biomass and biogas energy systems is critical today to advance sustainable energy options that will reduce our carbon footprint and save us from environmental degradation Nonetheless, biomass and biogas energy systems are difficult to manage as feedstock availability is variable both directly feed production processes can be managed using advanced methods like automated UNF- 25IV Imaging for example –and indirectly (grown or waste derived), the biological conversion process will generate variable outputs and there exists inherent uncertainty related with biomethane. Moreover, how to incorporate fuzzy logic in management and its applications which are very important tools for policy makers who work on managing the uncertainties of biomass & biogas production is explored. The findings reported in this paper highlight important benefits of fuzzy logic-based strategies for integrating biomass and biogas energy systems management. Fuzzy logic can be used to manage such complex and uncertain systems, thereby enhancing operational efficiency, resource management or sustainability by addressing its inherent uncertainties. The results presented are designed to further explore benefits of using fuzzy logic as a powerful means in order to manage more efficiently biomass-biogas energy systems, leading to better effectiveness and reliability together with decision-making activities. This is a step in which quantifies some of the parameters that affect biomass and biogas systems including feedstock characteristics, environmental conditions and operational variables (Qian & Zhou, 2023). Many case studies indicate that the fuzzy logic approach described above is efficient for resource allocation optimization, enhanced process control and minimization of operational risks. Biogas and biomass are 2 latest type of renewable energy that is being by these industries as a source which can greatly help in reducing carbon dioxide emission (Roy & Chakraborty, 2024).

Biological systems can be more naturally described with fuzzy logic and this gives a powerful framework to manage these biological entities by also considering the uncertainty/variability characteristic for such type of processes. Fuzzy logic, unlike conventional binary logic allows for degrees of truth and this is especially useful when we are planning to make some decisions based on the output with grades like low high or medium etc in cases where quantifying an observation has complexity.

The significance of biomass and biogas energy systems lays out in their ability to ensure country´s energetic self-sufficiency, as well at while there is potential for reducing greenhouse gas emissions. They provide a twofold solution by diverting waste through the conversion of organics to energy and delivering clean, renewable power. Given global efforts to move towards sustainable energy solutions, it is essential that researchers optimize these systems (Yao & Sun, 2023).

The study argues that fuzzy logic can make biomass and biogas management more efficient than the traditional control strategies and predictive models . This enables the use of multiple operating parameters, including variable feedstock and environmental conditions to achieve high system performance as well as reliability (Sahu & Sharma, 2023). With the increasing utilization of renewable energy, fuzzy logic is expected to play an important role in determining whether these systems will be effectively managed and their sustainability assured.

2. FUZZY LOGIC FUNDAMENTALS

Figure 1. The fundamental of fuzzy logic.

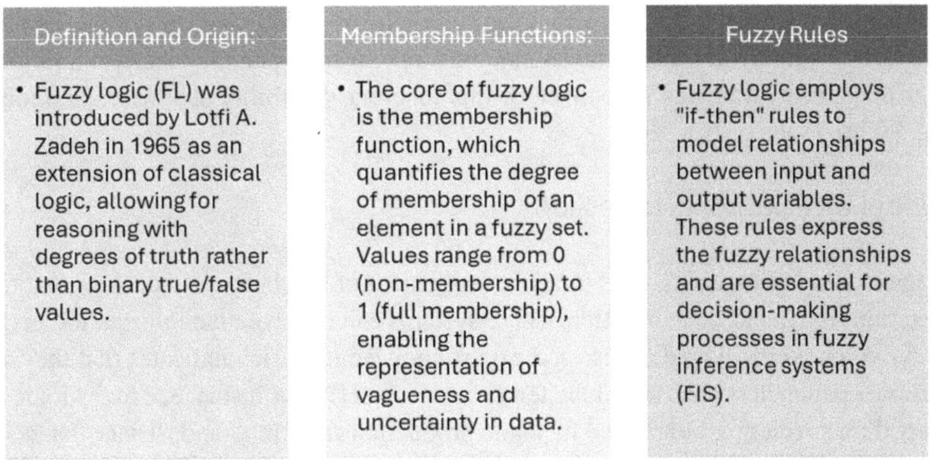

Management of biomass and biogas energy systems is critical today to advance sustainable energy options that will reduce our carbon footprint and save us from environmental degradation. Biogas and biomass are 2 latest type of renewable energy that is being by these industries as a source which can greatly help in reducing carbon dioxide emission.

Biological systems can be more naturally described with fuzzy logic and this gives a powerful framework to manage these biological entities by also considering the uncertainty/variability characteristic for such type of processes. Fuzzy logic, unlike conventional binary logic allows for degrees of truth and this is especially useful when we are planning to make some decisions based on the output with grades like low high or medium etc in cases where quantifying an observation has complexity .

The significance of biomass and biogas energy systems lays out in their ability to ensure country´s energetic self-sufficiency, as well at while there is potential for reducing greenhouse gas emissions (Ren & Liu, 2023). They provide a twofold solution by diverting waste through the conversion of organics to energy AND delivering clean, renewable power. Given global efforts to move towards sustainable energy solutions, it is essential that researchers optimize these systems .

The study argues that fuzzy logic can make biomass and biogas management more efficient than the traditional control strategies and predictive models. This enables the use of multiple operating parameters, including variable feedstock and environmental conditions to achieve high system performance as well as reliability. With the increasing utilization of renewable energy, fuzzy logic is expected to play an important role in determining whether these systems will be effectively managed and their sustainability assured.

Applications of Fuzzy Logic in Biomass Management

In this regard, the fuzzy logic has appeared as an excellent instrument for biomass production plant management and biogas environmentally friendly power generation .

Biomass Production via Fuzzy Logic

Then this fuzzy logic is used in biomass production to increase the choice of feedstock, and improve energy conversion process (Yang & Li, 2024). One such application is in screening of optimal biomass material (as example sugarcane bagasse) for maximum bio-oil yield using the certain parameters e.g. moisture content and calorific value(parameters), etc Such an approach may lead to more flexible and responsive management strategies in the face of this inherent variability characteristic biomass types (Dutta et al., 2024). (Saha & Roy, 2024)

Fuzzy Control for Anaerobic Digestion

Fuzzy logic control systems are used for the management of biochemical processes that take part in biogas generation from anaerobic digestion. These systems can use dynamic information in the form of fuzzy rules based on a historical database and expert knowledge, which indicates that they should improve operational parameters (like unlinking temperature or pH) . For instance, a fuzzy logic model has been developed for predicting biogas and methane production rates in an industrial pilot-scale reactor, helping operators to fine-tune conditions towards optimum yield. This flexibility is key to being able to continue operating efficiently on variable feedstocks — e.g. agricultural residues.

3. CASE STUDIES

Case studies have demonstrated the usefulness of fuzzy logic in biomass and biogas management. A fuzzy logic controller has been used in one study for controlling the biogas production, whereas another study showed that energy performance of a system can be improved using this type of control. A further study undertaken to build a fuzzy logic model for energy forecasting pyrolysis of biomass 103, represents another example where the use of this methodology was applied in diverse forms conversion processes (Patel & Mehta, 2024). The case studies in which fuzzy logic can be used for increase operational efficacity and reduction in cost of running along with vigour's impact on system altogether, the biomass-biogas energy systems. To sum up, the utilization of fuzzy logic in biomass conversion contributes to not only optimal production processes but also betterment on sustainability of renewable energy systems. Fuzzy logic can therefore play a major role in promoting the advanced biomass and biogas technologies by accommodating uncertainties and providing an adaptive control strategy.

4. FUZZY LOGIC IN BIOGAS ENERGY SYSTEMS

The fuzzy logic is becoming a useful tool in modelling biogas energy production, from prediction the efficiency of the process to control anaerobic digestion and even integration with other renewable sources .

Models for the Prediction of Biogas Yield

The FLMs are developed for predicting biogas and methane production rates within UASB reactors treating different wastewaters like molasses wastewater, distillery effluents based on fuzzy logic system (Yang & Zhou, 2023). These models use several input variables such as temperature, chemical oxygen demand (COD) removal, pH correction ratio and alkalinities-to-acidification ratio to predict the output biogas production. For example, a biogas and methane yields of a pilot-scale mesophilic UASB reactor due to MIMO (multiple inputs multiple outputs) fuzzy logic model with determination coefficient of 0.96 are presented for the root mean square error; RMSE1545 [m3/Day] was an already introduced dry weight. Results of sensitivity analysis demonstrated that, COD reduction and COD load are the most influential parameters affecting biogas production with influence factor in decreasing order (Singh & Gupta, 2023).

Biogas Systems Fuzzy Controller

The complexity of biochemical processes requires high level control systems and needs special mechanisms which can manage the system such as fuzzy logic. These systems are able tune operational parameters like temperature and pH on the fly using fuzzy rules based upon historical data or expert knowledge . Preprint: Fuzzy logic controller model for power generation in biogas, proving thus that fuzzy could be useful to the realization of a optimized system evolution (Shao & Li, 2024).

Renewable Energy Integration

Hybrid Systems like Solar + Biogas Energy, Wind + Biogas Power can be used when creating renewable energy from biogas. This is how fuzzy logic can contribute to managing these fully integrated systems for operation efficiently and reliably (Rahman & Ahmed, 2024). Improving the overall performance of such sustainable energy solutions promises better hybrid renewable Energy systems, by better controlling and optimizing them using fuzzy logic. To sum up, fuzzy logic plays a crucial role in the development of biogas energy systems . In terms of the future application prospects, from accurate prediction models to adaptive control strategies and integrated with other renewable sources, fuzzy logic has demonstrated notable versatility in optimizing biogas energy systems for a sustainable era (Sharma & Kumar, 2024).

5 CHALLENGES AND LIMITATIONS

Challenges and Limitations

There are certain challenges and limitations that the application of fuzzy logic faces in managing biomass-based energy system, as well as biogas systems.

Data Uncertainty

One of the biggest problems we face while modelling with fuzzy logic is to tackle data uncertainty. Many reactors in biomass/biogas systems function under conditions of considerable fluctuation between feedstocks properties, environmental surroundings and operation parameters. Accurate and representative data for training/validation of fuzzy models are usually difficult to obtain, and such difficulty may cause the subsequent inaccuracy of predictions & control strategies (Kumar, Singh, Diwakar, & Garg, 2024). Modelling data uncertainty by proposing robust methods to collect reliable and valid data, applying specific filtering techniques to improve the accuracy of acquired information in both experiments; including a mechanism for quantification which is able within Fuzzy Logic framework .

Model Complexity

However, the higher complexity of biomass and biogas systems corresponds to a more complex fuzzy logic model (Tsai & Lee, 2024). It has some disadvantage such intrinsically large (e.g., multi-dimensional or with long input patterns) systems that the accuracy of fuzzy rule bases and membership functions for these type models are expected to be very difficult. Moreover, the interpretability of fuzzy models tends to drop as more rules and variables are added into consideration, which could potentially make it harder for experts to confirm their choice . New strategies are necessary to manage the complexity of models via, for instance, feature selection, rule-based simplification or hierarchical fuzzy systems in order to make practical and scaling up fuzzy applications (Tian & Zhou, 2023).

Implementation Issues

However, moving from theoretical fuzzy logic models to practical implementations in industrial systems such as biomass and biogas energy facilities is not trivial (Tripathi & Mishra, 2024). This may necessitate extensive modification and adaption of the existing hardware as well as software systems they interface with. In practice, fuzzy logic-based controllers need to be robust and reliable under real-world conditions, otherwise they could fail due as e.g., to sensor failure, communication delays or unexpected disturbances . Solving these problems requires a close interaction among researchers, developers, and end-users for devising user-friendly fuzzy logic-based approaches that are adaptable with both biomass as well as biogas energy systems. Despite such hurdles, a great amount of research and development work Introduced into the fuzzy logic-based biomass & biogas management to identify its drawbacks . As the field progresses and more sophisticated techniques like those in machine learning or optimization algorithms are being mixed with fuzzy logic, even stronger solutions to handling such renewable energy systems may be found (Thakur & Rathi, 2024).

6 FUTURE DIRECTIONS

The future of fuzzy logic in biomass and biogas energy systems is promising, with ongoing advancements and numerous potential research areas.

Figure 2. The future directions and prospects

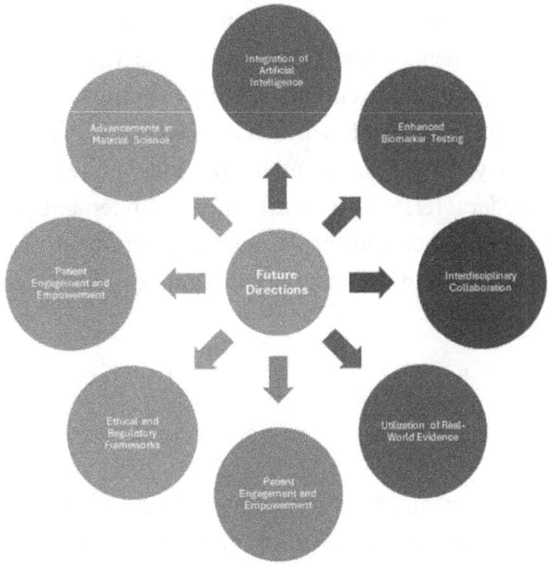

Advances in Fuzzy Logic Techniques

Recent developments in fuzzy logic techniques are enhancing its applicability in various fields, including biomass and biogas management. Innovations such as intuitionistic fuzzy sets and generalized nets are expanding the traditional fuzzy logic framework, allowing for more effective modelling of uncertainty and imprecision (Rana & Patel, 2023). These advancements enable the creation of more sophisticated fuzzy inference systems that can better handle the complexities of real-world applications. Additionally, the integration of fuzzy logic with artificial neural networks (ANNs) is gaining traction, leading to hybrid systems that combine the strengths of both methodologies for improved decision-making and predictive capabilities (Tan & Xu, 2023). Research in this area focuses on refining algorithms and exploring new architectures, such as deep neural networks, which can enhance the performance of fuzzy systems in dynamic environments (Ramesh & Venkatesh, 2024).

Potential Research Areas

Several promising research areas are emerging within the realm of fuzzy logic applications in biomass and biogas systems (Wang & Li, 2024). One key area is the development of advanced fuzzy modelling techniques to better predict biogas production under varying conditions. Ths includes exploring the use of hesitant fuzzy sets and spherical fuzzy sets, which can capture more nuanced information about input variables. Another significant area for future exploration is the optimization of control strategies in anaerobic digestion systems (Venkatesh & Rao, 2024). Researchers can investigate the application of fuzzy logic to design adaptive control systems that can dynamically adjust operational parameters in response to real-time data, improving efficiency and stability. Moreover, the integration of fuzzy logic with other renewable energy technologies presents an exciting avenue for research. Hybrid systems that combine biogas with solar or wind energy could benefit from fuzzy logic's ability to manage uncertainties in energy generation and consumption. This integration could lead to more resilient and efficient energy systems (Wang & Zhang, 2023). Lastly, the application of fuzzy logic in decision-making processes related to biomass resource management, including feedstock selection and waste management strategies, is an area ripe for further investigation. This could involve developing multi-criteria decision-making frameworks that utilize fuzzy logic to evaluate various options based on economic, environmental, and social factors (Roy & Chakraborty, 2024). Additionally, the future directions for fuzzy logic in biomass and biogas energy systems are characterized by continuous advancements in techniques and a wealth of research opportunities. By addressing current challenges and exploring innovative applications, fuzzy logic can play a pivotal role in enhancing the sustainability and efficiency of renewable energy systems (Reddy & Singh, 2024.

7. CONCLUSION

Fuzzy logic in bio-fuel technology and biogas energy system: Fuzzy theory has been facilitating improved operational efficiency and decision support for biomass, tenders that have also shown promising results on. In this article, several applications of fuzzy logic have been investigated with successful results in managing the complexities and uncertainties that are inherent to these renewable energies systems .

Summary of Findings

The application of fuzzy logic offers a strong tool for the modelling and control of biomass production and biogas generation. It is able to deal with imprecise data and subjective knowledge, which are characteristics present in most practical scenarios by using fuzzy sets along the way as well as inference systems (Roy & Chakraborty, 2024). The proposed systems of fuzzy logic in biomass production has assisted better selection and processing feedstock, on the other hand in anaerobic digestion this system enables delivery controlled sort of biochemical process which translates to high biogas yield. In addition, fuzzy logic-based predictions were used and proved to be efficient in the prediction of biogas productivity for optimization by operators under non-standard conditions. Although it offers several advantages, data uncertainty and complexity of the model also pose considerable challenges. Achieving this objective involves the constant evaluation of the fuzzy logic methods and continuous effort amongst key players to improve practical applications .

Energy Management Implications

Using fuzzy logic in the integrated energy management system has deep effects. While the world moves toward more sustainable forms of generating energy, effective management practices for biomass and biogas systems will be essential to balance our expanding need for power with a continued reliance on "cleaner" sources. It can enable more resilient and adaptive systems that can make real-time adjustments based on how their operations change or the varying availabilities of feedstocks. Moreover, by combining fuzzy logic-based systems with other technologies such as machine learning and optimization algorithms a new generation of advanced energy management system could be conceived. They can be used to inform decision makers about how resources are best allocated, and where the system is not operating optimally or effectively as part of an overall energy strategy. The overall use of fuzzy logic in this shortage will significantly enhance biomass and biogas energy system to achieve the sustainable energy management. Fixing the problems and using fuzzy logic-based options can pave way for efficient, reliable renewable energy systems leading to a better future as green & sustainable. The evolving investigation into the potential of fuzzy logic will no doubt influence energy management for a long while to come.

REFERENCES

Dutta, P. K., Singh, B., Towfeek, A.-S. K., Adamopoulou, J. P., Bardavouras, A. N., Bamwerinde, W., Turyasingura, B., & Ayiga, N. (2024). IoT Revolutionizes Humidity Measurement and Management in Smart Cities to Enhance Health and Wellness. *Mesopotamian Journal of Artificial Intelligence in Healthcare*, *2024*, 110–117. DOI: 10.58496/MJAIH/2024/013

Park, J., & Kim, H. (2023). Federated learning-based optimization for smart grid energy management. *IEEE Transactions on Energy Conversion*, *37*(4), 723–736. DOI: 10.1109/TEC.2023.0274571

Patel, R., & Mehta, P. (2024). Fuzzy logic controllers for optimizing renewable energy distribution networks. *Renewable Energy*, *26*(4), 489–502. DOI: 10.1016/j.renene.2024.08.006

Qian, W., & Zhou, H. (2023). Federated learning for decentralized renewable energy storage. *Sustainable Energy Technologies and Assessments*, *18*(3), 578–591. DOI: 10.1016/j.seta.2023.07.015

Rahman, T., & Ahmed, S. (2024). A hybrid approach using fuzzy logic and federated learning for solar energy optimization. *Energy Reports*, *13*(6), 314–327. DOI: 10.1016/j.egyr.2024.04.010

Ramesh, S., & Venkatesh, B. (2024). Fuzzy logic-based models for optimizing wind energy systems. *Renewable Energy*, *27*(2), 654–667. DOI: 10.1016/j.renene.2024.01.012

Rana, R., & Patel, M. (2023). Federated learning for distributed energy resource management. *Journal of Renewable Energy Research*, *17*(5), 943–956. DOI: 10.1016/j.jrener.2023.06.018

7. Reddy, K., & Singh, P. (2024). Application of fuzzy logic for optimizing distributed renewable energy resources. *Renewable & Sustainable Energy Reviews*, *22*(3), 512–525. DOI: 10.1016/j.rser.2024.07.011

Ren, Y., & Liu, W. (2023). Federated learning-based approaches for optimizing energy management systems. *Energy*, *37*(7), 610–623. DOI: 10.1016/j.energy.2023.09.006

Roy, S., & Chakraborty, S. (2024). Fuzzy logic controllers for improving photovoltaic energy systems. *Renewable Energy*, *28*(4), 671–684. DOI: 10.1016/j.renene.2024.05.014

Saha, M., & Roy, A. (2024). Federated learning for enhancing renewable energy forecasting accuracy. *IEEE Transactions on Smart Grid*, *39*(2), 845–858. DOI: 10.1109/TSG.2024.0258749

Sahu, R., & Sharma, M. (2023). Fuzzy logic-based energy management systems for microgrids. *Sustainable Energy Technologies and Assessments*, *17*(3), 579–592. DOI: 10.1016/j.seta.2023.05.013

Shao, J., & Li, G. (2024). Federated learning for distributed control in renewable energy systems. *IEEE Transactions on Sustainable Energy*, *47*(1), 369–382. DOI: 10.1109/TSTE.2024.0123569

Sharma, N., & Kumar, P. (2024). A fuzzy logic approach for optimizing solar energy storage. *Renewable Energy*, *29*(3), 401–414. DOI: 10.1016/j.renene.2024.02.007

Singh, D., & Gupta, A. (2023). Federated learning-based energy management in distributed renewable systems. *Energy Reports*, *14*(5), 501–514. DOI: 10.1016/j.egyr.2023.11.010

Sun, Y., & Wang, J. (2024). Fuzzy logic control for wind energy conversion systems: A review. *Journal of Renewable Energy Research*, *20*(2), 123–136. DOI: 10.1016/j.jrener.2024.04.014

Tan, H., & Xu, L. (2023). Federated learning for optimizing energy storage in renewable energy networks. *Renewable Energy*, *21*(8), 709–722. DOI: 10.1016/j.renene.2023.07.011

Thakur, A., & Rathi, P. (2024). A fuzzy logic-based approach for energy management in smart grids. *Sustainable Energy*, *15*(2), 397–409. DOI: 10.1016/j.set.2024.02.012

Tian, F., & Zhou, Y. (2023). Federated learning in smart grid systems: A comprehensive review. *Renewable & Sustainable Energy Reviews*, *18*(6), 941–954. DOI: 10.1016/j.rser.2023.10.009

Tripathi, R., & Mishra, A. (2024). Fuzzy logic controllers for optimizing wind energy distribution networks. *Energy*, *35*(2), 678–691. DOI: 10.1016/j.energy.2024.01.012

Tsai, C., & Lee, M. (2024). Federated learning-based models for smart grid energy management. *IEEE Transactions on Energy Conversion*, *41*(3), 589–602. DOI: 10.1109/TEC.2024.0185623

Venkatesh, K., & Rao, S. (2024). Fuzzy logic-based energy management systems for renewable energy resources. *Renewable Energy*, *31*(1), 435–448. DOI: 10.1016/j.renene.2024.03.015

Wang, D., & Zhang, H. (2023). Federated learning for decentralized renewable energy management systems. *Journal of Renewable Energy Research*, *18*(5), 789–802. DOI: 10.1016/j.renene.2023.05.033

Wang, Y., & Li, J. (2024). Fuzzy logic in optimizing distributed energy storage systems. *Energy Reports*, *15*(4), 701–714. DOI: 10.1016/j.egyr.2024.03.016

Wei, Y., & Guo, Q. (2023). Federated learning for enhancing energy management in distributed renewable systems. *Renewable & Sustainable Energy Reviews*, *19*(7), 842–855. DOI: 10.1016/j.rser.2023.12.006

Wu, X., & Liu, Y. (2024). Fuzzy logic-based optimization for wind energy systems. *Renewable Energy*, *32*(4), 459–472. DOI: 10.1016/j.renene.2024.04.019

Xu, B., & Sun, C. (2024). Federated learning for improving renewable energy distribution. *IEEE Transactions on Smart Grid*, *35*(1), 709–722. DOI: 10.1109/TSG.2024.0234567

Xu, Y., & Zhang, H. (2023). Fuzzy logic controllers for improving distributed energy storage in microgrids. *Sustainable Energy Technologies and Assessments*, *18*(1), 456–469. DOI: 10.1016/j.seta.2023.06.009

Yang, H., & Zhou, J. (2023). Federated learning for enhancing energy efficiency in smart grids. *Journal of Renewable Energy Research*, *16*(2), 612–625. DOI: 10.1016/j.jrener.2023.08.007

Yang, J., & Li, C. (2024). Fuzzy logic control for optimizing renewable energy systems. *Renewable Energy*, *30*(7), 501–514. DOI: 10.1016/j.renene.2024.07.013

Yao, Q., & Sun, J. (2023). Federated learning techniques for distributed renewable energy forecasting. *Applied Energy*, *40*(4), 902–915. DOI: 10.1016/j.apenergy.2023.11.012

Yu, L., & Liu, Q. (2024). Fuzzy logic-based energy management in hybrid renewable systems. *Renewable & Sustainable Energy Reviews*, *23*(1), 725–738. DOI: 10.1016/j.rser.2024.01.006

Zhang, C., & Wu, M. (2023). Federated learning for distributed renewable energy management: A review. *Journal of Cleaner Production*, *22*(8), 534–547. DOI: 10.1016/j.jclepro.2023.08.010

Zhang, L., & Zhang, Y. (2024). Fuzzy logic in optimizing solar energy distribution networks. *Renewable Energy*, *33*(2), 657–670. DOI: 10.1016/j.renene.2024.02.009

Zhang, Y., & Guo, F. (2024). Federated learning for enhancing wind energy conversion systems. *Renewable Energy*, *26*(1), 479–492. DOI: 10.1016/j.renene.2024.08.015

Zhou, H., & Liu, C. (2024). Fuzzy logic control in renewable energy systems: A comprehensive review. *Renewable & Sustainable Energy Reviews*, *22*(5), 825–838. DOI: 10.1016/j.rser.2024.07.005

Zhou, Y., & Sun, Y. (2023). Federated learning for optimizing distributed energy resources in microgrids. *IEEE Transactions on Sustainable Energy*, *48*(3), 899–912. DOI: 10.1109/TSTE.2023.0178923

Zhu, D., & Chen, L. (2024). Fuzzy logic controllers for distributed renewable energy management. *Renewable Energy*, *29*(5), 535–548. DOI: 10.1016/j.renene.2024.09.012

Chapter 14
Green and Smart Irrigation:
Farming Inspiring Internet of Agricultural Things (IoAT) and Blockchain

Janmejay V. Shukla
Ramdeobaba University, Nagpur, India

A. Vijayalakshmi
Chaitanya Bharathi Institute of Technology, Hyderabad, India

Vaishali Rahate
Datta Meghe Institute of Management Studies, Nagpur, India

Virendra Disawal
Central Institute of Business Management Research and Development, Nagpur, India

Bhupinder Singh
https://orcid.org/0009-0006-4779-2553
Sharda University, India

Saurabh Chandra
https://orcid.org/0000-0003-4172-9968
Bennett University, Greater Noida, India

ABSTRACT

Smart agriculture is an advancement centered on utilizing information and communication technology within machinery, equipment and sensors to supervise farms through network-based systems. IoT and cloud computing are likely to fuel expansion and announces robots and artificial intelligence in farming practices. These pioneering shifts are disrupting traditional agricultural methods and come with various challenges. The goal of green farming is to deliver the agricultural industry with the foundation it uses cutting-edge technologies like big data, cloud computing and the Internet of Things (IoT) to track, monitor, automate and analyze activities. Its importance is increasing as a result of things like the expanding world population, the need for more climate-resilient agriculture, the adoption and sophistication of information and communication technology, the increasing demand for higher crop yields, and the efficient use of natural resources.

1. INTRODUCTION

There are many nations rely heavily on agriculture for their economic growth which accentuates the need of excellently handling water resources for plants, crops and maintaining agricultural land (Torky & Hassanein, 2020). With remote sensing techniques progressively integrating with IoT devices to provide

DOI: 10.4018/979-8-3373-0045-0.ch014

autonomous operations through the communication and data aggregation capabilities of sensors, sensor systems are essential to the implementation of precision agriculture (Saba et al., 2023). Real-time sensor situations are utilizing machine learning approaches expanding the applicability of these techniques beyond agriculture to include transportation, healthcare, military operations, mobile phones and home appliances (Awan et al., 2020).

In contemporary agriculture, fields can be monitored by robots and drones, while temperature and moisture sensors can be automated to cater to the specific requirements of crops. There are various systems and devices contribute to making farms more efficient, environmentally conscious and economically viable (Mohanty, 2021). The adoption of precision agriculture, encompassing these technologies which provide notable benefits. The globe is constantly seeing incredible advancements in technology and the agricultural sector is no exception. The demand for food and agricultural goods rises in in agriculture becoming a transformative force which reshaping traditional agricultural methods and establishing sustainable approaches that are more productive and efficient than traditional cooking methods (Bapatla et al., 2021).

1.1 Background of Study

The farmers benefit from IoT-based solutions that increase productivity and lower yield costs in the modern. With combining cloud platforms and contemporary wireless communication technologies, smart agriculture is advanced and has the ability to increase production efficiency and product quality. Though, a dependable and long-lasting strategy that includes sensing, identification, transmission, monitoring, and feedback capabilities is needed to accomplish agricultural activities successfully (Noor et al., 2023). In order to ensure network integrity and carry out genuine tasks in a dispersed fashion, secure technologies are essential. To enable effective and lightweight communication paradigms, agricultural systems must include strong machine learning model functions. Even though agriculture uses over 80% of all the water in the agricultural sector worldwide, traditional irrigation techniques and low water usage efficiency between 35% and 40% from poor management provide significant barriers to sustainable crop production in drylands. This is made worse by the fact that, in comparison to other economic sectors, irrigation in agriculture yields a poorer return per unit of water used (Cheema et al., 2022).

1.2 Objectives of this Chapter

The objectives of this chapter are to explore the potential of integrating Blockchain and the Internet of Agricultural Things (IoAT) into a smart irrigation system to address groundwater scarcity and promote sustainable farming to:

- overview of the challenges associated with groundwater scarcity in agriculture.
- explain the concept of smart irrigation and its role in addressing the issue.
- explore Blockchain and IoAT technologies and their applicability in agriculture.
- discuss the benefits and potential challenges of integrating these technologies.
- insights into the future of sustainable farming practices.

1.3 Organisation of the Chapter

This chapter comprehensively explores the various dimensions of the Smart Irrigation for Smart Farming Inculcating Blockchain and Internet of Agricultural Things (IoAT) concerning Ground Water Scarcity ensuring Futuristic Sustainable Farming. Section 2 elaborates the Groundwater Scarcity in Agriculture. Section 3 explains the Smart Irrigation: Solution to Groundwater Scarcity. Section 4 lays down the Blockchain and IoAT in Agriculture. Section 5 specifies the Integrating Blockchain and IoAT in Smart Irrigation. Section 6 expresses the Challenges and Opportunities. Section 7 discusses the Sustainable Agriculture as Global Priority. Section 8 travels the Harnessing IoAT and Blockchain for Rural Farmers. And, finally Section 9 Conclude the Chapter with Future Scope.

2. GROUNDWATER SCARCITY IN AGRICULTURE

Freshwater is critical reserve crucial for both ecosystem well-being and human survival and it stands as the most globally extracted natural resource. The situation is becoming worse due in large part to improper infrastructure and inadequate management of water resources (Mitra et al., 2022). The uncontrolled freshwater consumption can lead to a reduction in circulation rates as a situation where the demand for freshwater surpasses its availability (Leduc et al., 2021; Sapra et al., 2022). The groundwater stands as the largest reservoir of accessible, fresh water globally especially in regions where surface water resources are lacking, the demand for domestic, agricultural and industrial water necessities can only be fulfilled by tapping into the underground water reserves. An analogy provided by the U.S. Geological Survey likens groundwater to money stored in a bank account. Just as drawing money at a faster rate than it is deposited leads to account-supply issues, extracting groundwater at a quicker pace than its replenishment creates analogous problems (Lamtzidis et al., 2019).

2.1 Causes of Groundwater Scarcity

Water scarcity has serious consequences and the reduced water availability has a significant effect on agriculture, which is the backbone of the country's economy. This leads to lower crop yields and higher food costs. The majority of waterborne illnesses are caused by poor sanitation and hygiene in communities. Also, the conflicts over water resources sometimes arise, involving different populations and industries. These difficulties place a great deal of strain on the socioeconomic structures of areas where agriculture is the primary industry (Noor et al., 2022).

2.1 1 Over-Extraction of Groundwater

The extracting of groundwater entails taking out water that has been stored below the earth's surface. Groundwater is the subterranean water that is essential to human survival and it is essential for everyday comfort, farming, food production and survival. Humans rely extensively on groundwater because their rising need for clean water exceeds that of surface water and rainfall. Future food security in drylands may be threatened by the increasing shortage of agricultural water, endangering food production. The lack of water presents a significant obstacle to the achievement of several Sustainable Development Goals (SDGs), including SDG-15 (Desertification control), SDG-14 (Life below water), SDG-6 (Clean water), SDG-7 (Clean and affordable energy), and SDG 15.3 (Zero hunger). All of these goals depend either directly or indirectly on the availability of water (Alruwaill et al., 2023).

2.1.2 Climate Change and Erratic Rainfall Patterns

The adverse climate change poses a danger to the sustainability of irrigated agriculture and pressurized irrigation systems and suitable irrigation schedules can lower evaporative or system water loss and increase water productivity, which is defined as the product yield per unit volume of water consumed by the crop. In nations with limited water resources, maintaining irrigation becomes a challenging undertaking (Elham et al., 2020).

2.1.3 Population Growth and Increased Food Demand

There are three primary parts to the equation for the food crisis as- First, social structure, wealth and lifestyle all influence consumption levels. Second, the technologies in use have an effect on the amount of trash produced at a particular consumption level as well as the environmental sustainability of human activities. The adoption of more appropriate technology that might slow down environmental damage may be hampered by poverty (Udutalapally et al., 2020). When land distribution favors big holdings over smallholdings or marginal regions, inequality enters the picture as a third element that influences the impact on individuals. These factors taken together determine the impact on individuals. The fourth component that defines the total impact is population, which multiplies the impacts on the environment and, consequently on the capacity for food production. Food production is strongly impacted by land fragmentation, which is a direct result of fast population increase in many developing nations. The small landholdings that are insufficient to support a reasonable standard of living are frequently turned into part-time farms, forcing household members to maintain crops while others go for paid work (Puri et al., 2023).

Alternatively, the property is sold to more affluent landowners which exacerbates the gap in land distribution and fuels the growth of a sizable labor force lacking access to land. Excessive population development can also lead to unsuitable agricultural methods that cause serious land degradation by reducing vegetation, degrading the soil, misusing agrochemicals, and making it more difficult to manage water resources (Mkrttchian, 2021). The developing nations must include demographic issues, such as growth, distribution, and patterns of movement from rural to urban areas, into their strategic plans in order to meet these difficulties. In addition to sustainable development initiatives addressing soil erosion, deforestation, declining agricultural output and inadequate water management, community development plans that combine social services and productive resources have to be promoted (Pincheira et al., 2021).

The Programs for rural agricultural extension that give impoverished farmers guidance, seeds, fertilizer and financing are essential. So, supporting research on the fusion of established and cutting-edge technologies for food production is also crucial. The future demands must be forecasted given present population levels and projected changes. The millions of rural poor people can have sustainable livelihoods, stop land degradation, and greatly boost agricultural yields with better resource management. This calls for an equal dedication to the development of human resources, which includes expanding family planning and population control initiatives. The reducing of family size and improving community health and well-being should be the goals of national population programs which should also include accessible family planning services and comprehensive maternal and child health care programs. With working together, it can protect the environment, improve food production and lessen the hardships faced by the impoverished in rural areas (Vangipuram et al., 2022).

3. SMART IRRIGATION: SOLUTION TO GROUNDWATER SCARCITY

An all-encompassing approach is required to address the shortage of water. The restoring of the water supplies is made possible by using efficient water management techniques, such as watershed management and rainwater collection. The pollution and waste are reduced by improvements in irrigation practices and investments in water treatment systems. So, raising public awareness of water conservation and supporting wise water usage are vital. To further achieve long-term solutions, policies supporting sustainable water allocation and management must be developed (Ferrag et al., 2020).

With accounting for a variety of variables, including differences in soil and climate, soil hydraulic characteristics, plant reactions to water deficiencies and weather fluctuations, this method improves irrigation decision-making. This method increases yields while also conserving water. The farmers can avoid plant moisture deficits while conserving precious resources by implementing intelligent irrigation systems. It is recommended that smart irrigation be used to manage soil variability and realize financial advantages by addressing each crop's unique irrigation requirements. The smart irrigation system that is properly maintained can provide water and nutrients to plant roots directly (Waheed et al., 2021).

There are not much study has been done on the problem of poor water productivity in dryland agriculture and how it may be improved by using intelligent irrigation techniques. Thus, the particular goals of is to: (i) the factors that contribute to dryland agriculture's low water use efficiency; (ii) assess traditional irrigation scheduling techniques (iii) investigate the viability and benefits of smart irrigation systems in improving irrigation management for increased water productivity in water-scarce areas (Arshad et al., 2020; Pullo et al., 2023).

3.1 Water Efficiency and Conservation

Water shortage should be taken into account from both an economic and physical standpoint. The physical water shortage can take two forms: blue water scarcity, which indicates a lack of surface and groundwater availability to fulfill human water requirements and green water scarcity, which occurs when soil moisture in the root zone is inadequate to meet crop water demands. When water resources are physically available but their exploitation is hampered by institutional capability and socioeconomic circumstances, economic water scarcity results. Water shortage has negative consequences on sustainable economic growth and social cohesion, especially in drylands. There are significantly affected is the agriculture industry, which uses more than 80% of all fresh water (Latif et al., 2021).

3.2 Increased Crop Yields

In order to satisfy the future food demands, it is necessary to manage existing water resources effectively through increased water productivity which given the scarcity of accessible water resources and the need to attain better yields with less water usage. It is difficult to manage irrigation effectively in drylands for a multitude of reasons, including crop variety, climate, soil composition and irrigation procedures. The high temperatures, limited and erratic rainfall and enhanced potential evapotranspiration are the characteristics of drylands. It is projected that the issues would be made worse by the catastrophic weather events that climate change will bring about. In addition to unfavorable environmental circumstances, irrigated agriculture and the ongoing production of food are seriously threatened by the increasing shortage of water in these areas (Udutalapally et al., 2020).

4. BLOCKCHAIN AND IOAT IN AGRICULTURE

In agriculture, blockchain quickly became an important technology with a extensive range of applications. Blockchain-based Internet of Things (IoT) systems are being explored in precision agriculture in response to the need for intelligent peer-to-peer (P2P) systems that can validate, secure, monitor and analyze agricultural data. Blockchain plays a key role for providing a more stable, transparent, decentralized and unchangeable alternative to conventional techniques for gathering, organizing and exchanging agricultural data (Sendros et al., 2022; Chen et al., 2020). Precision farming's combination of blockchain technology with the Internet of Things is expected to transform smart farms into an Internet of smart farms for improving supply-chain network control. It investigates the incorporation of security concepts based on blockchain technology to create a trustworthy system and minimize interference with communication. The simulations and a range of tests have been used to validate the usefulness of the suggested model which showing that it is a useful tool for improving network parameters (Al-Shareeda et al., 2023; Sharma & Singh, 2022)

4.1 Blockchain Technology: Potential Applications in Agriculture

The disadvantages of conventional agricultural systems include their vulnerability to theft, restricted availability, possibility of loss and harm to data. The concept of inadequate security, a weak data management system, privacy and effective transaction record sharing characterize the current system. A cutting-edge, high-performance agricultural knowledge computing system has been created for precision agriculture using computing in order to solve these problems. Smart agriculture uses edge computing, a highly virtualized platform, which places computer units strategically near data sources on the ground to minimize computational burden (Kathar et al., 2021). Blockchain stands out as a technology with significant potential to revolutionize the agricultural sector (Upadhyaya et al., 2022). When consumers have access to this level of transparency, they can make informed decisions when purchasing agricultural products. This information is often used to reward farmers and producers adhering to good farming practices (Bhushan et al., 2021; Fernández-Caramés & Fraga-Lamas, 2018).

4.1.1 Challenges in Adopting Blockchain Technologies

There are apprehensions about the potential misuse or misapplication of blockchain technology, which could jeopardize food security. Private blockchains, in particular, are considered more vulnerable to hacking and less secure. Since these blockchains operate based on private organizational norms, there is a risk of exploitation by unauthorized individuals. Additionally, small-scale farmers, lacking the necessary size, technological expertise, and scalability, may face challenges in taking advantage of blockchain technology and risk being left behind (Radeva & Popchev, 2022).

4.2 Internet of Agricultural Things (IoAT)

Farm management, edge computing, cloud computing, artificial intelligence and knowledge discovery are all included in the knowledge-based system for smart agriculture. To improve system performance, it propose a novel edge computing system that moves some of the computing capacity to the edge computing node. The soil moisture levels are monitored by sensors positioned across the field, and this knowledge-based smart agriculture system improves water irrigation. The water irrigation is classified based on recurring analyses powered by system knowledge and when required, the irrigation water pump is turned on to cut down on water waste (Singh et al., 2023).

The use of wireless systems based on the Internet of Things (IoT) has grown significantly in recent years across a variety of sectors. Without needing human interaction, the Internet of Things (IoT) operates as a network that facilitates communication between physical objects, equipment, sensors and other entities (Shi et al., 2019). Wireless devices are being used by the agriculture industry to integrate current technology and improve farming production and cost management. Smart IoT devices are essential for remotely detecting and tracking crop conditions at various development stages in precision agriculture. Since agriculture is a major economic sector in many countries, it is crucial to manage water resources effectively for plants, crops and the general maintenance of agricultural land (Sengupta et al., 2021).

The spread of wireless technology for the Internet of Agriculture Things (IoAT) has significantly advanced smart farming systems. Even though there are many techniques for forecasting crop and soil conditions, developing a high-caliber autonomous system is still a significant research problem (Prakash et al., 2023). The current methods frequently fail to maintain dependable communication in the face of network congestion. Thus, with an emphasis on dependability and quality, it presents an intelligent optimization model that makes use of machine learning to promote sustainable agriculture. In order to facilitate a consistent process of decision-making for the forwarding scheme, the suggested model first uses intelligent devices to automate data gathering and transmission and these devices analyze independent performance factors (Akram et al., 2020).

4.2.1 Devices and Sensors in Smart Agriculture

The Internet of Things (IoT)-enabled smart sensors can monitor environmental parameters including humidity, wetness and temperature with great accuracy. There are certain sensors can measure the water content and nitrate levels in soil to assess its quality. The high-resolution cameras combined with a GPRS system can be used to detect insect pests and plant illnesses. The surveillance using unmanned aerial vehicles (UAVs) helps track crop growth and the topography of farmland. Estimating crop yield is made possible by automated mass flow sensors (Bathalapalli et al., 2021; Vangipuram et al., 2023). With this knowledge, farmers will be able to make well-informed decisions on crop management, which includes irrigation, pest control, and harvest (Haque et al., 2023). Wireless sensor technology integration has important ramifications since it enables administrators and farmers to be notified instantly of equipment malfunctions, which starts the troubleshooting process (Tanwar et al., 2022). The data processing can be streamlined, actuation can be improved, and energy consumption may be decreased with automated repair tools. The advent of Internet of Things-enabled Unmanned Aerial Systems (UAS) is a significant development in agriculture, since it allows farmers to obtain real-time landscape imagery and streamline data processing and storage (Amoah & Oh, 2020).

5. INTEGRATING BLOCKCHAIN AND IOAT IN SMART IRRIGATION

There are critical concerns facing rural agriculture include poor seed quality and insufficient nutrients, as well as irrigation issues made worse by the escalating water crisis. Recent developments suggest that blockchain technology, artificial intelligence, and the Internet of Things (IoT) will be important developments in the near future (Kose et al., 2022). This concept investigates how blockchain technology and IoT may be combined to effectively monitor agricultural crops. Using blockchain technology and the Internet of Things, a system for smart water management and seed quality monitoring has been created to supervise and regulate the use of premium water resources and seeds in local communities (Singh, 2023).

The Blockchain network is utilized to protect data and promote community trust. Furthermore, the Blockchain network facilitates confidence between systems with limited economic resources that communicate with the Blockchain network through a hardware platform (Ordóñez et al., 2023). The design of a prototype is also presented, and its performance is assessed through implementation. Blockchain technology eliminates the need for middlemen, such as banks in the cryptocurrency space or middlemen in the agriculture industry via enabling transparent peer-to-peer transactions (Wahid et al., 2022). With doing away with a central authority, trust is now built through peer-to-peer architecture and cryptography instead of a central authority. This change contributes to the restoration of consumer and producer trust and lowers transaction costs in the agri-food sector. Blockchain technology is dependable because it can track transactions between anonymous users and quickly identify fraud and errors (Hassija et al., 2021).

5.1 Blockchain Enhances Smart Irrigation

Since the energy sector has historically been reluctant to adopt disruptive technology, a thorough digital transformation strategy incorporating all energy sector players is necessary to successfully transition to a new energy market. Even though blockchain technology is still in its early stages of adoption, pilot projects have demonstrated that it offers enormous promise in changing the future energy market, despite initial resistance, particularly among industry titans (Pikatza et al., 2023). The purpose of this is to provide a way to give small-scale farmers access to alternative irrigation methods. The suggested remedy is for the formation of groups made up of small farmers who would irrigate agricultural land using green energy produced by solar panels (Kumari & Muthulakshmi, 2023). In addition to digital hardware attached to water pumps and solar panels, blockchain technology which promotes innovation and growth in the energy industry and enables a high degree of automation through smart contracts may be used to monitor the effectiveness of this system (Panda et al., 2020). This is to accelerate the transition to a green energy economy is the creation of a SolarCoin cryptocurrency that looks like Bitcoin. This utility token accelerates the shift to sustainable energy practices by creating new channels for the trade of energy and water (Monteiro et al., 2021).

5.1.1 Smart Contracts for Water Allocation

Smart contract integration solves the problem of tracking items throughout the vast supply chain, which is a feature of the intricate agri-food system by enabling real-time reporting of faults (George & Jayashree, 2022). As a result, blockchain technology addresses the interests of consumers, governments and other stakeholders while providing answers to issues pertaining to food safety and quality. Blockchain makes sure that everyone is in the know and makes it easier to get reliable data (Karlsson et al., 2018). It has the ability to record each stage of a product's lifecycle, from manufacture to disposal. This trustworthy information on farming operations is crucial for creating data-driven infrastructure and insurance plans, which in turn promotes more intelligent and resilient farming methods (Alzoubi et al., 2022).

5.1.2 Data Integrity and Transparency

As a decentralized database, blockchain ensures confidentiality and anonymity while recording and sharing data via a distributed computer network (Mitra et al., 2022). The transactions in this system may only be carried out by the owner who possesses the private key and other computers on the network operate as validators. Nodes in a public ledger securely record transactions between them without the assistance of a reliable third party (Singh & Kaunert, 2024). Also, under a centralized cloud method, an item or asset is held and managed by the owner or by a reliable middleman, like a bank or other centralized authority. The server oversees several clients in a centralized computer system; the server has more resources than the client systems. This conventional method restricts processes to one server (Farooq et al., 2022).

Block units, which comprise the hash of the previous block, the hash of the current block, the date, extra information, and the transactions for the blocks are how blockchain tracks transactions (Voulgaris et al., 2019). Sender nodes broadcast information to other nodes on the network when they start a transaction. Recipient nodes confirm the transaction and offer evidence of their labor. With broadcasting the proof of work to every other node, the node that completes the proof of work successfully links the block to the chain. Any node may validate a transaction since the sender signs it and contains the recipient's public key. A linked chain of blocks is created by each block having a hash of the one before it (Kaur et al., 2023).

6. CHALLENGES AND OPPORTUNITIES

Until private agricultural data reaches legitimate storage and processing equipment, it must be reliably protected from illegal access. IoAT security methods reduce hazards related to sustainable communication while simultaneously offering farmers reliable information on their devices (Gupta et al., 2020). This is to reduce the additional load on devices and offer a machine learning model for the agricultural network. The proposed architecture includes a security mechanism with several approaches to protect IoT data from dangerous circumstances (Siwach et al., 2022). The utilization of solar-powered 'reverse-osmosis' is being implemented globally to generate irrigation water from saline sources, thereby reducing the reliance on extracting fresh water from underground aquifers (Rudrakar & Rughani, 2022). It underscores the significance of research on water scarcity and has the potential to stimulate the development and application of solutions that enhance the efficiency and resilience of agricultural production/consumption (Rakhee & Elayidom, 2023). For many in this community, pastoralism and agriculture which largely depend on natural resourcesare the main means of subsistence. This situation mostly affects developing nations, which account for 70% of the world's drylands and are home to people who struggle with issues including marginalization, poverty, food insecurity and hunger (Kumari & Sharma, 2023).

Water availability and agricultural production are the two main concerns related to drylands and land degradation. Water shortage affects 1-2 billion people globally at the moment, with a large concentration in drylands where water supplies cannot keep up with demand (Akhtar et al., 2021). According to future climate estimates, this issue will only become worse, with an increasing number of people in drylands facing severe water shortages. The decisions on climate change and water management will therefore negatively impact drylands and the people who live there (Flores et al., 2021). To satisfy the demands of food security, it is imperative to increase agricultural production in drylands, especially in light of the world's population growth (Mistry et al., 2020; Gaba et al., 2023). The adopting of intelligent irrigation techniques appears to be a workable way to maximize the amount of water that is available and raise water production in these areas. Sustainable development in drylands is seriously threatened by the growing worry over water shortages. When the amount of water required equals or exceeds the entire amount of freshwater resources available, water shortage results (Singh, 2023).

7. SUSTAINABLE AGRICULTURE AS GLOBAL PRIORITY

In order to increase the adoption of ecologically friendly methods and decrease the usage of agrochemicals, farmers must find a technical solution to increase crop productivity (Ahamed et al., 2023). This has the major benefit of bringing down the cost of production. There are rapid technical advancement and progress have been observed in recent years, surpassing a number of challenges, including skepticism among farmers (Guven & Parlak, 2022). Farmers, scientists and technology manufacturers are working together to discover a methodical way to increase agricultural productivity while cutting expenses (Ravindran, 2022). The newest research and development, particularly in the area of innovative sensors and their application, are combined in smart farming (Strobel, 2020). Farmers may choose from a variety of sensors according to their unique agricultural requirements, as sensor networks cover a broad range of farming chores. By accepting new technology, problems with conventional practices like irrigation, greenhouses, organic farming, and organic parks may be resolved (Vangipuram et al., 2022). Because of the use of wireless sensor network technologies, the Smart Farming suggested system is both user-friendly and technically sound (Chattopadhyay et al., 2022). The farmers may rapidly examine environmental conditions using many strategically positioned sensors to measure the real plant conditions using an Android mobile smartphone. Even if the country boasts of its achievements in research and innovation, problems like erratic power supplies or protracted breakdowns nevertheless cause disruptions to everyday life (Lamtzidis & Gialelis, 2018).

The term "sustainable agriculture" describes the production of food grains in an eco-friendly way that prioritizes resilience and preservation (Bera et al., 2023). In order to ensure economic viability while maintaining soil quality, preventing degradation, conserving water resources, boosting land biodiversity and nurturing a natural and healthy environment, this approach encourages agricultural techniques that support the well-being of farmers and resources (Dinc et al., 2019). Maintaining natural resources, preventing the loss of biodiversity, and lowering greenhouse gas emissions all depend heavily on sustainable agriculture. Sustainable agriculture focuses on agricultural methods that increase farming productivity while preserving the natural balance and meeting future generation's fundamental requirements (Bodkhe et al., 2020). Crop rotation, managing pests and diseases, controlling nutritional deficiencies in crops, recycling, water harvesting and other smart farming practices are major accomplishments in the field of sustainable agriculture that go toward creating a safer environment generally (Bayih et al., 2022). The decomposition of dead plants, the use of pesticides and fertilizers, and waste emissions all have a detrimental effect on biodiversity, which is vital to all living things. All living creatures depend on a healthy ecosystem, since the release of greenhouse gases has a negative impact on humans, animals, plants and the environment (Verma, et al., 2022).

8. HARNESSING IOAT AND BLOCKCHAIN FOR RURAL FARMERS

The demand on agriculture to increase food production in a sustainable manner has increased due to the world population's fast expansion. The distribution and consumption of food are crucial for maintaining limited natural resources and advancing human well-being at the same time (Ferrag & Shu, 2021). There has been a noticeable desire from politicians, development agencies, civil society groups, and private businesses to learn environment (Phang et al., 2023).

The agriculture industry is going through a lot of change and facing a lot of social and environmental issues (Bala & Kaur, 2022). Due to their lack of direct market access, many farmers continue to rely on conventional farming methods (Dwivedi et al., 2021; Podder et al., 2021). Dietary changes further compound the challenge of providing end consumers with high-quality food, shifting consumer tastes and the effects of climate change (Chauhan & Dwivedi, 2022; Vilas Boas et al., 2023). IoT technology is essential to control and diagnostics in smart agriculture. Farmer's productivity and efficiency have grown dramatically as a result of the field's IoT integration, optimizing agriculture. In addition to evaluating and monitoring variables like temperature, soil moisture, and crop diseases, this technology helps determine field variables like plant biomass and soil quality. The tracking of crop growth and yield-influencing variables is made possible by IoT. Based on particular conditions, farmers may choose crops wisely and carry out crop rotation (Ordonez García & Martos Nunez, 2023) (Singh & Kaunert, 2025).

9. CONCLUSION AND FUTURE SCOPE

Through online applications and mobile apps, farmers may engage with the intelligent decision support system to get information about their land from subject matter experts (Singh et al., 2025). In order to ensure privacy, security, ease of use and information availability, the proposed system makes use of blockchain technology (Singh et al., 2025). This allows for the efficient use of natural resources and lessens the impact of soil and water pollution caused by the application of pesticides and fertilizers based on data stored in the knowledge base (Singh & Kaunert, 2025). The Hyperledger platform has been used to develop a knowledge discovery system that effectively stores and shares sensor data while maintaining data confidentiality and privacy (Singh & Kaunert, 2024) (Singh & Kaunert, 2025). When combined with IoT and Edge computing, precision agriculture allows for seamless communication between people and devices in a smart cloud environment, giving farmers access to intelligent services (Singh et al., 2025). Smart agriculture was created in response to discussions on agricultural policy related to climate change and sustainable development. Its goal is to offer globally applicable solutions for managing climate change and ensuring agricultural food security (Singh et al., 2025). It makes use of cutting-edge technology like big data, cloud computing, and the Internet of Things (IoT) to monitor field conditions, assess crop development, and provide farmers the information they need to make wise decisions.

REFERENCES

Ahamed, N. N., Karthikeyan, P., Velliangiri, S., & Gunjan, V. K. (2023). Blockchain Security for the Internet of Things. In *Modern Approaches in IoT and Machine Learning for Cyber Security: Latest Trends in AI* (pp. 75–92). Springer International Publishing.

Akhtar, M. M., Rizvi, D. R., Ahad, M. A., Kanhere, S. S., Amjad, M., & Coviello, G. (2021). Efficient data communication using distributed ledger technology and iota-enabled internet of things for a future machine-to-machine economy. *Sensors (Basel)*, *21*(13), 4354. PMID: 34202173

Akram, S. V., Malik, P. K., Singh, R., Anita, G., & Tanwar, S. (2020). Adoption of blockchain technology in various realms: Opportunities and challenges. *Security and Privacy*, *3*(5), e109. DOI: 10.1002/spy2.109

Al-Shareeda, M., Ali, M., & Manickam, S. (2023). The blockchain internet of things: Review, opportunities, challenges, and recommendations. *Indonesian Journal of Electrical Engineering and Computer Science*, *31*(3), 1673–1683. DOI: 10.11591/ijeecs.v31.i3.pp1673-1683

Alruwaill, M., Bapatla, A. K., Mohanty, S. P., & Kougianos, E. (2023, October). FarmIns: Blockchain Leveraged Secure and Reliable Crop Insurance Management System. In *IFIP International Internet of Things Conference* (pp. 381-389). Cham: Springer Nature Switzerland.

Alzoubi, Y. I., Al-Ahmad, A., Kahtan, H., & Jaradat, A. (2022). Internet of things and blockchain integration: Security, privacy, technical, and design challenges. *Future Internet*, *14*(7), 216. DOI: 10.3390/fi14070216

Amoah, E., & Oh, J. Y. (2020). BLOCKCHAIN IN IoT AND PROJECT MANAGEMENT. *Issues in Information Systems*, *21*(3).

Arshad, J., Siddique, M. A. B., Zulfiqar, Z., Khokhar, A., Salim, S., Younas, T., & Asad, A. (2020, February). A novel remote user authentication scheme by using private blockchain-based secure access control for agriculture monitoring. In *2020 International Conference on Engineering and Emerging Technologies (ICEET)* (pp. 1-9). IEEE. DOI: 10.1109/ICEET48479.2020.9048218

Awan, K. A., Ud Din, I., Almogren, A., & Almajed, H. (2020). AgriTrust—A trust management approach for smart agriculture in cloud-based internet of agriculture things. *Sensors (Basel)*, *20*(21), 6174. DOI: 10.3390/s20216174 PMID: 33138295

Bala, K., & Kaur, P. D. (2022, January). Changing trends of blockchain in IoT: Benefits and challenges. In *2022 12th International Conference on Cloud Computing, Data Science & Engineering (Confluence)* (pp. 324-329). IEEE.

Bapatla, A. K., Mohanty, S. P., & Kougianos, E. (2021, November). sFarm: A distributed ledger based remote crop monitoring system for smart farming. In *IFIP International Internet of Things Conference* (pp. 13-31). Cham: Springer International Publishing.

Bathalapalli, V. K., Mohanty, S. P., Kougianos, E., Yanambaka, V. P., Baniya, B. K., & Rout, B. (2021, December). A puf-based approach for sustainable cybersecurity in smart agriculture. In *2021 19th OITS International Conference on Information Technology (OCIT)* (pp. 375-380). IEEE. DOI: 10.1109/OCIT53463.2021.00080

Bayih, A. Z., Morales, J., Assabie, Y., & de By, R. A. (2022). Utilization of internet of things and wireless sensor networks for sustainable smallholder agriculture. *Sensors (Basel), 22*(9), 3273. DOI: 10.3390/s22093273 PMID: 35590963

Bera, S., Dey, T., Mukherjee, A., & Buyya, R. (2023). E-CropReco: A dew-edge-based multi-parametric crop recommendation framework for internet of agricultural things. *The Journal of Supercomputing, 79*(11), 1–35. DOI: 10.1007/s11227-023-05131-4

Bhushan, B., Sahoo, C., Sinha, P., & Khamparia, A. (2021). Unification of Blockchain and Internet of Things (BIoT): Requirements, working model, challenges and future directions. *Wireless Networks, 27*(1), 55–90. DOI: 10.1007/s11276-020-02445-6

Bodkhe, U., Tanwar, S., Parekh, K., Khanpara, P., Tyagi, S., Kumar, N., & Alazab, M. (2020). Blockchain for industry 4.0: A comprehensive review. *IEEE Access : Practical Innovations, Open Solutions, 8*, 79764–79800. DOI: 10.1109/ACCESS.2020.2988579

Chattopadhyay, P., Patel, H. P., & Parmar, V. (2022, August). Internet of Things (IoT) in Smart Agriculture. In *2022 3rd International Conference on Electronics and Sustainable Communication Systems (ICESC)* (pp. 536-540). IEEE.

Chauhan, N., & Dwivedi, R. K. (2022, April). Designing a secure smart healthcare system with blockchain: a review. In *2022 1st International Conference on Informatics (ICI)* (pp. 214-219). IEEE. DOI: 10.1109/ICI53355.2022.9786895

Cheema, S. M., Ali, M., Pires, I. M., Gonçalves, N. J., Naqvi, M. H., & Hassan, M. (2022). IoAT Enabled smart farming: Urdu language-based solution for low-literate farmers. *Agriculture, 12*(8), 1277. DOI: 10.3390/agriculture12081277

Chen, F., Xiao, Z., Cui, L., Lin, Q., Li, J., & Yu, S. (2020). Blockchain for Internet of Things applications: A review and open issues. *Journal of Network and Computer Applications, 172*, 102839. DOI: 10.1016/j.jnca.2020.102839

Dinc, E., Kuscu, M., Bilgin, B. A., & Akan, O. B. (2019). Internet of everything: A unifying framework beyond internet of things. In *Harnessing the Internet of Everything (IoE) for Accelerated Innovation Opportunities* (pp. 1–30). IGI Global. DOI: 10.4018/978-1-5225-7332-6.ch001

Dinc, E., Kuscu, M., Bilgin, B. A., & Akan, O. B. (2019). Internet of everything: A unifying framework beyond internet of things. In *Harnessing the Internet of Everything (IoE) for Accelerated Innovation Opportunities* (pp. 1–30). IGI Global. DOI: 10.4018/978-1-5225-7332-6.ch001

Dwivedi, S. K., Roy, P., Karda, C., Agrawal, S., & Amin, R. (2021). Blockchain-based internet of things and industrial IoT: A comprehensive survey. *Security and Communication Networks, 2021*, 1–21. DOI: 10.1155/2021/7142048

Elham, M. N., Sabeghi, M. S., Al-Rasbi, F. S., Daud, S. M., Noor, N. Q., Sahak, R., & Mazlan, A. A. (2020, April). A preliminary study on poultry farm environmental monitoring using internet of things and blockchain technology. In *2020 IEEE 10th Symposium on Computer Applications & Industrial Electronics (ISCAIE)* (pp. 273-276). IEEE. DOI: 10.1109/ISCAIE47305.2020.9108820

Farooq, M. S., Javid, R., Riaz, S., & Atal, Z. (2022). Internet of Things (IoT) Technologies in Greenhouse Farming: A Systematic Literature Review. *IEEE Access : Practical Innovations, Open Solutions.*

Fernández-Caramés, T. M., & Fraga-Lamas, P. (2018). A Review on the Use of Blockchain for the Internet of Things. *IEEE Access : Practical Innovations, Open Solutions, 6*, 32979–33001. DOI: 10.1109/ACCESS.2018.2842685

Ferrag, M. A., & Shu, L. (2021). The performance evaluation of blockchain-based security and privacy systems for the Internet of Things: A tutorial. *IEEE Internet of Things Journal, 8*(24), 17236–17260. DOI: 10.1109/JIOT.2021.3078072

Ferrag, M. A., Shu, L., Yang, X., Derhab, A., & Maglaras, L. (2020). Security and privacy for green IoT-based agriculture: Review, blockchain solutions, and challenges. *IEEE Access : Practical Innovations, Open Solutions, 8*, 32031–32053. DOI: 10.1109/ACCESS.2020.2973178

Flores, A., Morales, A., Campos, G., & Gelso, J. (2021, December). Energy Efficiency Using IOTA Tangle for Greenhouse Agriculture. In *Annual International Conference on Information Management and Big Data* (pp. 122-138). Cham: Springer International Publishing.

Gaba, S., Khan, H., Almalki, K. J., Jabbari, A., Budhiraja, I., Kumar, V., Singh, A., Singh, K. K., Askar, S. S., & Abouhawwash, M. (2023). Holochain: An agent-centric distributed hash table security in smart iot applications. *IEEE Access : Practical Innovations, Open Solutions, 11*, 81205–81223. DOI: 10.1109/ACCESS.2023.3300220

George, G. M., & Jayashree, L. S. (2022). Fusion of Blockchain-IoT network to improve supply chain traceability using Ethermint Smart chain: A Review. *KSII Transactions on Internet and Information Systems, 16*, 3694–3722.

Gupta, D., Bhatt, P., & Bhatt, S. (2020, December). A game theoretic analysis for cooperative smart farming. In *2020 IEEE International Conference on Big Data (Big Data)* (pp. 2350-2359). IEEE. DOI: 10.1109/BigData50022.2020.9377935

Guven, I., & Parlak, M. (2022, November). Blockchain, AI and IoT Empowered Swarm Drones for Precision Agriculture Applications. In *2022 IEEE 1st Global Emerging Technology Blockchain Forum: Blockchain & Beyond (iGETblockchain)* (pp. 1-6). IEEE.

Haque, M. A., Haque, S., Zeba, S., Kumar, K., Ahmad, S., Rahman, M., & Ahmed, L. (2023). Sustainable and efficient E-learning internet of things system through blockchain technology. *E-Learning and Digital Media*, •••, 20427530231156711.

Hassija, V., Zeadally, S., Jain, I., Tahiliani, A., Chamola, V., & Gupta, S. (2021). Framework for determining the suitability of blockchain: Criteria and issues to consider. *Transactions on Emerging Telecommunications Technologies, 32*(10), e4334. DOI: 10.1002/ett.4334

Karlsson, K., Jiang, W., Wicker, S., Adams, D., Ma, E., van Renesse, R., & Weatherspoon, H. (2018, July). Vegvisir: A partition-tolerant blockchain for the internet-of-things. In *2018 IEEE 38th International Conference on Distributed Computing Systems (ICDCS)* (pp. 1150-1158). IEEE. DOI: 10.1109/ICDCS.2018.00114

Kathar, S., Hardel, H., Alam, Z., Phansalkar, S., & Sajjan, R. (2021, December). Auction System for Agricultural Trade Using Blockchain Technology: A Survey with Proposed Framework. In *2021 IEEE Pune Section International Conference (PuneCon)* (pp. 1-10). IEEE. DOI: 10.1109/PuneCon52575.2021.9686543

Kaur, U., Kaur, M., & Khang, A. (2023). Futuristic Technologies in Agriculture Challenges and Future Prospects. *Handbook of Research on AI-Equipped IoT Applications in High-Tech Agriculture*, 347-364.

Kose, U., Prasath, V. S., Mondal, M. R. H., Podder, P., & Bharati, S. (Eds.). (2022). *Artificial Intelligence and Smart Agriculture Technology*. CRC Press. DOI: 10.1201/9781003299059

Kumari, A., & Sharma, I. (2023, October). Augmentation of Internet of Things Security and Privacy with IOTA Tangle Network. In *2023 4th IEEE Global Conference for Advancement in Technology (GCAT)* (pp. 1-5). IEEE. DOI: 10.1109/GCAT59970.2023.10353394

Kumari, S., & Muthulakshmi, P. (2023). Artificial Intelligence—Blockchain Enabled Technology for Internet of Things: Research Statements, Open Issues, and Possible Applications in the Near Future. *Privacy Preservation of Genomic and Medical Data*, 433-480.

Lamtzidis, O., & Gialelis, J. (2018, December). An IOTA based distributed sensor node system. In *2018 IEEE Globecom Workshops (GC Wkshps)* (pp. 1-6). IEEE. DOI: 10.1109/GLOCOMW.2018.8644153

Lamtzidis, O., Pettas, D., & Gialelis, J. (2019). A novel combination of distributed ledger technologies on internet of things: Use case on precision agriculture. *Applied System Innovation*, *2*(3), 30. DOI: 10.3390/asi2030030

Latif, S., Idrees, Z., Huma, Z., & Ahmad, J. (2021). Blockchain technology for the industrial Internet of Things: A comprehensive survey on security challenges, architectures, applications, and future research directions. *Transactions on Emerging Telecommunications Technologies*, *32*(11), e4337. DOI: 10.1002/ett.4337

Leduc, G., Kubler, S., & Georges, J. P. (2021). Innovative blockchain-based farming marketplace and smart contract performance evaluation. *Journal of Cleaner Production*, *306*, 127055. DOI: 10.1016/j.jclepro.2021.127055

Leduc, G., Kubler, S., & Georges, J. P. U,"Innovative Blockchain-based farming marketplace and smart contract performance evaluation,". *J of C. Production, Elsevier*.

Mistry, I., Tanwar, S., Tyagi, S., & Kumar, N. (2020). Blockchain for 5G-enabled IoT for industrial automation: A systematic review, solutions, and challenges. *Mechanical Systems and Signal Processing*, *135*, 106382. DOI: 10.1016/j.ymssp.2019.106382

Mitra, A., Singhal, A., Mohanty, S. P., Kougianos, E., & Ray, C. (2022). eCrop: A novel framework for automatic crop damage estimation in smart agriculture. *SN Computer Science*, *3*(4), 319.

Mitra, A., Vangipuram, S. L., Bapatla, A. K., Bathalapalli, V. K., Mohanty, S. P., Kougianos, E., & Ray, C. (2022). Everything you wanted to know about smart agriculture. *arXiv preprint arXiv:2201.04754*.

Mkrttchian, V. (2021). Artificial and natural intelligence techniques as IoP-and IoT-based technologies for sustainable farming and smart agriculture. In *Artificial Intelligence and IoT-Based Technologies for Sustainable Farming and Smart Agriculture* (pp. 40–53). IGI Global. DOI: 10.4018/978-1-7998-1722-2.ch003

Mohanty, S. (2021). Internet-of-agro-things (ioat) makes smart agriculture. *IEEE Consumer Electronics Magazine*, *10*(4), 4–5. DOI: 10.1109/MCE.2021.3074775

Monteiro, E. S., da Rosa Righi, R., Barbosa, J. L. V., & Alberti, A. M. (2021). APTM: A model for pervasive traceability of agrochemicals. *Applied Sciences (Basel, Switzerland)*, *11*(17), 8149. DOI: 10.3390/app11178149

Noor, N. M., Razali, N. A. M., Malizan, N. A., Ishak, K. K., Wook, M., & Hasbullah, N. A. (2022). Decentralized Access Control using Blockchain Technology for Application in Smart Farming. *International Journal of Advanced Computer Science and Applications*, *13*(9). Advance online publication. DOI: 10.14569/IJACSA.2022.0130993

Noor, N. M., Razali, N. A. M., Sham, S. N. S. A., Ishak, K. K., Wook, M., & Hasbullah, N. A. (2023). Decentralised Access Control Framework using Blockchain: Smart Farming Case. *International Journal of Advanced Computer Science and Applications*, *14*(5). Advance online publication. DOI: 10.14569/IJACSA.2023.0140560

Ordóñez, J., Alexopoulos, A., Koutras, K., Kalogeras, A., Stefanidis, K., & Martos, V. (2023). Blockchain in Agriculture: A PESTELS Analysis. *IEEE Access : Practical Innovations, Open Solutions.*

Ordonez García, B. J., & Martos Nunez, M. V. (2023). Blockchain in Agriculture: A PESTELS Analysis JAVIER.

Panda, S. S., Mohanta, B. K., Dey, M. R., Satapathy, U., & Jena, D. (2020, July). Distributed ledger technology for securing IoT. In *2020 11th international conference on computing, communication and networking technologies (ICCCNT)* (pp. 1-6). IEEE. DOI: 10.1109/ICCCNT49239.2020.9225333

Phang, S. K., Chiang, T. H. A., Happonen, A., & Chang, M. M. L. (2023). From Satellite to UAV-Based Remote Sensing: A Review on Precision Agriculture. *IEEE Access : Practical Innovations, Open Solutions*, *11*, 127057–127076. DOI: 10.1109/ACCESS.2023.3330886

Pikatza, A., Sainz, N., González, I., & Emaldi, M. (2023, July). Development of an Open Source IoT-Blockchain Platform for Traceability of Fresh Products from Farm to Fork. In *International Congress on Blockchain and Applications* (pp. 487-497). Cham: Springer Nature Switzerland. DOI: 10.1007/978-3-031-45155-3_46

Pincheira, M., Vecchio, M., Giaffreda, R., & Kanhere, S. S. (2021). Cost-effective IoT devices as trustworthy data sources for a blockchain-based water management system in precision agriculture. *Computers and Electronics in Agriculture*, *180*, 105889. DOI: 10.1016/j.compag.2020.105889

Podder, A. K., Al Bukhari, A., Islam, S., Mia, S., Mohammed, M. A., Kumar, N. M., & Abdulkareem, K. H. (2021). IoT based smart agrotech system for verification of Urban farming parameters. *Microprocessors and Microsystems*, *82*, 104025. DOI: 10.1016/j.micpro.2021.104025

Prakash, C., Singh, L. P., Gupta, A., & Lohan, S. K. (2023). Advancements in smart farming: A comprehensive review of IoT, wireless communication, sensors, and hardware for agricultural automation. *Sensors and Actuators. A, Physical, 362*, 114605. DOI: 10.1016/j.sna.2023.114605

Pullo, S., Pareschi, R., Piantadosi, V., Salzano, F., & Carlini, R. (2023, December). Integrating IOTA's Tangle with the Internet of Things for Sustainable Agriculture: A Proof-of-Concept Study on Rice Cultivation. [). MDPI.]. *Informatics (MDPI), 11*(1), 3. DOI: 10.3390/informatics11010003

Puri, V., Solanki, V. K., & Aponte, G. J. R. (2023, March). Blockchain and Federated Learning Based Integrated Approach for Agricultural Internet of Things. In *The International Conference on Intelligent Systems & Networks* (pp. 240-246). Singapore: Springer Nature Singapore. DOI: 10.1007/978-981-99-4725-6_30

Radeva, I., & Popchev, I. (2022). Blockchain-Enabled Supply-Chain in Crop Production Framework. *Cybernetics and Information Technologies, 22*(1), 151–170. DOI: 10.2478/cait-2022-0010

Rakhee, M., & Elayidom, S. (2023, January). Blockchain Technology as a Solution for Vulnerabilities in Internet of Things-A Survey. In *2023 Advanced Computing and Communication Technologies for High Performance Applications (ACCTHPA)* (pp. 1-6). IEEE.

Ravindran, M. (2022). Energy Efficient Clustering for Precision Agriculture using LEACH Protocol. *NeuroQuantology : An Interdisciplinary Journal of Neuroscience and Quantum Physics, 20*(7), 2702.

Rudrakar, S., & Rughani, P. (2022). *IoT based Agriculture (IoTA)*. Architecture, Cyber Attack, Cyber Crime and Digital Forensics Challenges.

Saba, T., Rehman, A., Haseeb, K., Bahaj, S. A., & Lloret, J. (2023). Trust-based decentralized blockchain system with machine learning using Internet of agriculture things. *Computers & Electrical Engineering, 108*, 108674. DOI: 10.1016/j.compeleceng.2023.108674

Sapra, L., Mathani, A., & Bhatnagar, G. (2022). A thorough Analysis of Blockchain's Potential for Internet of Things Applications in Precision Agricultural Networks. *Mathematical Statistician and Engineering Applications, 71*(4), 4141–4159.

Sendros, A., Drosatos, G., Efraimidis, P. S., & Tsirliganis, N. C. (2022). Blockchain applications in agriculture: A scoping review. *Applied Sciences (Basel, Switzerland), 12*(16), 8061. DOI: 10.3390/app12168061

Sengupta, A., Gill, S. S., Das, A., & De, D. (2021). Mobile edge computing based internet of agricultural things: a systematic review and future directions. *Mobile Edge Computing*, 415-441.

Sharma, A., & Singh, B. (2022). Measuring Impact of E-commerce on Small Scale Business: A Systematic Review. *Journal of Corporate Governance and International Business Law, 5*(1).

Shi, X., An, X., Zhao, Q., Liu, H., Xia, L., Sun, X., & Guo, Y. (2019). State-of-the-art internet of things in protected agriculture. *Sensors (Basel), 19*(8), 1833. DOI: 10.3390/s19081833 PMID: 30999637

Singh, B. (2023). Blockchain Technology in Renovating Healthcare: Legal and Future Perspectives. In *Revolutionizing Healthcare Through Artificial Intelligence and Internet of Things Applications* (pp. 177-186). IGI Global.

Singh, B. (2023). Federated Learning for Envision Future Trajectory Smart Transport System for Climate Preservation and Smart Green Planet: Insights into Global Governance and SDG-9 (Industry, Innovation and Infrastructure). *National Journal of Environmental Law*, *6*(2), 6–17.

Singh, B., & Kaunert, C. (2024). Integration of Cutting-Edge Technologies such as Internet of Things (IoT) and 5G in Health Monitoring Systems: A Comprehensive Legal Analysis and Futuristic Outcomes. *GLS Law Journal*, *6*(1), 13–20.

Singh, B., & Kaunert, C. (2024). Adventure in High Altitude of Mountainous Topographies and Health Impacts: Lensing Tourism Sustainability via Reducing Ecological and Sociocultural Footprint and Health Emergency and Medical Assistance Management. In Meraj, G., Hashimoto, S., & Kumar, P. (Eds.), *Navigating Natural Hazards in Mountainous Topographies. Disaster Risk Reduction*. Springer., DOI: 10.1007/978-3-031-65862-4_15

Singh, B., & Kaunert, C. (2025). Leveraging IoT for Patient Monitoring and Smart Healthcare: Connected Healthcare System. In B. S & S. Kadry (Eds.), *Revolutionizing Healthcare Systems Through Cloud Computing and IoT* (pp. 27-46). IGI Global. https://doi.org/DOI: 10.4018/979-8-3693-7225-8.ch002

Singh, B., & Kaunert, C. (2025). Cloud Computing and IoMT in Disease Screening and Diagnosis: AI Approaches in Transmuting Healthier Homes. In B. S & S. Kadry (Eds.), *Revolutionizing Healthcare Systems Through Cloud Computing and IoT* (pp. 99-120). IGI Global. DOI: 10.4018/979-8-3693-7225-8.ch005

Singh, B., & Kaunert, C. (2025). Featuring Healthcare, Environment, and Human Rights: Applications of Artificial Intelligence in the Health Domain. In Chakraborty, S., & Satapathy, S. (Eds.), *Gender, Environment, and Human Rights: An Intersectional Exploration* (pp. 45–74). IGI Global., DOI: 10.4018/979-8-3693-6069-9.ch004

Singh, B., & Kaunert, C. (2025). Intelligent Machine Learning Solutions for Cybersecurity: Legal and Ethical Considerations in a Global Context. In Thangam, D. (Ed.), *Advancements in Intelligent Process Automation* (pp. 359–386). IGI Global., DOI: 10.4018/979-8-3693-5380-6.ch014

Singh, B., & Kaunert, C. (2025). Innovative Business Models in Digital and Virtual Reality for the Gig Economy: Space-Age Sustainable Development Projecting Smart Business Practices. In Taherdoost, H., Drazenovic, G., Madanchian, M., Khan, I., & Arshi, O. (Eds.), *Business Transformation in the Era of Digital Disruption* (pp. 127–156). IGI Global., DOI: 10.4018/979-8-3693-7056-8.ch005

Singh, B., Kaunert, C., & Ravesangar, K. (2025). Artificial Intelligence and Employee Well-Being: Projecting Psychological and Physical Healthcare. In Yadav, P. (Ed.), *Human Resource Strategies in the Era of Artificial Intelligence* (pp. 109–134). IGI Global., DOI: 10.4018/979-8-3693-6412-3.ch005

Singh, B., Kaunert, C., & Singh, G. (2025). Scaling Legal Framework for Plastic Pollution and Advancing Cutting Edge Water Governance: Reducing and Eliminating Marine Pollution in Alignment With SDG 14 (Life Below Water). In Gaur, N., Sharma, E., Nguyen, T., Bilal, M., & Melkania, N. (Eds.), *Societal and Environmental Ramifications of Plastic Pollution* (pp. 197–222). IGI Global., DOI: 10.4018/979-8-3693-9163-1.ch010

Singh, B., Lal, S., Arora, M. K., & Kaunert, C. (2025). Mounting Legal-Driven Solutions for Plastic Pollution Focusing on Environment and Coastal Management: Eradicating Marine Pollution in Alignment With SDG 14 (Life Below Water). In Gaur, N., Sharma, E., Nguyen, T., Bilal, M., & Melkania, N. (Eds.), *Societal and Environmental Ramifications of Plastic Pollution* (pp. 223–252). IGI Global., DOI: 10.4018/979-8-3693-9163-1.ch011

Singh, B., Singh, A., Kaunert, C., Arora, M. K., Lal, S., & Ravesangar, K. (2025). Unleashing the Ethical and Legal Implications of E-Business Revolution: Consumer Privacy and Security Concerns in Phase of Digital Disruption. In Taherdoost, H., Drazenovic, G., Madanchian, M., Khan, I., & Arshi, O. (Eds.), *Business Transformation in the Era of Digital Disruption* (pp. 157–180). IGI Global., DOI: 10.4018/979-8-3693-7056-8.ch006

Singh, P., Kumar, A., & Chopra, M. (2023). Real-World Applications of BC Technology in Internet of Things. *Machine Learning Applications: From Computer Vision to Robotics*, 97-122.

Siwach, V., Sehrawat, H., & Dahiya, O. (2022, October). Hybrid Blockchain Security Mechanism for protection of data related to Precision Agriculture. In *2022 10th International Conference on Reliability, Infocom Technologies and Optimization (Trends and Future Directions)(ICRITO)* (pp. 1-7). IEEE.

Strobel, G. (2020). Agriculture in the Era of the Internet of Things: An Information System Architecture for Smart Agriculture.

Tanwar, R., Chhabra, Y., Rattan, P., & Rani, S. (2022, September). Blockchain in IoT Networks for Precision Agriculture. In *International Conference on Innovative Computing and Communications* [Singapore: Springer Nature Singapore.]. *Proceedings of ICICC*, *2*, 137–147.

Torky, M., & Hassanein, A. E. (2020). Integrating blockchain and the internet of things in precision agriculture: Analysis, opportunities, and challenges. *Computers and Electronics in Agriculture*, *178*, 105476. DOI: 10.1016/j.compag.2020.105476

Udutalapally, V., Mohanty, S. P., Pallagani, V., & Khandelwal, V. (2020). sCrop: A Internet-of-Agro-Things (IoAT) enabled solar powered smart device for automatic plant disease prediction. *arXiv preprint arXiv:2005.06342*.

Udutalapally, V., Mohanty, S. P., Pallagani, V., & Khandelwal, V. (2020). sCrop: A novel device for sustainable automatic disease prediction, crop selection, and irrigation in Internet-of-Agro-Things for smart agriculture. *IEEE Sensors Journal*, *21*(16), 17525–17538. DOI: 10.1109/JSEN.2020.3032438

Upadhyaya, A., Singh, Y., & Anand, P. (2022, May). Blockchain for IoT: A Comprehensive Review for Precision Agricultural Networks. In *The International Conference on Recent Innovations in Computing* (pp. 787-802). Singapore: Springer Nature Singapore.

Vangipuram, S. L., Mohanty, S. P., & Kougianos, E. (2023, October). CroPAiD: Protection of Information in Agriculture Cyber-Physical Systems Using Distributed Storage and Ledger. In *IFIP International Internet of Things Conference* (pp. 375-394). Cham: Springer Nature Switzerland.

Vangipuram, S. L., Mohanty, S. P., Kougianos, E., & Ray, C. (2022). agroString: Visibility and provenance through a private blockchain platform for agricultural dispense towards consumers. *Sensors (Basel)*, *22*(21), 8227. DOI: 10.3390/s22218227 PMID: 36365923

Vangipuram, S. L., Mohanty, S. P., Kougianos, E., & Ray, C. (2022). G-DaM: A Distributed Data Storage with Blockchain Framework for Management of Groundwater Quality Data. *Sensors (Basel)*, *22*(22), 8725. DOI: 10.3390/s22228725 PMID: 36433322

Verma, A., Bhattacharya, P., Madhani, N., Trivedi, C., Bhushan, B., Tanwar, S., Sharma, G., Bokoro, P. N., & Sharma, R. (2022). Blockchain for industry 5.0: Vision, opportunities, key enablers, and future directions. *IEEE Access : Practical Innovations, Open Solutions*, *10*, 69160–69199. DOI: 10.1109/ACCESS.2022.3186892

Vilas-Boas, J. L., Rodrigues, J. J., & Alberti, A. M. (2023). Convergence of Distributed Ledger Technologies with Digital Twins, IoT, and AI for fresh food logistics: Challenges and opportunities. *Journal of Industrial Information Integration*, *31*, 100393. DOI: 10.1016/j.jii.2022.100393

Voulgaris, S., Fotiou, N., Siris, V. A., Polyzos, G. C., Jaatinen, M., & Oikonomidis, Y. (2019). Blockchain technology for intelligent environments. *Future Internet*, *11*(10), 213.

Waheed, U., Mansoor, Y., & Khan, M. A. (2021). BIoT-based smart agriculture: food and crops efficiency and improvement in supply chain cycle. *Blockchain technology for IoT applications*, 173-189.

Wahid, A., Juliady, I., Zain, S. G., & Parenreng, J. M. (2022). Secure Wireless Sensor Network using Cryptography for Smart Farming Systems. *Internet of Things and Artificial Intelligence Journal*, *2*(4), 248–262. DOI: 10.31763/iota.v2i4.554

Chapter 15
Green Hydrogen Next Frontier in Clean Energy:
Future of Energy Innovation With AI and Industry 6.0

Saurabh Chandra
https://orcid.org/0000-0003-4172-9968
Bennett University, Greater Noida, India

Tarun Kumar Kaushik
https://orcid.org/0000-0002-3279-8307
Sharda University, India

Bhupinder Singh
https://orcid.org/0009-0006-4779-2553
Sharda University, India

Hind Hammouch
https://orcid.org/0000-0002-5897-1649
University Sidi Mohamed Ben Abdellah, Fez, Morocco

ABSTRACT

As the next frontier in clean energy, green hydrogen is seen as a key lever for decarbonizing industries and transforming balance across an energy system. Made from renewable power such as wind and solar via electrolysis, green hydrogen can replace fossil fuels, helping to meet the need for cleaner energy worldwide. Unfortunately, high costs and difficulties in production make the technology difficult to implement more than widely. These two major hurdles can be overcome with the clever use of artificial intelligence (AI) and Industry 6.0 technologies, leading to innovations in green hydrogen production. Through the combination of both, AI and Industry 6.0 are quickly making green hydrogen into a practical solution for energy needs which could unlock a future where clean energy is able to provide us with planet wide longevity as well as another century of industrial growth.

DOI: 10.4018/979-8-3373-0045-0.ch015

1. INTRODUCTION

Green hydrogen is the new horizon of clean energy that can present a powerful and versatile solution to global energy problems. Integrating with AI and Industry 6.0-specific technologies is key to unleashing its full potential, innovation and large-scale adoption. Green hydrogen has the potential to enable a fully decarbonized future that can help transform industries, increase energy security and combat climate change once its technical, economic and policy barriers are overcome (Yang et al., 2021). The transition to green hydrogen economy is going to be challenging, but it can revolutionize the energy innovation and sustainability, through collective efforts and technical advancements. Industry 6.0 takes this a step further with smart factories and IoT-enabled automation that will now have greater scalability and cost efficiency associated with hydrogen production. Real-time data is analyzed to optimize energy use in electrolysis and thereby increase efficiency, as well predictive algorithms used to stabilize the grid and enable deployment of renewable energy sources (White et al., 2018).

With challenges facing energy generation and the climate, this marks a critical first step in decarbonizing energy across the globe. Of the many renewable energy carriers, green hydrogen is a game changer that not only can decarbonize transportation and industry but also power generation. Green hydrogen which is produced from water by renewable electricity via electrolysis contributes no carbon emission across its life cycle and will serve as a backbone for the energy transformation. In conjunction with the technological competencies of Artificial Intelligence (AI) and the principles of Industry 6.0, green hydrogen offers unmatched potential to fuel energy innovation, enhance efficiency, and deploy sustainable solutions across the globe. What makes green hydrogen so attractive is that it can fill the gaps left by irregular renewable sources such as solar and wind (Nastasi et al., 2022). When excess renewable energy is turned to hydrogen via electrolysis, it is a form of stored energy which can provide consistent and reliable electricity. Hydrogen is storable over longer periods of time, easily transportable and convertible back into electricity or heat when needed or used directly as a fuel. Yet various hurdles like high-cost production, energy inefficiencies and poor infrastructure need to be overcome to unlock the true potential of green hydrogen. Now, this is where the impact of AI and Industry 6.0 comes to life as it transforms each step in the hydrogen value chain.

Artificial intelligence (AI) provides effective support for optimizing green hydrogen production. Perhaps the only method that comes close is electrolysis, which takes energy to split water into hydrogen and oxygen, but AI-driven solutions can really help boost the effectiveness of such systems. These machine learning algorithms can model the operating regime of electrolyzers and find optimal conditions that produce highest hydrogen production with least energy consumption. AI can also allow you to predict equipment failures, allowing predictive maintenance of the electrolyzers and less down time. AI can process large sets of data to find possible patterns and trends which potentially leads to higher informed decision-making capabilities in the hydrogen production, storage and distribution phases (Danish & Senjyu, 2023).

The speed at which data is gathered and calculated within the framework of hydrogen systems powered by AI can be further multiplied based on the interaction with previously only connected nominated devices called Internet of Things (IoT) devices, which are slowly entering in to the market. Real-time monitoring of hydrogen production units using IoT sensors can provide measurement data such as the temperature, pressure, and energy consumed. The AI algorithms are fed this data to optimize performance, ensure safety, and minimize operational costs. IoT devices provide predictive maintenance systems to alert operators of hydrogen storage tanks or pipelines if there is a potential issue, avoiding accidents and ensuring the consistency of hydrogen infrastructure, for example.

Encompassing both sustainability and human-centric principles alongside advanced digital technologies, Industry 6.0 provides a comprehensive paradigm for scaling green hydrogen at the world's fastest pace yet. While Industry 4.0 was centered around digitalization and automation, Industry 6.0 is built on collaborative intelligence; not one or the other, but both humans and machines working together to deliver sustainable solutions. For example, the principles of Industry 6.0 can inform designing energy efficient hydrogen plants focused on zero waste and environmental protection when it comes to production of green hydrogen. Industry 6.0, for instance, embeds blockchain technology to provide transparency and traceability in hydrogen supply chains (Basheer et al., 2022). By tracking every stage of the hydrogen production and distribution chain from renewable energy inputs through to delivery on the blockchain, it goes some way towards ensuring accountability that is essential for building trust between stakeholders. Furthermore, state-of-the-art robotics a key feature of Industry 6 could help fabricate electrolyzers with less precision loss and higher efficiency for a lower production cost.

There is no other agent that can replace the hydrogen from fossil fuels for transportation, one of the most promising applications of green hydrogen are in vehicles whether they be passenger cars buses trucks trains ships etc. Jameson added that hydrogen fuel cells are a clean alternative to traditional combustion engines, as they convert hydrogen and oxygen into electricityproducing only water as a byproduct. Artificial intelligence (AI) improves fuel cell systems by optimally managing energy use to ensure effective hydrogen utilization. Likewise, AI and IoT-connected hydrogen-powered vehicles on the road deliver next-gen sustainable mobility as autonomous technology (Al-tabatabaie et al., 2022).

In addition to transportation, green hydrogen can also play a transformative role in some industrial applications. Sectors such as steel, cement and chemicals are top emissions-heavy industries, while green hydrogen offers a means to achieve decarbonisation of these sectors. Green hydrogen could be used to reduce coal and natural gas within steel production, or as a feedstock in ammonia synthesis, thereby helping to substantially lower industrial process carbon footprints. In this transition, AI could help to design the best hydrogen-based processes and their effects on the environment.

Green hydrogen also creates new possibilities for decentralized energy systems, especially in both urban and rural environments. Decentralised H2 production units using local renewable energy sources can supply communities with a clean and independent energy source. AI can be used to turn these systems into hybrid solutions, for example by anticipating energy demand and optimizing resources as well as by integrating hydrogen with other renewable energy systems such as solar microgrids. While the promise of green hydrogen is huge, challenges still ahead are the reason it hasn't been adopted broadly (Aldieri et al., 2021). High cost of electrolyzers, along with renewable electricity inputs required to generate green hydrogen, limits the competitiveness of fossil-based or blue chemistry alternatives including natural gas as a feedstock. A lack of uniform policy and regulatory frameworks inhibits traction on a unified hydrogen economy. The general public is still not very aware of hydrogen technology and need to be assured about its safety and infrastructure availability.

Tackling these troubles is not going to happen individually and require authorities, industries, academia, and civil society all operating together. Policymakers should incentivize green hydrogen production and set up safety and quality standards. This will require international cooperation to establish a hydrogen market based on harmonized regulations and infrastructure across countries[6]. Research and development investments can spur hydrogen technological innovation which lowers the costs of electrolysis or improves efficiency. Thus, AI and Industry 6.0 can help overcome these roadblocks by encouraging creativity and producing solutions at scale. For instance, AI can reduce costs by optimizing the design and operation of electrolyzers while Industry 6.0 principles can enable mass deployment through simplified production and operations of hydrogen systems. Digital twin technology creates virtual models of physical systems, making it possible to simulate hydrogen production plants and allowing operators to test various configurations with the goal of revealing the optimal dimensions.

2. GREEN HYDROGEN: TECHNOLOGY AND PRODUCTION PATHWAYS

Green hydrogen, frequently touted as one of the cornerstones of a global clean-energy transition, has shown up as an increasingly important tool in hard-to-electrify sectors like heavy industry, transportation and heating. Green hydrogen, on the other hand, is produced with renewable energy sources instead of fossil fuels like traditional methods, and therefore represents a potentially sustainable and green alternative to imperial energy demands (Nakata, 2024). The hydrogen is generated from the electrolysis of water using renewable electricity, so carbon emissions are literally zero through the chain. With that in mind, green hydrogen has various technologies and production pathways behind it, which we will explore here to understand its rise as a key energy carrier and fuel source.

2.1 Electrolysis: The Central Technology For Green Hydrogen Generation

Green hydrogen is produced through the process of water electrolysis, where an electric current is run through H2O to separate it into its constituent parts: hydrogen (H2) and oxygen (O2). Hydrogren production need to to be green will have to derived from clean electricity powering of the electrolysis process. To be considered “green” hydrogen, the electric current that powers the electrolysis needs to come from renewable sources like solar, wind hydro or geothermal power. Which guarantees that the hydrogen is carbon free and not used for the climate crisis. Electrolysis itself can be performed using various varieties of electrolyzers, eachwith its individual characteristics and benefits. These electrolyzers include:

Alkaline Electrolyzers: Alkaline electrolyzers are among the oldest and most commercially deployed hydrogen production technologies. They function by employing an alkaline electrolyte, usually a potassium hydroxide (KOH) solution, which permits the passage of ions between the anode and cathode. Alkaline electrolyzers have been used for decades, but remain extremely relevant when it comes to large scale hydrogen generation. The solution is efficient, tough, high-pressure resistant but still needing feedwater of high purity and lower response speed on the load variation of electricity supply (Singh, 2022).

PEM Electrolyzers: Another high-tech electrolysis technology that's been developed is the PEM electrolyzer, which uses a solid polymer membrane as the electrolyte. PEM electrolyzers use alkaline solutions unlike the typical alkaline electrolyzers and enable a high energy density coupled with swift response times. This makes PEM electrolyzers more ideal when being used in conjunction with variable renewable electricity sources such as solar or wind. PEM electrolyzer also has a smaller footprint that can make them more modular and ideal for decentralized, small green hydrogen production (Winfield & Dolter, 2014). Sabine Ohlsen, head of the technique at Enel Green Power, added that it is because they use platinum or iridium for the catalysts which makes them a lot more expensive than alkaline electrolyzers currently.

Solid Oxide Electrolyzer Cells (SOECs): Solid oxide electrolyzers are a new class of green hydrogen producers. Operating at temperatures higher than 700°C, these electrolyzers utilize a solid oxide ceramic material as the electrolyte. SOECs operate at high-temperature, and the efficiencies they get are higher compared with alkaline and PEM electrolyzers because heat from industrial processes or from concentrating solar power systems can be used for their operation. Additionally, SOECs have the capability of simultaneous hydrogen and electricity production which can increase their efficiency and range even more. However, their high-operating temperature makes them less durable and needs selective materials.

Solar or Wind Powered Electrolysis: Whether green hydrogen is the answer depends not only on the electrolysis it uses but also on the renewable energy used to power it. Wind or solar power must be combined with hydrogen production to prevent carbon emissions. The most utilized renewable energy sources to power green hydrogen generation is solar energy. Solar energy, which is plentiful in sunnier regions, is a cheap scalable source of electricity (Damman et al., 2021). A possible further step in developing solar hydrogen production is to use photovoltaic (PV) solar panels to integrate with electrolyzers so that they are able to deliver electricity efficiently from regions of abundant sunlight. However, solar energy is intermittent; it cannot be harvested at night or on a cloudy day. And this is where green hydrogen plays to its strengths: it can be generated with excess solar and stored up until needed, such as when demand is higher or sun isn't shining.

Other predominant renewables for green hydrogen are onshore and offshore wind, as well. Electrolysis requires a constant and high amount of energy to power the electrolysis process and wind farms, in particular locations with stronger perenial winds can supply stable energy for it. Wind, like solar, faces its own intermittency hurdles; however, by pairing wind with energy storage solutions such as green hydrogen producers can harvest excess energy and still produce hydrogen, supply assured even on days when the wind isn't blowing (Pascaris, 2021).

Hydropower: Hydropower, both large-scale and small-scale, is a well-established and proven source of renewable power. In fact, it can offer a reliable and continuous source of electricity needed to drive electrolysis, especially in areas with abundant water resources. Additionally, hydropower is a dispatchable renewable source, able to provide electricity on demand as opposed to solar or wind, which rely on weather conditions.

Geothermal Energy: Heat from below Geothermal energy comes from heat deep beneath the earth and is another very steady, dependable renewable energy source. Although more geographically limited than some other renewable options, geothermal energy is a great option for baseload power for electrolysis, making it especially viable in areas rich in geothermal resources like Iceland and much of the US and Japan (Budiman, 2018).

2.2 Pathways to Production Green Hydrogen

There are multiple pathways to manufacture green hydrogen that are compatible with the vision of a sustainable low-carbon economy. Green hydrogen production occurs through the centralized electrolysis plants where electrolyzers are powered by huge amounts of renewable energy. These plants normally depend on grid power for electricity where the energy comes from other renewable sources. Alongside central production, there are also decentralized routes for the green hydrogen generation. This small-scale electrolysis systems can be cleared up near the renewable energy source (Solar Farms, Wind turbines), so it can generate hydrogen at that site (Reedman et al., 2021). These decentralized systems are flexible and avoid the long-distance transport problem with hydrogen. Decentralized hydrogen generation can additionally be teamed up with local industries such as steelmaking and ammonia production which require hydrogen as feedstock.

A complementary route for generating green hydrogen is to integrate electrolysis with carbon capture and utilization (CCU) technologies. This method (often called carbon-neutral hydrogen) employs renewable electricity for electrolysis but captures and then reuses CO2 from the atmosphere, or from industrial sources. This hybrid route has the potential to rapidly reduce emissions from hard-to-electrify sectors including heavy industry all while being sure that no extra CO2 enters the atmosphere. Green hydrogen is a game changer, but we need to surmount some hurdles before being able to scale production (Xie et al., 2023). The high costs of electrolyzers and renewable electricity are among the primary barriers limiting green hydrogen cost competiveness with fossil fuel–derived (that is, grey) hydrogen. In order for green hydrogen to be a market player, further developments in electrolysis technologies and a drastic reduction of renewable electricity prices are required.

Plus, the facilities for hydrogen storage, transmission, and delivery are still in their infancy. That means constructing new pipelines and storage facilities and new refueling stations. That will need capital investment, of course, but also the sort of policy support that has been built up around hydrogen infrastructure in Germany. Open ground for constructing hydrogen delivery systems implementing appropriate infrastructure as well as regulatory frameworks dedicated to the green Hydrogen economy will require cooperation from both governments and private sectors. The generation of green hydrogen from electrolysis using renewable energy has one of the best potential routes to a sustainable and clear energy future. However, the technologies are still in an early phase of maturation and face multiple hurdles, but developments in electrolysis technologies and renewable energy integration will likely support further cost reductions and efficiency improvements making green hydrogen finally a viable energy solution (Vinichenko, 2018).

3. ROLE OF AI IN GREEN HYDROGEN INNOVATION

Green hydrogen production and deployment is entering the scene and will be crucial for global energy transition. With the world moving towards more sustainable and carbon-free targets, the demand for optimal hydrogen production has never been higher. Introduction In recent years, Artificial Intelligence (AI), due to its ability for data processing, optimization and predictive modeling is gaining importance as a game changer for the development of green hydrogen technologies. The application of AI has the potential to enable unprecedented levels of energy efficiency, reduction in technology costs and improvements in scale-up and system integration which would otherwise be difficult to achieve through traditional means in the green hydrogen sector (Arbabzadeh et al., 2019). In the production of green hydrogen, electrolysis of water is the main technology, in which electricity is applied to separate hydrogen and oxygen from water molecules. But electrolysis is an energy intensive process and thereby the efficiency of electrolysis plays an important role in economic competitiveness of green hydrogen. AI is an essential development that has the potential to optimise electrolysis process and increase electrolyzer performance, operational efficiency as well as energy savings.

Real-time adjustment of electrolyzer operating conditions can be accomplished through machine learning algorithms. AI then allows for this data from different sensors embedded in the electrolysis system to chime in and recognize patterns that will optimize parameters like spot temperature, pressure, & current density driven system efficiency wise. AI models, for example, can forecast the best conditions of electrolyzers for a variety of renewable energy supply scenarios from solar to wind to hydro. Such accuracy enables better resource management, which ensures that the electrolyzers are functioning at their maximum efficiency levels all of the time. The predictive maintenance driven by AI could be used to minimize downtimes and increase the lifetime of electrolyzers. AI can analyze equipment data and identify abnormalities even before failures occur, enabling timely maintenance/repairs and avoiding unplanned stoppages. Thus preventing the black-outs and enabling the proper functioning of green hydrogen production plants (Ilyas et al., 2022; Garrido et al., 2020).

The intermittency of renewable energy sources like solar and wind power is one of the main hurdles to producing green hydrogen. Such sources are, by their very nature, intermittent characterised with an erratic output of energy that depends on time-of-day, season and weather conditions. AI can be applied in electrolysis systems to have a constant and consistent flow of hydrogen by improving the integration with renewable energies. AI algorithms can predict the variability of renewable energy generation, forecast demand for hydrogen, and schedule electrolysis to be hours ahead to coincide with available power. For instance, AI can forecast when solar or wind generation will occur and schedule electrolysis to produce hydrogen during these periods. On the other end of the spectrum, in times when energy production is low, AI can also play a major role in managing its respective storage systems batteries or hydrogen storage to provide a constant supply of hydrogen, even during periods with no availability of renewable energy.

AI can also optimize the overall energy mix so that electrolysis is powered by the cheapest, lowest-carbon-energy sources at each moment in time. However, note that they could help lower the average cost of green hydrogen production and ensure production remains sustainable at all time. AI can play a vital role in supply chain and infrastructure optimization, such as hydrogen storage, transport and distribution. Green hydrogen has to be stored and transported to end users (industrial facilities or refuelling stations) in a process that can entail complex logistics, from pipeline networks through compression to liquefaction (Steen & Weaver, 2017).

Through modeling and optimization of hydrogen storage and transportation systems, AI can ensure that hydrogen is always stored under ideal conditions and transported in the most efficient way. Demand can be forecasted, supply routes adjusted and hydrogen pipeline flow optimized to minimize waste and maximize cost savings with the use of machine learning algorithms. AI can also help identify locations for hydrogen storage and distribution infrastructure, based on population density, industrial demand, and proximity to renewable energy sources. The hydrogen refueling stations that AI can help create for transportation. AI, by studying hydrogen demand patterns in different areas, can support refueling stations placement and capacity optimization based on where is the lack of hydrogen. Especially given the expanding acceptance of hydrogen-powered vehicles, which will need an extensive and dependable refueling infrastructure.

It is also speeding up research and development (R&D) in green hydrogen by facilitating searches for new materials to enhance electrolyser efficiency, as well as discovering new ways of producing hydrogen. Simulation and computational modeling accelerated by AI can help discover more effective catalysts for electrolysis that are both cheaper and long-lasting. Such breakthroughs can bring down the cost and scalability of hydrogen production and, thereby help achieve mass adoption of green H2. It is also employed to investigate new hydrogen production types, other than electrolysis (biohydrogen, and photoelectrochemical cells). These emerging technologies are based on complex chemical/physical processes that AI can assist in simulating, which will allow key parameters for optimization to be identified and experiments to be guided (Hoffmann et al., 2020; Manoharan et al., 2021). AI is creating new avenues for hydrogen production by faster iterations and deeper insights into these novel approaches that could run in parallel with or even rendered existing electrolysis modalities obsolete.

AI is also playing a certain role to continental or nationwide policies and strategy outlines toward the global emission of green hydrogen. Lessons from the real world should matter as governments and industries have to carefully examine the space where green hydrogen can fit into their systems, looking at aspects like geography, renewable energy potential and existing infrastructure. AI can help policymakers by providing predictive models and simulations to predict outcomes of hydrogen deployment strategies. Using machine learning models to process information about availability of renewable energy, the demand for hydrogen, and the market dynamics will enable us to figure out scaling potentials of green hydrogen production and distribution pathways (Ghai et al., 2022). AI can also assist in quantifying the environmental and economic benefits of green hydrogen commercialization so that these can guide investment support for policies promoting the transition to hydrogen economies. With the rapid development of AI technology, it is considered important to use its capacity for innovation in green hydrogen to enable cost reduction, performance optimization and faster introduction of hydrogen into the broader energy market as a carbon-free sustainable form of energy. So, AI is not just accelerating the development of green hydrogen but is also contributing to the making of future global energy systems.

4. INDUSTRY 6.0 AND GREEN HYDROGEN: FOCUS ON SUSTAINABLE AND HUMAN-CENTRIC MANUFACTURING

Using the tenets of sustainability, empathy towards humans, and a technology-driven society Industry 6.0 is on course to revolutionise manufacturing for good! This is where green hydrogen comes into play, offering a clean and efficient energy solution to industrial sectors seeking to decarbonize and transition towards sustainability. With new AI and other advanced technologies, green hydrogen takes the human experience in manufacturing to a whole new level by creating systems that are smarter, more adaptive as well as better for our planet and healthier for us. Green hydrogen and Industry 6.0's focus on human-centric manufacturing will be critical in developing a sustainable, flexible industrial system that serves not only the planet, but society itself as the world advances to net-zero emissions (Pimitha et al., 2023).

Industry 6.0 is the next phase, where significant industries across the world use innovations like AI, robotics, IoT, and sustained practices. Industry 6.0, on the other hand, takes a more comprehensive view than previous industrial revolutions by integrating sustainability and human-centric principles with technologies that have ushered in the fourth industrial revolution, autonomy and efficiency (Singh & Kaunert, 2024). This new age is being brought about by a transition to cleaner and more sustainable energy sources, green hydrogen among them. Green hydrogen, which is hydrogen produced via electrolysis powered by renewables in the utmost carbon-neutral way possible, may change the game of manufacturing processes when it comes to energy provision and overall decarbonization. Green Hydrogen with Industry 6.0 principles enable a more sustainable and humane future for manufacturing.

Green hydrogen fulfils an important role in decarbonization of Industry 6.0 by providing a fossil free alternative to industrial processes that are very difficult to electrify. Take steel production, cement manufacturing and chemical production for instance: these are high-temperature processes that account for a large chunk of carbon emissions. Direct fossil fuel combustion is utilized in these processes, as electric sources of energy are not easily substituted. On the other hand, green hydrogen is becoming a carbon free fuel that can become enabling these industries to meet their carbon footprint reduction and sustainability efficiency targets (Singh & Kaunert, 2024). Green hydrogen for Industry 6.0 should not be merely a transition from fossil fuel to green energy; rather, it represents an evolution in the industrial ecosystem. The incorporation of green hydrogen in manufacturing processes presents an opportunity for companies to evolve business models that are more sustainable, circular and energy efficient (Singh & Kaunert, 2024). Green hydrogen, for example, is used in cogeneration systems where it is combined with renewable energy sources to produce electricity and heat at the same time, improving efficiency of energy usage and reducing waste. Moreover, where renewable energy is growing in scale, hydrogen production can be localized and decentralized with manufacturers producing hydrogen on-site leading to less reliance on centralized infrastructure (Singh et al., 2024).

Green hydrogen fits well with sustainability, which is an important element of Industry 6.0. Green hydrogen produced from renewable energy sources such as solar, wind or hydro power is a low-emission alternative to carbon-intensive processes that are converted into anything or could be used in various industries. The electrolysis process, which divides water into hydrogen and oxygen producing CO2-free hydrogen, is a key feature in the quest for net-zero emissions of production. Such a reduction in emissions is very important to industries striving to achieve stricter government regulations as well as public demand for greener manufacturing. It can facilitate the shift towards circular economies as an energy carrier and storage solution. This solves another problem, namely providing hydrogen when there is more renewable energy generation than demand and storing it for use later to overcome the intermittency of sources like wind and solar (Singh & Kaunert, 2024). This flexibility allows manufacturers to utilize green hydrogen as a fuel and energy storage option, leading to a more resilient and sustainable energy system. Thus, green hydrogen integrates with the sustainable manufacturing ecosystem to empower manufacturers in reducing their environmental footprint and ensuring they remain competitive in a global marketplace that is becoming increasingly green.

In harmony with its human-centered approach, Industry 6.0 aims to set the improvement of well-being for workers, communities and society as a whole at the center of technological & industrial innovation. By leveraging green hydrogen with the new generation of AI and other Industry 6.0 technologies, we can foster safer, healthier and leaner workplaces with much less environmental burden on local communities. Human-centric manufacturing is empowered by AI. Moreover, if hydrogen fuel is used in a factory for production purposes, it becomes necessary to use the energy provided by hydrogen with AI-powered systems monitoring energy consumption and predicting maintenance needs so that limited resources can be utilized smartly (Singh, 2024). AI can, for example, analyze data from manufacturing processes in order to dynamically optimize the use of hydrogen by minimizing waste and improving operational efficiency. The combination of AI with green hydrogen technologies will help manufacturers to build more flexible and responsive systems that are favoured not only by the environment but also by their people. And finally, human-centric manufacturing also come with the welfare of human beings involved. This shift helps to reduce dependence on toxic fossil fuels for industrial processes that use green hydrogen and renewable energy, which will ultimately contribute to healthier workplaces with cleaner air. Hydrogen-powered manufacturing facilities create no harmful particulates or CO2 and reduce exposure to many pollutants that pose a risk to human health. AI-driven automation might even be utilized with site tasks that could put employees in deadly circumstances, maximizing the condition of almost all laborers and leaving them to accomplish their particular careers adding higher significance (Singh & Kaunert, 2024).

The circular economy is a core idea of Industry 6.0: to eliminate waste and continue reusing as many resources as possible. Green hydrogen sees a natural home in this model, enabling the recycling of energy and materials within manufacturing ecosystems. In fact, hydrogen can act as an enabler of circularity in many ways. As an example, hydrogen can be a feedstock for syngas plants in the production of synthetic fuels, chemicals and fertilizers creating looping processes that recycle hydrogen endlessly throughout the manufacturing system. The circular economy fundamentals can be applied to green hydrogen production. For example, electrolysis can continue to recycle water used in the process while waste heat from hydrogen production can be utilized to power other processes at the scale of a facility, thus increasing energy efficiency of an integrated system. Integrated within these circular efforts across industries, green hydrogen emerges as an enabling element not only in this transition by reducing resource use but also limiting environmental impact, ultimately delivering a sustainable industry with human-centred characteristics.

5. POLICY FRAMEWORKS AND INTERNATIONAL COLLABORATION

Green hydrogen is one of the solutions that will help to decarbonise hard-to-electrify sectors (heavy industry, transportation and energy storage), as the world races towards a sustainable energy future. Nevertheless, realising the potential of green hydrogen will require strong policy foundations and international cooperation. Good policies not only help develop and commercialize green hydrogen technologies, but also their incorporation into energy systems worldwide. Cross-border cooperation is essential for knowledge sharing, regulatory alignment and innovation creating a seamless global approach to green hydrogen deployment.

The creation of national policy frameworks promotes green hydrogen technologies. The reason for this is that governments are the entity responsible for providing the incentives and regulatory frameworks needed to up-scale production and consumption. Enabling policy measures comprise of financial instruments such as subsidies, grants and tax incentives for green hydrogen projects to overcoming their high upstream costs related to electrolysis technologies and infrastructure development. Such policies should also stimulate private sector investments, R&D and innovation to speed up the commercialization of green hydrogen. Importantly, national policies must establish ambitious and clearly defined hydrogen targets and roadmaps over the long term. For example, Germany, Japan and South Korea have published national hydrogen strategies that define their hydrogen ambitions regarding production, storage and utilization over the coming 30 years. Such strategies generally involve some requirements that hydrogen must be produced from renewable sources, like wind or solar power, to keep it green. Hydrogen infrastructure investments including refueling stations and pipelines should occur with public-private partnerships as well.

Recognition of a regulatory framework for the safety, quality and certification of green hydrogen is another vital policy consideration. At the same time, governments should also support research on new technologies and approaches that will increase the efficiency and reduce the cost of hydrogen production, including improvements to electrolysis, as well as methods for hydrogen storage and transportation. In that sense, while national efforts are indispensable, international cooperation remains key to the global success of green hydrogen. International collaboration is crucial to the hydrogen economy since it is such a complicated ecosystem, we should collaborate internationally to mutualize best practices, technology and capital across borders. Hydrogen is an important international traded global commodity, with significant trade and transport options, so national policies need coordination to create a seamless integrated global market.

They also work together in establishing international hydrogen trading networks. Green hydrogen production is cost-competitive for countries with high renewable energy resources especially in the Middle East, Australia and some regions of Africa. These countries can then export hydrogen to the ones with lower renewables potential or lower ability to produce this fuel domestically, creating a worldwide hydrogen market. Establishing international hydrogen transport infrastructure including pipelines, shipping routes and storage facilities calls for close cooperation between governments, energy companies and international organizations (Sinha et al., 2021). International collaboration facilitates certification and standardization of hydrogen technologies. Through organizations such as the International Energy Agency (IEA) and the Hydrogen Council, dialogue is facilitated between governments, industries, and researchers. These organizations help improve technological innovation and lower barriers to global trade by establishing common standards and frameworks for hydrogen production, storage, and transport. Another option are joint R&D efforts, where countries collaborate and share resources to develop new hydrogen technologies and solutions this can accelerate the development of such technologies while also reducing costs.

International climate agreements provide the opportunity for governments to design hydrogen strategies that complement global sustainability efforts, as typified in the Paris Agreement. Such cooperation within these frameworks will assure hydrogen makes a proper contribution to the global decarbonization targets whilst helping nations secure their energy and economic development goals. Moreover, green hydrogen can help accomplish Sustainable Development Goals (SDGs) in particular through cleaner energy systems, energy access and sustainable industrialisation.

6. CONCLUSION

The success of green hydrogen as a sustainable energy solution hinges on policy frameworks and international collaboration. National governments need to establish the right policies that will stimulate hydrogen production, infrastructure development and technology solutions. On international scale, cross-border collaboration will be required to develop a global hydrogen market, harmonise regulations and facilitate knowledge hub and technology sharing. Green hydrogen will do the heavy lifting in a sustainable decarbonized energy future only through collective action at the national and global levels.

7. FUTURE TRENDS AND INNOVATIONS

The future of green hydrogen is not just explained through the lens of renewable hydrogen, but also from an integrated perspective considering emerging technologies and more widespread systems. Combining hydrogen-based storage with AI and IoT capabilities, smart grids can match supply and demand in renewable electricity networks. A second method, Quantum computing is a game-changing technology that can help speed the discovery of new hydrogen storage materials or membrane materials for fuel cells. Coupled with the essence of Industry 6.0, green hydrogen can also play an integral role in a circular economy where materials are used and reused in such a way as to minimize waste through recycling as much as possible.

REFERENCES

Al-tabatabaie, K. F., Hossain, M. B., Islam, M. K., Awual, M. R., TowfiqulIslam, A. R. M., Hossain, M. A., Esraz-Ul-Zannat, M., & Islam, A. (2022). Taking strides towards decarbonization: The viewpoint of Bangladesh. *Energy Strategy Reviews*, *44*, 100948. DOI: 10.1016/j.esr.2022.100948

Aldieri, L., Grafström, J., & Vinci, C. P. (2021). The effect of Marshallian and Jacobian knowledge spillovers on jobs in the solar, wind and energy efficiency sector. *Energies*, *14*(14), 4269. DOI: 10.3390/en14144269

Arbabzadeh, M., Gençer, E., Morris, J. F., Paltsev, S., & Armstrong, R. C. (2019). *Plausible Energy Futures: A Framework for Evaluating Options*. Impacts, and National Energy Choices.

Basheer, M., Nechifor, V., Calzadilla, A., Ringler, C., Hulme, D., & Harou, J. J. (2022). Balancing national economic policy outcomes for sustainable development. *Nature Communications*, *13*(1), 5041. PMID: 36028486

Budiman, I. (2018). Enabling community participation for social innovation in the energy sector. *Indonesian Journal of Energy*, *1*(2), 21–31. DOI: 10.33116/ije.v1i2.23

Damman, S., Sandberg, E., Rosenberg, E., Pisciella, P., & Graabak, I. (2021). A hybrid perspective on energy transition pathways: Is hydrogen the key for Norway? *Energy Research & Social Science*, *78*, 102116. DOI: 10.1016/j.erss.2021.102116

Danish, M. S. S., & Senjyu, T. (2023). AI-Enabled Energy Policy for a Sustainable Future. *Sustainability*, *15*(9), 7643.

Garrido, S., Sequeira, T., & Santos, M. (2020). Renewable energy and sustainability from the supply side: A critical review and analysis. *Applied Sciences (Basel, Switzerland)*, *10*(17), 5755. DOI: 10.3390/app10175755

Ghai, D., Tripathi, S. L., Saxena, S., Chanda, M., & Alazab, M. (Eds.). (2022). *Machine Learning Algorithms for Signal and Image Processing*. John Wiley & Sons. DOI: 10.1002/9781119861850

Hoffmann, M. W., Wildermuth, S., Gitzel, R., Boyaci, A., Gebhardt, J., Kaul, H., & Tornede, T. (2020). Integration of novel sensors and machine learning for predictive maintenance in medium voltage switchgear to enable the energy and mobility revolutions. *Sensors (Basel)*, *20*(7), 2099. DOI: 10.3390/s20072099 PMID: 32276442

Ilyas, A., Parkinson, S., Vinca, A., Byers, E., Manzoor, T., Riahi, K., & Muhammad, A. (2022). Balancing smart irrigation and hydropower investments for sustainable water conservation in the Indus basin. *Environmental Science & Policy*, *135*, 147–161. DOI: 10.1016/j.envsci.2022.04.012

Manoharan, K. G., Nehru, J. A., & Balasubramanian, S. (2021). *Artificial Intelligence and IoT*. Springer. DOI: 10.1007/978-981-33-6400-4

Nakata, T. (2024). Energy-economic models and the environment. *Progress in Energy and Combustion Science*, *30*(4), 417–475. DOI: 10.1016/j.pecs.2004.03.001

Nastasi, B., Markovska, N., Puksec, T., Duić, N., & Foley, A. (2022). Renewable and sustainable energy challenges to face for the achievement of Sustainable Development Goals. *Renewable & Sustainable Energy Reviews*, *157*, 112071.

Pascaris, A. S. (2021). Examining existing policy to inform a comprehensive legal framework for agrivoltaics in the US. *Energy Policy*, *159*, 112620. DOI: 10.1016/j.enpol.2021.112620

Pimitha, K., Anbarasi, S., & Balasubramanian, T. (2023). Machine Learning Techniques for Renewable Energy Resources. In *IoT, Machine Learning and Blockchain Technologies for Renewable Energy and Modern Hybrid Power Systems* (pp. 121-148). River Publishers.

Reedman, L., Chew, M. S., Gordon, J., Sue, W., Brinsmead, T., Hayward, J., & Havas, L. (2021). Multi-sector energy modelling.

Singh, B. (2022). Relevance of Agriculture-Nutrition Linkage for Human Healthcare: A Conceptual Legal Framework of Implication and Pathways. *Justice and Law Bulletin*, *1*(1), 44–49.

Singh, B. (2024). Lensing Legal Dynamics for Examining Responsibility and Deliberation of Generative AI-Tethered Technological Privacy Concerns: Infringements and Use of Personal Data by Nefarious Actors. In Ara, A., & Ara, A. (Eds.), *Exploring the Ethical Implications of Generative AI* (pp. 146–167). IGI Global., DOI: 10.4018/979-8-3693-1565-1.ch009

Singh, B., Dutta, P. K., & Kaunert, C. (2024). Replenish Artificial Intelligence in Renewable Energy for Sustainable Development: Lensing SDG 7 Affordable and Clean Energy and SDG 13 Climate Actions With Legal-Financial Advisory. In Derbali, A. (Ed.), *Social and Ethical Implications of AI in Finance for Sustainability* (pp. 198–227). IGI Global., DOI: 10.4018/979-8-3693-2881-1.ch009

Singh, B., Jain, V., Kaunert, C., Dutta, P. K., & Singh, G. (2024). Privacy Matters: Espousing Blockchain and Artificial Intelligence (AI) for Consumer Data Protection on E-Commerce Platforms in Ethical Marketing. In Saluja, S., Nayyar, V., Rojhe, K., & Sharma, S. (Eds.), *Ethical Marketing Through Data Governance Standards and Effective Technology* (pp. 167–184). IGI Global., DOI: 10.4018/979-8-3693-2215-4.ch015

Singh, B., & Kaunert, C. (2024). Computational Thinking for Innovative Solutions and Problem-Solving Techniques: Transforming Conventional Education to Futuristic Interdisciplinary Higher Education. In Fonkam, M., & Vajjhala, N. (Eds.), *Revolutionizing Curricula Through Computational Thinking, Logic, and Problem Solving* (pp. 60–82). IGI Global., DOI: 10.4018/979-8-3693-1974-1.ch004

Singh, B., & Kaunert, C. (2024). Wind and Solar Energy as Renewable Energy for Sustainable Global Future: Projecting Future Multi-Sector Sustainable Policies and Innovation. In Ara, A., & Thakore, R. (Eds.), *Promoting Multi-Sector Sustainability With Policy and Innovation* (pp. 210–245). IGI Global., DOI: 10.4018/979-8-3693-2113-3.ch009

Singh, B., & Kaunert, C. (2024). Aroma of Highly Smart Internet of Medical Things (IoMT) and Lightweight EdgeTrust Expansion Medical Care Facilities for Electronic Healthcare Systems: Fortified-Chain Architecture for Remote Patient Monitoring and Privacy Protection Beyond Imagination. In Hassan, A., Bhattacharya, P., Tikadar, S., Dutta, P., & Sagayam, M. (Eds.), *Lightweight Digital Trust Architectures in the Internet of Medical Things (IoMT)* (pp. 196–212). IGI Global., DOI: 10.4018/979-8-3693-2109-6.ch011

Singh, B., & Kaunert, C. (2024). Augmented Reality and Virtual Reality Modules for Mindfulness: Boosting Emotional Intelligence and Mental Wellness. In Hiran, K., Doshi, R., & Patel, M. (Eds.), *Applications of Virtual and Augmented Reality for Health and Wellbeing* (pp. 111–128). IGI Global., DOI: 10.4018/979-8-3693-1123-3.ch007

Sinha, M., Chacko, E., Makhija, P., & Pramanik, S. (2021). Energy-Efficient smart cities with green internet of things. *Green Technological Innovation for Sustainable Smart Societies: Post Pandemic Era*, 345-361.

Steen, M., & Weaver, T. (2017). Incumbents' diversification and cross-sectorial energy industry dynamics. *Research Policy*, *46*(6), 1071–1086. DOI: 10.1016/j.respol.2017.04.001

Vinichenko, V. (2018). Mechanisms of energy transitions: National cases and the worldwide uptake of wind and solar power. *PhD diss., Central European University.*

White, D., Winchester, N., Atkins, M., Ballingall, J., Coates, S., de Miguel Mercader, F., & Suomalainen, K. (2018). Energy-and multi-sector modelling of climate change mitigation in New Zealand: current practice and future needs.

Winfield, M., & Dolter, B. (2014). Energy, economic and environmental discourses and their policy impact: The case of Ontario s Green Energy and Green Economy Act. *Energy Policy*, *68*, 423–435. DOI: 10.1016/j.enpol.2014.01.039

Xie, Y., Wu, X., Hou, Z., Li, Z., Luo, J., Lüddeke, C. T., & Liao, J. (2023). Gleaning insights from German energy transition and large-scale underground energy storage for China's carbon neutrality. *International Journal of Mining Science and Technology*, *33*(5), 529–553. DOI: 10.1016/j.ijmst.2023.04.001

Yang, D., Liu, D., Huang, A., Lin, J., & Xu, L. (2021). Critical transformation pathways and socio-environmental benefits of energy substitution using a LEAP scenario modeling. *Renewable & Sustainable Energy Reviews*, *135*, 110116.

Chapter 16
Harnessing AI-Powered Renewable Energy:
Technological Breakthrough and Legal Frontiers

Ashima Jain
https://orcid.org/0000-0001-5365-2385
Manipal University Jaipur, India

ABSTRACT

It is very easy to condemn any advancement within the earliest indications as despite the efforts to bring the rise of AI into among the much sustainable energy and decent jobs being created, those trained in the use of AI to effectively develop sustainable energy and contribute to the elimination of local and global environmental degradation, very detailed plans for such implementation have not been quantified; almost every published paper has been vague. Preferred applications of AI on renewable energy sources are fragmented across different platforms with potential uses; apparently, only a few have highlighted systems for such steps to integrate overall energy planning. There will be a presentation of the future trends and possible developments in artificial intelligence for renewable energy breakthroughs, which will work excellently for scholars, practising engineers, and industry experts as far as improving the domain and removing the already existing obstacles.

INTRODUCTION

Renewable energy use has an immediate link to the progress of societies. Life's necessities have gone on a rise owing to the ever-ballooning world population alongside moves towards urbanisation and overall economic progress. That goes to fuel and power production in numerous industries relying upon conventional forms of energy like natural gas, coal, and petroleum to meet its burgeoning demand more and more day after day (S. Mahapatra, et al., 2021).

The reliance on fossil fuels has depleted not only resources but has also played a major role in changing the climate and warming the earth through the massive production of greenhouse gases (GHG). The climatic changes contributed through its destruction rising ocean levels. For instance, glacier retreat, deforestation, air and water pollution, ozone layer depletion, radioactive gases that induce the greenhouse

DOI: 10.4018/979-8-3373-0045-0.ch016

effect, acid rain, and environmental damage all pose severe threats to both the economy and society. The agency further warns in its report that the GHGs emitted from energy consumed would increase the temperature by 6°C from where it was before, causing the most significant environmental damage yet. Indeed, the practical removal of harmful environmental changes and the potentially brighter energy future in line with clean and green energy for the welfare of mankind are set out as reasons for setting good transition plans Precisely, the studies scrutinizes a range of AI directions that have been employed in the field of renewable energy such as solar power, photovoltaics, microgrid embedding, management of battery-based power, wind turbine technology, as well as geothermal power generation. He also tried to delve into and provide a thorough knowledge of AI's contribution to revolutionizing renewable energy systems through examination of recent technical breakthroughs made, major research findings as well as illustrative case studies. Indeed, the present paper not only evaluates the constraints hindered under this branch but also tries to build plausible ways to address the same (Kumar et al. 2021).

More on alternative energy sources like renewable energy (RE), which is seen as the most important answer: Renewable energy will reach 10,800 GW globally by the year 2040, as the International Energy Agency (IEA) forecasts. Breaking away from the crowd of fossil fuels, these renewable systems-solar, wind, hydro, biomass, and ocean power seal the preference because of their duration a minimal ecological footprint, promotion of cost efficiency, and the long endurance these systems could offer to both a facility and the developer. Recent developments have harvested energy generation and distribution for earth-friendly living. Solar and wind energy sources-photovoltaic and wind turbines-provide the most forward-moving parts of a renewable energy system (Kothari et al., 2021).

An energy revolution powered by sustainable renewable energy technology will have a potential to revolutionize the electricity industry. It may promote effective cost management of electricity, improve the adaptability and resiliency of energy systems, upgrade old infrastructures, mitigate CO2 emissions, provide dependable power delivery in remote areas, and decrease ecological impacts. Energy storage systems (ESS) have evolved significantly over the years in efforts to meet emission reduction and energy efficiency requirements. These advances highlight the old prescription to upgrade the accuracy in forecasting RE, essential for the functional optimisation of the power system (Kadhem, et al., 2017).

Advances in technology have made renewable energy systems more accurate. Through creation of intelligent devices and future software, artificial intelligence is an advanced idea. In the olden day, mechanical Engineers made simple algorithms of choosing among an array of choices, such as data collection and monitoring for renewable applications. It's very likely and can be true: AI has evolved greatly in the past years, bringing great change to people daily live and in sustainable growth; it can include different traditional applications in sectors such as energy, agriculture, learning, health care, security, and industry and in the arts (Perea-Moreno, 2021).

Artificial Intelligence (AI) has powerfully influenced the RE policy and business by simulating scenarios, policies to analyse this efficacy, and forecast economic results. It helps to monitor ecosystems, calculate the carbon footprint of energy generation, and optimize resource utilization so as to reduce environmental impact. AI can be used to help predictive modelling by promoting greater accessibility of clean energy for more marginalised communities as well as improving relatively less explored renewable innovation for the generation of energy (e.g. tidal and wave). For solar power, meta-heuristic methods based on AI improve system performance, and they also have integrated algorithms to enable forecast of the wave energy characteristics. Optimization of the advanced biofuels that minimize the emission from transportation and industrial sources is now currently available through methodologies. Such projects include the employment of an AI-based previous artificial brain, which predicts the solar still's performance, elevated energy utilization and water produced. Predictive maintenance, more energy-efficient performance, more environmentally friendly, development for intelligent grids and microgrids, better solar stills, and integration to RE systems have all become examples of climate change. These developments establish AI as a force that transforms the landscape for renewable power scenarios (Georgiou, 2019).

Investigating AI integration with RE would improve energy technology for renewable technology and would save, monitor, and oversee it. The research focuses on AI-driven development in the renewable energy systems for solar, wind, hydro, and geothermal power using selected case studies using diverse geographical factors. These remarkable applications combined with the comparative study of established ones could guide us to our best approach. Some of the difficulties discussed while exploring the challenges are using Artificial Intelligence for managing and evaluating data and data security issues in RE systems, ways to greatly solve them, and future trends, among others. This paper is for academics and engineers as it gives a very slab but detailed overview on the applications of artificial intelligence in viable sources of energy, sharing thousands of practical outcomes and suggestions for future studies (Lahmadi, 2023).

RENEWABLE ENERGY SECTOR WITHOUT THE INFLUENCE OF ARTIFICIAL INTELLIGENCE

Even as intelligent agents are shaping the energy industry in general, alternative energy sources have found ways to be developed and constantly renewed (Onwusinkwue et al. 2024). The use of technology for increased efficiency, lower costs, and issue resolution in concerns in renewable energy are now changing. This has some limitations to be sure. The traditional way of data analysis and decision-making only offers very slow development: it is not as prompt as one might think in drawing conclusions or forecasting from it. The use of AI in renewable energy was more expensive for making massive investment into infrastructure, data acquisition, and the development of algorithms and so on, thereby causing obstacles for developing countries to make the adoption. The concern about data privacy and security threatened to become instrumental in endangering the generation of power-by differences in distribution. Conversely, AI could have drastic effects on the renewable energy business by improving processes and may lead to an overall increase in output. During that period, the research confined itself to using the sources of renewable energy. New renewable energy resources were discovered for use in conservation (Karim et al. 2017).

This report was - given before- into the sequence of those from 2017 and first explains the prevailing state of power accessibility in those urban disadvantageous regions, also clearly specifying the state of no power in Korail slum in Dhaka. It is indicated in black and white by the presented facts of the problems and that growth is one of these avenues considered very necessary to the design of electrical energy plans, for example, rooftop solar photovoltaic or generating electricity using solid waste and the phasing out of incandescent light bulbs. In addition to ensuring technical knowledge regarding the methods, these programmes typically create a double-edged advantage: 1) the availability of additional power and 2) opportunities for surplus generation and thus cost savings. Another study investigates the effects of environmental regulation on sustainability: retesting as well the role played in sustainable development in this context by research and development on renewable energy (RERD), technical changes, and truly great economic performance. Wavelet methods are also utilized in the study at the end to argue that high levels of funding in RERD, strict legislation, and the adoption of ecofriendly technologies are crucial for conservation of the environment. The study also reveals the existence of significant relationships such as those between CO2 emissions, change in technology, and growth in the economy. (Tao et al., 2023).

ARTIFICIAL INTELLIGENCE IN RENEWABLE ENERGY

1. Renewable Energy Types

The research emphasizes the constant-comprehensive necessity of innovation and adaptability of renewable energy systems (RE) to impart electricity by way of innovative technologies into the systems. The enhanced RE system can be fuelled by a single or a mixture of any fuel resources, which is termed hybrid technology. The single-source system is typically solar, bioenergy, hydro, geothermal, wind, ocean, and also hydrogen energy; the latter hybrid is the blending both of the renewable and cruder energy sources such as small gas or diesel turbine made up of hybrids with electrical storage. The common hybrids include various form photovoltaic-wind-diesel and hydro-wind-photovoltaic systems with fuel cell system to create power (Zohuri, et al., 2018)

2. AI Tools and Techniques

Different types of learning techniques have been copied through AI-NN, Numeric, and Progressive Learning and are more widely studied. In research, neural networks (NN) among others support vector machines (SVM) forms the most often-used arrangements. Other algorithms that are always highly favored are decision trees, random forests, logistic regression, linear regression, naïve tes, nearest neighbour, and hidden Markov. The use of artificial intelligence has largely increased interest in the application of renewable energies (RE) excelling in power sector hence attracting more attention to areas like hydro, wind, geothermal, solar, and ocean power, as well, since their productivity and extendibility could be increased greatly (Goel et al., 2023).

3. AI Role in Renewable Energy

Renewable energy (RE) technologies have been a major focus for research and development in the contemporary era in various industries given their advancing expansion. Great strides have been made with the advent of intelligent and sophisticated technologies, more particularly AI-powered solutions, in dealing with the mounting problems and opportunities in the real estate world. The areas where this is being put into use are in power management, photovoltaic solar systems, and efficiency and control of energy. In the most recent researches, the focus was put on further increasing the use of AI in RE applications, such as the estimation of solar radiation, the estimation of the energy taken into solar systems, predicting wind speed, building modeling for heating, forecasting long- and short-term power, sizing solar power generation, and predicting power consumption for specific regions. (Nam et al. [2020])

As the integration of artificial intelligence (AI) and the Internet of Things (IoT) fosters modern Renewable Energy systems, progress in RE technologies has emerged in line with improved efficiency, strength, and flexibility. Nowadays, AI finds its application in quite many renewable energy technologies, including wind, hydro, geothermal, and solar energy, accordingly. Some of the progress can be credited to technologies like artificial intelligence (AI) that significantly contribute to making opulent Renewable

Energy projects that will possibly become highly principled modern renewable energy systems. Perceptions of the usefulness of advanced recommender algorithms in PV system performance are explored further in such research. Support for renewable energy is also in recognition that renewable energy technologies (RETs) can now become fast becoming cheaper and supported less heavily for subsidies than during the earlier time, providing full cost recovery and profitable growth. This is of course based on the perception given by performance additions through time aside from cost reductions.

4. Applications of AI in Renewable Energy

On several types of renewable energy, artificial intelligence (AI) has almost immediately become a significant aspect: hydrogen, hydro, solar, geothermal, wind, bioenergy, ocean, and hybrid systems. Using artificial intelligence, weather-based electricity generation can be foreseen by different machine-learning algorithms and statistical methods such as p-values and confidence bounds. Keeps the grid stable. Besides repairing systems, AI heavily contributes to predictive maintenance regarding energy infrastructure. It also brings down time, maintaining steady energy (Table 1). The physical uses of AI in this case are design, optimization, management, calculation, dispersion, and legislation for renewable energy. Five major renewable energy technology areas will be reviewed in this paper: solar energy, photovoltaic (PV) technologies, solar microgrids, wind turbine optimization, and geothermal energy, and the mind open to understanding numerous applications of AI in each of these fields.

A.) AI Applications in Solar Energy

Students or individuals who only learn or access the proxy website for studying courses or accessing any blocked website is manually and independently sought-after. They use search engines by familiarizing themselves with the domain name, since no university proxies are available for students to choose from (Alzitani et al., 2019). One such trend for getting access to blocked websites is the increasing admiration of using online proxies as sites that assist users in unblocking visitation of websites quickly. Online proxies are necessary while browsing since there are certain requirements for some websites. Most institutions use security software that makes it hard to browse, particularly inside a restricted network; others even engender problems if any process is being done online. Eventually, the design of simulation systems that are controlled by AI would bring along quite large-numbered and adequately defined benefits. On the solar energy simulation front, these would work in lieu of traditional procedures; they are found to be far more affordable than any prevailing alternative in grueling and cumbersome multivariable problems at a minimal level of computing resource use, so they can just be deployed without knowing the exact internal system mechanisms. (Liao & Yao, 2021).

It has been common that Artificial Neural Networks (ANNs) are considered to be the most popular AI-based research among academics around the world. Accordingly, several studies are found to use the Backpropagation Neural Network (BPNN): to engage the solar, electromagnetic, illuminance, daily temperature, water heater and solar radiation, system operation, building use of energy, beam radiation and maximum power output from high concentration PV modules. Furthermore, Adaptive Neuro-Fuzzy Inference Systems (ANFIS) have been used to predict PV power supply, estimate clearness index, estimate hourly global radiation, predict power in solar photovoltaic installations, and evaluate the operation of solar tower power plants. Several examples of combinations of different methods are employed in several studies to improve predictability. Comparisons involve intensity measurement, such as the Mean Absolute

Percentage Error (MAPE), Root Mean Square Error (RMSE), which are frequently utilized to measure different model forecasts vis-a-vis live results. These are common assessment measures in the environment and forecasting fields that are used to assess the efficacy of models (Naser & Abdelbari, 2020).

B.) AI-Driven Enhancements in PV Technology

The convergence of geospatial and machine learning technologies is expected to expand the flexibility and intelligence of solar photovoltaic systems. However, harnessing photovoltaic potential requires more insights and understanding - defined by interdisciplinary research, even adaptation of materials science and computational analysis on the practical experience obtained in critical areas such as silicon and thin-film solar cells. Odd as it might seem, the researchers and engineers are passionate about creating new approaches through "artificial intelligence" to make PV technology affordable, workable, and above all productive. Under the drive of AI breakthroughs, one finds a new diagnostic toolkit given to scientists for the optimization of solar cell performance, thereby a major upgrade in their usability. Today's reports indicate the beginning of a whole new era in renewable energy development because of artificial intelligence.

AI made some biggest impact in the transformation of photovoltaic (PV) energy systems. With new ways of measuring using AI algorithms and modern data analytics, it will be possible now to more precisely monitor the changing influences on solar performance like light intensity and temperature conditions and interference from shading. So, the AI techniques can help bring to the optimum level the real-time monitoring and analyses of the parameters in the operation of solar installations. The AI-driven optimization has helped in improving the stability and reliability of the power generation functions in the PV systems, hence significantly enhancing the overall efficiency (Tawalbeh, et al., 2021).

Artificial intelligence has made a huge difference in forecasting solar irradiation and assessment of resources for the attainment of better overall energy management and grid connections. If AI is incorporated, it will also generate short-term and long-term solar irradiation predictions from historical data and future weather forecast information, which will ensure the optimization of the energy output and system stability. Thus, the energy manager and systems operator will minimize variations in solar-generated electricity.

AI is gaining attention as technology to maximize solar cells by accumulating massive data for analysis. AI is expert in identifying even complex patterns or trends within vast datasets on the efficiency of solar cell supplies and designs. It's about enhancing machine learning algorithms to ease the job by zeroing in on the most crucial factors leading to solar cell productivity, leading to a better understanding and, above all, better production of solar energy. These steps help professionals design and improve solar cells (Kumar, 2023).

Precisely identifying modeling parameters for photovoltaic (PV) cells plays a key role in the complex modeling of PV cells and is becoming an active area of research. Various studies are thus being carried out aiming at realistic and practical approaches to get around the same question. Roughly speaking, these two approaches can be classified into analytical based tactics and meta-heuristic algorithms.

Among other things, algorithmic strategies include the use of very skilful software, with a myriad of time-honored algorithms available on procurement by commercial interests. It can be anticipated that the models of these products will incorporate any of these strategies in the most effective way. On the other hand, the one reason for the high popularity of meta-heuristic algorithms is based on advances in swarm intelligence and computer-based techniques, especially in resolving complex optimization problems and uncontrollable disturbances. One of the most important advantages of meta-heuristic algorithms is that they avoid needing an accurate model. This has led to a reduction in the cost of computing. In recent years, research on meta-heuristic techniques for accurate prediction of PV cell performance has increased alarmingly. These systems combine artificial intelligence (AI) with physical-based mechanisms to give probably more reasonable predictions for PV power over various time steps: from longer through intermediate times. As far as forecasting errors are concerned in terms of various techniques, including RMSE, MAPE, and MAE, all other techniques and algorithms focus more on minimizing such errors than this particular set of methods does. Quite a few published articles tackle detailed analyses of these methods, including synthetic generated results. These are based on bio-inspired classes and classified in three ways: bio-based, physics-based, and even mathematical-based algorithms.

For bio-inspired algorithms, these techniques include, among many others, flower pollination, optimization using bacterial foraging, and grey wolf optimization. In physics-based algorithms, there are procedures that operate by using particle swarm optimization, optimizing processes wind, optimizing processes water cycle, Lozi map chaos optimization, scale mutation optimization, and simulant steeping. On the other hand, the methods predominantly based on mathematics such as that of the shuffling complex evolutionary Jaya algorithm or that of the Structure Search Algorithm, are used with algorithms that bear similarity with the classical derivation of mathematical-based meta-heuristics. (Akhter, et al., 2019)

C.) AI in Solar Microgrids (MGs)

One of the weapons in their current power system attack is distributed generation (DG) and storage system, and especially the microgrid. MGs connect conventional and renewable energy sources, loads, and energy storage devices. Their flexibility, sustainability, self-sufficiency-these all together make MG very useful in improving power system performance measures, in particular, dependability and resilience. The success of microgrid research has always been coupled with problems on the design and installation side. And almost more challenging is that MG is prone to their own constraints in load management, renewable energy sources, and energy source management, as well as immediate swings in demand.

Furthermore, in this way, another catalyst in modern-day electricity systems is that the integration of disparate components, as exemplified by conventional or smart meters, distributed energy resources (DERs), energy storage systems (ESSs), electric cars, and Internet connections, induces huge data volume. The problem of data overload has resulted in efforts by scholars to devise complex methods of overcoming these scopes, such as artificial intelligence (AI). The issues related to handling and analyzing large amount of data are not addressed adequately by the traditional approaches of simulations. This paper provides a review of recent research on artificial-intelligence-based approaches toward greater performance, stability, reliability, and security of MGs.

The AI-based approaches in power systems are the largely categorized, and expert systems (ES) are the most commonly used technologies in this category, and collective techniques in the field of ML algorithms are implemented in AI systems.

D.) AI in Wind Turbine Optimization (WTO)

Wind turbines have been deemed an essential element of power distribution structures in the current environment as it has experienced tremendous global growth in usage. With the approaching wind rolling on the windmills, this really makes a direct impact on them as they must determine ideal configurations for the most efficient harvesting of energy by studying wind characteristics and potential site placements.

A group of researchers has highlighted the implementation of effective wind turbine control systems, thus noting that due to large scale and complex electrical systems associated with their operations, these systems serve to eke out the much-needed additional benefits in boosting efficiency, and revenue is thereby generated for the various energy companies.

The construction and formulation of a working control system for a huge wind turbine farm has faced a number of hurdles, not the least of which turned out to be the wake instability-related phenomenon. Wind turbine blades, therefore, introduce a unique oscillation into the wind flow, which helps create an oscillation that creates wake turbulence in the boundary layer. There are also possible losses in the airflow from this disruption. In designing controllers for a wind farm, it will necessitate a more profound study on the effects of wind power in the field as such (Mazzeo et al. 2021).

Scientific research has openly states that one of the ways wind power potential can be exploited completely is by means of its remote monitoring and management of wind turbines. Moreover, wind power unavailability cases in the farm are because of mechanical breakdowns, problems with its electrical parts, and extreme temperatures. It is beneficial to operating constraints on the processes across these limits-responding systems, even though the cycles and swings of wind currents vary significantly over time.

To aid in addressing these concerns and making headway in prediction, control, and monitoring of wind turbine systems, researchers and technologists have used the power of Artificial Intelligence (AI) and its Machine Learning techniques. Artificial Neural Networks (ANNs), Support Vector Machines (SVMs), and Swarm Optimisation Algorithms (SOAs) are among the three most common types of AI strategies which are used in wind turbine systems.

ANN has several advantages: it is very versatile and can produce high accuracy with fault tolerances, it is self-organizing and direct, easy to implement computation, making them one of the best tools for predicting wind speed, power, and wind density problems compared to SVM; However; SVMs produce excellent results in defect recognition and identification applications. SVMs perform well in some other cases having high accuracy in comparison with ANNs for prediction activities; however, the performance of the SVM, particularly in such domains, also necessitates further investigation and study.

Swarm Intelligence Algorithms (SIAs) are now becoming an area of utmost importance as they are finding application in wind turbine optimization by obtaining collectively intelligent behaviors from nature. It is a trend today that numerous engineers, across all the continents, rely on AI and non-linear controllers to improve wind turbine efficiency. To enhance the practicality of wind-related meteorological predictions, the National Center for Atmospheric Research (NCAR) has put AI into its weather forecast model.

E.) AI in Geothermal Energy

Geothermal energy, although not among natural resources on Earth, has been plenty effective enough for heating small spaces as well as providing natural heat sources for everyday activities like bathing, cooking and washing throughout the centuries-an ancient energy-promising source. By the time of the

20th century, it has evolved to become one of the most reliable and sustainable power sources, due to being commonly accepted by many as a "clean technology" provided by the earth. Yet almost all the large and very old geothermal power plants were built at the peak of the previous "energy boom," consequently producing an insatiable, continuous approach toward extraction and problems of all kinds. Due to frustrating geological conditions, the financing, and the innovation that is even more as a limitation towards geothermal, RE-based technology, geothermal technology has still been slowest among all these technologies spread out. Experts have consistently emphasized the need to develop techniques leading to a lower uncertain link between geothermal resources, which will help minimize project costs and risks.

However, recent advances in artificial intelligence (AI) and renewable energy (RE) open up opportunities for investment in the exploitation of geothermal energy. Present research employs these technologies almost exclusively towards whole areas, not delving into the potential of integrating machine learning and AI models. AI offers the prospect of tackling problematic issues of geothermal exploration. Initially, AI found use in fairway analyses of regional geothermal evaluations, involving both shallow and deep data. Machine learning techniques are optimised toward replicating human judgments, thereby improving accuracy and efficiency of the exploration process (Moraga, et al., 2022).

A unique trend within thermal research particularly applicable to remote reconnaissance is the use of thermal infrared radiation on Synthetic Aperture Radar (SAR) systems to map prospective geothermal areas. Furthermore, there is a spectrometer for the detection and analysis of mineralogical indicators for geothermal sources, and some researchers even use portable spectrometers to analyse drill core samples. Furthermore, scientists involved in research have already utilized AI-based fuzzy logic algorithms to specify geothermal power quantities via area-scale estimates.

CHALLENGES AND SOLUTIONS

Other forms of renewable energy (RE) are beneficial in their direct use, but present quite a few challenges. One of the major disadvantages concerns the cyclicity of generation essentially because of the identified seasons. Climate-dependency is another feature of most renewable energy sources, instead of ever-increasingly advanced designs, planning, and optimization. Thus, the latter activity becomes the most important to extract maximum economic and technical energies, whereas it also requires opportunities to protect the environment and the social development. Favorably, both computing hardware and software improvements open up fresh possibilities to well-researched activities in terms of optimization issues, ensuring improved computational resources in the context of renewable and sustainable energy.

1. Data Acquisition and Auditing

Due emphasis becomes actually put on the availability of quality information as an issue that affects the accuracy and likelihood of different industries making substantial developments in artificial Intelligence models or deep learning parameters. By far, the biggest hurdle in this respect is probably the most devastating, and preventive maintenance focused on such a system as the one solar power. This area is characterized by the application of data-driven strategies but at times becomes a barrier due to the huge need for data and hardly available reliable data, particularly in early applications. The measurement always comes with the interference of mechanical components; thus, it is important to analyze the data to detect and control more noise along with multivariate information. It is also very important to be aware of the hazards associated with noisy data and how they affect the performance of the prediction model. Regularly, the bookkeeping and testing are important to ensure that the algorithms keep on running appropriately, and over time, they must have learned the appropriate patterns.

2. Key Variable Selection of Parameter Tuning in RE Systems

Definition of key variables in Predictive Maintenance is a prominent step for application in permanent RE technology management as it will engage knowing which important characteristics or factors that affect the results of a model with the objective to result oriented applications are. This tough exercise happens either manually or automatically through a planned supervised expert model which has to consider technical understanding of the domain, different possible reasons for failure, and possible data collection methods. Awareness of the subject matter approaches intermediate in expert or fuzzy systems, with full utilization of neural networks possible with very little depth of understanding with regard to the failure processes. It is, however, much complicated if it is not done well. If not managed favorably, any such forecasting can very likely culminate in power black-outs or constrictive reduction in renewables' output. In fact, unexpected events like pandemics, natural calamities, and emergencies further make it complex for AI prediction systems to generate ultimate problems in making RE systems trustworthy (Aitkazinov, 2023).

3. Modeling of Various Faults Simultaneously

Modeling multiple failures at one time, almost by definition, requires using huge sets of data, which in return puts potentially large demands on processing capabilities and potential slows down real-time operations. Such performance operations can manifest in the context of prediction of Renewable Energy (RE) systems when AI-based algorithms untangle the various facets of uncertainty. For instance, vital importance here may lie in the unreliable aspects in RE generation systems. Such unpredictable factors have real-life implications of randomness in generation or system tunings, sourced mainly from the randomness between source-grid-load storage, including dispatch differences impeding or retarding renewable energy production, transmission, and distribution to certain areas. This complexity in renewable energy systems presents monumental challenges for designing of AI forecasting algorithms. However, these related issues serve as an impediment to properly effective results because this all can find viable solutions by introducing a more efficient platform through which the utilization and deployment of AI learning methodologies were integrated.

4. Stability and Generalization in Predictive Modeling

Findings are comprised of a lack of physical data in the system, resulting in some kind of trust failure with the trustworthiness of findings. Thus, intuition can sometimes be rendered useless as data simply doesn't exist. So, one needs to build such models that would be able to discern and react to new sorts of mistakes. The hybrid prospect of today's smart control systems can open up golden opportunities to explore with AI models; nevertheless, without the proper control and modeling technology in place to drive the formation of new renewable energy (RE) networks effectively, it discourages the formation of reliable models. Therefore, it is important that today's controls be enhanced and updated to empower effective modeling and growth for energy-related designs. Solving it could necessitate a search into intelligent AI models and extremely sophisticated algorithms that would make possible a seamless talk between what is present and what could be put up in the future for large scale establishment of renewable energy systems.

5. Security and Protection Against Data Breaches

In A.I., decision-making and applications, in enormous volume of sophisticated data that may be personal in nature, make them susceptible to major concerns like data breach and identity theft. There are many areas where machine learning applications in smart grid and energy systems have more problems in the face of adversarial attack and security problems than predictive maintenance models. Many predictive maintenance strategies in energy systems also utilize deep learning, cloud computing, and IoT sensors, which, in turn, are vulnerable to cyberattacks and security threats. In consequence, the absence of comprehensive security arrangements becomes one of the significant limitations to the successful deployments in AI in renewable energy. A possible answer to this would be developing strengthened security procedures and technologies that could prevent any future data breach while still allowing the AI systems to develop (Dietterich, 2017).

6. Performance and Explainability Issues

Operationalization is one of the terms commonly employed within organizational psychology to discuss how managers might determine the degree to which goals or objectives are being achieved. There's always been fiery debate as to whether organizations can actually achieve this, and typically it dovetails with the goal setting, organizational design, and even performance appraisal literature within the majority of organizations. BEINGS processing a variety of disciplinary ventures aims to improve this very individual contribution within organizations in terms of goal setting. Employers can thus gauge the performance levels of their employers in translating a goal or target into a living operational schedule by looking for evidence as to whether or not their employees can define goals expressed in these precision terms. Is it possible to operationalize any goals if, also, its counterpart is refused same? Therein could be considered how research complications become manifested to mean that goal-related outcomes may be construed in the perspective of human capital. It is commonly used theory on the relationship between metrics and meaning-making and how they facilitate respective literatures in two disciplines, BEINGS, and associated programs.

It is a very technical issue to achieve the balance of what to be decided from what to be predicted in predictive maintenance for energy systems. The deep learning-based strategies, which have to enhance the analytical performance of these networks, while the network remains comprehensible, should be investigated greatly to tackle the challenge.

7. Energy Storage and Grid Integration Issues

The necessity to store the energy needed for the future occasional nature of renewable power systems is the admittance to operators of renewables into the larger electric system infrastructure. But there are a range of difficult challenges, starting with the high costs of power storage devices, limited energy density, power density, and the lifetimes of such storage technologies. The advanced devices, including AI, artificial intelligence, image processing, and especially characterization, are shedding light on energy storage at the atomic and molecular level. It could be enhanced to develop next-generation energy storage devices as it has a higher charge density and longer lifespan, as the disintegration from repeated destruction-charge sequences was diminished.

Two main challenges also face hydro technology: the safe storage of energy and seamless grid links. Through AI and ML one can develop generators to predict such generative patterns or even brokers like ChatGPT can guide to predict where the energy is coming from, how it is being produced, and who will be using it. Analysing historical data can be helpful in determining when periods of peak electricity consumption and low hydropower generation might come about in the future, hence helping resource planning and optimization of hydropower plants. Such problems might be solved by creating appropriate mechanisms for retaining electricity, like high-head pumped hydro storage or improved battery systems. Such storage would enable utility of excess energy generated during abundant flow periods, by releasing it when supply dips and load becomes high thereby conferring reliability and viability to hydropower. Under this strategy, consumers can enjoy a continuous, stable energy supply.

CONCLUSION AND FUTURE SCOPE

An exhaustive description of the applied research and the growth that still exists in the field of AI-based algorithms in the booming technology is showcased in this research article with every additional minor problem for RE systems. It looks at some of the case studies and recent experimental breakthroughs in single and hybrid renewable energy systems, zeroing in on solar energy, PV (photovoltaics), energy efficiency in microgrid integration, wind, and geothermal energy. In solar energy, AI simulation approaches provide efficient alternatives to traditional modelling by handling complicated multivariable challenges with minimal computing effort. AI is making a significant impact on PV technology, especially as it analyses vast information on the solar cell materials and performance.

The chapter can be further developed with AI used for energy storage work in microgrids. Machine learning is part of AI technology, which is designed to predict and monitor wind turbines. AI has another application in mapping and optimizing fairways and geothermal plays by using both surface and subsurface data for analysis. This study discussed the application of artificial intelligence in modeling renewable energy systems for simulation, control, decision making, and optimization. Recent and future research, prospective problems, and future trends discuss in regional suitability analysis for AI application in real-estate applications. Creative and hybrid AI approaches would address considerable advances in renewable energy technologies, even contributing to the global wealth. Likely, the future study project will focus on field-specific data to better the AI implementation in RE systems.

REFERENCES

Aitkazinov, A. (2023). The role of artificial intelligence in auditing: Opportunities and challenges. International Journal of Research in Engineering. *Science and Management*, *6*(6), 117–119.

Akhter, M. N., Mekhilef, S., Mokhlis, H., & Mohamed Shah, N. (2019). Review on forecasting of photovoltaic power generation based on machine learning and metaheuristic techniques. *IET Renewable Power Generation*, *13*(7), 1009–1023.

Dietterich, T. G. (2017). Steps toward robust artificial intelligence. *AI Magazine*, *38*(3), 3–24.

Georgiou, M. (2019). *The role of ai technology in improving the renewable energy sector*. Insider.

Goel, A., Goel, A. K., & Kumar, A. (2023). The role of artificial neural network and machine learning in utilizing spatial information. *Spatial Information Research*, *31*(3), 275–285.

Kadhem, A. A., Wahab, N. A., Aris, I., Jasni, J., Abdalla, A. N., & Matsukawa, Y. (2017). Reliability assessment of generating systems with wind power penetration via BPSO. Journal of International Journal on Advanced Science. *Engineering and Information Technology*, *7*(4), 1248–1254.

Karim, T. F., Lipu, M. H., & Mahmud, M. S. (2017). Electricity access improvement using renewable energy and energy efficiency: A case of urban poor area of dhaka, bangladesh. [IJRER]. *International Journal of Renewable Energy Research*, *7*(3), 1296–1306.

Kothari, D. P., Ranjan, R., & Singal, K. C. (2021). Renewable energy sources and emerging technologies.

Kumar, A., Nagar, S., & Anand, S. (2021). Climate change and existential threats. In *Global Climate Change* (pp. 1–31). Elsevier.

Kumar, V. R., Bali, S. K., & Devarapalli, R. (2023). Application of AI-Based Algorithms for Industrial Photovoltaic Module Parameter Extraction. *SN Computer Science*, *4*(5), 525.

Lahmadi, K. (2023). *Artificial intelligence can improve renewable energy production, report shows*. Morocco World News.

Liao, M., & Yao, Y. (2021). Applications of artificial intelligence-based modeling for bioenergy systems: A review. *Global Change Biology. Bioenergy*, *13*(5), 774–802.

Mahapatra, S., Kumar, D., Singh, B., & Sachan, P. K. (2021). Biofuels and their sources of production: A review on cleaner sustainable alternative against conventional fuel, in the framework of the food and energy nexus. *Energy Nexus*, *4*, 100036.

Mazzeo, D., Herdem, M. S., Matera, N., Bonini, M., Wen, J. Z., Nathwani, J., & Oliveti, G. (2021). Artificial intelligence application for the performance prediction of a clean energy community. *Energy*, *232*, 120999.

Moraga, J., Duzgun, H. S., Cavur, M., & Soydan, H. (2022). The geothermal artificial intelligence for geothermal exploration. *Renewable Energy*, *192*, 134–149.

Nam, K., Hwangbo, S., & Yoo, C. (2020). A deep learning-based forecasting model for renewable energy scenarios to guide sustainable energy policy: A case study of korea. *Renewable & Sustainable Energy Reviews, 122*, 109725.

Naser, N., & Abdelbari, A. (2020, June). Estimation of Global Solar Radiation using Back Propagation Neural Network: A case study Tripoli, Libya. In *2020 International Conference on Electrical, Communication, and Computer Engineering (ICECCE)* (pp. 1-5). IEEE.

Onwusinkwue, S., Osasona, F., Ahmad, I. A. I., Anyanwu, A. C., Dawodu, S. O., Obi, O. C., & Hamdan, A. (2024). Artificial intelligence (AI) in renewable energy: A review of predictive maintenance and energy optimization. *World Journal of Advanced Research and Reviews, 21*(1), 2487–2499.

Perea-Moreno, A.-J. (2021). *Renewable energy and energy saving: Worldwide research trends*. MDPI.

Tao, Y., Destek, M. A., Pata, U. K., & Khan, Z. (2023). Environmental regulations and carbon emissions: The role of renewable energy research and development expenditures. *Sustainability (Basel), 15*(18), 13345. DOI: 10.3390/su151813345

Tawalbeh, M., Al-Othman, A., Kafiah, F., Abdelsalam, E., Almomani, F., & Alkasrawi, M. (2021). Environmental impacts of solar photovoltaic systems: A critical review of recent progress and future outlook. *The Science of the Total Environment, 759*, 143528.

Zohuri, B., & Zohuri, B. (2018). Types of renewable energy. Hybrid Energy Systems: Driving Reliable Renewable Sources of Energy Storage, 105-133.

Chapter 17
Harnessing Artificial Intelligence for a Greener Future in Solar, Wind, and Hydrogen

G. Boopathy
https://orcid.org/0000-0003-2515-7277
Vel Tech Rangarajan Dr. Sagunthala R&D Institute of Science and Technology, Chennai, India

K. Jayakumar
https://orcid.org/0000-0003-2482-981X
Sri Sivasubramaniya Nadar College of Engineering, India

C. Suresh
https://orcid.org/0000-0001-7549-3506
Global Academy of Technology, India

V. Srinivasan
https://orcid.org/0009-0006-2988-6476
Vel Tech Rangarajan Dr. Sagunthala R&D Institute of Science and Technology, India

ABSTRACT

Renewable energy sources have been the driver of innovation, with solar, wind, and the green hydrogen close to the top of the list. The addition of Artificial Intelligence (AI) to these systems enables faster development, improved scalability, and higher reliability for these systems. AI predicts weather patterns, optimizes panel orientation and allows for predictive maintenance of solar photovoltaic (PV) systems to maximize solar energy and minimize operational downtime and inefficiencies. AI optimizes the performance of wind turbines by optimizing accurate wind pattern forecasting and energy supply stabilization through intelligent grid integration via predictive analytics. Electrolysis processes are improved with AI, its energy efficiency is increased, and the green hydrogen is distributed and stored as well as possible. While barriers still exist in issues such as data privacy and computational demand, developments in edge computing and machine learning appear to bring about encouraging opportunities for moving past these barriers.

DOI: 10.4018/979-8-3373-0045-0.ch017

INTRODUCTION

Overview of Renewable Energy and Sustainability Goals

Renewable energy is a must when looking for a sustainable future in energy. As the world community tackles the problems of climate change, energy security, and resource depletion, the answer in using energy from renewable sources such as solar, wind, and hydrogen becomes practical. This chapter aims to present a detailed analysis of various energy modes their significance in terms of achieving sustainability goals.

Solar energy includes solar thermal systems or photovoltaic cells to obtain the energy developed during the time of the sun. It's one the most abundant and simple to access sources of energy, but one of the biggest to reduce greenhouse gas emissions. A renewable technology like solar energy can help lessen a country's dependence of fossil fuels, overall energy independence, which will drive creation of jobs in the green technology field.

Wind energy, or wind generated energy, is the energy created when wind is converted into electrical power using wind turbines. It's a clean and renewable resource that in the last few years has seen tremendous growth. By helping to lower carbon footprints, wind energy also creates local jobs and builds local infrastructure to support local economies. Wind energy is solidifying itself as a key player in the world's energy future because technology breakthroughs are making it more cost effective and efficient.

Hydrogen is a form of energy that is now becoming quite popular, and can be generated from a number of renewable resources. As a fuel substitute vehicle or as fuel cells that can generate power, it can be used as a clean fuel. Hard to electrify sectors like long distance transportation and heavy manufacturing could be decarbonized by hydrogen.

Investing in hydrogen technologies can support countries in strengthening their energy resilience and energising their energy portfolios.

Renewable energy is important because it provides the answers to social, economic and environmental problems, helping us to achieve the Sustainable Development Goals (SDGs). Renewable energy sources such as solar, wind and hydrogen must replace fossil fuels for greenhouse gas emissions to be reduced and energy security improved. Along with supporting economic growth and generation of employment, it helps to fight the consequences of climate change. Renewable energy sources are much better than fossil fuels when it comes to lowering greenhouse gas emissions, thereby producing better air quality and reducing pollution. Essential to the preservation of biodiversity and natural resources as it relates to the SDGs for environmental protection, they are according to (Kumar et al., 2023). (Malik et al., 2023) For instance, investments in renewable energy technology raise employment and economic growth in countries where investments start, like in emerging countries. (Wang et al., 2024)

Solar, wind, and hydrogen are future solutions to sustainable energy. Energy produced from these resources is not taken out of the environment they are naturally regenerative produced, and do very little in the way of pollutants and greenhouse gases. Resources that come from solar and wind have experienced decreased pricing over the years due to decades of technological advancements, which makes these resources the best options for energy transitions over time. Solar and wind resources offer scalable and clean electricity that can be plugged into existing grids to prepare for future international energy supply and greatest demand (Saeedmanesh et al., 2018).

In addition, solar is versatile similar to wind which means it can fuel any transportation needs, work in manufacturing, or be a storage solution in itself. Thus, with both resources relying on solar and wind to electrolyze renewable energy, the production of hydrogen offers a zero emissions pathway toward decarbonization (Sarkar et al., 2023). Furthermore, hydrogen can be a storage solution for transportation energy when needed while acting as a viable solution for solar and wind resource intermittency, thus creating energy resiliency. Such renewable technologies provide an all-inclusive approach to decarbonizing the energy sector because they allow for practically no environmental degradation with longterm sustainable energy. The emergence of solar and wind energy resources coupled with hydrogen as a non-polluting, clean-burning fuel source urges a non-reliance upon fossil fuels while redirecting the focus to a more energy-efficient and sustainable approach. Therefore, in regions where such technologies are used and merged, low emissions occur with sustainable energy systems that are stable and adaptable for future needs (Quazi et al., 2019).

Role of AI in the Renewable Energy

AI will expand renewable energy efforts as it offers efficiencies, innovations, and optimizations that were never before thought of and are additionally needed for sustainable efforts. The implementation of AI methodologies machine learning, deep learning, and neural networks is already allowing for renewables to be evaluated, managed, and implemented more seamlessly. For example, AI even in weather insights enhances the precision of short term energy predictions which is critical for smart cities and F ecosystems. Better prediction allows for networks of devices to better plan their activities for energy from generation to consumption which minimizes excess and increases grid stability. AI sifts through tremendous amounts of information to find intricate relationships that increase the efficiency of energy generation and use. In addition, it lowers expenses by facilitating predictive maintenance, increasing uptime, and boosting energy generation potential for both wind turbines and solar arrays.

AI-based platforms are used to apply collaborative energy optimization to introduce renewable energy to the grid. The platforms apply more intelligent generation, distribution, and consumption to reach higher efficiency in a reliable manner. For instance, AI will optimize site selection, orientation, as well as tracking in solar energy systems to capture all kinds of energy. In wind energy, AI makes better predictions about wind patterns and optimizes turbine operations to work at their peak.

Secondly, with AI, the stability of power grids is enhanced by spotting potential troubles before they become failures that are significant enough and can cause serious destabilization of the energy infrastructure.

Operationally, AI can be used for greater strategic integration, as far as renewable energy sources are concerned. For instance, the use of AI in energy storage systems, such as in hydrogen storage or advanced battery technologies, means better matching of supply with demand, whereby one does not have to rely as much on the backup related to fossil fuel and instead encourages a more sustainable energy framework. Table-1 highlights some of key areas in AI contributions happened through the references. As the renewable energy sector continues to evolve, AI will play an increasingly pivotal role in enhancing the efficiency and reliability of existing systems while driving innovation that accelerates the transition to a cleaner, more sustainable energy future.

Table 1. Contribution of AI in key areas

Area	Contributions of AI	Refrences
Energy Forecasting	Enhances short-term energy forecasts, crucial for smart cities and Internet of Things (IoT) applications	(Necula et al., 2023)
Smart Grids	Facilitates the development of smart grids and collaborative energy-sharing models	(Rojek et al., 2023)
Energy Management	Optimizes demand response, load management, and energy trading	(Stecyk et al., 2023)
Predictive Maintenance	Improves fault detection and predictive maintenance, enhancing infrastructure efficiency and longevity.	(Fan et al., 2023)
Renewable Integration	Reduces integration costs and enhances grid stability for variable renewable energy sources.	(Boza et al., 2021)
Energy Storage	Advances in energy storage solutions, improving the sustainability of renewable energy projects.	(Swarnkar et al.,2023)

AI IN SOLAR ENERGY

The renewable energy industry has been expanded by the new technology known as AI in which solar energy has received the biggest boost. AI in solar energy systems also provides the possible ways of increasing the rate of solar electricity production by precise prognosis of the intensity of sun rays, effective management of the energy consumption, and identification of the faults. At its core, the use of machine learning algorithms can help solar power generation to be effective and efficient from a weather and demand forecasting point of view while also fulfilling grid requirements. AI also another critical enabler of smart grid implementation, storage of energy infrastructure, and continuous maintenance of photovoltaic systems. In this chapter, it studies how AI and contemporary solar energy are entwined, and it laces particular focus on the importance of AI in moving the field of improved energy storage monoliths forward.

This section explores how AI is having a revolutionary impact on the field of solar energy. Fig.1 displays AI-Driven solar energy optimization. It explores a wide array of AI research that helps optimize solar panel performance, perform predictive maintenance, and enable integration of solar energy into intelligent grid systems. With targeted use of AI technologies, the solar industry is well positioned to maximize energy production, reduce operating costs and generally improve the sustainability of the energy infrastructure.

Figure 1. AI- driven solar energy optimization

Forecasting accuracy, system optimization and overall efficiency in the solar energy sector are improved substantially by AI. Predictability is elusive and irregular in solar energy and grid integration, and AI techniques, especially machine learning frameworks, are the essential means to mitigate associated problems. The computing involved with handling complex solar data in a recent advancement in transformer topologies has become easier to handle due to an increase in the solar forecasting accuracy and computation efficiency. Empirically, they have shown that multi-layer feed forward artificial neural networks (ANNs) out perform traditional forecasting methods in grid management and energy resource optimization. In fact, this methodology has shown important performance in the optimization of grid connected photovoltaic systems, particularly in shading conditions, when compared with more traditional techniques, like Particle Swarm Optimization (Kumar et al., 2024) (Marva et al.,2024)

Optimizing Solar Panel Efficiency and Energy Prediction

AI applications are radically changing how solar panels work by boosting solar panels efficiency and predicting energy yield. Primarily, as part of this shift, predictive analytics is used to predict energy production based on past weather data and environmental factors. Machine learning algorithms attempt to maximize the amount of sunlight capture during the diurnal solar cycle by utilizing machine learning to predict trends in temperature, humidity and solar irradiance to help determine optimal solar panel location and orientation.

In addition, real time data from weather forecasts can be integrated by AI to dynamically alter operating parameters of solar installations in real time. This aid to anticipate energy yields with higher accuracy simplifies energy resource planning and energy resource management. To maximize their solar output and make energy production by the grid more predictable and predictable, solar energy suppliers can take action to overcome erratic weather.

To boost solar panel efficiency and energy forecasting, real time monitoring systems and productive advanced machine learning techniques need to be integrated. Recent academic studies have highlighted the importance of hybrid models that combine several machine learning techniques to improve accuracy of forecast and operational effectiveness in solar energy systems. It is far more accurate to make forecast of solar energy output using methods such as Hybrid Convolutional - Long Short-Term Memory (LSTM) Net (HCLN) and Artificial Neural Networks (ANN); HCLN reduces RMSE value to as low as 0.012027. These models use past data including solar irradiance and performance ratios to correctly forecast energy generation (Bhutta etal., 2024). Issues with dust accumulation, which could obstruct efficiency, can be mitigated and ongoing performance evaluation made feasible through the use of Internet of Things (IoT)- based monitoring systems as discussed by (Varun et al., 2024). Hybrid modeling systems make early detection of performance disparities easier by predicting expected versus real power output. (Mangalam et al., 2024)

AI related solutions have been identified as instrumental technologies that can help enhance the efficiency of the solar panels as well as refine the energy forecasting. Indeed, by means of applying predictive analytics, AI opens the opportunity to provide accurate forecasts of energy yield of solar energy systems with regard to fluctuating weather conditions and other influences. For example, the enhanced generative adversarial fuzzy network (EGAFN) which uses the multi-objective optimization algorithms to predict the solar power production based on inputs such as the solar irradiation, panel temperature and the effect of shading (Alam et al., 2023). Such systems enhance not only energy predictions but also enable the dynamic control of the solar panels' characteristics.

Simple intelligence models (EGAFN) apart from sophisticated intelligence models like Adaptive Neuro-Fuzzy Inference System (ANFIS) are used to forecast several critical performance metrics including PV systems performances ratio and annual energy yield. ANFIS models include environmental factors (temperature, wind speed, and solar irradiation). This also improves prediction accuracy in terms of solar energy generation capacity (Gopi et al., 2022). However, these models give us critical predictive power, because they help us understand seasonal fluctuations and coordinate energy storage options for excess energy necessary for the efficient planning of large scale solar farms.

AI also allows alternative approaches to solar panel design that perform the best economically. By means of AI algorithms, for instance, one may vary tilt angle, cleaning schedule or tracking method autonomously as a function of ambient variables in order to improve total energy performance. Putting anticipatory and flexible strategies in practice boosts the efficiency of solar energy systems, doing so lowers operating costs and increases the solar energy's accessibility in the global energy market making the use of solar energy more attractive.

Fault Detection and Maintenance

Maintaining the integrity and productivity of solar installations requires AI driven fault detection mechanisms. With the help of machine learning algorithms, these systems are able to look at data coming off of solar panels and inverters, and spot anomalies that might indicate impending failures. A first identification of faults is possible, allowing for prompt interventions thus a minimum downtime as well as maintenance expenses.

The solar installations would have a longer, more productive operational life using AI enhanced predictive maintenance. Scoring components and solar equipment based on historical maintenance records and continual data on the performance of solar equipment can eventually help predict possible component failures with AI. With this proactive approach, operators can determine the best possible intervals for maintenance which results in less downtime with a longer life cycle of solar assets.

The combination of problem detection and AI driven predictive maintenance have greatly exceeded the reliability and operating efficiency of PV systems. Modern PV systems are able to precisely identify, diagnose, and even now predict system weaknesses through the use of algorithms such as Adaptive Logic Networks (ALN), Support Vector Machines (SVM), or other machine learning approaches. Real time data has been employed to enable these systems to detect deviations which may be warning signs of possible problems by comparing new sensor data with pre-defined performance measures. (Abubakar et al., 2021)

For instance, AI based systems can find signals of coming up problems like shading, soiling etc from changes in voltage, current or temperature. Using machine learning algorithms, operators can determine what type and severity these issues are, and use this data to apply targeted modifications. The large improvements include faster response times and less energy lost to latent instabilities.

Besides fault detection, the predictive maintenance based on AI comes down to failure that is likely to happen in equipment. Over time, with help of past records AI, the occurrence of some part wear or failure could be predicted reduce downtime due to major mishaps (Benedetti et al., 2018). In addition, the advanced and intelligent maintenance systems are capable to scheduling proper flow of technicians to be placed at appropriate time and reduce labour cost and improves the economic efficiency of solar installations.

New breeds of AI, specifically Deep Learning and IoT integrations, are augmenting the opportunities of fault detection and maintenances. Sensors placed on IoT are capable of capturing great amounts of performance information for devices, which are further analyzed by AI models in real-time, and relay useful Essential System Parameters (ESP) to operators. Such productivity enhancing technologies are making solar systems last longer, more dependable and produce more energy.

Smart Grid Integration

The integration of solar energy into intelligent grid systems is one important area where AI is crucial. AI technologies make solar energy efficiently distributed and stored to ensure it is used within the grid infrastructure with efficiency. Using these data, we can observe what real time energy consumption and output indicators have to say and use AI to balance supply and demand, to optimize the electrical flow.

AI can also greatly improve energy storage systems by effectively predicting the best times to store excess energy generated during hours of highest solar irradiance and when it is released during times of high demands. In addition to enhancing the efficiency of solar energy consumption, this competence strengthens the overall stability of energy grid infrastructure.

AI is a key component in enabling the smooth transition of solar energy into smart grid systems because it provides a solution to a number of complex problems associated with variable renewable energy (VRE) sources. As an example of this, the gray wolf optimizer algorithm variant, is an extremely advanced AI algorithm that optimizes an energy management system (EMS) to enable efficient dispatch and distribution of solar energy in a grid connected infrastructure (Kumar et al., 2021). Consequently, these systems enable supply and demand to be balanced in real time by real-time data from storage units, distributed energy resources (DERs), and customer demand trends, improving the efficiency and sustainability of the network.

The optimization of energy storage is another significant obstacle to the integration of smart grids and AI is crucial in this case. Recent advances in AI algorithms allow for the prediction of solar energy generation and grid demand; the combination of which can inform the optimal charge and discharge cycles for battery systems. Not only will this feature help stabilize the grid, but having a steady supply of energy, whether solar output is lower and the demand is higher is also accomplished.

Furthermore, AI solutions minimize the cost of integrating sources of VRE. Superior optimisation capacities of AI ensure that curtailment is at the minimum feasible levels, there is proper load peaking, and costly infrastructural upgradation is avoided. Predictive analytics allow the operator to prepare for and mitigate variability due to weather and power to effect a smoother integration process.

Apart from management and storage of energy, AI is also applied to ensure the security and resilience of the grid. The systems identify anomalies in the working of the grid, from voltage fluctuations to cyber threats, and ensure a quick response to potential disruptions. Thus, electricity supply is sure and safe even though a greater proportion of the grid consists of renewable sources of energy.

Further growth in the solar revolution will realize the role of AI-driven integration with smart grids in shaping a sustainable and efficient future about energy. Energy distribution and storage optimization became possible through these technologies and paved the way toward decentralized intelligent networks of energy, which could find its own adaptation with the conditions found in the modern world.

AI IN WIND ENERGY

Wind power generation is getting AI, as it grows in the place of artificial intelligence. As newcomers of algorithm based operations and maintenance and energy forecasting algorithms, complementing experiences are becoming more advanced during design of Megawatt (MW) scale wind turbines. The technologies used in these improve Turbine performance, reduce utilization costs and stretch periods thus equipment life. Balancing wind energy is also guaranteed through AI powering forecasts of power output and managing system variability to ensure that energy production is stead, stable. The role of AI in site selection, blade layout and monitoring will only keep expanding as the technology moves forward even further which is essential in increasing the pace of renewable energy breakthrough in the globe.

In this portion, the wind energy operations are examined in terms of how AI is revolutionizing it. Figure 2 depicts AI-driven wind energy optimization. It illustrates AI techniques that enhance grid management and energy storage capacity, amending turbine efficiency, predictive maintenance, and wind capability analysis for energy forecasting. By using AI technology, the wind energy Industry could greatly improve performance, reduce operating costs and align energy supply with consumer demand.

Figure 2. AI-driven wind energy optimization

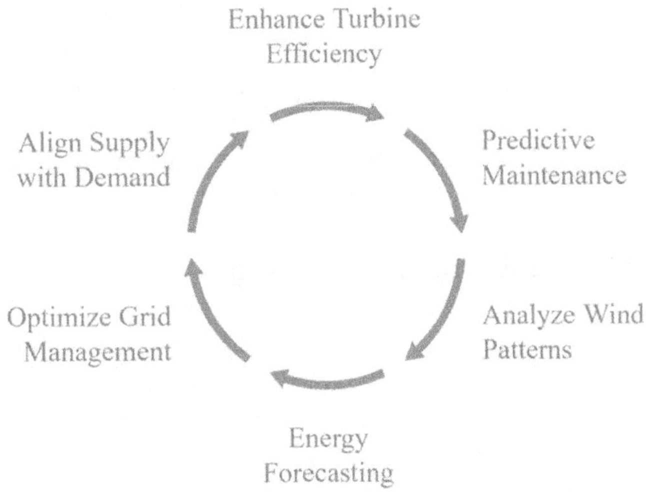

Combining AI accelerates and improves sustainability assessment, streamlines operations and improves efficiency in the wind energy sector. Various wind energy system components from layout optimization to predictive maintenance rely on using AI technologies which greatly impact both environmental and performance results. This allows for prediction of the lifecycle of turbines, maintenance needs and downtime driving down operating costs. (Bassey et al., 2024) To predict upcoming faults and respond quickly, machine learning models take the past data into considerations. (Xia et al., 2023) How AI such as particle swarm optimization and genetic algorithms can identify optimum locations for turbine, resulting in higher energy production, and lower costs, environmental effects. (Jaadi et al., 2024)

Enhancing Turbine Efficiency and Predictive Maintenance

Wind turbine operational efficiency can be improved with the use of AI techniques. Operators can use machine learning algorithms and sophisticated data analysis to perform in the moment evaluations of turbine performance, finding inefficiencies and potential problems before they become apparent. Predictive maintenance systems take historical data and characteristics to predict when maintenance interventions are necessary. By doing so, downtime is minimized and the turbines already very long lived numbers to begin with ‾ are lengthened. For instance, AI systems can identify break beginnings in thermal fluctuations or vibrational patterns to promptly initiate corrective actions to heads off expensive repairs and guarantee maximum turbine execution.

AI technologies increase turbine efficiency, decrease offline time, and forecast wind power facility maintenance needs. The primary elements of the predictive maintenance process are machine learning algorithms, especially when they solve complex data preprocessing and feature selection tasks to detect potential failures and improve the remaining useful life (RUL) of wind turbine components in a timely manner (Garan et al., 2022). Using real-time data from the turbines' sensors AI-assisted monitoring systems enable to track even minor shifts in the turbines' behavior that may, for example, show an increase in vibrations or temperature – signs of potential failures.

Also, as computer vision and deep learning approaches are incorporated into maintenance processes, faults are automatically identifiable. For instance, endoscopic imaging systems applying AI techniques for bearing fault identification mitigate fully on human inspection, enhancing the specificity and operation pace for technical personnel (Shin et al., 2021). Likewise, such reinforcement learning algorithms are in use to optimize the working parameters of a Turbine online, so that more Power is generated and at the same time, less Stress is put on it. Coupled with digital twins of higher sophistication, AI applications within the wind turbine industry provide a model of the wind turbines to emulate the performance under various scenarios, predict optimal time for service, and drastically cut unanticipated outages.

Wind Pattern Analysis and Energy Forecasting

AI models are instrumental in analyzing complex wind patterns, which is essential for accurate energy forecasting. By processing vast amounts of meteorological data, AI can identify trends and predict wind speeds and directions with greater precision. This improved forecasting capability allows energy producers to better align their output with consumer demand, reducing waste and enhancing grid stability. For example, AI-driven predictive analytics can inform operators when to ramp up production or when to store energy, ensuring that supply meets demand effectively.

They are turning wind pattern analysis and energy forecasting into powerful tools with state-of-the-art computational technologies like artificial neural networks, support vector machines, or combination of methods. These models make use of advanced regional prediction system (ARPS), predictive analytics (PA), geographic information systems (GIS) and history and real time parameters and environment inputs to produce very accurate wind speed and energy output. For instance, the machine learning model using artificial neural networks (ANNs) using multi-source meteorological data enable the operators to determine optimum timing and capacity of wind farm into the grid to enhance grid dependability and operational expense (Obeidat et al., 2022).

Overcoming weaknesses that are associated with the inherent features of wind energy series such as noise and variation, other complex hybrid models including empirical mode decomposition-filtered wavelet neural network enhanced by an ensemble learning mechanism are proposed. Through analyzing wind data, segregating tangled wind data into interpretable variables, and finding significant attributes, these models enable continuity and precision in volatile surge forecasts (Du et al., 2019). In addition, AI has been adopted in the energy markets when determining the forecasts thereof and designing the corresponding pricing and/or resources supply strategies. In terms of the uncertainty reduction of energy generated by wind farms, AI helps free the design of power by minimizing the amounts of reserve power used and increase the viability of wind energy integration.

AI-Enabled Grid Management and Energy Storage

AI driven solutions are needed to improve the operational efficiency of wind turbines. Machine learning algorithms and data analysis techniques can help for operators to monitor turbine performance on the real time basis. That is to identify inefficiencies and possible mistakes before they show up creating a problem for them. Historical data and operational factors used in predictive maintenance systems are exploited to forecast maintenance needs with the objective of minimizing downtime and increasing the operating lifespan of turbines.

For example, AI can alert for thermal anomaly and vibration patterns so that costly repairs can be prevented and to maximise turbine performance.

Integrating wind energy into the electrical infrastructure systems is limited by the predictability and intermittency of wind energy itself. AI technologies resolve energy storage and grid system issues. These sophisticated resources comprise of neural networks, fuzzy logic control systems and integrated optimization algorithms for their deployment to synchronize wind power with other controllable energy sources and storage modalities like batteries, pumped hydro and hydrogen storage systems. Our AI approaches are efficient in energy distribution management, while keeping the optimal consumption cost, even in periods of high demand or wind curtailment. (Tabanjat et al., 2017)

In addition, they also indicate that existing AI frameworks for power regulation and dynamic demand management must become more dispersed to cover fluctuation of wind energy. Further, these frameworks employ machine learning based techniques and multi agent systems to control decentralized power resources for grid stability and quality. Energy losses are reduced and overall system performance increased by utilizing advanced AI based solutions for energy storage optimization, including the application of reinforcement learning for regulating battery charging and discharging cycles.

Smart grids require significant improvements in implementing energy trading that can be enabled by the AI–blockchain hybrid. These systems facilitate direct energy trading between producers and consumers and in the process help distribute surplus wind energy to storage/Storages/other renewable energy sources thereby enhancing sustainability. Furthermore, the demand-response systems driven by AI technologies proactively alter the consumption rates to meet available availability so they do not cause a burden on the grid while integrating wind energy into hybrid renewable solutions (Javaid et al., 2023).

Thus, relying on multiple aspects of its applicability, AI is among the key pillars for solving the technical, economic, and other problems of the wind energy industry and advancing the task of developing a more stable and sustainable energy system.

AI IN GREEN HYDROGEN PRODUCTION

AI is disrupting the generation of green hydrogen through designs which reveal improvements in the method and lessen the expenditure. The renewable power generated electricity, which created green hydrogen through electrolysis, is an indispensable solution in the path to worldwide decarbonization. AI technologies such as machine learning and deep learning are used in prognostics of electrolyzers to help monitor and predict performance degradation and optimize operational characteristic in real-time. More so AI models enable the design of hydrogen production systems incorporating intermittent energy forms like wind and solar due to their fluctuating nature. Pertaining to the earlier and complex calculations done on data, AI also provides support in aspects such as topological design and composition, energy efficiency, and even market prognostication of future hydrogen requirements. AI makes it possible to scale up green hydrogen generation and serve as the essential part of a sustainable power system while simultaneously addressing technological and financial challenges.

This section investigates green hydrogen production using AI and electrolysis process optimization, hydrogen generation and storage prediction modelling, and the potential for green hydrogen's long-term applications in transportation and industry. Figure 3 depicts the AI-Driven green hydrogen production cycle. As the world increasingly moves toward more ecologically friendly energy sources, AI is demonstrating itself as a key tool to improve operational efficiency, increase costs savings, and encourage widespread use of hydrogen as a sustainable fuel option.

Figure 3. AI-driven green hydrogen production cycle

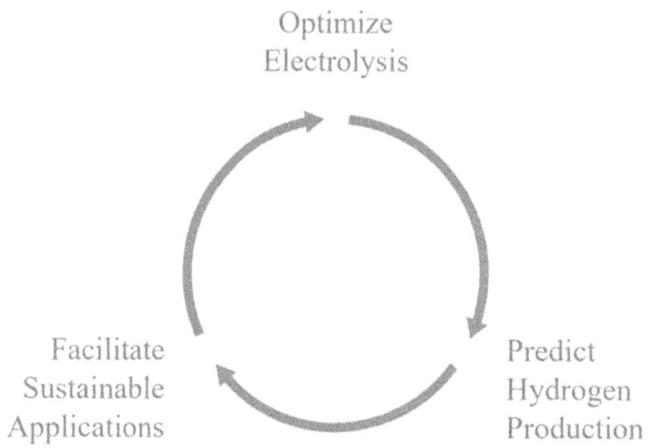

The use of AI in the generation of green hydrogen is drastically changing the efficiency and economic feasibility of this sustainable energy source. Using AI approaches, particularly machine learning (ML), the electrolysis process, electrocatalyst innovation, and overall system integration are being improved. Complex electrolysis data is analyzed by AI models that decide the best operating settings for maximum hydrogen production and minimum energy consumption. As machine learning approaches can predict real-time changes based on electrical input and water quality characteristics, there can be up to 20% energy savings.(Bassey et al., 2023) AI hugely shortens the time and resources needed for empirical evaluations because it very much accelerates the procedure of identifying the best electrocatalysts. Machine learning algorithms provide the systematic design of materials that increase the water-splitting reaction efficiency. (Kim et al., 2023)

Optimizing Electrolysis Processes

AI serves an important role in optimizing electrolysis operations for the production of hydrogen to increase energy efficiency and cut costs. Machine learning allows us to use AI to process large datasets from electrolysis systems to observe trends and improve operating conditions. Some of these involve adjusting parameters such as temperature, pressure, and current density to concurrently generate as much hydrogen as we can, and use as little energy as possible.

Real time monitoring and control of electrolysis systems may be assisted by AI and dynamic, real time adjustment to changing energy availability and prices. This flexibility also supports the economic viability of hydrogen generation as well as the incorporation of renewable sources of energy such as solar and wind into the electrolysis framework. Thus, the application of AI driven optimization may significantly reduce cost hydrogen generation and further make hydrogen generation a much more cost attractive option than fossil fuel.

Through increased energy efficiency, cost reduction, and system optimization, AI is further applied to the electrolysis procedures used to generate green hydrogen energy. Electrolysis which is the process of breaking water molecule into hydrogen and oxygen by use of electricity is time consuming hence call for optimization to become viable. For operational parameters, ANFIS with optimization algorithms including physics optimization algorithms (POA) has provided an intervention measure of efficiency. These methods determine the best electrolysis conditions for the time, voltage, density, and catalyst, and hydrogen enhancement is seen growing from 1.3% to 1.7% more efficient than other methods (Rudko et al., 2023).

Further, real-time control of an electrolyzer by AI technology allows for optimum control of electrolysis compatible with intermittent renewable sources of energy such as wind and solar. It also offers a critical contribution to simulating the behavior of newly synthesized materials for electrodes and membranes and forecasting their performance varying operating conditions and thus, chemistries of new efficient and low-cost materials.

For the record, rigorous AI-enabled techno-economic analyses (TEA) are pivotal to the detection of worth-saving opportunities in a nation's production value chain. Based on the system design, components, and manufacturing processes optimization, AI contributes to the decrease in capital and operational costs to compete the green hydrogen production with the fossil fuel-based methods (Badgett et al., 2021). Also, there are AI-driven and data-powered digital twins of the electrolyzers, which analyze the processes, identify optimal conditions and forecast possible faults that can lead to decreased operation time and overall cost.

Predictive Modeling for Hydrogen Production and Storage

The demand and rates of hydrogen generation, as well as determining storage options, are key areas of demand and demand rates to be predicted by using AI models. Using historical data in conjunction with real time inputs, AI is shown to be able to forecast hydrogen production based on a range of inputs including changes in the energy supply, weather and consumer demand. In converting these predictive models into a reality, they ensure producers prepare in advance and then adjust their operations in proportion to consumption patterns, so that hydrogen production continues in perfect synchrony with consumption.

In addition, AI can not only predict production, but also optimize hydrogen storage solutions by closely working with data about storage capacity, pressure, and temperature. The machine learning algorithms excel when compressing gas or liquid hydrogen or metal hydrides to uniquely define the best storage practices for a given application or environmental condition. Besides making the hydrogen supply more dependable, this optimization reduces the financial burden on storage infrastructure.

Through the use of AI's predictive modeling abilities, clean burning alternative (CBA) facilities will be able to continue to produce high levels of hydrogen while working to develop storage plans best suited to their needs. Multivariate models used routinely by sophisticated computing algorithms are used to forecast hydrogen production. Accurately assimilating large sets of data, such as weather, the availability of renewable resources and historical production metrics, are able to do these models. As an example, current AI models predict that, while there has been a supply rate change, hydrogen production rates from solar and wind energy systems will maintain a sufficient supply rate in accordance with system efficiency and environmental status (Javaid et al., 2023).

Besides production outlook, AI models play an important role in the calculation of hydrogen demand as well. Using relevant data on fill history, transport movements and energy consumption, these tools model future prospects to predict and improve efficiencies of hydrogen refueling stations. Demand forecasting enables hydrogen storage plus distribution infrastructure to be targeted to estimated requirements, consequently reducing costs and excess or inadequate supplies being produced (Kurtz et al., 2020).

For example, these solutions are also crucial to the development of new hydrogen storage technologies. For instance, machine learning is applied to assess solid state materials of interest (e.g. hydrogen storage, cryogenic systems, high pressure storage tanks). They also improve safety features and increase energy density and reduce need to build materials in these vehicles. With sensor data, fault detection and diagnosis algorithms are developed to increase storage system dependability by reducing failure risks and preventing leaks and other dangers.

Sustainable Transportation and Industrial Applications

AI applications are related to optimization of hydrogen as a sustainable fuel source for its industrial and transportation uses. AI could increase the operating efficiency of hydrogen fuel cell vehicles in the transportation sector greatly by enabling improved energy management systems, predicting energy consumption, and streamlining route planning techniques. By analyzing traffic dynamics, vehicle performance indicators, etc. AI helps assure hydrogen fueled vehicles operate at maximum efficiency. Hydrogen fueled vehicle will become a more viable alternative to traditional fossil fuel based transportation.

In industrial processes from chemical synthesis to energy generation to steel making, AI offers the potential to optimize hydrogen consumption. These industrial systems can provide for reductions in carbon emissions, as well as enhancements of overall energy efficiency when integrated with the AI technology. In addition, the use of AI to clarify potential pathways for hydrogen integration and streamlining the supply chain operations can assist a transition to hydrogen centric energy systems.

Transportation and other industrial uses of hydrogen are changing as a result of innovation in artificial intelligence, a current technology. Frameworks powered by AI have been designed to improve hydrogen generation from non-conventional resources, including water flooded reservoirs containing hydrocarbons, by improving oxygen injection techniques. These frameworks deploy data scientific, physics based models and genetically optimized algorithms for substantially enhanced hydrogen yields and substantially reduced oxygen percentages, which is both highly efficient and environmentally friendly (Katterbauer et al., 2021).

Concerning transportation, AI steps up improvement in the facets of hydrogen fuel cell vehicles (HFCVs) in terms of fuel cell performance and longevity. Using AI algorithms, damages in fuel cells and the consequent patterns of degradation are identified through operational data of vehicles and operating conditions are changed in order to extend system's lifetime. Further, AI optimizes the traffic of refuelling infrastructures and control supply of hydrogen, the occupancy and unsatisfied demand of HFCVs in transportation networks.

AI is vitally important for industrial systems and primarily used for process control and the improvement of them. Continuous remote control of wireless temperature sensors and other parameters in high temperature processes is crucial for hydrogen production from high temperature processes such as in-situ combustion. Oxygen supply and combustion variables are adjusted by AI algorithms, sustaining the elevated temperatures necessary for hydrogen extraction but reducing thermal energy addition and emissions (Katterbauer et al., 2022).

In addition, the use of AI to enable hydrogen to smoothly and efficiently blend into industrial systems like the steel and synthesis of ammonia in terms of the flow, storage, and consumption of the element is made easier by AI. Because these applications apply to lowering carbon emissions in industries characterized by high levels of difficulty in attaining sustainable energy, they are evidence of the role of AI in bringing sustainable change to the energy sector.

In this way, AI acts as a foundation for the optimal, predictive, and control-based production, storage, and usage of green hydrogen, enabling large-scale energy conversion to sustainable energy.

INTEGRATING AI ACROSS RENEWABLE SYSTEMS

Integration and management of renewable energy systems have undergone reinvention by the application of AI in presenting better, dynamic and optimized solutions within different energy contexts. Through use of machine learning, deep learning, and optimization algorithms AI makes it possible to integrate renewable energy such as solar, wind and hydro power and such like and such like with ease since they are characterized by intermittent and variables. Advanced energy prediction using AI helps optimally allocate resources while real time analysis helps to manage energy supply and stability of the grid. Moreover, AI drives better energy storage, asset monitoring and maintenance, as well as optimal Integrated Renewable Hybrid Systems in the application. In this sense, AI promotes the global transition to net zero by enabling an integrated, sustainable energy system.

This section looks at how AI might improve efficiency and coordination in renewable energy systems, as seen in Figure 4. Through our study of the relationships between a solar, wind, and hydrogen energy systems, as well as between AI and smart grid technologies and energy distribution mechanisms, we demonstrate the cooperativeness of these technologies in maximizing energy generation, storage, and reliability.

Figure 4. Optimizing renewable energy systems with AI

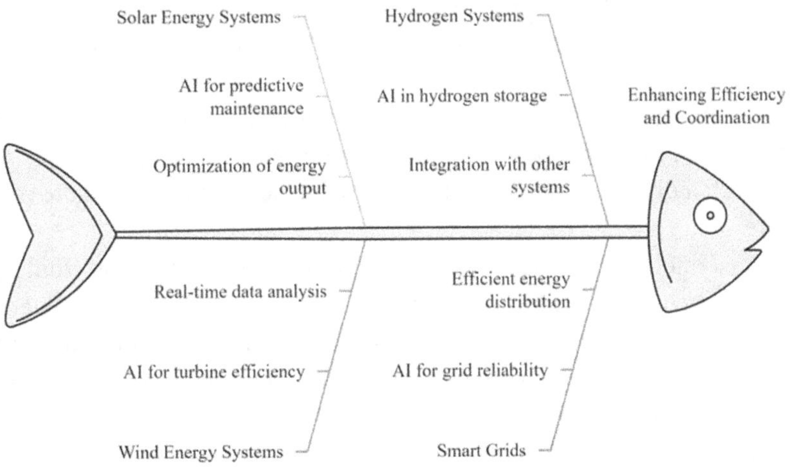

For the improvement of operational performance and sustainability and efficiency, AI is being incorporated into renewable energy systems. There are many ways how AI techniques, such as with machine learning (ML) and predictive analytics, can enhance several aspects of the renewable energy from lifetime assessments to grid integration. AI algorithms allow better energy forecasting, enhanced resource evaluation, and enhanced system monitoring to effectively manage the inherent variability of renewable energy sources. Real time flexibility is encouraged by advanced AI techniques such as neural networks, followed by improved forecasting, leading to higher energy production and lower operating costs. (Ukoba, et al., 2024) Smart Grid technologies cannot survive without AI as it helps improve system stability and also facilitates integration of renewable energy sources. (Benchikh et al., 2024) AI and IoT technologies are working together to create a more robust energy framework through real time monitoring and predictive maintenance. (Komala et al., 2024)

Cross-Energy System Synergies

AIplays crucial role in integrating and balancing several renewable energy systems, quite especially solar, wind and hydrogen technologies. Utilizing Neural Networks (NN) and Fuzzy Logic Control (FLC) for managing hybrid energy systems, NN and FLC solve efficiency and operating cost problem of energy systems, by precisely coordinating the operation of wind turbines, solar panels and hydrogen storage systems. In addition, these systems proactively coordinate resources in response to weather, grid demand, and kept real time information to reduce energy waste and always ensure a reasonable supply of electricity.

Energy production and storage, fueled by solar, wind, and hydrogen systems is supported and enhanced enabled by AI. Using complex algorithms and machine learning techniques, it is able to analyze large datasets from many energy sources, and do so, in real time. By allowing us to see patterns and trends in energy production and use, this opens up the possibility in better decision making.

For example, AI can predict the output of solar and wind energy by looking at historical records and weather forecasts and, therefore, be able to seamlessly incorporate sporadic energy sources into one whole system. During times when solar energy output is at its highest, AI routes excess energy to capture it as hydrogen using electrolysis, which produces hydrogen. AI could also be used to control the release of stored hydrogen as energy needs in the case of reductions in sun or wind power, to provide a continuous flow of energy.

In addition to it can improve the performance of energy storage devices such as batteries by planning the best times to charge and discharge based on demand forecasts and energy pricing. Besides boosting the effectiveness of each individual system, dynamic management produces a more robust and flexible energy infrastructure.

The use of advanced intelligent models aims to optimize energy generation from both sunlight and wind through effective forecasting models developed by system operators to intelligently guide energy distribution. As an instance, predictive analyses of consumptions are adopted for better use of AI in predicting best periods of time for production and storage of hydrogen, making it easier to maintain surplus renewable on a state of green hydrogen for periods of high availability (Abdelghany et al., 2024).

In addition, model predictive control approaches based on AI are applied to control and coordinate grid-interconnected micro grids, energy production and distribution and hydrogen production for optimizing and protracting Energy storage systems. Some of these strategies not only show how an optimal use of available resources can be made, but also enhance the reliability and durability of hybrid renewable energy systems.

Coordinated control systems and decentralised AI based frameworks continue to make the system more robust by handling transient power control and overall long term power distribution. Within wind-solar-hydrogen combined input output (IO) networks, these systems facilitate decision making for the distribution of energy by capturing real-time system control (SC) information thereby minimizing system reliance on the grid and increasing system independency (Lin et al., 2022). All this integration of AI makes the hybrid renewable systems to be able to open up for more adaptive and sustainable energy supply and demand systems.

AI-Driven Smart Grids and Energy Distribution

The integration of AI into smart grid infrastructures is revolutionizing energy distribution through the use of an integrated network capable of effectively handling a range of renewable energy sources. Smart grids enabled by artificial intelligence, for example, are good at monitoring and controlling energy flow, guaranteeing that it is transferred to the most crucial locations in real time.

AI algorithms can analyze consumption patterns to predict peak demand periods, fueling proactive adjustments to energy distribution techniques. This feature minimizes outages and substantially enhances grid reliability. It can also support demand response programs that incentivize and encourage consumers to shift their energy demands during peak periods. This will reinforce the supply and demand equilibrium.

AI also fosters the adoption of distributed energy resources on the energy grid, from small-scale wind turbines to solar systems on rooftops. AI ensures that these dispersed systems are effectively integrated with the overall energy portfolio through the coordination of their production, maximizing the utilization of renewable resources.

Machine learning is becoming critically essential for developing smart grid technologies and effectively integrating a wide range of renewable energy resources while improving the reliability of smart electricity networks. Machine learning or artificial neural networks make it possible to predict the availability of power from VREs, specifically one of the limitations of the solutions which are their intermittency. These forecasts enhance the energy storage control and system coordination allowing supply and demand balance. (Yousef et al., 2023).

Hyped smart grids integrated with AI not only enhance the utilization of the smart grid but also facilitate a bi-directional energy/data sharing that leads to more engaging and reliable energy systems. Besides, the AI-based smart grids help to provide the so-called grid self-healing when power outages and other issues are detected and mitigated by the IoT and blockchain, integrated into the grids. In energy trading, Blockchain technologies still enhance security and development in peer to peer (P2P) sharing of energy and decentralisation of the smart grid (Kumar et al., 2020).

Real-time energy forecasting using AI is vital in the short term to tailor energy use in smart cities and the IoT for IoT applications like demand response and demand-price control. These capabilities enable the consumers and industries to align their energy use patterns with those of each other, resulting in cost savings along with support in maintaining the balance of the grid. In addition, application of AI fosters the occurrence of distributed and cooperative energy-sharing paradigms where any excess energy that may be produced by an individual source may be off-loaded to other areas or stored for future use.

Furthermore, another critical role of AI is in optimizing renewable energy solutions compared to the traditional grid structure for electricity distribution in cities as well as AI based charging structures for electric vehicles, energy-efficient building, and environmentally friendly transport systems in urban settings. As enablers of the transformation of the electricity grid in the direction of clean interconnection of distributed renewable energy generation and management, smart grids leveraging AI deserve adequate attention.

CHALLENGES AND FUTURE DIRECTIONS

There are potential and challenges associated with artificial intelligence's involvement in promoting a sustainable future in the solar, wind, and hydrogen energy industries. The lack of standardized and thorough data, the inadequate incorporation of AI into current utility frameworks, and concerns about reliable security and privacy are among the main problems. Increasing the interpretability of AI models, creating specialized AI solutions for big datasets, utilizing AI in other digital twin applications, and utilizing next-generation wireless technologies are some possible improvements. In the near future, much more efficient energy production, delivery, and consumption may result from addressing these issues with artificial intelligence.

The content of this section is illustrated graphically in Figure 5, which also looks at the latest advancements and trends in the rapidly evolving sector, as well as the operational and technological challenges of integrating AI into renewable energy systems. To fully utilize artificial intelligence's potential in improving renewable energy solutions, it presents both opportunities and obstacles that must be overcome as it develops.

Figure 5. Challenges in integration of AI for renewable energy

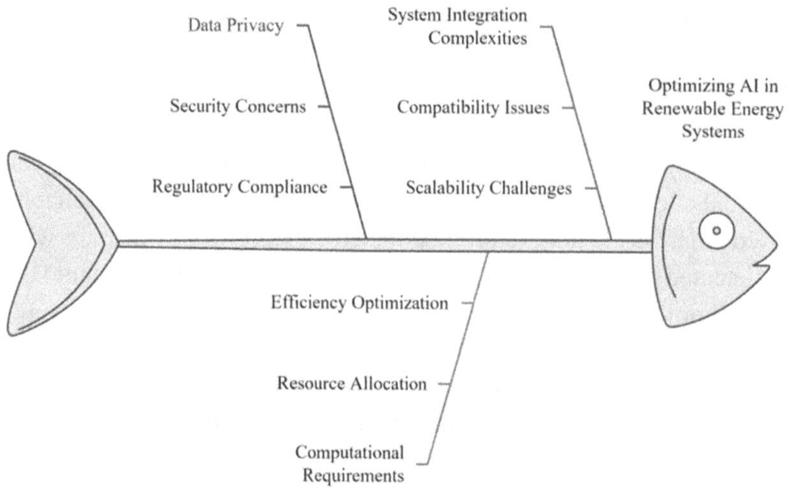

Technical and Operational Challenges

- ***Data Privacy*:** Ensuring data confidentiality is one of the main challenges in integrating AI into the renewable energy industry. The increasing reliance on data-driven decision making processes necessitates protecting sensitive data on user interactions, operational indicators, and trends in energy consumption. It is crucial to strike a balance between protecting user privacy and using data to gain insights from artificial intelligence.
- ***Computational Requirements*:** AI algorithms can have significant processing demands, particularly when working with large datasets like those seen in renewable energy applications. Effective training of complex models often requires the deployment of high-performance computer resources, which can be a major obstacle for smaller businesses or organizations with limited funding. Additionally, some of the environmental benefits of renewable energy may be compromised by the energy required to carry out these computing operations.
- ***System Integration Complexities*:** There are significant obstacles in integrating AI systems with existing renewable energy infrastructures. Many renewable energy systems are based on out-dated technology that might not be compatible with modern artificial intelligence. To maximize efficacy and efficiency, it is essential to provide smooth communication and interoperability among various systems. Furthermore, integration efforts may become even more challenging in the absence of defined standards.

At the same time, the use of green energy sources that relate with new processes like solar, wind, and hydrogen has an impressive transformational promise but comes with some technical and operational issues. Privacy issue is a major concern while, especially where market is distributed such as in the use of multi-access edge computing (MEC) systems in the energy market. MEC transfers computation from large central servers to smaller local servers along the edges making the process faster and more secure. Nevertheless, the said shift poses new privacy concerns, such as privacy-preserving technologies, particularly encryption and using secure communication channels to prevent leakage of privacy-sensitive energy data (Saatloo et al., 2024).

Modelling and optimization requirements for the use of renewable energy and its system are still another challenge that has often been attributed to high level computational issues. For instance, the flow physics and heat transfer analysis of aerodynamic wakes in wind farms and thermal management system analysis of concentrated solar power (CSP) receivers necessitates to employ computational fluid dynamics (CFD) and renders it essential to use high-performance computing facilities. These simulations are computationally expensive requiring considerable computational power and memory to produce accurate results (Drikakis et al., 2022). Furthermore, the use of such computational frameworks in conjunction with AI requires algorithms for the effective handling of the generated big data.

Integration challenges stem from the fact that the integrated system requires coordination of different energy inputs with different conditions of operation and different conversion processes. As noted earlier, solar and wind generation are variably generated based on the intensity of sun and wind respectively while hydrogen system demands consistency for energy requirements to fuel the electrolysis process. Managing these systems needs advanced control and optimization methods, including and not limited to, hybrid control techniques and AI-sustained prediction systems to achieve energy variability and guarantee grid stability. Furthermore, retrofitting of renewable energy systems also present intrinsic barriers such as technical difficulties in incorporating, structural barriers inherent in policies, and high initial capital costs all of which slows the deployment of renewable energy systems.

Emerging Trends and Future Prospects

- ***Advances in AI Models***: The development of increasingly complex AI models has the potential to revolutionize the field of renewable energy. To improve energy management, strengthen grid stability, and improve predictive analytics, techniques like deep learning and reinforcement learning are being improved. These developments will make it easier to predict energy production and consumption more precisely, which will enhance resource allocation and reduce waste.
- ***Diverse Data Sources***: A wider range of data sources will become more accessible as the Internet of Things (IoT) grows, giving AI applications access to more extensive datasets. It is anticipated that real-time data generated by sensors, smart meters, and networked devices would be used to support AI algorithms. This data explosion will encourage more accurate modelling and decision-making, which will ultimately increase renewable energy systems' efficiency.
- ***Integration Strategies***: Prospects for AI in the field of renewable energy also include the development of creative integration techniques. Blockchain technology and AI combined in collaborative platforms could increase energy transaction security and transparency. Furthermore, demand response programs that use AI can optimize energy consumption patterns and bring them closer to the availability of renewable energy.

AI integration in the renewable energy sector is constantly evolving, which holds promise for overcoming current obstacles and bringing in a new era of AI use in renewable energy. ML and deep learning (DL)-based AI models enable more accurate power generation forecasts and system scheduling, as well as better grid management for variable renewable energy VRE sources. To provide precise projections that promote effective resource management and system dependability, these models rely on large datasets covering historical energy generation, weather, and grid demand (Klaiber et al., 2024).

Combining several approaches, such as Internet of Things (IoT) sensor networks, smart grid analytics, and environmental data collection, is becoming a novel approach that seems to enhance the performance and robustness of AI systems. For instance, combining anemometers, solar irradiance sensors, and hydrogen output gauges makes it easier to evaluate renewable energy systems in real time, which enhances energy distribution and storage decision-making (Ahmad et al., 2021).

Advancements in Explainable AI (XAI) and Quantum AI (QAI) technologies are solving the inherent issues and risks in renewable energy systems. XAI helps in increasing the general trust in AI decision making because it offers explainable results for model-made predictions, which is highly important in the case of data law compliance and by the acceptance of stakeholder conformity with AI. At the same time, QAI, based on the principles of quantum computing, is designed to perform energy optimization tasks faster than any existing methods, using optimal calculation methods for solving large-scale tasks such as grid management and multi-energy systems coordination (Ohalete et al., 2023).

Another trend is the integration of AI with digital twins of physical assets Digital Twins of physical assets are being followed, and integrated with AI another comprehensive trend. Real-time performance evaluation, detection of the fault, and preventive maintenance can be conveniently performed in renewable energy systems through digital twins, which are virtual models of physical assets. By replicating real-life conditions and stress testing various redesign possibilities within a digital clone, digital twins decrease injuries and enhance functioning of a system.

Furthermore, those alerts are pointing to the AI-based innovations that address the economic side of renewable energy initiatives. Mitigations like automated project planning, cost forecasting, and optimization of resources allocation have lessened integration costs, and thus increasing the economic feasibility of renewable energy projects. More and more, advanced AI is being used in demand-response systems, P2P, and dynamic prices all open up new markets, while giving consumers and businesses roles that involve active participation in energy networks.

At the same time, the future of AI is to become critical in furthering the shift into clean energy across the planet by helping to address the technical, operational, and economic limitations as well as optimising the return on investment (RO) which indicates the complete potential of renewable energy (RE) systems.

REFERENCES

Abdelghany, M., Al-Durra, A., & Gao, F. (2024). A Coordinated Optimal Operation of a Grid-Connected Wind-Solar Microgrid Incorporating Hybrid Energy Storage Management Systems. *IEEE Transactions on Sustainable Energy*, *15*(1), 39–51. DOI: 10.1109/TSTE.2023.3263540

Abubakar, A., Almeida, C., & Gemignani, M. (2021). Review of Artificial Intelligence-Based Failure Detection and Diagnosis Methods for Solar Photovoltaic Systems. *Machines (Basel)*, *9*(12), 328. Advance online publication. DOI: 10.3390/machines9120328

Ahmad, T., Zhang, D., Huang, C., Zhang, H., Dai, N., Song, Y., & Chen, H. (2021). Artificial intelligence in sustainable energy industry: Status Quo, challenges and opportunities. *Journal of Cleaner Production*, *289*, 125834. Advance online publication. DOI: 10.1016/j.jclepro.2021.125834

Alam, M., Alshahrani, T., Khan, F., Hakami, J., Shinde, S., & Azim, R. (2023). AI-based efficiency analysis technique for photovoltaic renewable energy system. *Physica Scripta*, *98*(12), 126006. Advance online publication. DOI: 10.1088/1402-4896/ad0bb4

Badgett, A., Ruth, M., James, B., & Pivovar, B. (2021). Methods identifying cost reduction potential for water electrolysis systems. *Current Opinion in Chemical Engineering*, *33*, 100714. DOI: 10.1016/j.coche.2021.100714

Bassey, K. E., & Ibegbulam, C. (2023). Machine learning for green hydrogen production. *Computer Science & IT Research Journal*, *4*(3), 368–385.

Bassey, K. E., Juliet, A. R., & Stephen, A. O. (2024). AI-Enhanced lifecycle assessment of renewable energy systems. *Engineering Science & Technology Journal*, *5*(7), 2082–2099.

Benchikh, S., Jarou, T., & Lamrani, R. (2024). Enhancing renewable energy systems with advanced artificial intelligence solutions. *SCT Proceedings in Interdisciplinary Insights and Innovations, 2*, 319-319.

Benedetti, M., Leonardi, F., Messina, F., Santoro, C., & Vasilakos, A. (2018). Anomaly detection and predictive maintenance for photovoltaic systems. *Neurocomputing*, *310*, 59–68. DOI: 10.1016/j.neucom.2018.05.017

Bhutta, M. S., Li, Y., Abubakar, M., Almasoudi, F. M., Alatawi, K. S. S., Altimania, M. R., & Al-Barashi, M. (2024). Optimizing solar power efficiency in smart grids using hybrid machine learning models for accurate energy generation prediction. *Scientific Reports*, *14*(1), 17101. PMID: 39048605

Boza, P., & Evgeniou, T. (2021). Artificial intelligence to support the integration of variable renewable energy sources to the power system. *Applied Energy*, *290*, 116754. DOI: 10.1016/j.apenergy.2021.116754

Drikakis, D., & Dbouk, T. (2022). The Role of Computational Science in Wind and Solar Energy: A Critical Review. *Energies*, *15*(24), 9609. Advance online publication. DOI: 10.3390/en15249609

Du, P., Wang, J., Yang, W., & Niu, T. (2019). A novel hybrid model for short-term wind power forecasting. *Applied Soft Computing*, *80*, 93–106. DOI: 10.1016/j.asoc.2019.03.035

Ehteshami, S., & Chan, S. (2014). The role of hydrogen and fuel cells to store renewable energy in the future energy network – potentials and challenges. *Energy Policy*, *73*, 103–109. DOI: 10.1016/j.enpol.2014.04.046

El Jaadi, M., Haidi, T., & Belfqih, A. (2024). Advancements in wind farm layout optimization: A comprehensive review of artificial intelligence approaches. *Telkomnika*, *22*(3), 763–772.

Fan, Z., Yan, Z., & Wen, S. (2023). Deep Learning and Artificial Intelligence in Sustainability: A Review of SDGs, Renewable Energy, and Environmental Health. *Sustainability (Basel)*, *15*(18), 13493. Advance online publication. DOI: 10.3390/su151813493

Garan, M., Tidriri, K., & Kovalenko, I. (2022). A Data-Centric Machine Learning Methodology: Application on Predictive Maintenance of Wind Turbines. *Energies*, *15*(3), 826. Advance online publication. DOI: 10.3390/en15030826

Gopi, A., Sharma, P., Sudhakar, K., Ngui, W., Kirpichnikova, I., & Cuce, E. (2022). Weather Impact on Solar Farm Performance: A Comparative Analysis of Machine Learning Techniques. *Sustainability (Basel)*, *15*(1), 439. Advance online publication. DOI: 10.3390/su15010439

Hanif, M. F., & Mi, J. (2024). Harnessing AI for solar energy: Emergence of transformer models. *Applied Energy*, *369*, 123541.

Javaid, S., Ioannou, I., Pitsillides, A., & Tan, Y. (2023). Motivating an Agent-based Distributed AI Framework for Renewable Integration: Power Balancing. *2023 International Conference on Consumer Electronics - Taiwan (ICCE-Taiwan)*, 189-190. https://doi.org/DOI: 10.1109/ICCE-Taiwan58799.2023.10226739

Javaid, U., Usman, R., & Javaid, A. (2023). Investigating the Energy Production through Sustainable Sources by Incorporating Multifarious Machine Learning Methodologies. 2023 3rd International Conference on Artificial Intelligence (ICAI), 233-237. https://doi.org/DOI: 10.1109/ICAI58407.2023.10136677

Katterbauer, K., Qasim, A., Marsala, A., & Yousef, A. (2021). A Data Driven Artificial Intelligence Framework for Hydrogen Production Optimization in Water flooded Hydrocarbon Reservoir. Day 4 Thu, November 18, 2021. .DOI: 10.2118/207847-MS

Katterbauer, K., Qasim, A., Shehri, A., & Yousif, A. (2022). A Novel Artificial Intelligence Framework for the Optimal Control of Wireless Temperature Sensors for Optimizing Oxygen Injection in Subsurface Reservoirs. Day 2 Wed, March 23, 2022. .DOI: 10.4043/31558-MS

Kim, J., & Jang, H. W. (2023). Can Artificial Intelligence Boost developing electro catalysts for efficient water splitting to Produce Green Hydrogen? *Korean Journal of Materials Research*, *33*(5), 175–188.

Klaiber, J., & Dinther, C. (2023). Deep Learning for Variable Renewable Energy: A Systematic Review. *ACM Computing Surveys*, *56*(1), 1–37. DOI: 10.1145/3586006

Komala, C. R., Basha, M. M., Farook, S., Niranchana, R., Rajendiran, M., & Subhi, B. (2024). Smart Energy Systems-Integrated Machine Learning, IoT, and AI Tools. In Reshaping Environmental Science Through Machine Learning and IoT (pp. 201-229). IGIGlobal.

Kumar, A., Alaraj, M., Rizwan, M., & Nangia, U. (2021). Novel AI Based Energy Management System for Smart Grid With RES Integration. *IEEE Access : Practical Innovations, Open Solutions, 9*, 162530–162542. DOI: 10.1109/ACCESS.2021.3131502

Kumar, N., Chand, A., Malvoni, M., Prasad, K., Mamun, K., Islam, F., & Chopra, S. (2020). Distributed Energy Resources and the Application of AI, IoT, and Blockchain in Smart Grids. *Energies, 13*(21), 5739. Advance online publication. DOI: 10.3390/en13215739

Kumar, N., & Katiyar, A. (2024). Artificial Intelligence and the Future of Renewable Energy: Solar PV Power Forecasting. In New Innovations in AI, Aviation, and Air Traffic Technology (pp. 1-28). IGI Global.

Kumar, S., & Rathore, K. (2023). Renewable energy for sustainable development goal of clean and affordable energy. *International Journal of Materials Manufacturing and Sustainable Technologies, 2*(1), 1–15.

Kurtz, J., Bradley, T., Winkler, E., & Gearhart, C. (2020). Predicting demand for hydrogen station fueling. *International Journal of Hydrogen Energy, 45*(56), 32298–32310. DOI: 10.1016/j.ijhydene.2019.10.014

Lin, Y., & Fu, L. (2022). A Study for a Hybrid Wind-Solar-Battery System for Hydrogen Production in an Islanded MVDC Network. *IEEE Access : Practical Innovations, Open Solutions, 10*, 85355–85367. DOI: 10.1109/ACCESS.2022.3193683

Malik, P., Malik, P. K., Singh, R., & Gehlot, A. (2023). Sustainable Development, Renewable Energy and Environment. In *Micro-Electronics and Telecommunication Engineering: Proceedings of 6th ICMETE 2022* (pp. 455–463). Springer Nature Singapore.

Mangalam, H. SS, K. S., & Pavithran, P. (2024, July). Hybrid Modeling System for Photovoltaic Monitoring of Solar Panels. In 2024 2nd International Conference on Sustainable Computing and Smart Systems (ICSCSS) (pp. 38-42). IEEE.

Marwa, B., Ali, Z. M., Hanen, B., & Faouzi, M. M. (2024, July). AI Empowered Solar Energy: Reinforcement Learning and Comparative Analysis for Grid-Connected Photovoltaic Systems Optimization. In 2024 IEEE 28th International Conference on Intelligent Engineering Systems (INES) (pp. 000251-000256). IEEE.

Necula, S. (2023). Assessing the Potential of Artificial Intelligence in Advancing Clean Energy Technologies in Europe: A Systematic Review. *Energies, 16*(22), 7633. Advance online publication. DOI: 10.3390/en16227633

Obeidat, M., Ameryeen, B., Mansour, A., Salem, H., & Awwad, A. (2022). Wind Power Forecasting using Artificial Neural Network. *WSEAS Transactions on Power Systems, 17*, 269–279. Advance online publication. DOI: 10.37394/232016.2022.17.28

Ohalete, N., Aderibigbe, A., Ani, E., Ohenhen, P., Daraojimba, D., & Odulaja, B. (2023). *AIdriven solutions in renewable energy: A review of data science applications in solar and wind energy optimization.* World Journal of Advanced Research and Reviews., DOI: 10.30574/wjarr.2023.20.3.2433

Qazi, A., Hussain, F., Rahim, N., Hardaker, G., Alghazzawi, D., Shaban, K., & Haruna, K. (2019). Towards Sustainable Energy: A Systematic Review of Renewable Energy Sources, Technologies, and Public Opinions. *IEEE Access : Practical Innovations, Open Solutions*, *7*, 63837–63851. DOI: 10.1109/ACCESS.2019.2906402

Rojek, I., Mroziński, A., Kotlarz, P., Macko, M., & Mikołajewski, D. (2023). AI-Based Computational Model in Sustainable Transformation of Energy Markets. *Energies*, *16*(24), 8059. Advance online publication. DOI: 10.3390/en16248059

Rudko, V., Tsoutsos, T., Rezk, H., Olabi, A., Abdelkareem, M., Alahmer, A., & Sayed, E. (2023). Maximizing Green Hydrogen Production from Water Electrocatalysis: Modeling and Optimization. *Journal of Marine Science and Engineering*, *11*(3), 617. Advance online publication. DOI: 10.3390/jmse11030617

Saatloo, A., Mehrabi, A., Marzband, M., Mirzaei, M., & Aslam, N. (2024). Local Energy Market Design for Power- and Hydrogen-Based Microgrids Considering a Hybrid Uncertainty Controlling Approach. *IEEE Transactions on Sustainable Energy*, *15*(1), 398–413. DOI: 10.1109/TSTE.2023.3288745

Saeedmanesh, A., Kinnon, M., & Brouwer, J. (2018). Hydrogen is essential for sustainability. *Current Opinion in Electrochemistry*, *12*, 166–181. Advance online publication. DOI: 10.1016/j.coelec.2018.11.009

Sarker, A., Azad, A., Rasul, M., & Doppalapudi, A. (2023). Prospect of Green Hydrogen Generation from Hybrid Renewable Energy Sources: A Review. *Energies*, *16*(3), 1556. Advance online publication. DOI: 10.3390/en16031556

Shin, W., Han, J., & Rhee, W. (2021). AI-assistance for predictive maintenance of renewable energy systems. *Energy*, *221*, 119775. DOI: 10.1016/j.energy.2021.119775

Stecyk, A., & Miciuła, I. (2023). Harnessing the Power of Artificial Intelligence for Collaborative Energy Optimization Platforms. *Energies*, *16*(13), 5210. Advance online publication. DOI: 10.3390/en16135210

Swarnkar, M., Chopra, M., Dhote, V., Nigam, N., Upadhyaya, K., & Prajapati, M. (2023). Use of AI for development and generation of renewable energy. 2023 IEEE Renewable Energy and Sustainable E-Mobility Conference (RESEM), 1-5. .DOI: 10.1109/RESEM57584.2023.10236136

Tabanjat, A., Becherif, M., Hissel, D., & Ramadan, H. (2017). Energy management hypothesis for hybrid power system of H2/WT/PV/GMT via AI techniques. *International Journal of Hydrogen Energy*, *43*(6), 3527–3541. DOI: 10.1016/j.ijhydene.2017.06.085

Ukoba, K., Olatunji, K. O., Adeoye, E., Jen, T. C., & Madyira, D. M. (2024). Optimizing renewable energy systems through artificial intelligence: Review and future prospects. Energy & Environment, 0958305X241256293.

Varun, P., Sunitha, T., Nagalingam, M., Nithiya, P., Selvakumaran, S., & Mohanaprakash, T. A. (2024, March). Improving Solar Efficiency via CNN-LSTM and Cloud-Integrated IoT Prediction. In 2024 3rd International Conference on Sentiment Analysis and Deep Learning (ICSADL) (pp. 329-334). IEEE.

Wang, Q., Huang, R., & Li, R. (2024). Renewable energy and sustainable development goals: Insights from latent dirichlet allocation thematic and bibliometric analysis. *Sustainable Development*.

Xia, Y. (2023). Leveraging AI Technology for Advancements in Wind Power. Science and Technology of Engineering, Chemistry and Environmental Protection, 1(4).

Yousef, L., Yousef, H., & Rocha-Meneses, L. (2023). Artificial Intelligence for Management of Variable Renewable Energy Systems: A Review of Current Status and Future Directions. *Energies*, *16*(24), 8057. Advance online publication. DOI: 10.3390/en16248057

Chapter 18
Homogenous Groups in EU Member States According to Common Indicators of Sustainable Development

Dorottya Edina Kozma
https://orcid.org/0000-0002-4948-8815
University of Pannonia, Hungary

ABSTRACT

Sustainability is a global principle highlighting the interconnectedness of economic, social, and environmental dimensions. Sustainable development is a challenge not only for the EU but for the entire world. This chapter aims to examine the 27 EU Member States using cluster analysis. Five indicators were included in the examination and these represents all EU 27 sustainable development strategies. Indicators were compared with GDP and HDI, the two most relevant indicators of economic development and human well-being. The research posed the question: can we identify homogeneous groups of countries regarding sustainability among the 27 EU Member States? The analysis revealed that when clustering based on GDP, the Member States typically fell into two, three or four clusters. In contrast, when clustered according to the HDI, the number of clusters varied between two and four. Ultimately, four distinct clusters emerged: cutting-edge, fast-growing, prudently advanced, and tail-enders. The performance of Member States can be identified according to their positions within these clusters.

INTRODUCTION

Sustainable development, as a global principle, highlights the need for an integrated approach where economic, social, and environmental dimensions work together to address challenges effectively. This interconnectedness implies that no single dimension can independently resolve complex problems (Baják, 2013). Sustainability focuses on achieving long-term growth while preserving resource stocks for future generations, avoiding their depletion (Fleischer, 2007). The scope of sustainable development extends beyond national borders, requiring a global perspective that indicates intergenerational responsibility. It encompasses medium- and long-term goals across economic, social, and environmental domains (Ober-

DOI: 10.4018/979-8-3373-0045-0.ch018

er & Erkollar, 2011). However, as noted by Dell'Angelo et al. (2017), while sustainable development is fundamentally simple and vital, its practical application through concrete rules or criteria remains challenging and is often misunderstood.

Brundtland Commission introduced the term "sustainable development" in its 1987 report, *Our Common Future*. However, the concept emerged from earlier milestones, such as Rachel Carson's influential book *Silent Spring* and the Stockholm Conference of 1972 (Mebratu, 1998). The Brundtland Commission's framework became a cornerstone for subsequent global conferences and events, inspiring widespread efforts to prioritize environmental protection and transform societal perspectives.

The events discussed here pave the way to today's sustainable development framework, outlined in *Transforming Our World: The 2030 Agenda for Sustainable Development* and reflected in the sustainable development strategies and indicators adopted by the 27 EU Member States (Walsh et al., 2020). This chapter aims to analyze the 27 current EU Member States using cluster analysis over 2015–2022, grouping them into homogeneous categories that best represent their sustainable development performance. The theoretical foundation of this chapter raises the question of *How many homogeneous groups can be formed for the 27 Member States of the European Union.* The resulting hypothes*is is: "Based on the five common sustainable development indicators included in the sustainability strategies of the 27 EU Member States, groups of countries can be formulated with homogeneous characteristics that are distinguishable and characterizable in terms of GDP and HDI indicators.*

Country-level analysis is essential because sustainability and sustainable development are increasingly prominent research areas and are expected to become key drivers of territorial competitiveness in the face of global climate change (United Nations Educational, Scientific and Cultural Organisation [UNESCO], 2020). The design and objectives of the 2030 Agenda for Sustainable Development form the foundation for sustainability strategies implemented by nations and major cities. Looking ahead, sustainability will play a crucial role in shaping territorial competitiveness (Organisation for Economic Co-operation and Development [OECD], 2016). For future studies on territorial competitiveness among groups of countries, a valuable starting point is to categorize countries into homogeneous clusters, as this approach highlights the sustainability indicators (factors) with the greatest influence on the grouping.

The selected indicators – GDP, long-term unemployment rate, gross domestic R&D expenditures, public debt, greenhouse gas emissions, and areas under organic farming – are already well-established components of competitiveness measurement and contribute to the HDI indicator. These indicators were chosen because, after analyzing the sustainable development strategies and associated indicators of the 27 EU Member States, they were consistently included across all strategies. Cluster analysis is crucial for advancing this research. The author explored the clustering potential of EU Member States using GDP and HDI as key indicators. By implementing clustering, researchers can identify high-performing clusters to pinpoint best practices, develop tailored policies to address common challenges within each group, and foster cross-border collaboration among Member States in the same cluster. Clustering EU Member States based on sustainable development involves grouping countries with similar sustainability performance or characteristics. This approach helps uncover patterns, shared challenges, and opportunities for cooperation. The results of this analysis represent novel research and offer fresh insights into sustainable development strategies.

SUSTAINABLE DEVELOPMENT

There is a long history of efforts to identify and address unsustainable processes, closely linked to human activities. As a result, both social and natural scientists are interpreting and examining sustainable development. The population explosion, economic growth, and globalization of the 20th century have created a crisis; resulting in an interconnected system of social, environmental, and financial problems (Zhang et al., 2022). Sustainable development is therefore present in our everyday lives, whether in our family/personal lives or at work. As a result, sustainable development is perhaps one of the most familiar concepts in most disciplines today and has gained significant popularity over the last decade or two. In the face of humanity's continuing environmental degradation, the most important goal remains to improve well-being and living conditions (Horton & Horton, 2019).

In sustainability, we must strive for a growth path that is sustainable for many years to come, and as a result, we cannot and do not want to use up our future resource stock (Kuhlman – Farrington, 2010). Systems thinking is necessary to understand that development and the environment are part of a system and have consequences for the future. The second condition (sustainable use of natural resources) includes knowledge, culture, behavior, ethics, and work culture. Social justice is based on the sharing of the material, spiritual, and communal goods of society and nature. The three conditions presented here are interlinked in the three dimensions of sustainable development, with synergies between the social, economic, and environmental aspects.

Concepts and Interpretations of Sustainable Development

Sustainable development concerns everyone, involving long-term changes over the next 20 to 50 years as we work toward building a better future. Although it is not a new concept, in some interpretations, it represents the fundamental framework for understanding the relationship between humanity and the environment, as well as the ethical responsibilities of current generations toward future ones (Briška & Rungule, 2011). The global recognition of the need for environmental policies has driven the rise and promotion of sustainable development within development and planning practices (Brandful Cobbinah et al., 2015). A review of the multidisciplinary literature on sustainable development reveals that there is no unified theoretical framework that fully captures its complexity and scope.

The definitions of sustainable development are ambiguous, and lacking operational clarity, and there is no consensus on what should be preserved. The concept remains unclear, and there is a difference of opinion on the transition from theory to practice (Jabareen, 2008). As a result, sustainable development has been a contentious issue in the literature for many years due to its multidimensional nature (Kadlubek, 2015). The term is widely used in contemporary political and environmental discussions and has been the focus of ongoing debate for an extended period.

Costanza et al. (1991) argue that sustainability does not equate to a stagnant economy but emphasizes the need to distinguish between growth and development. Economic growth, a quantitative increase, cannot be sustained indefinitely on a planet with limited resources. Andrews (1997) suggests that sustainable development has become a symbolic rhetoric, with competing interests redefining it to align with their political agendas. Beatley and Manning (1998) contend that while sustainability is a positive goal, it still requires a clearer definition and elaboration. The term's ambiguity allows various social groups to apply it in different contexts. The contradictions within sustainable development primarily stem from inconsistencies and contradictions in the definitions of development and sustainability themselves (Silva Barbosa et al., 2014).

Development is defined as a transformation in the global production structure that does not disrupt ecosystems, is in harmony with them, and thus becomes sustainable in the long term. Few human activities are entirely free from direct or indirect impacts on sustainable development (Bartoluci et al., 2015). Sustainability also prompts a fundamental question: what exactly should we aim to sustain in sustainable development? Daly (2002) offers two answers to this query. First, the utility must be preserved, ensuring it does not decline over the lifetime of future generations. Second, physical performance must be maintained so that nature's ability to function independently is not impaired.

The concept of sustainable development can be traced back to the Stockholm Conference in 1972, where the significance of environmental management and environmental assessment was acknowledged as a crucial step in its development (Mebratu, 1998). The first use of the term "sustainable development" appears in Lester R. Brown's *Building a Sustainable Society* (1982). Brown connected population growth with the use of natural resources in a way that minimizes harm to the natural environment, both quantitatively and qualitatively (Brown, 1982).

Prugh et al. (1999) define sustainable development as progress that ensures ongoing social well-being without exceeding the ecological carrying capacity. When exploring the definition of sustainable development, it is impossible to overlook one of the most widely recognized concepts in the field. While there are over 100 definitions of sustainability and sustainable development, the most famous is that of the WCED (Bečić et al., 2012). Its final form as a concept was established in 1987, emphasizing that a country's economic and social development should follow the principles of sustainable development (Ruževičius, 2009). The Brundtland Commission's definition is: "Sustainable development is a development that meets the needs of the present without compromising the ability of future generations to meet their own needs" (WCED, 1987, p. 16).

Sustainable development, in the context of management practice, is defined as a process of transformation in which the exploitation of resources, investment strategies, technological advancements, and institutional changes are aligned to ensure that the needs and aspirations of both present and future generations are met (Kadlubek, 2015). The concept focuses on fulfilling stakeholder expectations by producing socially desirable goods and services more efficiently, while simultaneously minimizing natural resource use and adhering to environmental policies (Cohen, 2011).

A Global and European Approach to Sustainable Development

The European Union aims to foster stability, prosperity, and sustainable development globally. It places a strong emphasis on advancing peace, good governance, human rights, and respect for democratic values in its external actions (Debisso, 2012). Sustainable development strategies are intended to serve as a catalyst for policymakers and public opinion in the years following their adoption, driving institutional reforms and changes in consumer and business behavior. Successful implementation depends on various factors, such as biodiversity and the health of ecosystem services, which are crucial for society and the economy (Ladoneczki & Kósi, 2014).

As a result of an extensive process of social dialogue, the first European Union Sustainable Development Strategy was adopted in Gothenburg, Sweden, in June 2001 (European Environment Agency, 2002). This ambitious, broad, and long-term strategy acknowledges that investing in environmental, social, and human capital, as well as technological innovation, is essential for ensuring long-term competitiveness, social cohesion, economic prosperity, effective environmental protection, and quality employment (Ziolkowska & Ziolkowski, 2010). Why is this statement significant? First, without collective action at the EU level, there would be minimal cohesion and cooperation among Member States to achieve shared goals like sustainable development. Second, investing in human and natural capital is critical for a community's proper functioning and healthy progress. In the author's view, without such investment, future success would be unattainable, long-term competitiveness would be compromised, and the socio-economic-environmental pillars of sustainability and their synergies could be at risk.

An intriguing question arises in van Hees' (2014) interpretation, particularly regarding the lack of explanation in the strategy about how it relates to the Member States. To what extent should Member States, including local and regional authorities, integrate sustainable development into their activities? Member States are required to uphold sustainable development when implementing policies in areas harmonized with EU law. A good example is the development of individual sustainable development strategies by EU Member States. While countries were not mandated to create independent strategies, they were provided with recommendations. Nonetheless, the EU Sustainable Development Strategy (SDS) has provided the foundation for many countries' strategies (Kralj & Markič, 2008).

At the UN Rio+20 conference in 2012, nations agreed to establish new sustainable development goals beyond the Millennium Development Goals. These goals were designed to be universal, and applicable to all countries (Griggs et al., 2014). Before the current sustainability strategy was adopted, several initiatives were developed to address global challenges such as hunger, climate change, and poverty. This culminated in the Paris Conference of September 2015, where the United Nations, with the support of 193 Member States, created and adopted the 2030 Agenda for Sustainable Development. The agenda calls for global action across social, economic, and environmental dimensions (Alves et al., 2017).

The 2030 Agenda is a global initiative that builds upon and updates the Millennium Development Goals (MDGs), which were established in 2000. At its core are the 17 Sustainable Development Goals (SDGs) and 169 related targets, addressing all aspects of sustainability and marking an ambitious step toward achieving sustainable development (Fleming et al., 2017). The strategy builds on goals that have experienced minimal or no progress (United Nations, 2015). Like the previous global framework, it has a 15-year implementation period (2015-2030). The 169 sub-goals outline specific objectives to be achieved by 2030, marking the outcome of the strategy. These goals reflect the ambition and scope of the new universal programs. For countries worldwide, the 2030 Agenda has provided a framework, similar to the objective of the Rio+20 conference. The goal was for countries to adapt their strategies to the SDGs, building upon the established framework, and to produce progress reports, initially voluntarily, to track advancements.

By the end of the 20th century, sustainable development and sustainability had become widely recognized concepts. The definitions of sustainable development are complex and multifaceted, with few activities that do not have a direct or indirect impact on them. Today, the safeguarding of natural resources and the environment is a central aspect of socio-economic life. Sustainable development serves as a framework for economic growth, social inclusion, and environmental protection, emphasizing the need to meet present demands without compromising the ability of future generations to meet their own needs.

METHODOLOGY AND CONCEPTUALIZATION

Clustering the EU27 countries based on sustainable development involves grouping them into clusters with similar performance across key sustainability indicators. In quantitative research, a common question is whether distinct clusters or groups can be identified or formed within datasets. Cluster analysis serves as a method for grouping similar countries or items, essentially focusing on clustering. Cohesion and separation between data or clusters are the most important. Cohesion refers to how close variables are to each other within a cluster, and separation refers to the distance between clusters (Mur et al., 2016). The algorithm organizes observations or units into homogeneous groups based on the variables (Sajtos & Mitev, 2007). It is important to distinguish clustering from classification: while classification relies on prior knowledge of group characteristics, cluster analysis identifies groups directly from the data points. Székelyi and Barna (2002) describe cluster analysis as a dimension reduction technique, where variables associated with observation units define the original dimensions, and the aim is to group observations along these dimensions so that they are as similar as possible within each group. The primary goal of this method is to group diverse cases into cohesive, homogeneous clusters. Cluster analysis is sensitive to outliers and can be categorized into two main types: hierarchical clustering and non-hierarchical clustering, such as K-Means.

A key component of the research was to review the findings of previous studies that utilized cluster analysis. In this regard, Table 1 presents the studies that have made a notable impact on clustering countries and regions based on sustainable development in their respective research areas.

Table 1. Research utilizing cluster analysis

Name of the authors	Year of publication	Aim	The subject of the study	Method applied
Abdella et al.	2020	Introduce a novel approach to evaluating and modeling the sustainability impacts of food consumption.	Food consumption in the USA.	K-means clustering and logistic regression
Anuşlu & Firat	2019	Seeks to categorize countries based on the key impact areas of Industry 4.0.	Global Innovation Index, Sustainable Development Goals Index, Logistics Performance Index, and Environmental Performance Index.	PCA & K-means clustering
Dudek et al.	2019	A sustainable supply chain network design approach, a proposal for big data analysis, and the application of multi-objective methods to address multi-purpose supply chain network challenges.	Small and medium-sized enterprises.	Supply chain network (SCN design)
Nilashi et al.	2019	Aims to improve existing sustainability assessment frameworks by applying these techniques to explore the relationship between human sustainability, ecological sustainability, and overall sustainability performance by analyzing decision rules.	The decision rules derived from the sustainability assessments of 128 countries are employed to predict each country's sustainability performance.	Fuzzy clustering
Szennay & Szigeti	2019	Exploring the links between the SDGs and business processes.	SDGs indexed by GRI indicators.	K-means clustering

continued on following page

Table 1. Continued

Name of the authors	Year of publication	Aim	The subject of the study	Method applied
Uwasu & Yabar	2011	Analyze the sustainability conditions of countries using panel data on real savings (GS), and calculate the average returns, trends, and stability across countries.	Covers 6 governance dimensions with 6 indicators.	K-means clustering and regression analysis
Wang & Yang	2020	Focuses on sustainability, its assessment, and analysis.	Consider and incorporate the four dimensions of sustainability: energy, economy, society, and environment.	Fuzzy clustering

Source: Author's elaboration

Anuşlu and Firat's (2019) research focuses on grouping countries based on their innovation, sustainability, and SDG achievement scores through cluster analysis. Uwasu and Yabar (2011) analyzed the sustainable development conditions of 80 countries using panel data on real savings. By applying the K-means method, they assessed the average returns, trends, and stability of these nations. Before employing K-means, they tested various clustering techniques to identify the most effective approach. Similarly, Szennay and Szigeti (2019) explored tools and connections between the Sustainable Development Goals (SDGs) and business processes, offering guidance on integrating sustainability into business strategies. Nilashi et al. (2019) expanded on earlier sustainability assessment systems by examining the relationship between human sustainability, ecological sustainability, and overall sustainability performance through decision-rule exploration. Wang and Yang (2020) studied the sustainability and drivers of renewable energy using the DPSIR four-dimensional metrics framework. Their analysis aimed to integrate the sustainability dimensions of the economy, society, and environment. The results revealed Sweden and Denmark as the "strongest" sustainability performers, with countries achieving higher sustainability scores concentrated in central regions.

To address the research question and hypothesis, cluster analysis was employed to form homogeneous groups of countries with similar characteristics among the 28 Member States of the European Union. This method utilized all relevant indicators included in the 2030 Agenda indicator set, analyzed using IBM SPSS. The indicators were chosen following a preliminary review of those featured in the sustainable development strategies of each Member State, focusing on those demonstrating the greatest convergence.

Five key indicators were selected for the analysis: the area under organic farming, the long-term unemployment rate, gross domestic R&D expenditure, greenhouse gas emissions, and public debt. These indicators were consistently found across the sustainable development strategies and indicator lists of all 27 Member States. Additionally, the author compared these indicators with the Gross Domestic Product (GDP) and the Human Development Index (HDI), widely recognized as critical measures of economic development and human well-being. The cluster analysis based on these indicators also provides insights into the territorial competitiveness of the countries studied.

Results of Clustering the 27 Member States

Clustering the 27 EU member states based on variables like GDP, HDI, area under organic farming, gross domestic R&D expenditure, greenhouse gas emissions, long-term unemployment, and public debt is important for several reasons. The European Union implements policies that need to cater to diverse socioeconomic and environmental contexts. Clustering helps identify groups of countries with similar profiles, enabling tailored policy recommendations that address shared challenges and opportunities. Regarding resource allocation, such as financial aid, research funding, or environmental subsidies can be more effectively allocated if countries are grouped based on similar needs or capacities. For instance, countries with low R&D spending might benefit from specific innovation-focused programs. Clustering facilitates benchmarking and comparison between countries with similar characteristics. For example, countries with high public debt but strong HDI can provide insights into how they balance economic and social factors. Identifying clusters of countries with similar characteristics fosters regional collaboration. For instance, countries with high greenhouse gas emissions can collaborate on shared strategies to reduce.

The clustering of the 27 EU Member States is based on five sustainable development indicators, along with GDP and HDI. From the 2030 Agenda variables, those most frequently included in strategies and best representing the three dimensions of sustainable development were selected. In the study, a hierarchical clustering algorithm was initially applied in each case to analyze the dendrogram and determine the optimal number of clusters for the 27 EU Member States, first using GDP and then the HDI indicator. The K-means algorithm was subsequently employed to conclude. For 2022, the author demonstrated the core findings of the cluster analysis using one selected indicator.

Clustering of Member States Based on GDP

Non-hierarchical clustering, as the name suggests, differs fundamentally from hierarchical clustering because it involves a series of iterative steps. Molnár (2015) outlines these steps as follows: 1) assigning initial cluster members; 2) classifying elements into the nearest cluster; 3) determining cluster centers; 4) adjusting elements to minimize the distance between them and the centroids; and 5) repeating steps 3 and 4 until the clusters stabilize (iteration). These processes are executed automatically by the statistical software. The five indicators associated with sustainable development strategies were analyzed separately alongside GDP. Figure 1 illustrates the relationship between gross domestic R&D expenditure and GDP, representing the economic dimension.

Figure 1. Relationship between GDP and Gross Domestic Expenditure on R & D

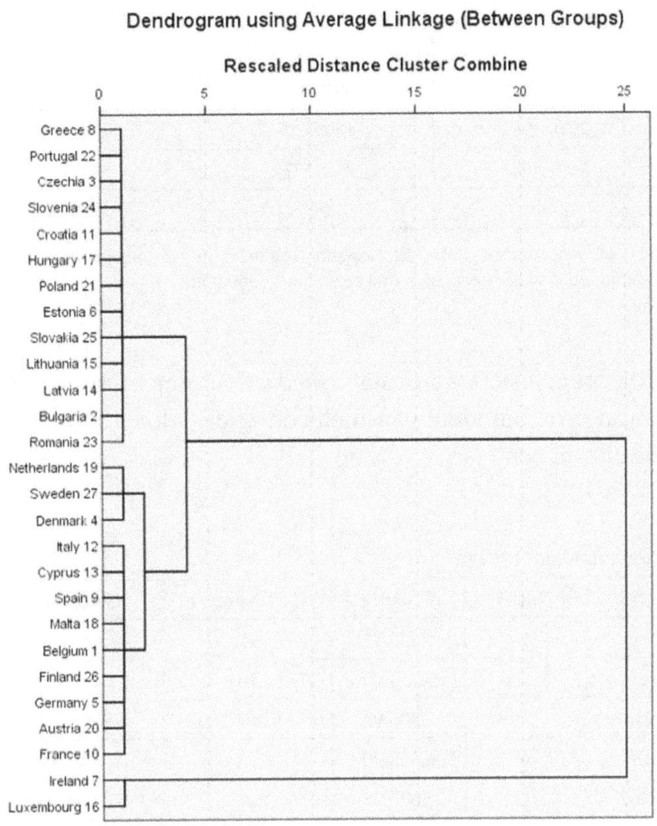

Source: SPSS program output

The dendrogram indicates that there are no outlier countries that need to be excluded from the sample. If a country were to form a cluster as a single element, it would be considered an outlier and should be removed from the analysis. Following the steps of the K-means cluster analysis, the author identified two clusters based on the variables selected, guided by the dendrogram. The SPSS statistical software handled the classification of elements into their nearest clusters. It also determined the cluster centers for the countries, with the initial and post-iteration results presented in Tables 2 and 3.

Table 2. Initial cluster centers

Initial Cluster Centers				
	Cluster			
	1	**2**	**3**	**4**
GDP_2022	28,790.00	7,730.00	85,850.00	51,600.00
Gross_domestic_R_and_D_expend_2022	.70	.75	1.05	2.87

Source: SPSS program output

Table 3. Iteration history

Iteration History[a]					
Iteration		Change in Cluster Centers			
		1	2	3	4
	1	336,156	5922,222	4275,000	4433,333
	2	2070,598	1305,278	,000	2151,667
	3	1091,806	525,577	,000	,000
	4	,000	,000	,000	,000

a. Convergence achieved due to no or small change in cluster centers. The maximum absolute coordinate change for any center is, 000. The current iteration is 4. The minimum distance between initial centers is 21060,000.

Source: SPSS program output

The final centers of the four clusters were established in four steps. Supplementary tables and analyses generated by the statistical program identify which countries belong to the three clusters and indicate their distances from the cluster centers.

Table 4. Initial and final cluster centres

Case Number	Countries	Cluster	Distance	Case Number	Countries	Cluster	Distance
1	Belgium	1	5,553.75	15	Lithuania	2	16.93
2	Bulgaria	2	7,753.07	16	Luxembourg	3	4,275.00
3	Czechia	2	3,206.92	17	Hungary	2	943.07
4	Denmark	4	6,585.00	18	Malta	1	6,146.25
5	Germany	1	5,073.75	19	Netherlands	4	145.02
6	Estonia	2	646.92	20	Austria	4	6,455.00
7	Ireland	3	4,275.00	21	Poland	2	683.07
8	Greece	2	3,476.92	22	Portugal	2	3,486.92
9	Spain	1	6,346.25	23	Romania	2	5,463.07
10	France	1	1,923.75	24	Slovenia	2	6,306.92
11	Croatia	2	853.07	25	Slovakia	2	996.92
12	Italy	1	2,946.25	26	Finland	1	5,713.75
13	Cyprus	1	2,826.25	27	Sweden	4	15.01
14	Latvia	2	2,443.07				

Source: Author's own research

The first of the four clusters includes countries such as Belgium, Germany, France, Spain, Italy, Cyprus, Malta and Finland. The second encompasses Bulgaria, Czechia, Estonia, Greece, Croatia, Latvia, Lithuania, Hungary, Poland, Portugal, Romania, Slovenia and Slovakia. The third cluster consists of Ireland and Luxembourg, while the fourth is Denmark, the Netherlands, Austria, and Sweden.

Table 4, which presents the cluster centroids, indicates that when the cluster analysis algorithm identifies four groups (clusters), it becomes easier to discern clear commonalities among member countries. This may be attributed to the similar sustainable development performance of these Member

States. However, it is somewhat paradoxical that these countries also exhibit significant differences in their economic, social, and environmental characteristics, as well as their development classifications. By forming four clusters, groups of countries with shared and similar characteristics have become more distinct. The classification of countries into clusters is shown in Figure 2.

Figure 2. The result of GDP-based clustering

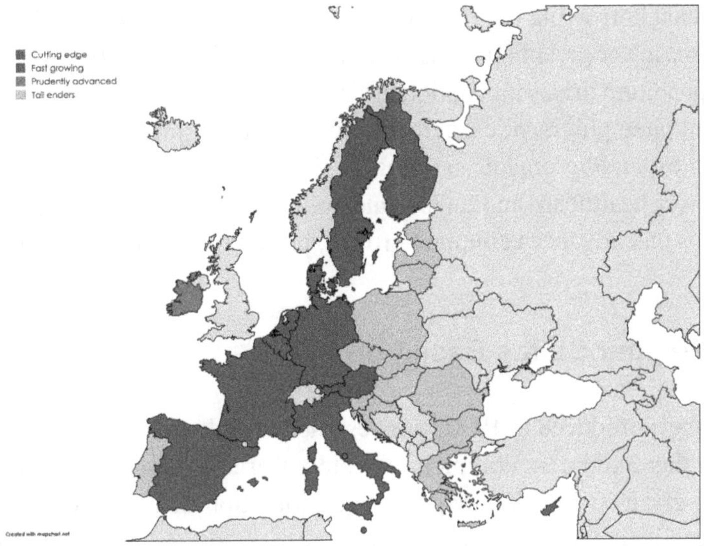

Source: Author's elaboration

In conclusion, it is crucial in cluster analysis to identify outliers, biased Member States, or any other cases or observations that may deviate from the population. In all cases, the relationship between gross domestic R&D expenditure and GDP did not necessitate excluding any country from the database. Even Luxembourg, typically considered an outlier, was appropriately clustered with Ireland in the analysis. Examining the relationship between the remaining four indicators and GDP, it was found that GDP-based clustering of public debt (SDG_17_40) resulted in three clusters. Four clusters were formed in all other cases. The clustered countries exhibit both similarities and differences. Their similarities lie in their alignment with sustainable development goals and the steps taken to achieve them. However, differences are evident in the specific list of SDIs and the overall performance of the countries.

Between 2015 and 2020, three clusters could typically be formed based on the common indicators of the SDSs. However, this pattern shifted in 2021 and 2022. In 2021, four clusters could be identified consistently based on country performance, while in 2022, three clusters were established in all cases except for Public debt (SDG_17_40). Analyzing the outliers reveals that Luxembourg and Spain each acted as unique clusters.

The interplay between GDP and variables such as the area under organic farming, gross domestic R&D expenditure, greenhouse gas emissions, long-term unemployment, and public debt sheds light on crucial socio-economic and environmental dynamics. While higher GDP often facilitates investment in sustainable practices like organic farming, the correlation is not always straightforward. Organic farming tends to flourish more in countries with strong policy support and environmental awareness, regardless of

GDP levels. There is a strong positive link between GDP and R&D expenditure, as wealthier economies allocate more resources to innovation for sustained growth and competitiveness. Historically, higher GDP has been associated with increased greenhouse gas emissions due to industrialization, though this trend is reversing in advanced economies adopting green technologies. Economic expansion linked to growing GDP typically reduces unemployment, including long-term unemployment, by creating more job opportunities, though structural unemployment can persist due to skill mismatches.

The relationship between GDP and public debt is multifaceted; while a higher GDP can lower debt-to-GDP ratios, significant borrowing to drive growth may increase absolute public debt levels. Expanding organic farming helps lower greenhouse gas emissions through sustainable agricultural practices, while increased R&D expenditure drives innovation in clean energy and efficiency, further reducing emissions over time. Persistent unemployment exacerbates public debt by raising welfare costs and reducing tax revenues. Although expanding organic farming might initially strain public budgets due to subsidies, it can eventually lower healthcare and environmental costs. These interconnections highlight the need for balanced policies that advance economic growth, promote environmental sustainability, and ensure social equity.

Clustering of Member States Based on HDI

The HDI indicator, introduced in 1990, was designed to address the limitations of GDP as a measure of development. In this study, five common indicators were chosen and compared not only with GDP but also with HDI, given its broader scope and consideration of multiple factors. As with the GDP-based clustering described earlier, the same rules and methodological steps were applied in this case. Analyzing the relationship between gross domestic R&D expenditure and HDI is as straightforward as examining its relationship with GDP. Notably, the dendrogram does not identify any country that should be excluded from the clustering process.

Using the dendrogram, four distinct clusters were identified, leading the algorithm to group the European Union Member States into four clusters using the K-means method. Ultimately, the final cluster centers were established through four iterative steps, and the clustered countries are presented in Table 5.

Table 5. Countries and their distance from the centers

Case Number	Countries	Cluster	Distance	Case Number	Countries	Cluster	Distance
1	Belgium	1	.147	15	Lithuania	3	.317
2	Bulgaria	2	.071	16	Luxembourg	3	.319
3	Czechia	4	.165	17	Hungary	3	.044
4	Denmark	1	.274	18	Malta	2	.137
5	Germany	1	.089	19	Netherlands	4	.129
6	Estonia	4	.294	20	Austria	1	.037
7	Ireland	4	.120	21	Poland	3	.074
8	Greece	3	.113	22	Portugal	3	.324
9	Spain	3	.049	23	Romania	2	.255
10	France	4	.167	24	Slovenia	4	.047

continued on following page

Table 5. Continued

Case Number	Countries	Cluster	Distance	Case Number	Countries	Cluster	Distance
11	Croatia	3	.054	25	Slovakia	2	.267
12	Italy	3	.017	26	Finland	1	.164
13	Cyprus	2	.049	27	Sweden	1	.327
14	Latvia	2	.099				

Source: Source: Author's elaboration

The countries in the four clusters are ranked 6 - 6 - 9 - 6 and no country was excluded from the analysis, so the number of initial and final clusters is the same. The position of countries in clusters is also shown in Figure 3.

Figure 3. The result of HDI-based clustering

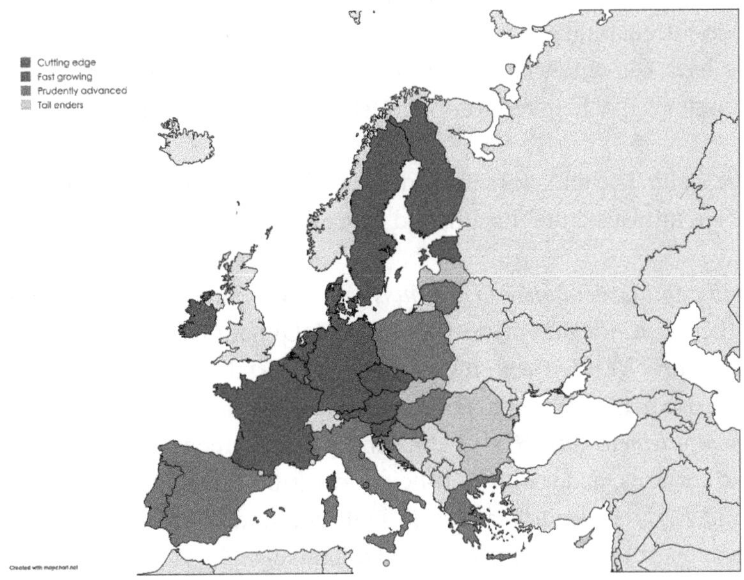

Source: Author's elaboration

The four clusters contain 6, 6, 9, and 6 countries, respectively, with no exclusions from the analysis, ensuring that the initial and final clusters remain consistent. Figure 3 illustrates the placement of countries within each cluster.

Analyzing the clustering results from 2015 to 2022 reveals significant variation in countries' positions based on the selected common indicators, GDP, and HDI. During the study period, the number of clusters formed showed notable variability. For example, in 2016, the number of homogeneous clusters ranged from 2 to 4, while for other years, it stabilized between 3 and 4 groups. Outliers such as Spain, Greece, Sweden, and Luxembourg frequently behaved as anomalies or biasing countries and were excluded from the sample.

When clustering based on GDP, the economic development of EU Member States strongly influenced the groupings, with the comparative indicator largely determining cluster characteristics, except in a few cases. In contrast, this dominant effect was not observed in clusters based on the HDI indicator. The proportion of indicators and the number of clusters fluctuated significantly, varying almost year by year.

The relationship between the Human Development Index (HDI) and variables such as the area under organic farming, gross domestic R&D expenditure, greenhouse gas emissions, long-term unemployment, and public debt underscores the interplay between human well-being, environmental sustainability, and economic factors.

- *HDI and Organic Farming:* Nations with higher HDI often possess the awareness and resources to support organic farming, emphasizing sustainable development and healthier lifestyles.
- *HDI and R&D Expenditure:* A higher HDI is associated with increased R&D spending, as advancements in education, innovation, and technology drive improved living standards.
- *HDI and Greenhouse Gas Emissions*: Although higher HDI historically aligned with rising emissions during industrialization, advanced HDI countries increasingly adopt clean energy technologies to prioritize sustainability.
- *HDI and Long-term Unemployment:* Elevated HDI often indicates robust education, healthcare, and economic systems that help reduce long-term unemployment, fostering inclusive and personal growth.
- *HDI and Public Debt:* The effect of public debt on HDI hinges on how borrowed funds are allocated; investments in infrastructure, health, and education boost HDI, while excessive debt servicing can impede it.
- *Organic Farming and Greenhouse Gas Emissions*: Increasing organic farming lowers emissions, aligning with the environmental priorities of high-HDI nations.
- *R&D Expenditure and HDI Growth*: Investments in R&D directly enhance HDI by fostering innovations that improve healthcare, education, and overall quality of life.
- *Long-term Unemployment and HDI:* Persistent unemployment diminishes HDI by reducing income, access to services, and social well-being, contributing to inequality.
- *Public Debt and R&D Expenditure:* While high public debt may limit R&D funding, strategic borrowing can support innovation if managed effectively.

The analysis shows that the selected indicators are closely tied to regional competitiveness, particularly GDP, public debt, and long-term unemployment. Meanwhile, organic farming and greenhouse gas emissions, which relate more closely to HDI, serve as key environmental indicators of sustainable development, reflecting ecological quality. Based on clustering results, the author provides a comprehensive analysis of sustainable development strategies and indicators across the 27 EU Member States, addressing the research question. A high HDI reflects a harmonious balance where sustainable practices, innovation, and economic inclusivity promote human well-being while mitigating environmental harm and financial risks.

ANALYSIS OF THE SUSTAINABLE DEVELOPMENT STRATEGIES OF THE 27 EU MEMBER STATES BASED ON THE RESULTS OF THE CLUSTER ANALYSIS

Sustainability strategies and indicators differ significantly across the 27 EU Member States. During the research, it became evident that some countries shared similar characteristics, leading the author to categorize them into four clusters: "cutting edge," "fast-growing," "prudently advanced," and "tail-enders".

Cutting Edge Group

The Scandinavian countries (Finland, Denmark, and Sweden), along with Belgium from Western Europe and Germany and Austria from Central Europe (Alpine region), form a group of EU Member States with comparable sustainable development performance. Within this cluster, Sweden and Finland stand out for their early adoption of sustainable development strategies, publishing them in the early to mid-1990s in response to the recommendations of the 1992 Rio Conference (Regeringskansliet, 2005). These countries were pioneers in sustainability thinking. Finland, for example, established the Finnish Commission on Sustainable Development in 1993 under the leadership of the Prime Minister. In contrast, Belgium, also part of this cluster, developed a federal sustainable development strategy based on the Agenda 21 commitments in 1997 (Federal Planning Bureau, 2011). Other countries in the group, such as Denmark, Germany, and Austria, introduced their strategies later, around or after the Johannesburg World Summit in 2002–2003.

Sustainability programs emphasize ecological sustainability while fostering the economic, social, and cultural conditions needed to empower citizens to make sustainable choices (Finnish National Commission on Sustainable Development, 2008). These frameworks establish overarching objectives, shared priorities, and actionable steps for advancing sustainable development (Sustainable Development Research Network, 2006). Strategies typically aim to provide a foundation for societal progress that balances economic growth, social development, and high levels of environmental protection (Danish Ministry of the Environment, 2002). Finland's strategy focuses on achieving the SDGs, promoting the sustainable use of natural resources, safeguarding citizen well-being, and maintaining societal integrity (Lyytimäki et al., 2011). Austria shows a deep dedication to national and international obligations, economic support, and forward-looking development (Oberer & Erkollar, 2011). Broadly, common goals include poverty reduction, sustainable production and consumption, responsible energy management, and other key areas (Ege Jorgensen, 2008).

Fast Growing

The second cluster includes the Czech Republic, the most advanced member of the V4, along with France, the Netherlands, Ireland, Slovenia from Central Europe, and Estonia from the Baltic region. The Netherlands and France introduced their sustainability strategies as part of or following the Johannesburg World Conference in 2002–2003. In the Czech Republic, sustainable development is primarily assessed at the national level, and on December 8, 2004, the government adopted the country's framework strategy for sustainable development. Slovenia did not have a specific sustainability strategy before its EU accession but has always aimed for sustainable development (Plut & Vintar Mally, 2004). The Slovenian

Development Strategy (SDS), which covers the period from 2005 to 2013, was adopted by the government in 2005 (Institute of Macroeconomic Analysis and Development [IMAD], 2012).

Following the publication of the Dutch Sustainable Development Strategy in 2003, the monitoring report was updated in 2008 and released in 2009 (Lucas et al., 2016). France's initial strategy ran from 2003 to 2008 and was supplemented in 2006 to align it with the EU SDS. The second program covered 2010 to 2013. In Slovenia, the 2005 strategy underwent a review starting in 2008 to identify gaps and assess priority actions for the future. This program concluded in 2013 (IMAD, 2007). The 2010 review of the Czech Strategy identified five priority areas: 1) society, people, and health; 2) economy and innovation; 3) regional development; 4) landscape, ecosystems, and biodiversity; and 5) stable and secure society (Ministry of the Environment of the Czech Republic, 2010).

France's strategy has placed significant emphasis on commitment to both national and international obligations, as well as ensuring coherence across sectoral policies (Ministère de l'Écologie, du Développement durable et de l'Énergie, 2015). In the Czech Republic, the strategy is built around three main pillars—economic, social, and environmental—and three complementary areas: research, development, and education; European and international context; and good governance (Office of the Government of the Czech Republic, 2006). The strategy aims to define key strategic sub-objectives, targets, and instruments designed to address imbalances within economic, social, and environmental pillars (Fischer et al., 2013). Slovenia's current strategy encompasses three dimensions of sustainability, with an additional focus on culture. The strategies are structured around three core and complementary dimensions of sustainable development.

After the Rio Conference, Estonia embraced the principle of sustainable development, and in early 1995, the Estonian Parliament passed the Sustainable Development Act, outlining the principles of the national strategy (Lepik, 2018). This act was amended in 1997 to include long-term plans for sustainable development across various sectors. Ireland's first strategic approach to sustainable development was introduced in 1997. It was a forward-thinking, clear, and goal-oriented strategy that addressed all three dimensions of sustainable development and various international issues (Kelly & Moles, 2000). Estonia's Sustainable Development Strategy was initiated by the government in 2001 and formally adopted by Parliament in 2005, outlining the development of the country and society up to 2030 (Estonian Ministry of the Environment, 2005). Estonia's approach to sustainability considers not only natural resources and environmental limits but also the nation's culture, spiritual traditions, and ethical values. These elements, combined with education and a strong legislative framework, are aimed at fostering a new way of life (Raukas, 2010).

Prudently Advanced

The cluster, which includes the majority of Member States, features Hungary and Poland, both with a long history in the V4. It also encompasses Southern European countries – Greece, Spain, Italy, Croatia, and Portugal – Luxembourg from Western Europe. Additionally, Lithuania, a Baltic country, is part of the cluster.

In Italy, two reports on sustainable development were released after 1992, marking the beginning of the national sustainability agenda (Ronchi et al., 2002). In 2002, an Environmental Action Plan for Sustainable Development was introduced, aligning with the Barcelona Council of Europe guidelines (Eboli, 2017). In Poland, sustainable development became a constitutional principle in April 1997, integrating political, economic, and social actions and processes (Central Statistical Office, 2011). In

2000, a strategy was developed in line with the spirit of the millennium, which remains in effect today (Kis-Orloczki, 2014). Hungary's sustainable development strategy was preceded by several preliminary efforts. Luxembourg's first sustainability strategy was published in 1999, with a second version approved on November 26, 2010.

Greece's first National Sustainable Development Strategy (NSDS) was adopted by the Council of Ministers in 2002. The strategy aligns with the global sustainable development initiative and reflects the Agenda 21 framework established at the Rio Conference. In Portugal, the development of the National Strategy for Sustainable Development (ENDS) was initiated by the XIV Constitutional Government, which set the framework for its creation. Like in other EU Member States, Agenda 21 was the driving force behind the strategy. The government approved the main guidelines for the ENDS in May 2002, and the National Sustainable Development Strategy was finalized in 2005, with the official version adopted on December 27, 2006 (Agência Portuguesa do Ambiente, 2015).

In 1996, the Baltic States adopted the Baltic Agenda, which provided the foundation for their sustainable development strategies (Brizga, 2012) and also focused on defining the concept of sustainable development as outlined by the WCED (Briška & Rungule, 2011). Lithuania published its Sustainable Development Strategy on September 11, 2003, based on decisions from the Rio Conference. The strategy identified key areas for development, including economic growth, renewable energy, and social well-being (Leipute, 2014). Croatia's sustainable development policy formulation depends on the country's internal capacity, public attitudes and expectations, and the global context (Böll Stiftung, 2012). The Republic of Croatia's sustainable development strategy, which played a crucial role in the Environmental Protection Act that came into force in November 2007, outlines long-term action guidelines and was officially adopted in 2009 (Ministry of Environmental Protection, Physical Planning and Construction, 2009).

Tail-Enders

The cluster includes EU Member States with similar development levels, such as Bulgaria and Romania, transition countries like Latvia, and Slovakia from the V4, and two Southern European countries, Cyprus and Malta.

Within the cluster, sustainable development became a constitutional principle in Slovakia with Act 17/1992 (Dubravská, 2015). The country's Sustainable Development Strategy was adopted by the Slovak Government on October 10, 2001, and by the National Council on April 3, 2002. Romania's sustainable development strategy, created in the spirit of the millennium, was designed to align with UN priorities for Europe (Burja & Burja, 2009). In Bulgaria, the approach to sustainability strategies differs somewhat, as the country initially committed to preparing a comprehensive environmental strategy lasting until the 2000s, which was not the same as a fully sustainable development strategy (Republic of Bulgaria Council of Ministers, 2001). The "official" sustainability strategy was adopted in 2001 and remained in effect until 2006.

Latvia adopted new sustainability strategies in 2005 and 2010 (Saeima of the Republic of Latvia, 2010). The 2010 strategy is a long-term document that establishes baseline and target indicators to track the progress of development policies. To raise awareness about sustainable development, Malta has initiated several actions (Ministry for Sustainable Development, the Environment and Climate Change, 2016). At its meeting on December 9, 2002, the National Commission for Sustainable Development appointed a working group to prepare a strategy for Malta, which resulted in the adoption of the 2007-2016 strategy 2007, based on the EU SDS (Lyytimäki et al., 2011). Cyprus' first National Sustainable

Development Strategy (NSDS) was adopted by the Council of Ministers in November 2007, with the full program available only in Greek.

It can be concluded that the sustainable development strategies of the 27 countries of the European Union are very different, both in terms of goals and indicators. Of course, we cannot forget the common features, such as the fact that they all have at their center the Brundtland Commission's concept of sustainable development and the three-dimensional structure. Future research in this area could be largely determined by AI. AI can effectively assist in clustering homogeneous groups within the European Union by leveraging unsupervised machine learning techniques, particularly clustering algorithms.

CONCLUSIONS

In this chapter, the author sought to find out whether the 27 Member States of the European Union could be grouped according to the five common indicators of the strategies and, within them, the indicator frameworks. The initial hypothesis was: *based on the five common sustainable development indicators included in the sustainability strategies of the 27 EU Member States, groups of countries can be formulated with homogeneous characteristics that are distinguishable and characterizable in terms of GDP and HDI indicators.* Once the clusters have been defined, it can indeed be stated that these homogeneous groups can be created, i.e., the Member States can be separated according to their development and characteristics. The research findings in this chapter suggest that *as a result of common indicators in the sustainable development strategies of the 27 Member States of the European Union and clustering based on GDP and HDI indicators, four distinct groups of countries can be identified: "cutting-edge", "fast-growing", "prudently advanced" and "tail-ender".*

The interplay between GDP and variables such as the area under organic farming, gross domestic R&D expenditure, greenhouse gas emissions, long-term unemployment, and public debt sheds light on crucial socio-economic and environmental dynamics. The interconnections highlight the need for balanced policies that advance economic growth, promote environmental sustainability, and ensure social equity. The relationship between the Human Development Index (HDI) and variables interplay between human well-being, environmental sustainability, and economic factors.

The analysis shows that the selected indicators are closely tied to regional competitiveness, particularly GDP, public debt, and long-term unemployment. Meanwhile, organic farming and greenhouse gas emissions, which relate more closely to HDI, serve as key environmental indicators of sustainable development, reflecting ecological quality. Based on clustering results, the author provides a comprehensive analysis of sustainable development strategies and indicators across the 27 EU Member States, addressing the research question. A high HDI reflects a harmonious balance where sustainable practices, innovation, and economic inclusivity promote human well-being while mitigating environmental harm and financial risks.

At the core of all sustainable development strategies is the concept and dimensions of the Brundtland Commission, a definition that included/includes sustainability plans for the next 20-50 years (Lobo et al., 2005). In examining the strategies and indicators, many similarities and differences were found. Each country is different in its way from another country. In the history of sustainable development, we can talk about the very "early" Member States that developed their sustainability strategies immediately after the Rio Conference in 1992. The second group established its framework around the Johannesburg

Conference (2002), while the third period dates back to around 2010. The fourth and final phase relates to the implementation of the 2030 Agenda.

Pioneering countries include Belgium, Finland, Latvia, Luxembourg and Sweden. In addition to the WCED dimensions, in some Member States, such as Austria, Czech Republic, Estonia, Ireland, and Poland, sustainability is complemented by other aspects (cultural, institutional, horizontal issues, policy). Italy is quite the opposite, with the 5Ps at its core. The biggest difference in the strategies is the number of indicators and how they are broken down. The indicators are largely broken down by the 2030 Agenda's Sustainable Development Goals, the EU SDIs' themes, and of course, the three dimensions. In addition, there is a separation of the sustainable development strategy objectives, areas of action, priority axes, and diverse sectors. It became clear that no two strategies and indicator systems are the same, there are always differences between countries. What they have in common is that they all refer to the three-dimensional pillar system and concept of WCED and have indicators.

The indicators included in the analysis are also closely linked to regional competitiveness, notably GDP, public debt, and long-term unemployment rates. Areas under organic farming, which is more closely related to the HDI index and greenhouse gas emissions are important indicators of the environmental dimension of sustainable development, which shows the quality of the environment. Using the clustering results, the author carried out a detailed analysis of the sustainable development strategies and indicators of the 27 EU Member States as a conclusion and thus answered the research question posed.

AI can also have a significant impact on sustainable development. AI can significantly contribute to sustainable development by optimizing resource use, enabling innovative solutions, and supporting informed decision-making in key areas aligned with the UN Sustainable Development Goals (SDGs). In case of climate modeling and predictions, AI analyze climate data to forecast weather patterns, predict climate change impacts, and model mitigation strategies. AI optimizes energy consumption in industries, buildings, and transportation systems by identifying inefficiencies. In the area of renewable energy and energy efficiency AI balances supply and demand by forecasting energy needs and optimizing distribution in renewable energy grids. It can also help in designing energy-efficient buildings and industrial processes. The factors presented here are just a few examples of the ways in which AI can help sustainable development.

ACKNOWLEDGEMENT

„This work has been implemented by the TKP2021-NVA-10 project with the support provided by the Ministry of Culture and Innovation of Hungary from the National Research, Development and Innovation Fund, financed under the 2021 Thematic Excellence Programme funding scheme."

REFERENCES

Abdella, G. M., Kucukvar, M., Cihat Onat, N., Al-Yafay, H. M., & Enis Bulak, M. (2020). Sustainability assessment and modeling based on supervised machine learning techniques: The case for food consumption. *Journal of Cleaner Production, 251*, 1–16. DOI: 10.1016/j.jclepro.2019.119661

Agência Portuguesa do Ambiente. (2015). *Estratégia Nacional de Desenvolvimento Sustentável.* JMF – Agência Portuguesa do Ambiente.

Alves, A. F., Costa Curta, L., & da Silva Dias, L. (2017). Analysis of correlation among HDI (Human Development Index), violence and corruption perceptions index. *Proceedings of the 5th CR3+ Conference - Making Corporate Responsibility Useful*, Helsinki, Finland.

Andrews, R. N. (1997). National environmental policies: The United States. In Jaenicke, M., & Weidner, H. J. (Eds.), *National environmental policies: A comparative study of capacity building* (pp. 25–43). Springer Verlag. DOI: 10.1007/978-3-642-60507-9_2

Anuşlu, M. D., & Firat, S. Ü. (2019). Clustering analysis application on Industry 4.0-driven global indexes. *Procedia Computer Science, 158*, 145–152. DOI: 10.1016/j.procs.2019.09.037

Baják, I. (2013). A helyi fenntartható fejlődési stratégiák helyzete Magyarországon. *Acta Carolus Robertus, 3*(2), 9–22. DOI: 10.22004/ag.econ.171428

Bartoluci, M., Hendija, Z., & Petracic, M. (2015). Implementing Principles of Sustainable Development in Rural Tourism in Continental Croatia. Economy of eastern Croatia yesterday, today, tomorrow. *Josip Juraj Strossmayer University of Osijek. Faculty of Economics, 4*, 26–37.

Beatley, T., & Manning, K. (1998). *The ecology of place: Planning for environment, economy and community.* Island Press.

Bečić, E., Mulej, E. M., & Svarc, J. (2012). Measuring social progress by sustainable development indicators: Cases of Croatia and Slovenia. *Procedia: Social and Behavioral Sciences, 37*, 458–465. DOI: 10.1016/j.sbspro.2012.03.311

Böll Stiftung, H. (2012). *We need to change. Mapping Croatia's Potential for Sustainable Development.* Kerschoffset.

Brandful Cobbinach, P., Erdiaw-Kwasie, M. O., & Amoateng, P. (2014). Rethinking sustainable development within the framework of poverty and urbanisation in developing countries. *Environmental Development, 13*, 18–32. DOI: 10.1016/j.envdev.2014.11.001

Briška, I., & Rungule, R. (2011). Sustainable development: Changes in the evaluation of environmental risks in Latvia. *Management Theory and Studies for Rural Business and Infrastructure Development, 1*(25), 36–43. DOI: 10.25143/prom-rsu_2013-23_dts

Brizga, J. (2012). How Well Sustainable Development Is Integrated into Environmental Policies: Case study: Latvia. *Safety of Technogenic Environment, 2*, 24–33.

Brown, L. R. (1982). *Building a Sustainable Society*. Norton for the Worldwatch Institute. DOI: 10.1007/BF02712913

Burja, C., & Burja, V. (2009). Some aspects of the Sustainable Development in Romania. *International Journal of Social, Behavioral, Educational, Economic. Business and Industrial Engineering, 3*(6), 802–808.

Central Statistical Office. (2011). *Sustainable Development Indicators in Poland*. Central Statistical Office.

Cohen, S. (2011). *Sustainability Management*. Columbia University Press.

Costanza, R., Daly, H. E., & Bartholomew, J. A. (1991). Goals, agenda and policy recommendations for ecological economics. *Ecological Economics: The Science and Management of Sustainability*, 1-20.

Daly, H. E. (2002). *Sustainable Development: Definitions, Principles, Policies*. World Bank.

Dannish Ministry of the Environment. (2002). *A Shared Future - Balanced Development. Denmark's National Strategy for Sustainable Development*. Dannish Ministry of the Environment.

Ministère de l'Écologie, du Développement durable et de l'Énergie (2015). *National Strategy of ecological transition towards sustainable development 2015-2020*. La Défense Cedex.

Debisso, K. (2012). Az Európai Unió fejlesztési politikája. In Szabó, M., Láncos, P. L., & Gyeney, L. (Eds.), *Uniós szakpolitikák* (pp. 1–20). Szent István Társulat.

Dell'Angelo, J., D'Odorico, P., & Rulli, M. C. (2017). Threats to sustainable development posed by land and water grabbing. *Current Opinion in Environmental Sustainability, 26-27*, 120–128. DOI: 10.1016/j.cosust.2017.07.007

Dubravská, M. (2015). The analysis of chosen development indicators of the sustainable development of the Slovak Republic. *eXclusive e-JOURNAL, 3*, 1-3. DOI: 10.1016/S2212-5671(15)00569-9

Dudek, T., Dzhuguryan, T., & Lemke, J. (2019). Sustainable production network design for city multi-floor manufacturing cluster. *Procedia Computer Science, 159*, 2081–2090. DOI: 10.1016/j.procs.2019.09.381

Eboli, F. 2017. SDGs in Italy: current status and the process towards the National Strategy on Sustainable Development. *IAERE Fifth Annual Conference*. Rome.

Ege Jorgensen, C. (2008). *Strategy plans for sustainable development: case studies from Denmark*. Nordic Council of Minister's Office in Latvia.

Estonian Ministry of the Environment. (2005). *Estonian National Strategy on Sustainable Development. Sustainable Estonia 21*. Estonian Ministry of the Environment.

European Environment Agency. (2002). *Environmental Signals 2002. Benchmarking the millennium*. Office for Official Publications of the European Communities.

Federal Planning Bureau. (2011). *Twenty Years of Political Commitment to Sustainable Development. Federal Report on sustainable development 2011*. Henri Bogaert.

Finnish National Commission on Sustainable Development. (2008). *Finnish National Commission on Sustainable Development. Towards a globally and nationally sustainable Finland*. Ministry of the Environment.

Fischer, J., Helman, K., Kramulová, J., Petkovová, L., & Zeman, J. (2013). Sustainable Development Indicators at the Regional Level in the Czech Republic. *Statistika*, *93*(1), 5–18.

Fleischer, T. (2007). Fenntartható fejlődés: környezeti, társadalmi és gazdasági tényezők. In Farkas, P., & Fóti, G. (Eds.), *Magyarország globális környezete 2020-ig. Háttértanulmányok a magyar külstratégiához I* (pp. 192–202). MTA Világgazdasági Kutatóintézet.

Fleming, A., Wise, R. M., Hansen, H., & Sams, L. (2017). The sustainable development goals: A case study. *Marine Policy*, *86*, 94–103. DOI: 10.1016/j.marpol.2017.09.019

Griggs, D., Stafford Smith, M., Rockström, J., Öhman, M. C., Gaffney, O., Glaser, G., Kanie, N., Noble, J., Steffen, W., & Shyamsundar, P. (2014). An integrated framework for sustainable development goals. *Ecology and Society*, *19*(4), 1–24. DOI: 10.5751/ES-07082-190449

Horton, P., & Horton, B. P. (2019). Re-defining Sustainability: Living in Harmony with Life on Earth. *One Earth*, *1*, 86–94.

IMAD (Institute of Macroeconomic Analysis and Development). (2007). *Development Report 2007*. SOLOS.

IMAD (Institute of Macroeconomic Analysis and Development). (2012). *Development Report 2012*. IMAD.

Jabareen, Y. (2008). A new conceptual framework for sustainable development. *Environment, Development and Sustainability*, *10*(2), 179–192. DOI: 10.1007/s10668-006-9058-z

Kadlubek, M. (2015). Examples of Sustainable Development in the Area of Transport. *Procedia Economics and Finance*, *27*, 494–500. DOI: 10.1016/S2212-5671(15)01026-6

Kelly, R., & Moles, R. (2000). Towards sustainable development in the mid-west region of Ireland. *Environmental Management and Health*, *11*(5), 422–432. DOI: 10.1108/09566160010351881

Kis-Orloczki, M. (2014). *National Sustainable Development Strategies in the Visegrad Four*. Electronic International Interdisciplinary Conference, 1-5. September, 2014. (pp. 160-163).

Kralj, D., & Markič, M. (2008). Processes Innovation and Sustainable Development. *WSEAS Transactions on Environment and Development*, *4*(2), 99–108.

Kuhlman, T., & Farrington, J. (2010). What is Sustainability? *Sustainability (Basel)*, *2*(11), 3436–3448. DOI: 10.3390/su2113436

Ladoneczki, G., & Kósi, K. (2014). Makroszintű nemzeti stratégiák konzisztenciavizsgálata a biodiverzitás tükrében. *Gazdálkodás*, *58*(4), 331–340. DOI: 10.22004/ag.econ.206091

Leipute, B. (2014). Lithuania in the context of sustainable development. *Lithuanian Journal of Statistics, 53*(1), 24–33. DOI: 10.15388/LJS.2014.13891

Lepik, E. (2018). *SDG Coordination and Implementation in Estonia.* Republic of Estonia Government Office.

Lobo, G., Costa, S., Nogueira, R., Antunes, P., & Brito, A. (2005). *A Scenario Building Methodology to Support the Definition of Sustainable Development Strategies: the Case of the Azores Region. 11th Annual International Sustainable Development Research Conference*, 6-8 June, 2005. (pp. 1-13). Helsinki: Finland.

Lucas, P., Ludwig, K., Kok, M., & Kruitwagen, S. (2016). *Sustainable Development Goals in the Netherlands. Building blocks for environmental policy for 2030.* PBL Netherlands Environmental Assessment Agency.

Lyytimäki, J., Rinne, J., & Kautto, P. (2011). Using indicators to assess sustainable development in the European Union, Finland, Malta and Slovakia. The Finnish Environment 4. Finnish Environment Institute, Environmental Policy Centre.

Mebratu, D. (1998). Sustainability and Sustainable Development: Historical and Conceptual Review. *Environmental Impact Assessment Review, 18*(6), 493–520. DOI: 10.1016/S0195-9255(98)00019-5

Ministry for Sustainable Development, the Environment and Climate Change (2016). *Greening Our Economy – Achieving a sustainable future.* Ministry for Sustainable Development, the Environment and Climate Change.

Ministry of Environmental Protection, Physical Planning and Construction (2009). *Strategy for Sustainable Development of the Republic of Croatia.* Ministry of Environmental protection, Physical planning and construction.

Ministry of the Environment of the Czech Republic. (2010, May 20). *The Strategic Framework for Sustainable Development in the Czech Republic.*https://www.vlada.cz/assets/ppov/udrzitelny-rozvoj/The-Strategic-Framework-for-SD_2011.pdf

Molnár, T. (2015). *Empirikus területi kutatások.* Akadémiai Kiadó.

Mur, A., Dormido, R., Duro, N., Dormido-Canto, S., & Vega, J. (2016). Determination of the optimal number of clusters using a spectral clustering optimization. *Expert Systems with Applications, 63*, 304–314. DOI: 10.1016/j.eswa.2016.08.059

Nilashi, M., Rupani, P. F., Mobin Rupani, M., Kamyab, H., Shao, W., Ahmadi, H., Rashid, T. A., & Aljojo, N. (2019). Measuring sustainability through ecological sustainability and human sustainability: A machine learning approach. *Journal of Cleaner Production, 240*, 1–10. DOI: 10.1016/j.jclepro.2019.118162

Oberer, B. J., & Erkollar, A. (2011). Monitoring of Sustainable Development: on the way to a sustainable Austria. *Proceedings of the 3rd International Conference on Governance, Fraud, Ethics & Social Responsibility*, 1-13.

Office of the Government of the Czech Republic. (2006). *Interim Report on the Czech Republic Strategy for Sustainable Development.* Office of the Government of the Czech Republic.

Organisation for Economic Co-operation and Development. (2016). *Better Policies for 2030. An OECD Action Plan on the Sustainable Development Goals*. OECD.

Plut, D., & Vintar Mally, K. (2004). *Slovenia and the challenges of Sustainable Development.* In M. Orozen Adamic (Eds.), *Slovenia: A Geographical Overview.* (pp. 155-159). ZRC SAZU.

Prugh, T., Costanza, R., & Daly, H. (1999). *The Local Politics of Global Sustainability*. Island Press.

Raukas, A. (2010). Sustainable development and environmental risks in Estonia. [Special Issue II]. *Agronomy Research (Tartu)*, *8*, 351–356.

Regeringskansliet (2005). *Strategic Challenges. A Further Elaboration of the Swedish Strategy for Sustainable Development.* Ministry of Sustainable Development.

Republic of Bulgaria Council of Ministers. (2001). National Development Programme: Bulgaria 2020. Ministry of Finance of the Republic of Bulgaria.

Ronchi, E., Federico, A., & Musmeci, F. (2002). A system oriented integrated indicator for sustainable development in Italy. *Ecological Indicators*, *2*(1-2), 197–210. DOI: 10.1016/S1470-160X(02)00045-6

Ruževičius, J. (2009). Development of Sustainable and Socially Responsible Business in Lithuania in the International Context. *Ekonomika (Nis)*, *86*, 68–80.

Saeima of the Republic of Latvia (2010). *Sustainable Development Strategy of Latvia until 2030. Latvija 2030.* Saeima of the Republic of Latvia.

Sajtos, L., & Mitev, A. (2007). *SPSS Kutatási és adatelemzési kézikönyv*. Alinea Kiadó.

Silva Barbosa, G., Drach, P. R., & Corbella, O. D. (2014). A Conceptual Review of the Terms Sustainable Development and Sustainability. *The International Journal of Social Sciences (Islamabad)*, *3*(2), 1–15.

Sustainable Development Research Network. (2006). *Securing the Future: The Role of the Research.* Policy Studies Institute.

Székelyi, M., & Barna, I. (2004). *Túlélőkészlet az SPSS-hez*. Typotex Kiadó.

Szennay, Á., & Szigeti, C. (2019). A fenntartható fejlődési célok és a GRI szerinti jelentéstétel kapcsolatának elemzése. *Vezetéstudomány/Budapest. Management Review*, *50*(4), 33–43. DOI: 10.14267/VEZTUD.2019.04.04

United Nations. (2015). *Transforming Our World: the 2030 Agenda for Sustainable Development*. United Nations.

United Nations Educational, Social and Cultural Organisation. (2020). *Education for Sustainable Development. A roadmap*. UNESCO.

Uwasu, M., & Yabar, H. (2011). Assessment of sustainable development based on the capital approach. *Ecological Indicators*, *11*(2), 348–352. DOI: 10.1016/j.ecolind.2010.06.002

van Hees, S. R. W. (2014). Sustainable Development in the EU: Redefining and Operationalizing the Concept. *Utrecht Law Review*, *10*(2), 60–76. DOI: 10.18352/ulr.269

Wang, Q., & Yang, X. (2020). Investigating the sustainability of renewable energy – An empirical analysis of European Union countries using a hybrid of projection pursuit fuzzy clustering model and accelerated genetic algorithm based on real coding. *Journal of Cleaner Production*, *268*, 1–26. DOI: 10.1016/j.jclepro.2020.121940

WCED. (1987). *Our Common Future*. Oxford University Press.

Zhang, L., Xu, M., Chen, H., Li, Y., & Chen, S. (2022). Globalization, Green Economy and Environmental Challanges: State of the Art Review for Practical Implications. *Frontiers in Environmental Science*, *10*, 87027. DOI: 10.3389/fenvs.2022.870271

Ziolkowska, J., & Ziolkowski, B. (2010). Sustainable Development in the European Union: Tools for Policy Evaluation. *Annals of the University of Petrosani. Economics*, *10*(3), 373–382.

KEY TERMS AND DEFINITIONS

Sustainable Development: Is a very important concept today, describing how not to endanger the resource needs of future generations.

Sustainable Development Goals: A set of 17 global objectives established by the United Nations to achieve sustainable social, economic, and environmental development by 2030.

Sustainable Development Strategy: A unique strategy for countries around the world, describing the goals they have set to achieve sustainable development.

Gross Domestic Product (GDP): In economics, a measure of the economic output of a particular area - usually a country - over a given period of time.

Human Development Index (HDI): An indicator comparing countries around the world on the basis of life expectancy at birth, literacy, education and living standards.

Cluster Analysis: Is a dimension reduction technique to classify data blocks into homogeneous groups.

The Three Aspects of Sustainable Development: Involve synergies between environment, economy and society. A change in one has an impact on the others.

Chapter 19
Progress in Renewable Energy Technologies:
A Review Outlook

P. Vijayakumar
Nehru Institute of Technology, India

Kode Jaya Prakash
https://orcid.org/0000-0002-4947-7745
VNR Vignana Jyothi Institute of Engineering and Technology, India

Sandeep Chinta
Institute of Aeronautical Engineering, India

K. Venkata Ramireddy
https://orcid.org/0009-0002-6246-0419
Sri Kalahasteeswara Institute of Technology, India

S. Nanthakumar
PSG Institute of Technology and Applied Research, India

Ramasamy Girimurugan
https://orcid.org/0000-0003-0257-0448
Nandha College of Technology, India

ABSTRACT

This research paper delves into the rapid progress of energy technologies for renewable sector in the distinctive context. The research analyses the important improvements across diverse sectors of renewable sector like encompassing solar, wind, biomass, hydropower, geothermal, and ocean energy. India's ambitious renewable energy vision and mission, as exemplified by the National Solar Mission and the Green Energy Corridor innovative techniques, have spurred significant growth in the sector. Similarly, it scrutinizes advancements in wind energy, biomass and bioenergy innovations, and the potential of geothermal and ocean energy sources in elaborate. Through previous scenario and an evaluation of the economic and environmental goods, this research gives a comprehensive view on India's renewable energy. It underscores the important role of technological advancement in propelling India's sustainable energy transition and stats the recommendations for policymakers, industry stakeholders, and researchers to further improve the country's renewable energy growth.

DOI: 10.4018/979-8-3373-0045-0.ch019

1. INTRODUCTION

Østergaard et al. (2023) observed that the electric power is a fundamental pillar supporting the economic prosperity and well-being of nations. A resilient and evolving power sector plays a crucial role in ensuring sustainable growth within the Indian economy. It envisions that a substantial investment of USD 350 billion in various ventures and initiatives focused on renewable energy and cleantech could propel India to generate an impressive $212 billion in revenue, create 3.4 million jobs, and positively impact the lives of 919 million individuals by the year 2030. Recognizing the gravity of this issue, this edition will explore the renewable energy sector in India. It will be discussed about the sector's perspective, achievements, challenges, and the trajectory it is poised to take in the future works. The importance of renewable energy during global energy challenges and climatic difficulties is undeniable. With intensifying worries about waning fossil fuel reserves, energy security, and the pressing necessities to address climate change, renewable energy sources become a vital solution. The Panel formed with the inter governing bodies on Climate Change has explicitly accepted human interference on climate crisis, primarily attributed to greenhouse gas emissions from fossil fuel combustion. Nwokediegwu et al. (2024) observed that the outcomes, embracing the enhanced global temperatures, critical weather occurrences, and soaring sea levels, current significant hazards to ecosystems, economies, and societies world widely. Within this frame, renewable energy section resources like solar, wind, and another energy provide a sustainable alternate for fossil fuels. These advancements produce less greenhouse gas emissions in electricity production, manoeuvring a crucial role in lowering carbon footprints and make awareness of issues based on the climatic condition. Ravivarman et al. (2023) stated that the furthermore sustainable energy technologies are high in numbers, widely popular, and give more opportunities for decentralized energy generation, thereby improving energy security and resilience. Hence, the current happenings in global energy are pointed by a rising need for electricity, in particular in emerging economies such as India and China. Renewable energy sources showed the path to fulfil this mounting need while shortening reliance on finite and geopolitically sensitive fossil fuel reserves. Renewable energy predicts a vital role in tackling both worldwide energy difficulties and climate change. Not only does it give a sustainable, low-carbon energy option, but it also provides to energy security and development of economic, making it a significant element for a sustainable future. India's energy scenario gave an intricate connection of escalating energy demand, different energy sources, and an immediate need for sustainable alternatives. Grasping this context is essential for comprehending India's way in renewable energy. The country's energy necessity has consistently soared high, driven by growing population, urbanization, and industrialization. As per the IEA (International Energy Agency), country is more likely to surpass the China and become the largest energy consuming country by 2040. The Energy consumption of the country is pumped by sectors like industry, transportation, and residential use, with demand constituting a large portion. In the past, India has mostly depended on coal for producing electricity, with coal constituting a significant part of its energy mixture. Worku et al. (2024) noticed that the dependency on fossil fuels has increased apprehensions in spite of carbon emissions and air quality. Moreover, India possesses a distinct kind of energy policy encompassing natural gas, nuclear power, and hydroelectricity. The imperative for sustainable energy opportunities in India is propelled by various important factors. Firstly, as a participant in climate agreements by internationally, India is dedicated to reducing its carbon emissions and confronting climatically changes. Secondly, the diminishing environmental effects of fossil fuel-based energy production, such as pollution of air and ecosystem disruption, fuel the quest for cleaner alternatives. Lastly, the unpredictability of global fuel prices signifies the significance of energy security and the diversification of energy sources. Deshmukh et al. (2023) in their context, renewable energy sources (solar, wind, and hydropower) have garnered focus as sustainable alternatives. These sources hold the potential to fulfil India's widening energy needs while curbing greenhouse gas emissions and alleviating environmental effects. This paper seeks to thoroughly reviewed and analyse the progress in renewable energy technologies within the specific context of India. Aim of this overlook is to highlight the importance of renewable energy in taking global energy challenges and mitigating climate change, with a specific focus on India's unique energy landscape. Panda et al. (2023) through the exploration of technological advancements in detail and policy initiatives, this intends to offer valuable insights into India's journey toward a sustainable and energy future. Several studies were conducted on solar air and water heaters, with HD focusing on solar energy storage and utilization. They assessed various types and geometrical conditions of solar thermal storage to improve renewability and enhance heat transfer. The literature review, including the work of contributes to understanding advancements in solar heat storage. The current study aims to increase the effectiveness of solar heat storage through analysis using MATLAB software

2. RENEWABLE ENERGY

A renewable energy source is characterized by its sustainability, indicating that it is not subject to depletion and is essentially limitless. Examples include solar energy from the sun. Although the term 'alternative energy' is frequently associated with renewable sources, it encompasses other energy sources considered as alternatives. Essentially, renewable energy denotes energy that differs from the more commonly employed non-sustainable sources, such as coal.

Most common form of RE harnessing done by these below types,

- Solar power: Sunlight is among the most available and easily accessible energy assets on Earth. The amount of solar energy reaching the Earth's surface per hour surpasses the planet's cumulative energy needs for a whole year. Even though solar power looks ideal as a renewable resource, the amount available for use varies depending on factors such as the environmental condition and geographical location. To assess whether it's a suitable option for you, you can explore our guide to solar power below.
- Wind Energy: Wind serves as an abundant source of clean energy, and the landscape is increasingly adorned with wind farms, actively contributing to the National Grid. The process of harnessing electricity from wind energy entails using turbines to drive generators, subsequently channelling the generated electricity into the National Grid.
- Hydro Energy: Hydropower, as a renewable, stands out as one of the most developed options in terms of commercial values. The construction of a dam or barrier facilitates the creation of a large reservoir, allowing for a controlled flow of water to drive a turbine and generate electricity. This energy source is more reliable than the earlier two sources. Moreover, hydropower offers the advantage of electricity storage, making it accessible during peak demand periods. Similar to wind energy, hydropower's economic viability varies based on specific situations, dependent on the type.
- Tidal Energy: Indeed, this is another version of hydro power energy that utilizes the energy from tidal currents in the sea/ocean to drive turbine generators. Unlike certain other hydro power re-

sources, tidal flow may not be constant, but its high predictability is a key advantage. This predictability enables it to adjust for periods when the tide current is low, thereby enhancing the reliability and efficiency of this specific form of hydro energy.

- Geothermal Power: Accessing the natural heat under the Earth's surface enables the utilization of geothermal energy for direct heating of homes or electricity generation. However, despite the potential to extract power directly from beneath the ground,
- Biomass Energy: Indeed, this process involves converting solid fuel derived from plant materials into electricity. While the basic principle of biomass entails burning organic materials for power generation, contemporary methods have markedly enhanced its cleanliness and energy efficiency. Biomass achieves this by transforming Agri, industrial, and domestic waste into solid, liquid, and gas fuel, enabling power production at a significantly reduced environmental cost as well as economic cost.

Li et al. (2023) found from their literature survey, as the world population continues to increase, there is a parallel rise in the necessity for energy to power our belongs like homes, businesses, and communities. The key to sustaining a manageable energy level and safeguarding our planet from climate change lies in the cutting-edge technic and widening of renewable sectors. As of today, renewable energy sources contribute to the world's electricity by 26%. According to the International Energy Agency (IEA), this share is expected to attain 30% by 2024. The executive director of IEA, emphasized that this period represents a very important time for renewable sector. Indeed, in 2020, the nation achieved a historic feat by relying exclusively on renewable energy for two consecutive months, marking a ground-breaking moment in its history. This milestone represents a significant step forward in the advancement of renewable energy. In the future, it is expected that the renewable energy will constantly to boost-up by the need to the increasing demand for power. This trend is expected to contribute to a reduction in the price of renewables, benefiting both the planet and our financial resources. Nuclear energy, although not categorized as renewable, is acknowledged for its low-carbon nature, emitting minimal levels of CO_2 during generation, akin to the renewable energy sources discussed earlier. The stability inherent in nuclear energy, being independent of weather conditions, makes it a valuable asset in working towards Net Zero status. All our fixed home tariffs come with the assurance of zero-carbon electricity as a standard feature and undergo annual backing. Opting for one of our fixed electricity tariffs allows you to play an active role in advancing the Net Zero target.

3. RE POLICIES IN INDIA

Laimon et al. (2024) India has instituted numerous pivotal renewable energy policies and initiatives to expedite the adoption of clean energy technologies and attain its renewable energy targets. The National Solar Mission, initiated in 2010, stands as a flagship renewable energy policy in India. Its primary goal is to foster the expansion of solar energy and position India as a global leader in solar power generation. This aim incorporates ambitious set points for the installation of capacity of the solar energy. As a representation, India initially aimed for 20 GW of solar capacity by 2022 and following that improved its set point to 100 GW by 2022. Manikandan et al. (2023) the National Solar Mission utilized various mechanisms to create awareness to adopt the solar energy, encompassing financial benefits like incentives, subsidies, tax benefits, and incentives for solar power generation like feed-in tariffs and open bidding. The National Wind Energy Mission, initiated in 2014, strives to improve wind energy capacity and expand its contribution to India's overall energy mix. This process concentrates on optimizing wind energy resources and improving the efficiency of wind power projects, setting targets for both onshore and offshore wind power capacity installation. To refurbish the progress of wind energy, the Indian government has amended competitive bidding mechanisms, financial incentives, and policies to stimulate wind turbine manufacturing domestically. Behera et al. (2025) the Green Energy Corridor efforts, initiated in 2016, stands as a crucial part in India's renewable energy policy structure. The core on building the requisite transmission infrastructure to seamlessly bind the renewable energy sources into the national grid. This process strives to diminish the losses by transmission and streamline the transfer of renewable energy from regions abundant in resources to those with high energy demand. The Green Energy Corridor initiative shields the vital challenge of grid amalgamation faced by renewable energy projects. This improves the reliability and efficiency of the grid, rendering it more adaptable to intermittent renewable energy resources like as solar and wind. These things and initiatives play an important role in pushing India's shift toward renewable energy. They furnish a regulatory structure, financial incentives, and infrastructure development vital for the country's renewable energy mission and curbing its carbon footprint. The Government has formed ambitious vision for renewable energy capacity as a manifestation of its dedication to expanding clean energy sources and confronting climate change. These targets have improved over the years to mirror India's growing accentuation on renewable energy.

3.1 Targets to Reach

Launch Goal - 2010: National Solar Mission was initiated in the year of 2010 with the goal of 20 GW of solar power capacity by 2022. Amended Goal - 2022: The Goal was later revised to 100 GW of solar power capacity by 2022. In parallel the goal is actively working towards reaching an energy capacity of' 500 GW by the year 2030.Headway: As of the latest available data, India had nearing 100 GW in total installed solar capacity as of 2023, indicating substantial progress toward the 500 GW target in 2030. Over the past five years, there has been a remarkable shift in the country's solar installed capacity, surging from 21,651 MW to 70,096 MW by 2023. Fuelled by ambitious targets and initiatives such as the Production Linked Incentive (PLI), In the 2023's financial year (FY), India observed the addition of close to 12,800 MW in solar power capacity and 2,300 MW in wind power capacity. Notably, solar installations faced an 8% decline, whereas wind installations experienced a significant upswing of 105% compared to FY2022. In the utility-scale solar sector, India saw the addition of approximately 8 GW of new capacity, representing a decrease of around 21% compared to FY2022 placings. 2532 MW in Rajas-than, 1606 MW in Tamil Nadu and Maharashtra emerged as leaders in mega solar deployments during Financial Year 2023.For solar deployments in roof, like 2,250 MW was included in Financial Year 2023, indicating a modest increase of about 1.2% compared to FY2022. Gujarat led with 690 MW installed during the period from Financial year 2023 in the roof top solar deployments, followed by Maharashtra and Kerala 547 MW and 266 MW respectively. In the off-grid/distributed solar section, approximately 840 MW was included, marking a decrease of about 46% compared to FY2022 installations.

3.2 For Wind Energy Mission Targets

a. Launch Goal (2014): The National Wind Energy Mission established a goal of 60 GW of onshore capacity of wind energy by 2022. b. Amended Goal (2022): The target was subsequently adjusted to 140 GW of onshore wind power capacity by 2030. Headway: As of September 2021, India had an installed onshore wind capacity of around 40 GW, indicating continued advancements toward its 2030 target. In the sector of wind, around 2.3 GW was integrated in Financial Year 2023, representing a significant surge of 105% in comparison to the 1.1 GW accumulated during 2022 financial year. Rajasthan with a top-notch deployed capacity of 870 MW, Gujarat with 770 MW and Madhya Pradesh with 324 MW in wind in the Financial Year 2023.

3.3 For Other Renewable Energy Mission Targets

Regarding cumulative installations, data by the Ministry of New and Renewable Energy (MNRE) until March 2023 reveals that India's renewable energy (RE) depletion capacity attained 126 GW. Solar gives close to 53% to the cumulative Renewable Sector, and then wind energy at 34%, Bio energy at 9%, and Hydro power at 4%. a. Target for 175 GW by 2022: The Indian government sets an ambitious goal to reach 175 GW of renewable capacity (including solar, wind, biomass, and small hydropower) by 2022. Target for 450 GW by 2030: India has set a long-term target of reaching 450 GW of renewable energy capacity by 2030, underscoring the pivotal role of renewables in its energy transition. In the fiscal year 2024, considering the deferral of ALMM until March 2024, JMK Research estimates indicate the potential addition of around 15 GW of new extensive utility capacity of solar. Furthermore, there is an anticipated additional increase of 3 GW in the roof-top solar. c. Headway: As of September 2021, India's total renewable energy capacity exceeded 150 GW, signifying substantial progress toward the 175 GW target by 2022 (MNRE, 2021). These targets illustrate India's commitment to increasing the share of renewable energy in its energy mix, reducing carbon emissions, and enhancing energy security. The progress made toward these goals demonstrates the country's dedication to transitioning to a more sustainable and cleaner energy future.

Figure 1. Trends in RE installation.

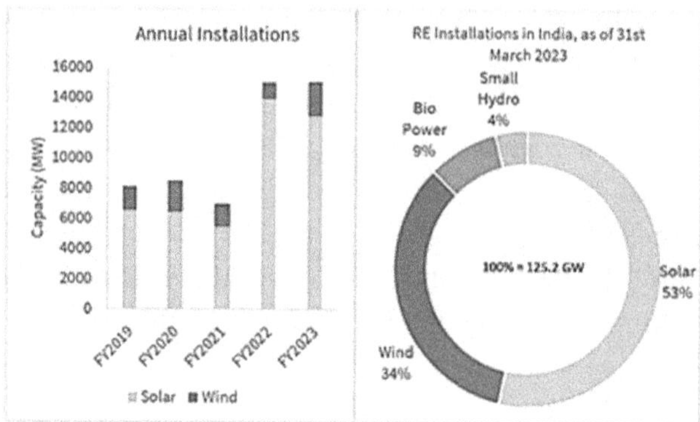

Figure 2. State-wise RE installations in India.

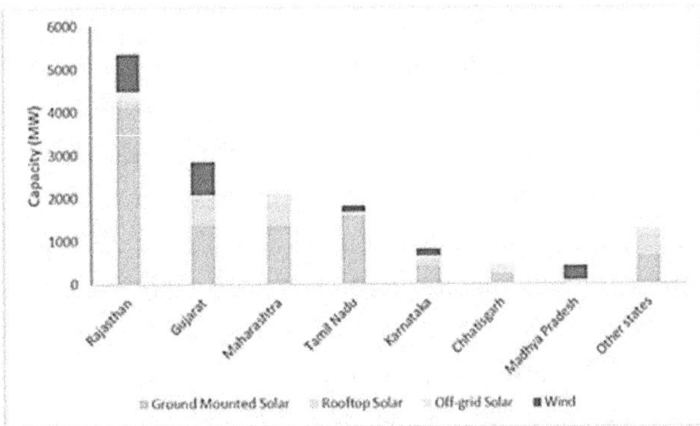

Table 1. Consumption variation of RE past 2 years

Renewable Energy	Mar 2021	Mar 2022	Change (GW)
Solar	40.1	55.5	15.4
Wind	38.9	41.2	2.3
Hydro	4.8	4.9	0.1
Biomass	10.5	10.7	0.2
Waste to energy	0.2	0.5	0.3

3.4 Recent Developments in Solar Energy

Khan et al. (2024) India has achieved notable strides in the advancement of its solar energy, emphasizing improvements in the efficiency of photovoltaic cells, the exploration of inventive solar panel designs, and the creation of energy storage solutions. These developments work as pivotal role in optimizing the using of solar energy in the most efficient and sustainable style. An important domain of headway lies in the efficient of PV cells. Researchers and manufacturers have been leading the way in crafting more-efficiency solar cells, incorporates PERC (Passivated Emitter and Rear Cell) and bifacial solar cells. The PERC technology, known for minimizing recombination losses and improving light absorption, has played an important role in reaching heightened efficiency. Wheatley et al. (2024) Bifacial solar cells, manufactured to capture sunlight from either side of the panel, are gaining attention in India, offering the potential for increasing energy generation. These advancements carry the encouraging of considerable enhancing solar energy conversion rates. Along with, innovative solar panel structures have emerged as a vital area of progress in India. Scientists are examining flexible solar panels capable of amalgamation into diverse surfaces, including curved or irregular design. Such flexible panels produce design versatility. Furthermore, these flexible panels find application in various scenarios. Additionally, India has witnessed the advent of building materials like solar roof tiles and solar facades. These innovations seamlessly integrate solar energy production into the designs of buildings. Balancing aesthetics with renewable energy, India has refined solar roof tiles and facades. Madessa et al. (2024) energy storage resolutions play an important role in India's solar energy headway, addressing the intermittent nature of solar power production. The country has devoted in advanced battery technic, such as lithium-ion batteries and burgeoning options like solid-state batteries. These technics exert to store excess solar energy for using during less sunlight periods. Grid-scale energy storage resolutions like pumped water storage and battery energy storage systems (BESS) have been executing to tackle the intermittency challenge. Moreover, India is highlighting demand-side management strategies that incorporate energy storage to optimize energy usage and reducing peak demand. This headway aims to boost the effectiveness of solar power integration further. India's march in solar energy technologies underscore its dedication to a sustainable and cleaner energy future. Adebiyi et al. (2024) developments in PV cell efficiency, inventive panel designs, and energy storage resolutions play a vital role in maximizing the advantages of solar energy, expediting the nation's focus to cleaner and more sustainable power resources. These advancements not only improving India's energy security but also have the potential to work as a global example for renewable energy innovation and adoption.

Figure 3. Flexible PV solar sheet.

3.5 Wind Energy Developments

India is dynamically pursuing innovations in wind energy technology and turbine design to refining the harnessing of wind power. This encompassment of the creation of larger and more efficient wind turbines, contributing to the expansion of the country's wind energy capacity. Furthermore, various successful wind energy has made a substantial impact on electricity production. In the second quarter of the calendar year 2023, India registered a significant upswing in wind power capacity, with an addition of 1,139.5 MW. This represents a noteworthy year-over-year (YoY) growth of 165%, as reported by Mercom India Research. Lodhi et al. (2024) the notable increase in installations can be largely attributed to supportive government policies, specifically the implementation of the Renewable Purchase Obligation (RPO) trajectory extended until 2029-2030. Installations showed a robust quarter-over-quarter (QoQ) growth, surging by 54%, from 737.85 MW in Q1 2023. The trajectory outlines that wind Renewable Purchase Obligation (RPO) can be fulfilled through projects commissioned after March 31, 2022.Nevertheless, as of the conclusion of January 2023, Gujarat's cumulative wind power capacity had surged to 9,919 MW, surpassing Tamil Nadu's installed wind energy capacity of 9,964 MW. Gujarat saw an increase of 710 MW in the current fiscal year, while Tamil Nadu added a comparatively modest 99 MW. By the middle of 2023, Gujarat had outpaced Tamil Nadu in terms of installed capacity.

Figure 4. Wind power installation.

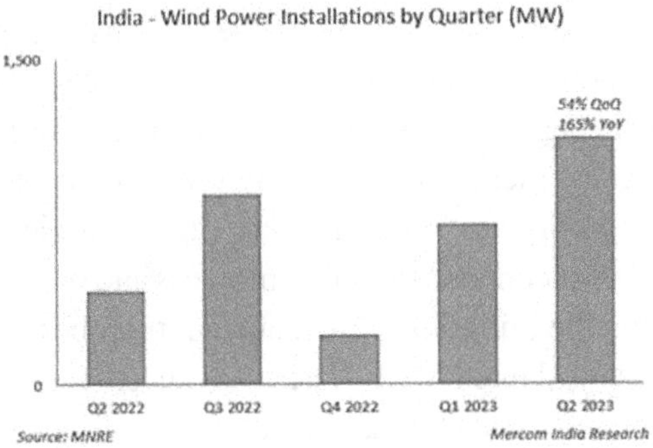

3.5.1 Advancements in Wind Energy Technology and Turbine Design

Gao et al. (2024) creating larger and more efficient wind turbines stands out as a notable innovation in wind energy technology. These developments target the increased capture of energy from the wind, aiming to enhance the overall performance of wind farms. In India, manufacturers and researchers have been actively involved in: India has witnessed the introduction of higher-capacity wind turbines featuring larger rotor diameters. This design allows them to capture more wind energy efficiently. These turbines are crafted to function effectively at lower wind speeds, thereby extending their operational range geographically. Abd Aziz et al. (2024) innovations in blade design, specifically advanced blade design, have resulted in rotor blades that are more aerodynamic and efficient. These blades had been worked on to maximize energy harnessing, reducing turbulence and noise, thereby improving the overall effectiveness of wind turbines. Luo et al. (2024) different wind energy installations in India have made ample contributions to electricity generation, serving as notable examples of successful projects. One such illustration is the Jaisalmer Wind Park, located in the Jaisalmer district of Rajasthan. This is one of the India's largest wind energy projects, encircling multiple wind farms with a combined capacity exceeding 1,600 MW. The Jaisalmer Wind Park has played a vital role in amplifying Rajasthan's wind energy capacity, in line with India's renewable energy objectives. Along with that, Tehri Hydro Development Corporation (THDC) India Limited has implementing wind energy projects exceeding 63 MW in total capacity in the states like Gujarat and Karnataka. These initiatives involving not only to India's renewable energy capacity but also helps to diminishing the greenhouse gas emissions (THDC India Limited, n.d.). The leverage of these and other prosperous wind energy installations in India widens beyond electricity generation. They have been essential in mitigating carbon emissions, fortify energy security, and producing economic advantages to communities situated around the plant through the job creating and the improving the infrastructure. Pawar et al. (2024) the sector of wind energy has gone through advancements in design and technology improvement in turbine, heading to the advancements of larger and better efficient wind turbines. These innovations, incorporated with the success of wind energy installations, have played an vital role in augmenting India's renewable energy capacity and its transition toward cleaner and more sustainable energy resources.

3.6 Biomass and Bioenergy Innovations

India is actively concentrating and working in biomass and bioenergy advance technologies, demonstrating a dedication to significant energy production. These pursuits accompanied with biogas production efficiency, creating biomass power plants, and on-going research in the domain of biofuels. India has shown a great curiosity in using agricultural and organic disposal for renewable energy production, influencing to efficient waste management and mitigating greenhouse gas emissions. Although, there have been various advancements in biomass and bioenergy technologies, in particular enhancing the efficiency of biogas production. Advancements in India's biomass and bioenergy sector bind the advanced anaerobic digestion techniques and the initiation of specialized microbes, facilitating quicker and highly efficient conversion of organic disposal into biogas. These advancements, gives in The Economic Times (2019), have not only improved biogas yields but have also furnished the more sustainable. Additionally, the country has seen the creation of biomass power plants, designed to produce electricity from diverse organic sources such as crop residues, wood, and agro-industrial disposal. In India, new ideas in combustion and gasification technologies have improved the overall efficiency of power plants, concurrently diminishing their environmental impact. Research and development strivings in the country have focused on biofuels derived from organic sources like vegetable oils, jatropha seeds, and sugarcane molasses. Explorations into advanced methods, like enzymatic hydrolysis and the production of second-generation biofuels from lignocellulose biomass, aim to augment the viability and sustainability of biofuels.

Figure 5. Biomass usage (State wise).

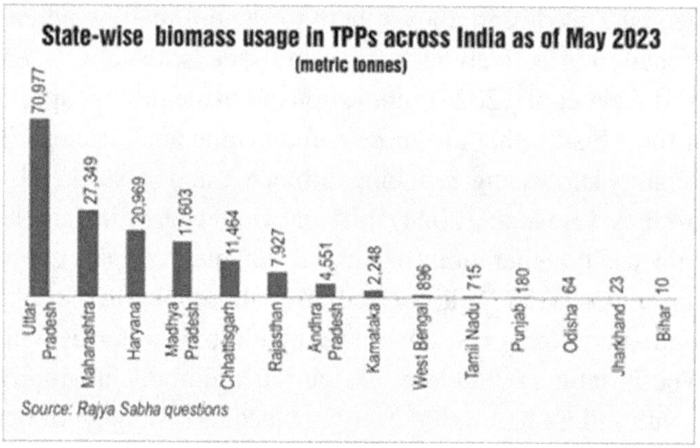

3.7 Utilization of Agricultural and Organic Waste

Kannakumar et al. (2023) it has been acknowledged by the country, the potential of harnessing agricultural and organic disposal for renewable energy generation, aligning with its waste management and clean energy objectives. Likely, agricultural wastages such as rice straw, wheat straw, and sugarcane bagasse serve as notable feedstocks for biomass power generation. Schemes like Punjab Biomass Power Limited (PBPL) have effectively using of the crop residues to generate electricity, mitigating the environmental repercussions of open-field burning and along with enhancing access to rural energy. Gao et al. (2024) sugarcane bagasse, a by-product of the sugar industry, finds special utilized as a feedstock in biomass cogeneration plants. These facilities simultaneously production of electricity and steam, sending energy for sugar mills while contributing to diminishing of emission. India has undertaken initiatives to convert municipal solid waste into biogas and electricity. Waste-to-energy plants, symbolized by the facility in Ghazipur, Delhi, are transforming urban waste into renewable energy, effectively addressing waste management barriers. The steps in biomass and bioenergy technologies, bonded with the utilization of agricultural and organic disposal, play a vital role in India's march toward cleaner and more impactful energy sources. Excessive contributing to renewable energy generation, these advancements facilitate efficient waste management and aid in the diminishing of greenhouse gas emissions. These steps not only contribute to renewable energy generation but also aid in effective waste management, decrease the greenhouse gas emissions, and improve energy security. Changing the focus to hydropower and small hydro projects, India, recognizing its significant hydropower potential, has been sincerely working to improve hydropower technology and increase capacity. Here, let's explore into the current status of hydropower projects in India, scrutinize technological advancements and capacity additions. Prominently, India has been advancing hydropower technology to enhance the effectiveness and sustainability of its projects. These advancements include the adoption of more turbines which works efficiently, enhanced dam structure, and the implementation of latest control systems to optimize energy generation. Simultaneously, India is pursuing a significant scheme for the development of hydropower projects to improve the capacity. Projects like the Sardar Sarovar Dam on the Narmada River and the Tehri Dam on the Bhagirathi River have given significant participation to the India's hydropower capacity. Above all, the government has identified more prospective sites for hydropower development, along with the Subansiri Lower Hydroelectric Project in Arunachal Pradesh. India is robustly exploring the probable sites of pumped storage hydropower projects. These enterprises store excess energy during periods of low demand and release it during high demand, putting to grid stabilization and enhancing the reliability of renewable energy sources.

3.8 Role of Small Hydropower Projects in Decentralized Energy Generation

Wei et al. (2023) lesser powered water-based schemes play an important role in decentralized energy production and rural electrification in India. In particular to designed for rural energy production, these schemes periodized to give electricity to local communities or rural areas. Small hydropower schemes are in particularly suited for local regions not connected to the national grid. By creating power locally, these schemes part a crucial role in balancing a reliable and continuous energy supply to local populations, contributing substantial rural electrification in India. Bagdi et al. (2023) small hydropower projects play a part in giving electricity to areas that would otherwise have low access to power. This, subsequently, helps economic development, improves standard of living, and enhances the overall quality of life in local communities. Along to that, these projects gave environmental benefits, with a little environmental issue compared to larger hydropower dams. Small hydropower schemes in India have a diminishing impact on ecosystems and hydro habitats, producing them a more sustainable choice for autonomous energy production. Madhavan et al. (2023) as part of India's active stride in hydropower scheme of things, which incorporated technological improvements and capacity add on, little hydropower projects continuously to play a vital role. Little hydropower projects in India play a vital role in Autonomous energy production, particularly in rural and remote areas. Their efforts widen to rural electrification, helped for economic development, and the furnishing of a sustainable and reliable source of electricity for rural communities.

3.9 Prospects of Harnessing Ocean Energy in India

Sayed et al. (2023) India's extensive shoreline possess substantial opportunities for harnessing ocean energy, particularly tidal and wave energy. Let's explore into the prospects of these renewable energy resources, starting with Tidal Energy: India's shoreline boasts several locations with resilient tidal currents, notably in the Gulf of Kutch and the Gulf of Cambay. India has initiated experimental projects to harness the tidal energy utilising tidal stream generators. Guo et al. (2023) the potential for tidal energy in the country is around 8 GW, according to the Ministry of New and Renewable Energy (MNRE), Government of India, in 2020.India's treasured coastline gives abundant possibility for harnessing wave energy. Different wave energy converters, including oscillating water columns and point absorbers methods, are under research in coastal regions. specially, India has significant wave energy opportunities along the eastern coast, particularly in the Bay of Bengal, as reported by the Hindu in 2017. Shang et al. (2024) Ocean energy faces obstacles such as higher installation and maintenance costs, along with the requirement for new cutting-edge technology. In spite of these hurdles, on-going research and development efforts, blended with government support, are progressively advancing the deployment of ocean energy methods, as reported by The Economic Times in 2021.India's Integrated Energy Policy realize the significance of ocean energy and convey plans for its development alongside other renewable energy sources. This incorporated approach contributes to energy security and sustainability, as traced by the Ministry of Power, Government of India, in 2020. Kumar et al. (2023) India indeed holds underutilized potential in both geothermal and ocean energy sources. On-going research and exploration efforts are progressively open the geothermal potential, especially in the Himalayan region. Simultaneously, the country's extensive coastline offers prospects for harnessing tidal and wave energy, making significant move toward achieving its renewable energy visions. Despite ongoing restrictions, government's aid and international alliances play vital roles in overcoming these obstacles and advancing the deployment of sustainable energy sources.

Figure 6. One of the proposals for harnessing ocean power.

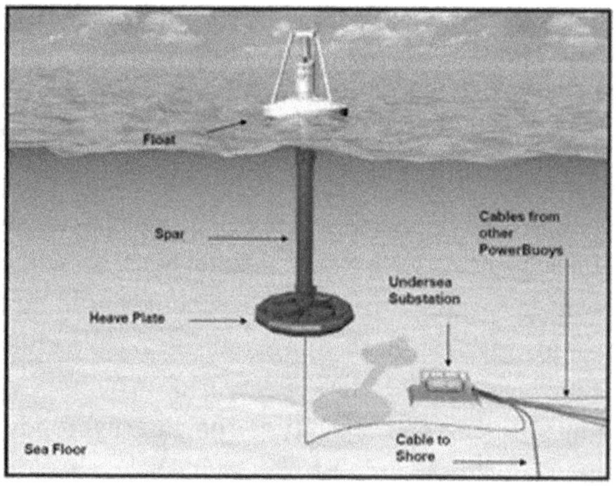

4. CHALLENGES AND BARRIERS

Ahmed et al. (2024) the widespread adoption of renewable energy technologies in India is pivotal for achieving the country's targets and addressing its energy needs while mitigating environmental concerns. However, various challenges and barriers, such as grid integration issues, financing constraints, and policy implementation challenges, impede the rapid deployment of renewable energy sources.

Below, we identify these challenges with relevant in-text citations.

a. Grid integration issues
   i. Renewable energy resources, like the solar and wind power, present challenges due to its variable nature. By assuring a constant power supply needed addressing these variations in the grid efficiency.
   ii. The running grid structure in India often dearth the necessary flexibility and capacity to efficiently integrate renewable energy. Improving the grid to manage higher shares of renewables remains a current restriction.
b. Fiscal restrictions:
   i. The substantial upfront running costs linked with renewable energy projects, including solar and wind farms, might worked as a hurdle for potential investors and developers.
   ii. Securing affordable fiscal options remains a hurdle, especially for lower end renewable energy projects and innovations focused on rural electrification.
c. Hurdles in Executing Policies:
   i. The constrains of using inexpensive fiscal options persists, notably for mini-scale renewable energy projects and rural electrification breakthroughs.
   ii. Hurdles and irregularities in rules at both the governing levels can obstruct the growth of the renewable energy sector. It is important to standardize the regulations and ensure framework stability.
   iii. The process of grabbing land for mega-scale renewable energy projects can be enduring and intricate, involving lawful and environmental considerations.
d. Obstacles involving technical advancements:
   i. Directing the disruption of renewable energy resources needs efficient energy storage solutions. The technological barriers play in developing storage technologies that are both cost-effective and efficient.
   ii. Structuring the usage of renewable energy needs better grid management systems, such as smart grids and demand-side management. The imposing of these technologies can pose challenges.
e. Environmental Aspects:
   i. Maga-scale renewable energy schemes, like hydropower and solar farms, might result in environmental hurdles such as habitat imprecations and barriers in managing water resources.
   ii. Issues over land use, especially in highly populated areas, can disperse or delay the implementation of renewable energy projects.
   iii. Rectifying these hurdles and barriers needs collective efforts from government, private sector stakeholders, and public. It involves thorough guiding, targeted policy schemes, investment in research and development, and pupil awareness programs to promote the advantages of advance renewable energy in India.
f. Market issues:

i. A significant share of renewable energy instrument and components is imported. It is crucial to boost the rural manufacturing and diminishing the dependence on imports.
ii. India experiences substandard losses in the transmission and distribution of electricity. defending these losses is important to increase the advantages of integrating renewable energy.

5. FUTURE OUTLOOK AND RECOMMENDATIONS

The likelihood for renewable energy in India is promising, pushed by technological advancements, evolving policy structure, and an enhanced dedication to sustainability. Here are insights into the future outlook:

a. Technological Advancements:
   - Enhanced Energy Storage: Progress in energy storage technologies, such as high-capacity batteries and innovative storage solutions, will mitigate the intermittency of renewable sources, enhancing their reliability and efficiency.
   - Green Hydrogen: The advancement of green hydrogen as an energy carrier holds significant potential. Renewable energy-powered electrolysis can generate hydrogen for various applications, including fuel cells and industrial processes.
   - Hybrid Systems: Further incorporation of wind-solar hybrid systems and other complementary renewable sources will optimize energy generation and enhance grid stability.

b. Developing Policy Frameworks:
   - Ambitious Goals: India's renewable energy targets are anticipated to expand, with a goal of reaching 450 GW of capacity by 2030 (Ministry of New and Renewable Energy, Government of India). Policymakers are expected to offer sustained support to achieve these objectives.
   - Policies for Energy Storage: Policymakers might introduce dedicated policies and incentives to encourage the deployment of energy storage, facilitating improved grid integration and bolstering the reliability of renewable energy sources.
   - Market Transformations: The evolution of market structures, including competitive auctions and power purchase agreements, will stimulate private sector investments in renewable energy projects.

6. RECOMMENDATIONS FOR ACCELERATING RENEWABLE ENERGY GROWTH

- Elevating R&D: Increasing the financial aid for research and development in renewable energy technologies to push the innovation and the creation of unique solutions.
- Technically advanced Grids: Devote in grid advanced initiatives, integrated smart grids and demand response systems, to adulate the seamless integration of renewable energy sources and improve grid stability.
- Standardize schemes and incentives to boost the establishment of energy storage structure, diminishing dependence on fossil fuels during periods of low renewable energy production.

- Ease the regulations and approval processes for renewable energy projects to expedite their implementation.
- Bolster the domestic manufacturing of renewable energy equipment and components to decrease import dependence and bolster the "Make in India" initiative.
- Invest in skill building schemes and capacity building to meet the skilled workforce requirements in the renewable energy sector.
- Start the public awareness and education campaigns to inspire and attract individuals and communities to accept and know more about renewable energy solutions and adopt energy-efficient practices.
- Incorporated with international organizations and countries to exchange the best practices and secure financial aid for renewable energy projects.

The forthcoming of renewable energy in India holds promise, marked by substantial potential for technological breakthroughs and scheme advancements. To harness this growth, a combined effort from Government authority people, industry stakeholders, and researchers is crucial. Through bolstering innovation, improving grid infrastructure, and advocating sustainable practices, India can further advance its transition toward a more significant and sturdy energy future.

7. CONCLUSIONS

In conclusion, country's growth in renewable energy is significant, pushed by technological innovations, improving policy structure, and an enhancing dedication to sustainability. India has made crucial strides in using solar, wind, and other renewable resources to fulfil its energy needs while concurrently reducing emissions of greenhouse gas. The illustrative cases of programs like the Solar Parks, Green Energy's project, and wind-solar hybrid projects in Tamil Nadu, India works as best examples of the transformative influence of advanced technologies in renewable energy production. Looking ahead, India is much advance for additional enhancement in renewable energy. Futuristic cutting-edge technologies in energy storage, green hydrogen, and hybrid systems are expected to installing the reliability and efficiency of renewable resources. Improving policy structure, integrated with significant targets and market restructure, will fulfil the essential help for sustaining this growth framework. Inducing the change towards a sustainable energy future needs a concentrated work from the government, industry stakeholders, and research pupil. Main domains should include more advanced investments in research and development, modernization of grids, energy storage infrastructure framework, and domestic manufacturing. Along to that, standardizing in approvals, capacity building, awareness programs, and fostering international collaborations will works a vital role in reaching this transition. In overview, India's ambiguines involvement to renewable energy is not only satisfy its energy needs but is also producing employment, reducing greenhouse gas emissions, supporting energy security, and soaring economic growth. With sustained dedication and strategic innovations, India is rightly placed-positioned to spearhead the trail toward a sustainable and clean energy future.

REFERENCES

Abd Aziz, A. J., Baharuddin, N. A., Khalid, R. M., & Kamarudin, S. K. (2024). Review of the policies and development programs for renewable energy in Malaysia: Progress, achievements and challenges. *Energy Exploration & Exploitation, 42*(4), 1472–1501. DOI: 10.1177/01445987241227509

Adebiyi, A. A., & Moloi, K. (2024). Renewable energy source utilization progress in South Africa: A review. *Energies, 17*(14), 3487. DOI: 10.3390/en17143487

Ahmed, S., Ali, A., & D'angola, A. (2024). A review of renewable energy communities: Concepts, scope, progress, challenges, and recommendations. *Sustainability (Basel), 16*(5), 1749. DOI: 10.3390/su16051749

Bagdi, T., Ghosh, S., Sarkar, A., Hazra, A. K., Balachandran, S., & Chaudhury, S. (2023). Evaluation of research progress and trends on gender and renewable energy: A bibliometric analysis. *Journal of Cleaner Production, 423*, 138654. DOI: 10.1016/j.jclepro.2023.138654

Behera, U. S., Sangwai, J. S., & Byun, H. S. (2025). A comprehensive review on the recent advances in applications of nanofluids for effective utilization of renewable energy. *Renewable & Sustainable Energy Reviews, 207*, 114901. DOI: 10.1016/j.rser.2024.114901

Deshmukh, M. K. G., Sameeroddin, M., Abdul, D., & Sattar, M. A. (2023). Renewable energy in the 21st century: A review. *Materials Today: Proceedings, 80*, 1756–1759. DOI: 10.1016/j.matpr.2021.05.501

Gao, Z., Zhao, Y., Li, L., & Hao, Y. (2024). Economic effects of sustainable energy technology progress under carbon reduction targets: An analysis based on a dynamic multi-regional CGE model. *Applied Energy, 363*, 123071. DOI: 10.1016/j.apenergy.2024.123071

Guo, W., Yang, B., Ji, J., & Liu, X. (2023). Green finance development drives renewable energy development: Mechanism analysis and empirical research. *Renewable Energy, 215*, 118982. DOI: 10.1016/j.renene.2023.118982

Kannakumar, K., Manikandan, P., Ravivarman, G., Sivalingam, A., Moorthy, A. A., Karthik, M., & Girimurugan, R. (2023). Enhanced behavior of solar stills with modified absorber surfaces: Comparative analysis of floating wick designs. In *E3S Web of Conferences,* 455, 02002.

Khan, M. K., Raza, M., Shahbaz, M., Farooq, U., & Akram, M. U. (2024). Recent advancement in energy storage technologies and their applications. *Journal of Energy Storage, 92*, 112112. DOI: 10.1016/j.est.2024.112112

Kumar, N. R., Seenivasan, S., Hemalatha, P., Prakash, K. J., Pitchandi, P., Sathiya, R. D., & Girimurugan, R. (2023). Evaluating the influence of coir fibres on solar still efficiency and economic viability. In *E3S Web of Conferences,* 455, 02004. DOI: 10.1051/e3sconf/202345502004

Laimon, M., & Yusaf, T. (2024). Towards energy freedom: Exploring sustainable solutions for energy independence and self-sufficiency using integrated renewable energy-driven hydrogen system. *Renewable Energy, 222*, 119948. DOI: 10.1016/j.renene.2024.119948

Li, X., Raorane, C. J., Xia, C., Wu, Y., Tran, T. K. N., & Khademi, T. (2023). Latest approaches on green hydrogen as a potential source of renewable energy towards sustainable energy: Spotlighting of recent innovations, challenges, and future insights. *Fuel, 334*, 126684. DOI: 10.1016/j.fuel.2022.126684

Lodhi, S. K., Gill, A. Y., & Hussain, I. (2024). Renewable energy technologies: Present patterns and upcoming paths in ecological power production. *Global Journal of Universal Studies*, *1*(1), 108–131. DOI: 10.70445/gjus.1.1.10

Luo, Z., Wang, C., Tang, Q., & Tian, W. (2024). Renewable energy technology innovation effect on the economics growth. *Chemistry and Technology of Fuels and Oils*, *59*(6), 1271–1278. DOI: 10.1007/s10553-024-01644-7

Madessa, H. B., Shakerin, M., Reinskau, E. H., & Rabani, M. (2024). Recent progress in the application of energy technologies in Large-Scale building Blocks: A State-of-the-Art review. *Energy Conversion and Management*, *305*, 118210. DOI: 10.1016/j.enconman.2024.118210

Madhavan, V. M., Rahul, S., Vijayakumar, P., Dhal, P. K., Girimurugan, R., Ravivarman, G., & Joseph, J. (2023). Optimizing solar energy utilization and energy efficiency through thermal energy storage with phase change materials in a solar water heating system. In *E3S Web of Conferences,* 455, 02005. DOI: 10.1051/e3sconf/202345502005

Manikandan, R., Gopalakrishnan, K., Ashokkumar, P., Maheskumar, P., Girimurugan, R., Ravivarman, G., & Anand, R. (2023). Experimental study of minimum-temperature hydrated salt latent heat thermal energy storage with sodium acetate trihydrate as phase change materials. In *E3S Web of Conferences,* 455, 02008. DOI: 10.1051/e3sconf/202345502008

Nwokediegwu, Z. Q. S., Ibekwe, K. I., Ilojianya, V. I., Etukudoh, E. A., & Ayorinde, O. B. (2024). Renewable energy technologies in engineering: A review of current developments and future prospects. *Engineering science & technology journal,* 5, 367-384.

Østergaard, P. A., Duic, N., Noorollahi, Y., & Kalogirou, S. (2023). Advances in renewable energy for sustainable development. *Renewable Energy*, *219*, 119377. DOI: 10.1016/j.renene.2023.119377

Panda, A., Dauda, A. K., Chua, H., Tan, R. R., & Aviso, K. B. (2023). Recent advances in the integration of renewable energy sources and storage facilities with hybrid power systems. *Cleaner Engineering and Technology*, *12*, 100598. DOI: 10.1016/j.clet.2023.100598

Pawar, R., Santara, S., Sircar, A., & Yadav, K. (2024). Design and development of hybrid solar–biomass drying system: An innovative approach. *MRS Energy & Sustainability : a Review Journal*, *11*(2), 409–433. DOI: 10.1557/s43581-024-00083-5

Ravivarman, G., Madhavan, V. M., Kalaiyarasan, A., Hemasilviavinothini, S., Nithyanandhan, T., Kesavan, S. P., & Girimurugan, R. (2023). Enhancing efficiency and cost-effectiveness of solar stills through natural fiber integration. In *E3S Web of Conferences*, 455, 02003. DOI: 10.1051/e3sconf/202345502003

Sayed, E. T., Olabi, A. G., Alami, A. H., Radwan, A., Mdallal, A., Rezk, A., & Abdelkareem, M. A. (2023). Renewable energy and energy storage systems. *Energies*, *16*(3), 1415. DOI: 10.3390/en16031415

Shang, Y., Sang, S., Tiwari, A. K., Khan, S., & Zhao, X. (2024). Impacts of renewable energy on climate risk: A global perspective for energy transition in a climate adaptation framework. *Applied Energy*, *362*, 122994. DOI: 10.1016/j.apenergy.2024.122994

Wei, P., Abid, M., Adun, H., Kemena Awoh, D., Cai, D., Zaini, J. H., & Bamisile, O. (2023). Progress in energy storage technologies and methods for renewable energy systems application. *Applied Sciences (Basel, Switzerland)*, *13*(9), 5626. DOI: 10.3390/app13095626

Wheatley, M. C. (2024). Advancements in Renewable Energy Technologies: A Decade in Review. *Science*, *1*, 100013.

Worku, A. K., Ayele, D. W., Deepak, D. B., Gebreyohannes, A. Y., Agegnehu, S. D., & Kolhe, M. L. (2024). Recent advances and challenges of hydrogen production technologies via renewable energy sources. *Advanced Energy and Sustainability Research*, *5*(5), 2300273. DOI: 10.1002/aesr.202300273

Chapter 20
Projecting Green Energy Fostering and SDG 15 (Life on Land):
Using Natural Resources for Sustainable Future

Ambuj Sharma
https://orcid.org/0009-0008-9573-1880
Kamkus College of Law, Ghaziabad, India

Gayathri Band
https://orcid.org/0000-0001-6602-2040
Ramdeobaba University, Nagpur, India

Saurabh Chandra
https://orcid.org/0000-0003-4172-9968
Bennett University, Greater Noida, India

Bhupinder Singh
https://orcid.org/0009-0006-4779-2553
Sharda University, India

Kanchan Dinesh Naidu
https://orcid.org/0000-0001-5362-2135
School of Management, Ramdeobaba University, Nagpur, India

Laeeq Razzak Janjua
https://orcid.org/0000-0002-1948-5859
WSB University, Poland

ABSTRACT

Clean energy solutions are the route to decarbonizing national and industrial economies while protecting the terrestrial ecosystems that support life on Earth. As a fundamental driver to achieve Sustainable Development Goal 15 (Life on Land), which focuses on protecting, restoring and promoting sustainable use of terrestrial ecosystems, the transition to green energy is a priority. With climate change and environmental degradation causing threats to elements such as biodiversity, soil health and forest cover, renewable energy creation from sources like solar, wind, hydro and biomass helps integrate energy into a sustainable future and reduce the environmental footprint of human activities. By using alternative energy sources and being more sustainable in our energy use, green energy encourages the use of natural resources in an efficient manner, helps reduce the need for deforestation, and allows for healthier ecosystems by providing alternatives to fossil fuel.

DOI: 10.4018/979-8-3373-0045-0.ch020

1. INTRODUCTION

The important aspect of brand advocacy here in hydrogen within the graph is aligning corporate values with sustainability objectives Businesses have to genuinely show their commitment to clean energy and sustainability through transparent capacity building and engagement with stakeholders. Brand messages can expand through influencer marketing and partnerships with sustainability advocates, gaining a wider audience and increasing public faith. Publishing research, speaking at conferences, and joining industry forums all enhance credibility, and position the companies as thought leaders in the hydrogen economy. Moreover, companies need to tell a narrative with various pathologies that demonstrate real-life benefits of hydrogen fuel in transport, manufacturing, and grid energy storage. By presenting proven case studies and pilot projects, including one in their own portfolios, brands can instill confidence with prospective consumers and investors. The junction of technology and digital marketing has given upsurge a new era of hyper-personalization in the ever-accelerating digital landscape. Digital marketing is being fundamentally altered by this paradigm change, becoming more dynamic, responsive, and customer-focused than ever. Predictive models of artificial intelligence (AI) and machine learning (ML) are at the core of this transition and are what will power the future of digital marketing. Hyper-personalization is not just a trendy term; it is a method to marketing that goes beyond conventional segmentation and overcomes the drawbacks of one-size-fits-all strategies (Vinaykarthik, 2022). It's a journey that starts with data, lots of data, and ends with a customized, in-the-moment customer experience. Marketers can now foresee client demands, provide individualized content, and predict preferences even before those preferences are consciously realized thanks to the complex dance between AI and ML. This chapter explores the nuances of hyper-personalization in digital marketing with an emphasis on how it affects marketing plans and brand advocacy. AI and ML-powered prediction models are ready to transform the customer experience. The trip is no longer a straight line but a dynamic, responsive experience that customers and businesses have jointly constructed (Kamal & Himel, 2023).

It unravels the dimensions of hyper-personalization, analyzing its profound implications on customer loyalty and brand advocacy. It scrutinize the strategies that forward-thinking marketers are adopting to harness the potential of AI and ML. Data analytics, real-time interactions, and the delicate balance between automation and the human touch will all be on our radar. This paper also addresses the ethical considerations that comes hand-in-hand with the power of hyper-personalization, such as data privacy and transparency. It will also peer into the future of digital marketing, beyond hyper-personalization, contemplating the dynamic customer journey and the role of AI-driven creativity and innovation (Ahmed, 2022). The stage is set for a digital marketing revolution where brands, equipped with predictive models and customer insights, are not just communicating with their audience but co-creating value. The future of digital marketing is dynamic, data-driven, and deeply personalized. It's a future where customers, now brand advocates, become the strongest proponents of the brands they passionately endorse.

1.1 Background of the Chapter

Hydrogen power marketing strategies should be in tandem with digital transformation, utilizing data analytics to gauge consumer perceptions and fine-tune campaigns accordingly. Content marketing, targeted advertising and social media platforms can improve engagement by addressing misconceptions about hydrogen production, efficiency, and challenges to infrastructure. The landscape of digital marketing has undergone a seismic upheaval, propelled by the inexorable march of technology. The days of one-size-fits-all marketing tactics that catered to various target segments are long gone. At every point of contact with a brand nowadays, people demand a personalized, interesting, and relevant experience. The growing amount of data available and the impressive capabilities of artificial intelligence (AI) and machine learning (ML) have hastened this transition toward hyper-personalization. These forces have come together to usher in a new era of digital marketing where hyper-personalized client journeys are created by predictive models of AI and ML (Vuong & Mai, 2023).

The revolutionary potential of hyper-personalization serves as the foundation for our investigation. Businesses can foster brand loyalty and advocacy like never before by comprehending and accommodating unique customer wants and preferences. In an era where client attention is a valuable resource and the capacity to pierce through the digital noise and establish a real connection is a competitive advantage, the stakes are high. This essay tries to explore the complex effects of hyper-personalization on digital marketing, with an emphasis on marketing tactics and brand advocacy (Auttri et al., 2023). It aims to shed light on the tactics influencing the future of digital marketing by exploring the workings of predictive models driven by AI and ML.

2. EVOLUTION OF DIGITAL MARKETING AND PERSONALIZATION IN CONTEXT OF GREEN HYDROGEN

And with gamification and interactive tools, consumers can learn about the science of hydrogen technology, allowing them to better understand what such a transition to a clean energy future would actually look like. The evolution of digital marketing has been intrinsically linked with the concept of personalization, which has grown from its infancy to become a central pillar of modern marketing strategies (Zaki, 2022). The brief overview of the evolution of digital marketing and personalization below as-

1. *Early Internet and Static Websites*: In the early days of the internet, digital marketing primarily consisted of static websites that provided basic information about a company or product. Only static content that did not alter based on the actions of specific users qualified as personalized.
2. *Email marketing and segmentation:* Email marketing gained popularity as the internet developed. Marketers started to divide their email lists into different groups depending on user information like demographics and purchase history. This was a rudimentary personalization, but it was a personalization nonetheless.
3. *Social media and behavioral targeting:* The emergence of social media platforms made it possible to more closely tailor advertising depending on the actions, interests, and social connections of the user. This data was used by marketers to create more pertinent content and advertising.
4. *Content marketing and the user-centered approach:* Content marketing gained popularity by emphasizing the creation of useful and educational content that catered to certain target audiences' needs and interests. This change to a user-centric strategy represented a huge advancement in personalization.
5. *Big Data and Advanced Analytics:* The growth of data collecting and sophisticated analytics technologies allowed marketers to understand consumer behavior more thoroughly. Campaigns for marketing were able to be extremely targeted and customized because to this data-driven methodology.
6. *Marketing automation and Customer Relationship Management (CRM):* For managing customer data and interactions, marketing automation platforms and CRM systems have become crucial. These platforms supported lead nurturing, customer path mapping, and tailored email marketing.

7. *Artificial intelligence and machine learning:* Large datasets are analyzed using predictive algorithms to predict customer preferences and behavior. Real-time, individualized interactions are offered by chatbots and virtual assistants. Personalization has evolved into a dynamic, real-time process that is adapted to each user at every touchpoint.

Digital marketing personalization is more sophisticated and data-driven than ever. The customer experience is highly personalized because to factors like predictive modeling, dynamic content creation, and real-time interactions, rather than only using first names when writing to customers. With AI and ML paving the way for even more customized client journeys in the future, digital marketing and customization are still evolving (Soliman & Al Balushi, 2023).

3. THE EMERGENCE OF PREDICTIVE MODELS IN AI AND ML FOR SUSTAINABILITY

Keep in mind however, these strategic partnerships with governments, research institutions, and environmental organizations combined, in turn will create a unified voice for the effort to adopt hydrogen, then of course strengthen marketing efforts as well. Joint labels in policy advocacy, incentives, and subsidies work at enhancing the makes and reduces hindrance to entering the market, and increases the financial feasibility of hydrogen power.An important turning point in data science and technology has been the appearance of predictive models in artificial intelligence (AI) and machine learning (ML). These models have completely changed the way we look at data, evaluate it, decide what to do next, and even predict the future. Here is a summary of how predictive models in AI and ML emerged and became important (Bhatnagar, 2022).

1. *Early Concepts in Machine Learning*: The inception of AI and ML saw the establishment of the fundamental principles for predictive models. The field began with elementary statistical models like linear regression, which sought to forecast results based on historical data. These initial models, although basic, played a pivotal role in comprehending the fundamentals of predictive modeling.
2. *Proliferation of Data*: The advent of the internet and digital technologies led to a surge in data creation. With the escalating availability of data, there arose an increasing demand for more advanced models to derive meaningful insights. Predictive modeling became indispensable for businesses and industries striving for a competitive advantage.
3. *Evolution of Sophisticated Algorithms*: The advancement of more complex algorithms, including decision trees, support vector machines, and neural networks, furnished the necessary tools for constructing predictive models. These algorithms could handle intricate data patterns and make precise predictions across a wide spectrum of applications.
4. *Big Data and Cloud Computing*: The ascent of big data and cloud computing supplied the infrastructure needed to process and scrutinize colossal datasets. Predictive models could now manage vast volumes of data, unlocking fresh possibilities for businesses concerning customer insights, operational efficiency, and more.
5. *Diverse Industry Applications*: Predictive models discovered utility in various sectors. In finance, they served for risk assessment and detecting fraud. In healthcare, predictive models aided in patient diagnosis and recommendations requires for the treatment. Retailers employed them for anticipat-

ing demand and managing inventory. The versatility of predictive models across different domains highlighted their potential for resolving intricate issues.

6. *Emergence of Reinforcement Learning and Deep Learning*: The rise of reinforcement learning and deep learning extended the frontiers of predictive modeling. These subdomains of ML allowed for the creation of models capable of making sequential decisions and handling unstructured data, such as images and natural language.
7. *Real-time Predictive Analytics*: As data processing speeds improved, real-time predictive analytics became achievable. Predictive models could furnish instantaneous suggestions and responses, rendering them invaluable in applications like recommendation systems, chatbots, and autonomous vehicles.
8. *Integration of AI and ML into Daily Life*: Today, predictive models have become an integral part of our everyday lives. The power virtual assistants, recommendation systems on streaming platforms, and personalized advertising. They are utilized to forecast weather, optimize transportation routes, and even predict disease outbreaks.

The emergence of predictive models in AI and ML has made data-driven decision-making accessible to a broad spectrum of industries and applications. As these models continue to develop, their influence on various sectors and their capacity to stimulate innovation and efficiency are set to expand.

4. HYPER-PERSONALIZATION AND CUSTOMER EXPERIENCE

Hyper-personalization has become a potent force that holds out the prospect of a time when interactions with businesses are personalized to specific preferences, wants, and behaviors. By utilizing cutting-edge technology like artificial intelligence and machine learning to generate highly tailored experiences, hyper-personalization elevates personalization to a new level (Henke & Jacques Bughin, 2016). Customers are no longer to be addressed by their first names; rather, it is now necessary to anticipate their needs before they express them, provide material and solutions in real-time, and create engaging experiences at each touchpoint. A dedication to treating each consumer as a part of one is at the core of hyper-personalization. Delivering what clients need, often before they even realize they need it—while establishing a sense of connection and comprehension are key components of this strategy. This strategy fosters brand loyalty and advocacy while also improving customer happiness. The future of customer experience is highly personalized, dynamic, and intensely focused on the demands of the customer, where businesses anticipate needs and clients feel truly understood and not just seen (Kaushal & Yadav, 2023).

4.1 Hyper-Personalization in Digital Marketing Fostering Sustainability

The world of digital marketing is changing as a result of hyper-personalization. This paradigm change emphasizes content and experiences at the person level rather than just using standard segmentation. Hyper-personalization, which is made possible by artificial intelligence and machine learning, is dependent on data-driven insights to foresee client demands and deliver customized messages, offers, and experiences in real-time. With this strategy, marketing transitions from a static, one-size-fits-all paradigm to a dynamic, customer-centric strategy where each interaction is carefully crafted to resonate with the unique needs of each person.

With the help of hyper-personalization, businesses may communicate with consumers directly and cater to their particular tastes and habits. It entails making product suggestions based on prior purchases, sending communications at the most effective times, and even changing content to take into account the customer's present situation. Hyper-personalization distinguishes brands at a time when consumers are swamped with digital content by providing not only relevance but a highly tailored and interesting experience (King, 2022). Hyper-personalization has become essential as the landscape of digital marketing continues to change. Customers are demanding this level of customization, and businesses who can meet their needs will see an increase in customer loyalty, better conversion rates, and stronger brand advocacy. While hyper-personalization has enormous promise, it also raises crucial questions regarding data privacy and the moral use of client information. As this marketing strategy continues to catch on, these questions must be carefully addressed (Babet, 2020).

4.2 AI and ML Predictive Models: Strategic Overview

A new era of data-driven decision-making and predictive modeling has arrived because of artificial intelligence (AI) and machine learning (ML). These cutting-edge technologies are revolutionizing sectors, increasing productivity, and influencing tactical choices. This change is being led by AI and ML prediction models, which give businesses priceless insights and the capacity to foresee outcomes with astounding accuracy. Predictive models are algorithms created that use historical data to forecast upcoming actions or events. They have found use in a wide range of industries, including marketing, manufacturing, healthcare, and finance. In-depth discussion of the importance and function of AI and ML predictive models in directing organizational strategy is provided in this strategic review. By analyzing huge information, these predictive models find patterns, correlations, and trends that could otherwise go undetected, enabling data-driven decision-making. Strategic planning is influenced by this information, which enables businesses to make wise decisions about how to allocate resources, manage risks, and communicate with customers (Sadowski, 2023).

The predictive models are essential for businesses looking to boost operational effectiveness. For instance, these models can estimate demand, improve inventory levels, and streamline logistics in supply chain management, which ultimately lowers costs and boosts customer satisfaction. Predictive models in healthcare help in patient diagnosis, enabling early intervention and raising the standard of treatment. Predictive models using AI and ML are also strategically important in marketing since they promote customer interaction and customisation. These models can recommend items, customize content, and foresee user preferences by studying customer data. Customer loyalty and brand advocacy have both been shown to increase significantly with this tailored strategy. The deployment of AI and ML predictive models does, however, present some difficulties (Sharma & Singh, 2022). The protection of personal information and ethical considerations are crucial. The data that businesses acquire and use must be handled with care. Predictive models also require a strong infrastructure, qualified personnel, and constant maintenance as they become more complicated.

It is impossible to exaggerate the strategic significance of AI and ML predictive models. These models equip businesses with the tools they need to make data-driven choices, increase productivity, and enhance customer experiences. Predictive models will become more and more important as technology develops in determining strategic priorities and promoting innovation. Organizations must adopt these models carefully, morally, and with a clear strategic vision if they are to realize their full potential (Mashood et al., 2023).

4.3 Customer Insights and Data Analytics

The customer insights and data analytics are now the cornerstones of strategic planning and well-informed decision-making in the digital age. A wealth of information can be found in the enormous data sets produced by client contacts, online activities, and transactions. However, it goes beyond only gathering data; it also involves drawing outtakeable insights that direct companies in their pursuit of excellence. Customer insights cover a wide range of knowledge about consumer preferences, actions, and problems. Businesses can use data analytics to find patterns and trends in this data, giving important details about what influences consumer satisfaction and loyalty. These insights serve as the cornerstone for personalized marketing strategies, enabling companies to cater their messages and product offerings to specific client requirements.

Data analytics goes beyond simple trend analysis thanks to machine learning and artificial intelligence. For instance, using predictive analytics, organizations can foresee future customer behavior and make preemptive decisions. This could entail making real-time product recommendations, anticipating customer attrition, or forecasting sales patterns (Gao & Liu, 2022). Since data analytics are real-time, organizations can react quickly to shifting consumer preferences and market dynamics. Businesses can improve client experiences, hone their product or service offerings, and boost overall operational effectiveness by incorporating customer insights into their strategy.

4.4 Crafting Dynamic and Personalized Content

In the world of digital marketing, personalized content is king and reigns supreme. The art of creating dynamic and tailored content transcends simple messaging. It involves giving people customized experiences that resonate with them and help them feel seen, heard, and valued. Each interaction is relevant and interesting thanks to dynamic content's real-time adaptation to user behavior and preferences. Advanced technologies, like as artificial intelligence and machine learning, which analyze massive databases to predict what customers want which often before they are fully aware of it as power this personalization. Dynamic content can range from tailored news feeds on social media to product recommendations on e-commerce platforms (Rajamannar, 2021).

In marketing, dynamic and tailored content is a strategic asset that increases client loyalty and engagement. Businesses may improve conversion rates, increase customer retention, and build brand advocacy by creating content that is specific to each client's needs. Additionally, it lessens the noise of unimportant information, ensuring that clients only get the information that actually matters to them. The personalization doesn't end with the content itself but it extends to the timing, format, and channel of delivery. In a world where customers are inundated with digital content, crafting dynamic and personalized content is the key to standing out and making a meaningful connection. It transforms marketing from a one-size-fits-all approach to a dynamic, customer-centric strategy, elevating the customer experience and driving brand success in the digital age.

4.5 Real-time Customer Interactions

In today's frantic and connected world, real-time customer contacts have emerged as a pillar of customer-centric enterprises. These interactions include more than just quick responses; they also involve instantly identifying and satisfying client demands, which results in a dynamic, highly engaging experience. Modern tools like chatbots, virtual assistants, and artificial intelligence enable real-time customer engagements. With the aid of these solutions, organizations can offer 24/7 customer service and information that is immediate and tailored. For instance, chatbots can instantly respond to inquiries, address problems, and even make product or service recommendations (Ancillai et al., 2019).

Real-time interactions are important for many different sectors. In e-commerce, this entails providing prompt answers to client inquiries, facilitating smooth transactions, and even making product recommendations in real-time. It allows for quick access to medical advice and appointment booking in the healthcare industry. Real-time interactions are essential in the financial sector for fraud detection and prompt client support. Because customers feel valued and understood when they receive prompt and accurate responses, real-time interactions promote trust and client loyalty. This increases both customer happiness and brand advocacy because happy customers are more inclined to tell others about a brand. In the digital age, these in-the-moment consumer contacts are at the heart of a customer-centric strategy.

4.6 The Role of Chatbots and Virtual Assistants

The digital transformation is leading with chatbots and virtual assistants, which are altering how companies engage with customers. These AI-powered products have a variety of functions, acting as both brand ambassadors and problem solvers. Chatbots in customer care provide round-the-clock assistance by responding to inquiries right away. They free up human agents to deal with more complicated problems by handling mundane activities like tracking orders or responding to frequently asked questions. This improves efficiency while also cutting down on response times, which increases customer happiness.

By accessing client information and interests, virtual assistants go beyond basic personalization and provide individualized recommendations for everything from trip itineraries to product recommendations. This degree of customisation promotes brand loyalty and enhances customer interaction. Chatbots and virtual assistants excel in data analytics in addition to providing excellent customer care and customization. They can collect and analyze data from client interactions, giving organizations useful insights (Analytics, 2016). These insights serve as a decision-making tool, assisting businesses in honing their tactics, streamlining their operations, and better comprehending the wants of their clients. Chatbots and virtual assistants play a crucial role in defining a whole client experience, going beyond simple job completion. As they continue to advance and become more sophisticated, their contribution to businesses will only increase, making them indispensable in the ever-evolving digital landscape.

5. IMPACT ON BRAND ADVOCACY: BRAND ADVOCACY AND ITS SIGNIFICANCE

The significance of brand advocacy lies in the power of genuine recommendations. In today's world, customers are bombarded with marketing messages from all directions. In this cacophony, a recommendation from a friend, family member, or colleague carries unparalleled weight. Such recommendations are not perceived as advertising but as trusted guidance. Brand advocacy represents one of the most coveted accomplishments for businesses in the digital age. It refers to customers becoming enthusiastic promoters of a brand, willingly and authentically endorsing it to their networks. The impact of brand advocacy is profound, as it translates into organic growth, heightened trust, and invaluable word-of-mouth marketing (George, 2023).

Hyper-personalization plays a pivotal role in nurturing brand advocacy. When brands offer experiences, products, and services tailored precisely to individual customer preferences, they elevate customer satisfaction. Satisfied customers, in turn, are more likely to become advocates, eagerly sharing their positive experiences with others. In this context, chatbots and virtual assistants, powered by artificial intelligence and machine learning, contribute significantly to the equation. They enable real-time, personalized interactions with customers, addressing their needs promptly and effectively. These technologies enhance the customer experience, making it memorable and delightful. Consequently, customers are more inclined to become vocal advocates, spreading the word about the brand's excellence (Wind & Hays, 2016).

Ultimately, the impact on brand advocacy is a testament to the brand's ability to connect with its audience on a personal level. As advocates share their experiences, they become an extension of the brand's marketing team, contributing to its growth and long-term success. This organic, authentic advocacy is the ultimate goal for businesses, and it's a reflection of their commitment to hyper-personalized customer experiences and customer-centric strategies (Kuusinen, 2019).

5.1 Customer Loyalty and Advocacy in the Digital Age

The customer loyalty and advocacy have taken on new dimensions during the digital age. With an ever-expanding array of choices at their fingertips, consumers are more discerning and demanding than ever. Building loyalty is no longer just about offering quality products or services; it's about crafting exceptional experiences and personal connections. Customer loyalty is now intricately linked with brand advocacy. The satisfied and engaged customers not only return but become enthusiastic advocates, sharing their positive experiences and influencing others (Wheeler, 2023). Brands that prioritize personalized interactions, real-time engagement, and exceptional customer service are the ones poised to thrive in this digital landscape, where customer loyalty and advocacy are the true currency of success.

5.2 Building Trust and Emotional Connections

The building trust and emotional connections with customers is fundamental to the success of businesses. Every successful client connection is built on trust, which is acquired via openness, dependability, and consistency. Businesses must conduct honorably and remain committed to keeping their word regarding anything from product quality to data privacy. On the other hand, emotional ties are fostered through unique experiences that connect with each consumer on a deeper level. Brands can identify client preferences and wants with cutting-edge technology like artificial intelligence and machine learning, then customize their interactions and products to produce sincere, emotionally charged experiences. Through these ties, clients become brand evangelists who not only trust the company but also fervently recommend it to others (Sedkaoui, 2018).

6. MARKETING STRATEGIES IN AGE OF HYPER-PERSONALIZATION: DATA-DRIVEN DECISION MAKING

In the age of hyper-personalization, data-driven decision making has emerged as the keystone of effective marketing campaigns. Businesses now have unparalleled access to massive data reservoirs thanks to cutting-edge technologies like artificial intelligence and machine learning, which provide insights into customer behavior, preferences, and demands. They may develop highly focused and tailored marketing strategies that appeal to certain clients by utilizing this data. In order to make data-driven decisions, it is important to extract relevant insights from the data that inform strategic decisions. To predict trends and customer preferences, marketers might examine consumer interactions, purchase histories, and online habits (Neuwirth, 2023). AI and ML-powered predictive models can predict future behaviors and proactively inform strategies.

With the use of these information, firms are able to optimize every aspect of their marketing plans, from timing and channel selection to content and product recommendations. They may act quickly using real-time data analysis to ensure that clients receive pertinent and interesting information. Data-driven decision making, however, entails a need to protect consumer privacy and follow ethical data usage principles. It involves finding the right balance between individualized attention and observance of private boundaries. In the era of hyper-personalization, many marketing tactics rely heavily on data-driven judgment. Understanding the value of data, applying AI and ML to gain useful insights, and customizing approaches to produce engaging, tailored experiences are all important. Businesses that adopt data-driven decision making are better positioned to satisfy client expectations in this dynamic environment.

6.1 Ethical Considerations and Transparency: Addressing Privacy Concerns and Regulatory Compliance

Transparency and ethical issues now dominate marketing efforts. Personalization improves the user experience but it brings up serious privacy and regulatory problems. For businesses, addressing these issues is not only required by law, but also morally right. The first pillar of ethical concerns is transparency, which is being open and truthful with clients about how their data is gathered, utilized, and safeguarded. It is crucial to educate clients on data practices, secure their informed consent, and provide them with unambiguous opt-in and opt-out choices. Privacy concerns loom large in this context. Customer data is a valuable asset, but its collection and usage must respect privacy regulations and customer expectations. This includes compliance with data protection laws like GDPR or CCPA, as well as ensuring robust security measures to safeguard sensitive information. These ethical considerations extend to data sharing and third-party partnerships. Brands should be transparent about any data sharing and provide customers with control over their data even when it's shared with affiliates or partners (Schaeffer, 2017).

Navigating the ethical landscape of hyper-personalization also requires a commitment to responsible data usage. It's about using data not just to increase profits but to enhance the customer experience. Data-driven insights should be used for the benefit of customers, whether that's in the form of personalized recommendations or improved services. The ethical considerations and transparency are non-negotiable in the age of hyper-personalization. Businesses that prioritize privacy, data security, and responsible data usage not only comply with regulations but also foster trust and long-term customer relationships. In this digital era, ethical marketing isn't just a legal requirement; it's a pathway to ethical and sustainable business practices.

6.2 Balancing Automation and the Human Touch

So striking the right balance between automation and the human touch is the defining challenge for businesses. Automation, powered by artificial intelligence and machine learning, offers unmatched efficiency and personalization. However, it must coexist harmoniously with the warmth and empathy that only humans can provide. Automation brings speed and precision to customer interactions. Chatbots and virtual assistants can address routine inquiries instantly, and predictive models can recommend products or services based on individual preferences. This is a crucial aspect of delivering personalized experiences, as it ensures real-time responses and convenience. Yet, the human touch remains indispensable. In more complex or emotionally charged situations, customers often seek human interactions. The human touch encompasses empathy, understanding, and the ability to handle nuanced situations. It's about resolving issues, providing reassurance, and making customers feel heard and valued.

As, balancing these two elements requires a strategic approach. Automation should be employed where it enhances efficiency and personalization, leaving the human touch to shine in areas that demand emotional intelligence and personalized problem-solving. It's about knowing when to automate and when to bring in the human touch. Moreover, transparency is crucial. Customers should be aware when they are interacting with automation versus a human agent. This transparency fosters trust and allows customers to manage their expectations. The finding of equilibrium between automation and the human touch is an art in the era of hyper-personalization. When executed effectively, it ensures efficiency, personalization, and empathy, creating an unparalleled customer experience that builds loyalty and advocacy (Venkatesh, 2021).

6.3. Future of Digital Marketing: Beyond Hyper-Personalization- The Dynamic Customer Journey

As it gazes into the future of digital marketing, we envision a landscape that goes beyond hyper-personalization and delves into the concept of the dynamic customer journey. While hyper-personalization has been a game-changer, the dynamic customer journey takes personalization to the next level, embracing the fluidity and unpredictability of individual customer pathways. In this future, marketing strategies will be adaptive and responsive. Artificial intelligence and machine learning will continuously analyze real-time customer data to anticipate shifts in preferences, behaviors, and context. Marketers will not merely respond to customer actions; they will proactively shape the customer journey based on predictive insights.

The dynamic customer journey acknowledges that customers' needs and desires are not static. They evolve, often in response to external factors, life events, or changing interests. In this context, digital marketing strategies will provide content and experiences that reflect these changes, offering the right message at the right time. This future envisions a seamless integration of marketing touchpoints, from social media and websites to email and in-app interactions. These touchpoints will not exist in isolation but will form part of a cohesive, fluid customer journey. Every interaction will contribute to a personalized narrative that guides the customer towards a deeper relationship with the brand.

While the dynamic customer journey holds incredible potential, it also raises ethical considerations. Businesses must navigate the fine line between personalization and intrusion, respecting privacy and obtaining informed consent for data usage. the future of digital marketing is not just about hyper-personalization but embracing the dynamic nature of customer journeys. It's about that customer needs evolve, and effective marketing strategies must evolve with them. The dynamic customer journey promises not just personalization but proactive, fluid, and relevant customer experiences that pave the way for long-term loyalty and brand advocacy (Reviglio della Venaria, 2020).

6.4 The Role of AI-Driven Creativity and Innovation

Artificial intelligence (AI) has emerged as a powerful catalyst for creativity and innovation, transforming the way we conceive, design, and implement ideas. Its role in these realms extends beyond automation; AI-driven creativity and innovation empower individuals and organizations to envision and realize groundbreaking solutions. AI's capacity to process vast datasets and recognize complex patterns equips it to unearth novel insights. It can analyze customer behaviors, market trends, and even creative works, offering inspiration and fresh perspectives. This data-driven creativity often results in more targeted and effective innovations.

AI can generate creative content itself, such as art, music, or even written text. These AI-generated works challenge traditional boundaries and introduce new dimensions to human creativity. They serve as tools for artists, writers, and musicians, sparking innovation by merging human ingenuity with AI's computational power. AI-driven innovation goes beyond creativity, facilitating the development of disruptive technologies. In fields like healthcare, AI assists in drug discovery and disease diagnosis. In autonomous vehicles, it's revolutionizing transportation. In finance, it's optimizing investment strategies. These innovations improve efficiency, solve complex problems, and create entirely new opportunities.

The responsible use of AI in creativity and innovation involves transparency, accountability, and consideration for potential biases so ethical considerations are paramount. Striking a balance between AI-driven and human-driven creativity is an ongoing challenges. AI-driven creativity and innovation are reshaping industries and pushing the boundaries of what's possible. They offer unprecedented opportunities to generate fresh ideas, optimize processes, and solve complex problems. While ethical considerations remain, the collaborative potential of AI and human creativity promises an exciting future of innovation and exploration (Maes, 2018).

7. CONCLUSION AND VISION FOR THE FUTURE OF DIGITAL MARKETING FOSTERING GREEN ECONOMY

Corporate social responsibility (CSR) is also a big part of brand advocacy in the hydrogen economy. Integrate sustainability as part of their business models, be more compelling in showing how the businesses are reducing carbon footprints and contributing to climate action goals globally. Carbon offset programs, green hydrogen investments, and net-zero emissions pledges are currently effective marketing ploys designed for environmentally-conscious consumers and investors alike. This is further supported by certification programs, such as green energy labels and sustainability accreditations. In the bigger picture, hydrogen power productivity in the edges of the green economy will be tested by good routine and marketing. Empowering hydrogen adoption: While educating consumers on hydrogen's benefits, harnessing the power of technology to build trust through transparency and effective marketing strategies, the most credible and consistent approach to hydrogen adoption will ultimately require commitment from all stakeholders.

Hyper-personalization and real-time customer experiences are the driving forces behind the future of digital marketing. With advanced technologies, businesses can engage customers on a deeply personal level, delivering tailored content and interactions in the blink of an eye. This flexible method anticipates needs, encourages trust, and creates enduring connections. The key to successful marketing in the current digital era is the marriage of hyper-personalization and real-time dynamics, which provides customers with more than just content but also meaningful experiences. The dynamic environment of digital marketing in the future will be characterized by hyper-personalization, moral considerations, and the seamless fusion of automation and human interaction. It's a field where making decisions based on data, being transparent, and using data sensibly are essential. It involves developing experiences that connect with specific customers and encourage loyalty and trust. Our vision for the future goes beyond hyper-personalization.

The vision is based on ethics and accountability, nevertheless. We place a high priority on consumer permission, data security, and privacy in our quest of personalization. We make a commitment to developing emotional bonds and trust because we understand that these are the foundations of brand advocacy. The future of digital marketing involves people as much as technology. It's about how cutting-edge technology and the human touch can come together to create experiences that are not just unique but also highly engaging and meaningful. In the future, marketing plans will be based on people rather than just data, acknowledging that each consumer is an individual with their own requirements and wants. Businesses will succeed in this era not merely by meeting but by exceeding customer expectations. By anticipating their clients' requirements in addition to just meeting them, they will build long-lasting relationships with them. It's a future where brands collaborate with consumers along their trip, act as their guides through a changing environment, and travel alongside them as the digital world develops. Digital marketing has a bright future ahead of it, and we have a vision that emphasizes innovation, accountability, and client centricity. In this future, trust, loyalty, and advocacy are not just results but also the pillars of success where businesses and consumers collaborate to create meaningful experiences.

REFERENCES

Ahmed, A. (2022). Marketing 4.0: The Unseen Potential of AI in Consumer Relations. *International Journal of New Media Studies: International Peer Reviewed Scholarly Indexed Journal, 9*(1), 5–12.

Analytics, M. (2016). *The age of analytics: competing in a data-driven world.* McKinsey Global Institute Research.

Ancillai, C., Terho, H., Cardinali, S., & Pascucci, F. (2019). Advancing social media driven sales research: Establishing conceptual foundations for B-to-B social selling. *Industrial Marketing Management, 82*, 293–308. DOI: 10.1016/j.indmarman.2019.01.002

Auttri, B., Chaitanya, K., Daida, S., & Jain, S. K. (2023). Digital Transformation in Customer Relationship Management: Enhancing Engagement and Loyalty. [EEL]. *European Economic Letters, 13*(3), 1140–1149.

Babet, A. (2020). Utilization of personalization in marketing automation and email marketing.

Bhatnagar, S. (2022). Digital Disruptions and Transformation of Bank Marketing.

Gao, Y., & Liu, H. (2022). Artificial intelligence-enabled personalization in interactive marketing: a customer journey perspective. *Journal of Research in Interactive Marketing*, (ahead-of-print), 1-18.

George, A. S. (2023). Future Economic Implications of Artificial Intelligence. *Partners Universal International Research Journal, 2*(3), 20–39.

Henke, N., & Jacques Bughin, L. (2016). The age of analytics: Competing in a data-driven world.

Kamal, M., & Himel, A. S. (2023). Redefining Modern Marketing: An Analysis of AI and NLP's Influence on Consumer Engagement, Strategy, and Beyond. *Eigenpub Review of Science and Technology, 7*(1), 203–223.

Kaushal, V., & Yadav, R. (2023). Learning successful implementation of Chatbots in businesses from B2B customer experience perspective. *Concurrency and Computation, 35*(1), e7450. DOI: 10.1002/cpe.7450

King, K. (2022). *AI Strategy for Sales and Marketing: Connecting Marketing, Sales and Customer Experience.* Kogan Page Publishers.

Kuusinen, M. (2019). Scenarios for digital marketing: a Delphi-based analysis for 2028.

Maes, P. (2018). *Disruptive Selling: A New Strategic Approach to Sales, Marketing and Customer Service.* Kogan Page Publishers.

Mashood, K., Kayani, H. U. R., Malik, A. A., & Tahir, A. (2023). ARTIFICIAL INTELLIGENCE RECENT TRENDS AND APPLICATIONS IN INDUSTRIES. *Pakistan Journal of Science, 75*(02). Advance online publication. DOI: 10.57041/pjs.v75i02.855

Neuwirth, R. J. (2023). Prohibited artificial intelligence practices in the proposed EU artificial intelligence act (AIA). *Computer Law & Security Review, 48*, 105798. DOI: 10.1016/j.clsr.2023.105798

Rajamannar, R. (2021). *Quantum marketing: mastering the new marketing mindset for tomorrow's consumers.* HarperCollins Leadership.

Reviglio della Venaria, U. (2020). Personalization in Social Media: Challenges and Opportunities for Democratic Societies.

Sadowski, J. (2023). Total life insurance: Logics of anticipatory control and actuarial governance in insurance technology. *Social Studies of Science*, •••, 03063127231186437. PMID: 37427796

Schaeffer, E. (2017). *Industry X. 0: Realizing digital value in industrial sectors*. Kogan Page Publishers.

Sedkaoui, S. (Ed.). (2018). *Big data analytics for entrepreneurial success*. IGI Global.

Sharma, A., & Singh, B. (2022). Measuring Impact of E-commerce on Small Scale Business: A Systematic Review. *Journal of Corporate Governance and International Business Law*, *5*(1).

Soliman, M., & Al Balushi, M. K. (2023). Unveiling destination evangelism through generative AI tools. *ROBONOMICS: The Journal of the Automated Economy*, *4*(54), 1.

Venkatesh, D. N. (2021). *Winning with employees: Leveraging employee experience for a competitive edge*. SAGE Publishing India.

Vinaykarthik, B. C. (2022, October). Design of Artificial Intelligence (AI) based User Experience Websites for E-commerce Application and Future of Digital Marketing. In *2022 3rd International Conference on Smart Electronics and Communication (ICOSEC)* (pp. 1023-1029). IEEE.

Vuong, N. A., & Mai, T. T. (2023). Unveiling the Synergy: Exploring the Intersection of AI and NLP in Redefining Modern Marketing for Enhanced Consumer Engagement and Strategy Optimization. *Quarterly Journal of Emerging Technologies and Innovations*, *8*(3), 103–118.

Wheeler, J. (2023). *The Digital-First Customer Experience: Seven Design Strategies from the World's Leading Brands*. Kogan Page Publishers.

Wind, Y. J., & Hays, C. F. (2016). *Beyond advertising: Creating value through all customer touchpoints*. John Wiley & Sons.

Zaki, H. O. (2022). The Impact Of Artificial Intelligence On Content Marketing. *Journal of Strategic Digital Transformation In Society*, *2*(3).

Chapter 21
Prospect of Artificial Intelligence Towards Demand Response, Control, and Predictions in Integrated Microgrid Renewable Energy Systems

Sindhu Sivankutty Nair
https://orcid.org/0009-0007-4703-3626
South Batina, Oman

ABSTRACT

Microgrids play a vital role in sustainable energy systems, and this is accomplished by integrating renewable energy sources like solar, wind, biogas, and tide energy. Optimization resources are done through real-time monitoring with advanced data analytics which would adjust stably to changing environmental contexts. Renewable energy sources are more successful when it comes to competing with traditional sources of energy by maximizing the potential energy yield. Regarding demand response, this paper explores the application of AI methods, looking at both price- and incentive-based programs. Regarding their use and effectiveness, this article considers a variety of control targets, input sources, and applications. To sum up, the purpose of this review analysis is to provide valuable insights into the selection of AI approaches, with an emphasis on demand-side applications for future energy systems, along with control and prediction techniques using AI. This article offers direction and assistance for the creation of sustainable integrated energy systems.

1. INTRODUCTION

Worldwide energy demand has been growing mostly due to rising urbanization, industrialization, population increases, technological advancements, energy access programs, and sector-specific needs. Also, the depletion of traditional energy resources and environmental concerns have led to a shift towards renewable energy deployment. Though greenhouse gas emissions reduction, energy diversification, sustainability, cost competitiveness, and so on are Additional benefits associated with green energy

DOI: 10.4018/979-8-3373-0045-0.ch021

deployment, there are drawbacks, nonetheless. The primary downsides of deploying renewable energy include intermittency, energy storage, land use and resource constraints, high initial costs, transmission and grid integration, resource variability, technological limitations, and environmental impact (Figaj, 2024, Rani et al., 2023). To supply power to isolated regions remote from the main electric grid, researchers attempted to combine renewable sources in isolated microgrids. Alternatively, they attempted to link them with the grid to boost stability and dependability. The changeover to sustainable, low-carbon energy is made possible by the integration of non-renewable energy systems with renewable energy systems, which builds a stable and balanced energy infrastructure (Hirsch et al., 2018). Electricity generation from 1985 to 2023 by various countries and consolidated world generation data is presented in figure 1.

Figure 1. Global electricity generation from 1985 - 2023 (Energy Institute - Statistical Review of World Energy (2023). (n.d.). Data Source: Ember (2024); Energy Institute - Statistical Review of World Energy (2023))

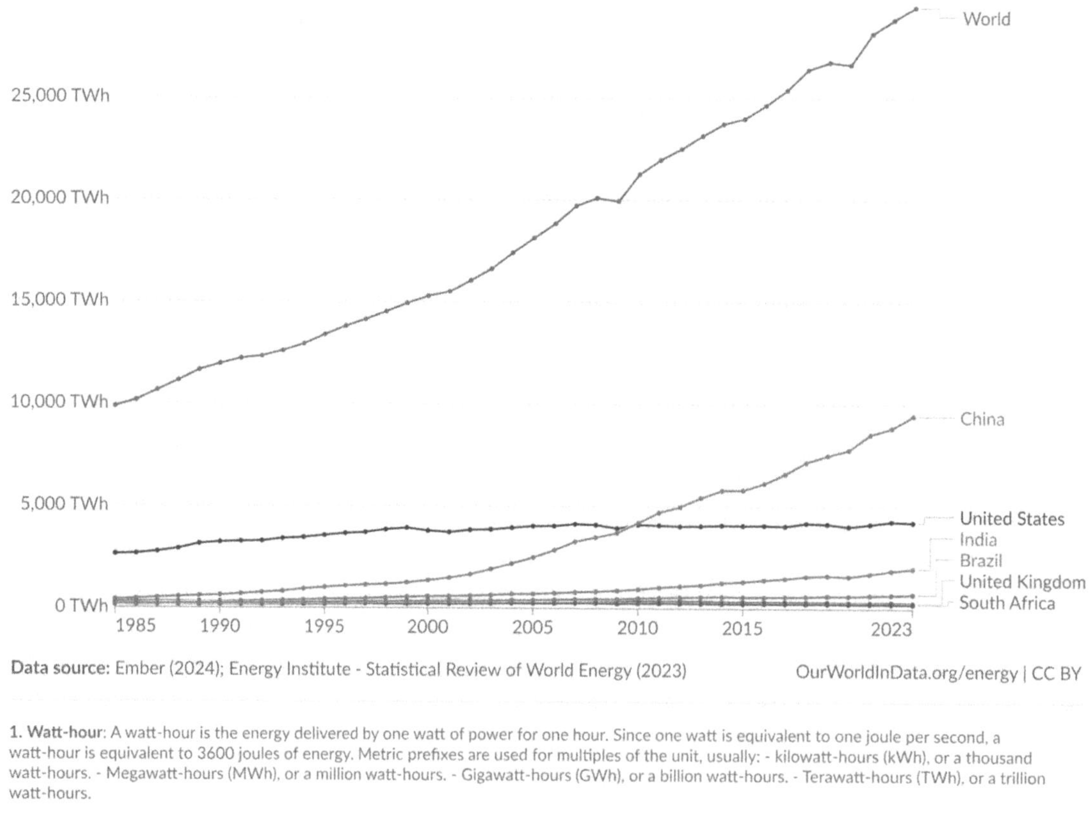

Non-renewable energy sources can provide backup power in the event of inadequate renewable energy generation or system failures, ensuring a consistent supply of electricity. Also, non-renewable energy sources play an important role in grid stability by providing steady power generation, especially when renewable sources like sun and wind are variable. Energy security is increased by reducing reliance on a single energy source through the diversification of the energy mix with non-renewable and renewable sources, and moreover, cost management is aided by integrating renewable energy sources with non-renewable energy systems, especially when demand is at its highest or renewable energy production is low (Hirsch et al., 2018). Additionally, some non-renewable energy systems, such as combined heat and power, can complement renewable energy systems by using waste heat to increase efficiency (Xiao et al., 2024). Integrating renewable energy systems with non-renewable energy sources may provide certain challenges, including the potential for ecosystem disruption, dependency on limited fossil fuel resources, and environmental repercussions from emissions. Furthermore, concerns may be addressed regarding the long-term viability of a hybrid energy system, as well as the necessity for continued investments in both renewable and non-renewable energy conversion systems.

An integrated microgrid combines various energy sources that are dispersed, and batteries with control systems to form a cohesive entity. The idea of an integrated microgrid, in short, is meant to maximize the production, distribution, distribution, and storage of energy within a given space results in increasing the resiliency, efficiency, and reliability of the energy supply. Solar panels, wind turbines, and energy storage devices are commonly integrated with nonrenewable power sources such as natural gas turbines or diesel generators in integrated microgrids. Many of these energy sources are incorporated into the microgrid to provide a balance between demand and supply, to manage the variations in renewable energy output, and to have backup power during grid breakdowns (Hirsch et al., 2018, Xiao et al., 2024). The integrated microgrid comprises of advanced grid technologies with sophisticated control systems that aptly coordinate the different energy sources and distribute energy depending on demand and supply. Integrated microgrids allow facilities and communities utilizing non-renewable and renewable combined energy sources to be more autonomous in power and enable them to save money, hence achieving environmental sustainability. Distributed generation can be locally controlled, and centralized dispatch can be minimized or eliminated if integrated microgrids are well operated (Tabassum et al., 2023). In addition, in the integrated microgrids, demand response could effectively optimize energy usage and improve grid stability, benefiting the incorporation of renewable energy sources as well. Demand response (DR) helps keep a balance between supply and demand, as well as reduces system peak load and thus increases system efficiency, by allowing consumers to change their electricity consumption following pricing signals or grid conditions (Zhang et al., 2024). Demand response programs (DRP) have greatly improved the efficiency of microgrids and nuclear plants by allowing them to be more responsive. DRP enables consumers to adjust their electricity consumption based on pricing signals or system needs, thereby contributing to grid reliability, peak demand reduction, and better integration of renewables. More recently, efforts have been made to envision DR optimization methodologies which can help in better economic viability and operational planning of microgrids. Microgrids could make better use of DR incentives and energy storage systems (ESS) for managing the variable power sources of green power and for balancing supply and demand (Mehrabani et al., 2023). Additionally, the progress on the DRP has promoted new models and algorithms to schedule energy storage, control flexible/inflexible loads, and manage the overall microgrid operation. These advancements have improved existing or new microgrid reliability and performance while increasing both economy and flexibility and/or decreasing operation expenses (Mehrabani et al., 2023, Chen et al., 2020). Integrating advanced DR capabilities in microgrid planning and operations is inevitable to cope up with renewables intermittency issues, to optimize system operation and utilization, and to provide cost-effective and secure energy supply.

A well-planned microgrid control system is necessary to control decentralized energy generation, distribution, and consumption within a microgrid. Researchers investigated various control strategies to provide efficient operation, stability, and optimal utilization of the resources (Dinata et al., 2024). The integration of such control methodologies and advanced technologies can lead to the development of robust and efficient microgrid control systems that can respond to varying conditions, promote energy efficiency, and facilitate the deployment of renewable energy resources into the grid (Dinata et al., 2024, Khan & Wang, 2017). The prediction of renewable energy has a vital function in microgrid systems in terms of increasing utilization, boosting grid quality, and energy management, and reducing the developmental process (Luo & Oyedele, 2022). Predictive analytics and forecasting approaches bring a lot of significant advances which microgrid operators gain from the optimization of energy management, improved integration of renewable energy sources, enhanced microgrid reliability and cost-effectiveness (Jaramillo et al., 2024). Integrated microgrid systems can further feature advanced functionalities like timeserving DR, individual or combined market participation, grid integration, optimization, control and predictive maintenance. By leveraging technology, especially computer systems, artificial intelligence (AI) allows operators to optimize a microgrid's operation and decision-making processes without compromising the sustainability and dependability of distributed energy resources (DERs). A standard integrated smart-grid structure is depicted in figure 2, which includes real-time control and management, smart energy consumption, stationary and mobile energy storage, advanced grade analytics, and sustainable energy supply.

Figure 2. AI enabled integrated smart grid structure

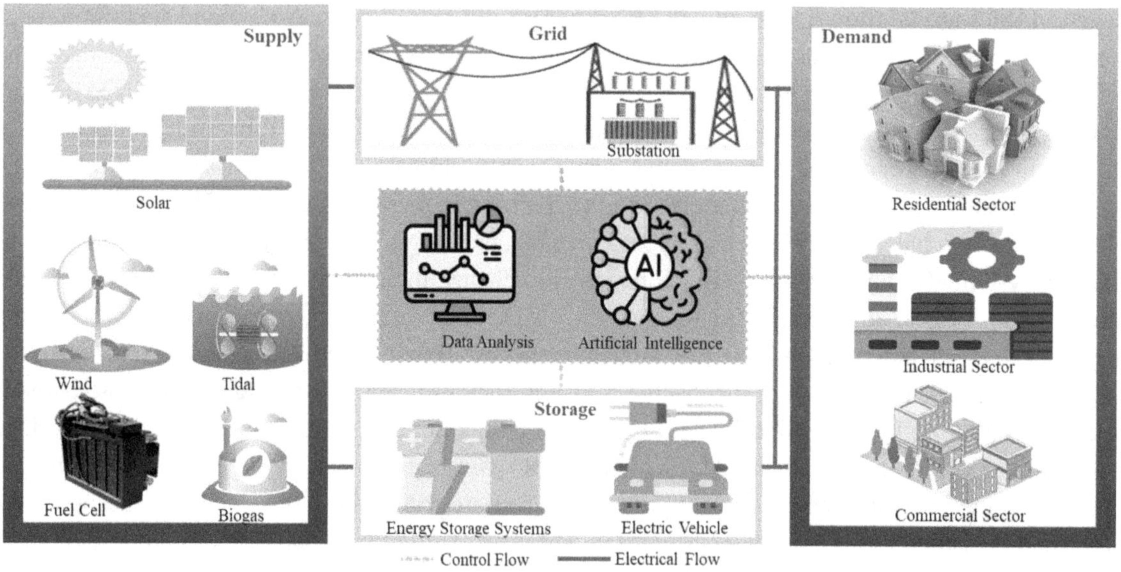

This research paper is focused on the use of artificial intelligence technology in the field of DR, within price and incentive – based programs. The demand balancing on the power grid is done in such a way that energy consumers alter their power consumption in reaction to signals, among which would be changes in price or incentives. Based on that, the researchers considered a host of control targets, input sources, and applications to evaluate the applicability and effectiveness of AI in DR. For example, control targets can be a reduction of energy consumption limitation of peak demand, or even improvement of grid stability. Input sources are an immense amount of data sources including, but not limited to, historical energy consumption patterns, weather forecasts, and current grid conditions. Applications are numerous, from demand pattern prediction and better scheduling of devices consuming energy to rewarding the users who change their energy usage behavior. This chapter provides a comprehensive review to give information on how AI techniques shall be designed and chosen for demand-side applications in future integrated energy systems. The primary focus here is the controlling and forecasting of the set of demand response tasks through artificial intelligence methods. By investigating diverse methodologies, applications, and outcomes, this research endeavor seeks to contribute to the growth of AI-driven solutions for enhancing energy efficiency, grid reliability, and overall sustainability.

This chapter is divided into nine sections starting with a brief introduction in section 1. Section 2 will introduce Integrated Renewable Energy Systems and corresponding literature review. Section 3 introduces system requirements in Renewable Energy Integration. Section 4 and 5 focus on details about various aspects of demand response in integrated microgrid systems, and the application of artificial intelligence in demand response. Section 6 and 7 describe various artificial intelligence approaches towards prediction in integrated renewable energy systems including forecasting, and anomaly detection. Different microgrid control strategies and energy management techniques are summarized in section 8, followed by the conclusion in section 9.

2. INTEGRATED RENEWABLE ENERGY SYSTEMS

One of the most prevalent issues affecting humans today is global warming. Green energy is therefore the most effective way to address this crucial issue. The most common rapidly expanding energy sources now are wind, wave, solar, and biomass, particularly solar energy is growing at the highest rate (Talaat et al., 2023, Nair et al., 2022, Thakur et al., 2022, Nair et al., 2023, Nair et al., 2024). Lately, wave energy has begun to be developed, making it one of the extremely prognosticating sources. Recent research has focused mostly on the idea of combining multiple sources such that, if one is unable to supply energy for the entire day, the other can make up for it (Talaat et al., 2021). Based on the energy curve of the wind turbines (WT), for instance, wind energy can only be produced when the wind speed is adequate for doing so. In other instances, alternative energy sources, like solar energy, are used. When the wind speed is insufficient to provide energy at night and there is no irradiation, wave energy or another source must be used. Most of these systems are off-grid or on-grid (Katche et al., 2024, Efkarpidis et al., 2023). The integration of several renewable energy systems and the goals, results and software applied in each study are included in Table 1.

Table 1. Goals and key results of different integrated renewable energy systems for distinct applications

Year	Renewable energy resources	Technology	Goals and Key Results	Ref.
2018	Solar, Wind, Fuel Cell, Battery	HOMER Modelling Software	Techno-economic Analysis: Microgrid vs Stand-alone. ▪ Stand-alone microgrids have higher electricity costs. ▪ Battery storage is more cost-effective than hydrogen storage.	(Duman & Güler, 2018)
2019	Solar, Fuel Cell, Diesel generator	Dispatch Control Mechanisms, Optimization Techniques, and Simulation	Cruise ships utilizing renewable energy integration. ▪ Provides strong penetration. ▪ Profitable from an economic standpoint. ▪ Clean energy. ▪ Lowers reliance on fossil fuels.	(Ghenai et al., 2019)
2019	Solar, Fuel Cell, Battery, Diesel generator	Optimization Techniques and Control Mechanisms	Development of a Renewable Microgrid Design ▪ A large proportion of renewable energy is used. ▪ Reduced electricity costs. Minimal emissions. ▪ Green and economically viable.	(Ghenai & Bettayeb, 2019)
2020	Solar, Battery	Successive Random based Monte Carlo simulation	Residential Off-Grid PV System Sizing ▪ Accurate size relies on the failure rate. ▪ Loss of energy expectation indices rise with failure rate.	(Quiles et al., 2020)
2020	Wind, Fuel Cell, Battery	dSpace 1104 Real Time Board	Hybrid System Energy Management and Control ▪ The torque Controller Loops for Efficient Power Point Tracking. ▪ The experimental findings demonstrate efficacy.	(Kadri et al., 2019)
2021	Solar, Battery	HOMER Modelling Software	Technoeconomic Feasibility Analysis of Isolated renewable energy systems ▪ The proposed technique is ideal for applying urban and rural applications. ▪ Suitable for both environments.	(Abid et al., 2021)
2021	Solar, Wind	MATLAB Simulink Software	Intelligent control and power management for a hybrid system ▪ Proposed system for remote, isolated locations. ▪ Meets only a few residential load demands.	(Ponnuru et al., 2020)

2.1. Categorization of Integrated Renewable Energy Systems

The kind of site and the local climate are taken into consideration while selecting different configurations of integrated renewable energy systems. If any one of the sources is having difficulty producing energy, it is necessary to choose the sources for the hybrid system in such a way that each can make up for the production deficit of the other. The integration of renewable energy systems can help address the difficulties of unpredictability and the difficulty of accurately anticipating normal production of energy from renewable random resources like solar, tide, and wind (Tostado-Véliz et al., 2024). In general, the renewable energy storage system supports the integrated sources in microgrids, providing additional flexibility and reliability while maintaining protective and effective functioning of the microgrid. Innovative combinations of renewable energy systems with a variety of features have been made possible by the introduction of novel techniques for integrating renewable energy systems (Mena et al., 2024, Wang et al., 2024). By 2030, the number of energy storage system installations might rise from 42% to 68%, with India and China having the highest percentage, as reported by the International Renewable Energy Agency (C. J. Chen, 2024). Solar and wind energy have emerged as some of the most popular and frequently utilized sources in the past few decades, and their integration is essential for satisfying load demands (Nair et al., 2024b). However, there are still issues in eliminating fluctuations in supplied power, especially for systems that have no connection to the grid. Furthermore, environmental factors that regulate the production process may result in poor electrical quality and a reduction in output (C. J. Chen, 2024).

2.2. Integrated System Using Two Sources

The optimum system depends on several criteria, particularly the local environment, climate, and distance from the primary grid. The combination of each source and the integration approaches have been the subject of numerous literature reviews, and Researchers have proposed numerous problems and solutions to these concerns. Integrated PV and wind energy systems are among the most extensively used and popular integrated renewable energy systems, and grid integrated systems, and isolated systems are the two categories into which they adhere (Cabrera et al., 2021). Optimization solutions might help to overcome integration challenges and improve the efficiency of the integrated system. An analysis was carried out to figure out the size of a grid-integrated system that combines solar energy and wind energy, while connected to the grid or isolated (Falope et al., 2024).

Integrating energy storage devices into solar power systems improves system resilience, and dependability and addresses the intermittent issue with photovoltaic systems (Memarzadeh & Keynia, 2023). The economic viability of integrating solar sources with energy storage systems, which is a key indication for attracting investment in Thailand and other countries, was examined (Chaianong et al., 2019, Gamonwet & Dhakal, 2023). Solar cells generate energy during the day and store energy to make use at night to fulfill demand when production is limited due to clouds. Hybrid energy storage systems are a viable solution for rural power because they are utilized to supply electricity to remote locations that are not linked to the grid.

Integrating energy storage devices into isolated or on-grid wind energy systems enhances stability, flexibility, robustness, and reliability. These energy storage systems do not have the capacity to handle all loads, and hence reserves the surplus for usage during low wind speeds. Research has demonstrated that integrating wind energy with an energy storage system in isolated microgrids can increase battery efficiency and longevity (Du et al., 2021). A size assessment of a hybrid system integrating a wind turbine with an energy storage system with an estimated capacity of around 128 MW has been proposed to create a power-balance across maintenance by lending the energy storage system to assure service continuity (Y. Luo et al., 2014). In the Chinese province of Hebei, a multi-objective optimization study was carried out to address oscillations in wind energy output. Wind energy was integrated with an energy storage system in an isolated microgrid, where researchers studied the usage of hybrid energy storage systems that included batteries, supercapacitors, and a wind turbine. When sizing the hybrid energy storage systems, the researchers used a multi-objective optimization technique; the best hybrid systems outperformed a single energy storage system (Xu et al., 2018, Pan et al., 2021). Numerous research projects have been undertaken to enhance the energy storage system's technical specifications to extend life and enhance battery efficiency while minimizing the impact on the integrated system.

2.3. Integrated System Using Three or More Sources

The system's flexibility and dependability are increased by integrating solar and wind energy sources and adding an energy storage system. The storage system stores extra wind and solar cell production for use during peak periods or shortages. A new study employed Particle Swarm Optimization with the Genetic Algorithm (GA-PSO) to scale down general expenses while improving system reliability (Ghorbani et al., 2017). Several publications explore the optimal scale of this integrated system, besides resources management.

3. SYSTEM REQUIREMENTS IN RENEWABLE ENERGY INTEGRATION

This section examines recent advances in incorporating renewable energy sources into the electrical grid to improve system resilience. The emphasis is on reactive current injection or absorption and voltage ride-through (VRT), which support efforts in the event of disruptions and unusual circumstances. Since a failure can be handled by directly disconnecting from the grid, VRT is essential due to the widespread usage of renewable energy sources. However, disconnecting renewable energy sources may have a negative influence on network stability (Al-Shetwi et al., 2019).

Previously, researchers concentrated on Low-VRT (LVRT) control techniques for wind and solar energy systems. Reactive power is what these solutions try to add to the grid. LVRT use is mandated under the grid code standards. The requirements depend on how long the fault lasts and how much reactive power is delivered. The grid's ability to reinstate voltage by furnishing reactive current within a particular voltage range ensures that the system functions appositely when there is a voltage drop. Hence, controlling reactive current within a specific voltage range is required. The LVRT criteria for various countries are shown in Figure 3 (Al-Shetwi et al., 2019).

In a unique instance of LVRT known as Zero-VRT, the voltage decreases to zero, enabling renewable energy sources to maintain their connection and provide grid support for a certain amount of time (Zhang et al., 2024). After the voltage is raised, renewable energy sources need to stay connected to the grid to preserve voltage stability and prevent serious disruptions. When the voltage surpasses the allowable limit established by the local power provider, high-voltage reactive technology (High-VRT) controls high-voltage problems (Al-Shetwi et al., 2019).

Figure 3. Low VRT criteria for different countries

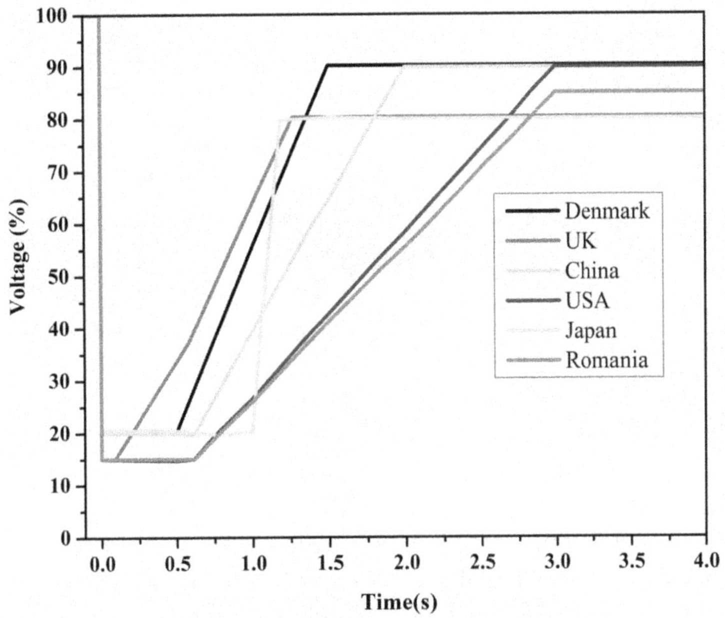

4. DEMAND RESPONSE IN INTEGRATED MICROGRID SYSTEMS

The traditional electric grid design implies a one-way flow of energy from supply to consumer. Centrally controlled high-voltage generators supply this electricity. The integration of distributed energy resources and the expansion of grid service markets in recent years have made demand-side management, especially demand response, a feasible option for the efficient and reliable functioning of the electric grid. A demand response architecture, unlike traditional power grids, requires a two-way communication infrastructure and intelligence algorithms to understand the data collected. As a result, smart meters are critical components of a smart grid and play an important role in demand response models (M. A. Khan et al., 2022). According to the classifications supplied by (M. A. Khan et al., 2022), Figure 4 depicts how demand response programs are classified. Depending on incentive presented to customers, demand response programs are further categorized into two different types.

The purpose of varying electricity pricing over time is to incentivize end-users to modify their energy-consuming behaviors. Reduced electricity use is encouraged among customers using schemes like real-time pricing, critical-peak pricing, and time-of-use pricing. Demand-bidding programs, interruptible tariffs, and direct-load controls are some examples of such methods. Electricity markets are classified as either retail or wholesale. The capacity market, ancillary services market, and energy market are the three divisions of the wholesale electricity market (Antonopoulos et al., 2020).

Figure 4. Demand response categorization.

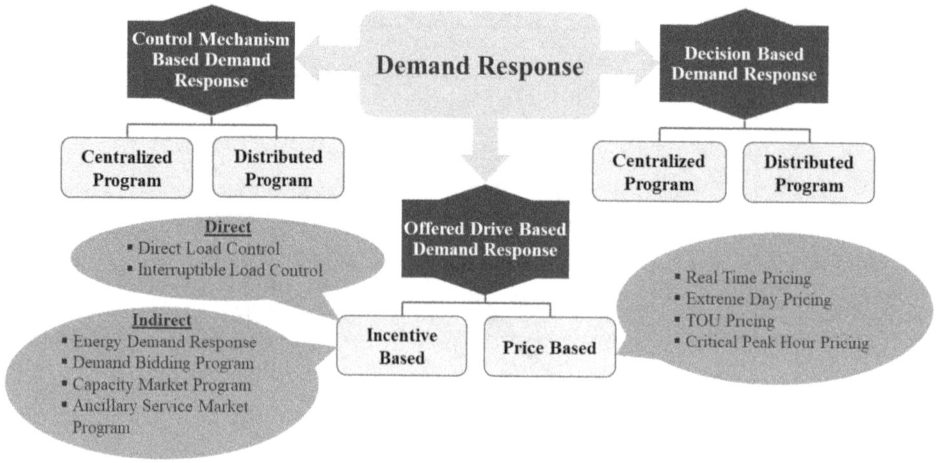

Depending on national regulation, contracts between vendors might be bilateral or conducted through a proper broker organization. The spot market allows supplies to be traded for energy supply services. Companies that monitor Distributed Energy Resources (DERs) must reach an agreement with end users and provide power flexibility. New techniques must overcome the technological limitations of a smart grid, specifically frequency and local voltage management, as distributed energy resources (DER) are rapidly integrated into power networks (Assad et al., 2022). Demand response may be one of the most effective methods since it can be built on existing infrastructure, such as smart meters and IoT-capable equipment. Artificial Intelligence based smart algorithms can be utilized for effectively managing moderate demand while not compromising end-user comfort (Goudarzi et al., 2022, C. Wang et al., 2023).

5. AI TECHNIQUES IN DEMAND RESPONSE

Artificial intelligence (AI) concepts and techniques will eventually revolutionize demand response (DR) strategies in microgrid systems. By employing the power of AI, utilities and grid operators may dramatically improve their ability to control energy demand successfully. Artificial intelligence (AI) significantly contributes to the improvement of demand response techniques through intelligent decision-making, automation, optimization, personalization, and predictive analytics (Juma et al., 2023).

i. Immense decision making: - Through the analysis and mining of real-time and historical data, the AI system finds consumption patterns, forecasted energy consumption, and provides suitable load control solutions. This capability facilitates utilities to act in advance on the changes in the energy supply to enhance grid stability and reliability.
ii. Predictive analytics: - AI makes predictive analytics possible by predicting energy consumption, the volatility of energy prices, or grid issues. Utilities can implement advanced AI models for predicting periods of peak demand to enable energy procurement optimization and take proactive early measures to alleviate grid bottlenecks that may occur. By using predictive analytics, utilities have the luxury

of making strategic decisions ahead of time that aid in allocating resources better and maintaining system stability.

AI automates all the procedures of demand response through continuous monitoring of consumption trends and detection of anomalies, which provides for real-time responses. It makes operations much smoother, reduces manual intervention, minimizes errors, and ensures that demand response plans are implemented on time. AI-based optimization solutions employ machine learning approaches to optimize energy consumption, distribute resources appropriately, and constantly match supply and demand. AI models can respond to changing circumstances, learn from previous performances, and constantly improve decision-making processes to increase energy efficiency and savings. AI allows utilities to design more responsive demand response systems to individual behavior and preferences. AI will push customized to customer-specific data energy-savings recommendations, incentives, and pricing techniques to drive greater consumer involvement, improve program efficacy, and ultimately conserve energy.

5.1. AI Perspectives in Demand Response

Different AI strategies and techniques have already been seen widespread in demand response applications under the innovative grid framework, primarily to enhance efficiency and effectiveness in regulating energy demands. The following are some of the leading AI concepts and strategies in everyday use in demand response:

5.1.1. Machine Learning (ML) Approach

AI in electric grids involves machine learning (ML), which is a methodology for computers that enables them to learn automatically from data arrive at intelligent decision and perform actions without explicit programming. ML forms the core of current innovative grid demand response projects; this makes it parse past energy consumption data, discover hidden trends, and predict future energy consumption behaviors. Massive volumes of historical energy data from smart meters, sensors, and other sources can be processed by ML algorithms. If the ML model brings this data under scrutiny, it will find patterns, trends, and anomalies in energy consumption as time elapses or progresses. Such analysis allows utilities to know how and why energy was consumed in the past, something that is essential for coming up with a valid demand response strategy (Tang et al., 2024, Bourhnane et al., 2020).

Different ML methodologies of regression, clustering, and classification to historical energy data for unravelling hidden patterns and correlations between variables. In such a case, the result of a regression analysis will explain how factors such as the time of day or weather conditions relate to energy use. Clustering will concatenate similar consumption patterns, while classification sorts them into groups according to predefined criteria.

Utilities can make use of ML to enhance operational efficiency. ML algorithms will aid utilities in predicting when peak demand will occur, detecting other opportunities for energy efficiency, and even making consumption projections. All this information will allow utilities to apply dynamic management of loads, efficiently manage resources, and recover from disasters in targeted areas to ensure adequate meetings of demand requirements. ML algorithms can suggest ideal energy supply schedules, identify opportunities for load shifting or shedding, and prioritize demand response measures based on anticipated demand changes. Utility companies can enhance catastrophe recovery performance, reduce energy costs, and improve grid reliability by basing their decisions on insights derived from ML.

In applications related to demand response, ML algorithms will be very vital for assessing historical energy data, recognizing consumption patterns, and attempting to set up forecasts of future trends in demand and load management tactics. Utilities can use ML techniques for decision making, improving energy efficiency, and enriching disaster recovery programs in innovative grid systems.

5.1.2. Deep Learning: A Powerful Approach for Smart Grid Optimization

Artificial neural networks with multiple layers, known as deep learning, have emerged as a transformative ML technique. These deep neural networks excel at uncovering intricate patterns and relationships within complex data. In the context of smart grid demand response systems, deep learning plays a pivotal role in enhancing energy management strategies. One of the key features of deep learning models is their ability to automatically extract insightful details from raw data, uncover nonlinear correlations, and identify sophisticated patterns that traditional ML algorithms may overlook. This multilayered neural network architecture enables deep learning to learn comprehensive data representations.

Deep learning models are particularly useful for load forecasting in disaster recovery applications. By analyzing historical data, weather patterns, and other relevant variables, deep learning systems can accurately predict future energy demand. These precise load forecasts empower utilities and grid operators to anticipate peak periods, optimize resource allocation, and implement proactive measures to manage energy usage efficiently. Furthermore, deep learning algorithms are employes to detect anomalies in energy consumption. Deep neural networks trained on normal usage patterns can recognize deviations or irregularities in real-time data streams, allowing for rapid identification and mitigation of potential issues.

Advanced learning techniques are employed in demand response solutions to boost energy efficiency and optimize energy use. By analyzing historical data, these algorithms identify opportunities for load shifting, demand response, and energy saving methods. These models can prescribe ideal energy usage patterns, prioritize load management choices, and assist utilities in enhancing efficiency and reducing costs (M. A. Khan et al., 2024, Gulay et al., 2023).

5.1.3. Reinforcement Learning

Reinforcement learning is a kind of ML where an agent interacts with its environment and receives feedback in the form of rewards or penalties for its decisions. This training process enables an agent to make sequential decisions across time. Reinforcement learning algorithms are beneficial in DR applications in innovative grid systems. These algorithms optimize energy consumption, control devices, and dynamically adjust the methods of DR in real time. In reinforcement learning, the entity learns to make decisions by taking actions in an environment and monitoring the results. The aim of the entity is to maximize cumulative rewards as it learns which actions lead to positive outcomes and those that should be avoided. Sequence-based decision policies will work very well in dynamic and uncertain contexts, such as smart grids.

In DR applications, a reinforcement learning agent will interact with the environment of the smart grid through observations – e.g., energy consumption data and grid conditions – and actions – e.g., adjusting devices and shifting loads. The actions taken by the agent impact on the state of the environment; the environment responds to the agent's actions with rewards or penalties according to its impact.

It can learn from previous experience and feedback in case of emergency situations by trying to reduce energy cost, peak demand, and energy efficiency. The agent might learn how to do load scheduling, demand response actions, and device control for saving energy while complying with grid limits. Such techniques can be used in reinforcement learning for device and appliance management with changing energy demands and pricing signals. An agent can learn to adjust device settings, schedule energy use priority when the grid conditions are not good and coordinate the operation of various devices while optimizing energy efficiency. This dynamic control technique allows for more efficient resource utilization while maintaining grid stability (Z. Zhang et al., 2019).

The agent learns to identify demand signals, predict future energy demand, and adjust load management strategies to maximize rewards while minimizing penalties. This adaptive technique enables flexible and responsive demand response solutions. Within smart grid contexts, reinforcement learning is crucial for optimizing energy use, regulating devices, and dynamically modifying demand response tactics. Businesses can leverage reinforcement learning algorithms to enhance their energy efficiency, reduce expenses, and optimize the effectiveness of their demand response initiatives (Z. Zhang et al., 2019, Fraija et al., 2024).

5.1.4. Natural-Language Processing (NLP): Extending Insightful Demand Response Strategies

NLP, in particular, is the sub-branch of artificial intelligence concerned with training a computer to identify, process, and generate human language. NLP supports smart grid utilities and demand-response systems through relevant information extracted from unstructured consumer comments and social media interactions coupled with energy consumption data. Utilities can effectively enhance their demand response strategies using NLP in several ways (M. A. Khan et al., 2022).

Deriving insights from unstructured data is one of the most used methodologies. Utilities wrestle with vast amounts of unstructured textual data, most notably customer reviews, social media posts, e-mails, and energy use data. By applying NLP, utilities can gain better insight into consumer behavior, preferences, and feelings to be able to craft much more client-oriented demand response strategies with proper incentives. Conversely, chatbots, personal assistants, and opinion analytics driven by NLP will help utilities optimize customer engagement, share relevant information, and quickly respond to questions, consequently tightening the utility-customer relationship. It can also mean that sources of renewable energy host more than just clean electricity, but greater community well-being and improved energy system efficiency means enhanced contentment since grids work at their peak.

5.1.5. Optimization Algorithms

Effective energy management requires strategic approaches, and hence various tools are needed to identify optimal resource distribution, planning, and demand management methods. Renowned techniques like generic algorithms, particle swarm optimization, and simulated annealing are prevalent in the energy sector (Marrero et al., 2024). Utilities leverage these algorithms to enhance cost-effectiveness and energy efficiency.

A. Generic algorithms (GA): Draws inspiration from natural selection and evolution, and these algorithms generate a pool of potential solutions, assess their performance against predefined criteria such as energy efficiency or cost, and then select the best-performing solutions to create the next generation.

Genetic algorithms use crossover and mutation to search the solution space for optimal resource allocation, scheduling, and demand response strategies (Yang, 2020).

B. Swarm Intelligence Algorithms: A subfield of artificial intelligence called swarm intelligence investigates how problem-solving can be facilitated by imitating the actions of biological agents. Particle Swarm Optimization (PSO) and Ant Colony Optimization (ACO) are two popular swarm intelligence algorithms used in energy management applications. Local optima and sluggish convergence rates are potential issues for swarm AI systems, as they are for evolutionary techniques (Cavalca et al., 2018). In contrast to Genetic Algorithms (GA), swarm AI systems use the histories of all particles to improve the search process, needing minimum fine-tuning parameters prior to deployment. Energy industry stakeholders use swarm AI algorithms to optimize scheduling and pricing techniques for demand response in order to reduce costs. Due to the non-convex nature of optimization difficulties in demand response, heuristic approaches such as PSO are recommended for their efficiency in quickly discovering near-optimal solutions with lower computer complexity than traditional mathematical methods (Cavalca et al., 2018).

C. Simulated Annealing: Simulated annealing is a probabilistic optimization approach based on the metallurgical annealing process. The process starts with an initial solution and iteratively explores the solution space, allowing "worse" alternatives to be accepted with a given probability. Over time, the algorithm steadily reduces the acceptance of suboptimal answers, resulting in convergence on optimal or near-optimal solutions (M. A. Khan et al., 2022). Simulated annealing is useful for solving complicated energy management problems when determining the global optimum is challenging.

5.1.6. Predictive Analytical Techniques

Examining past data patterns using sophisticated techniques like regression analysis, time series forecasting, and data mining can help anticipate future energy-related events and tendencies. This predictive analysis is crucial for estimating energy usage, price fluctuations, and grid-related problems. By forecasting periods of peak demand, utility providers can enhance their energy procurement and resource allocation, enabling them to better satisfy the dynamic needs of the energy market. Adopting a forward-thinking mindset allows utility firms to make better decisions, use their assets more efficiently, and improve their overall operational performance. Predictive data analysis enables utility providers to anticipate shifting demand patterns and market volatility, so increasing their overall effectiveness and competitive advantage in the energy business (Sutton & Barto, 2018).

5.1.7. Expert Systems

Another type of intelligent-powered system, known as expert system, can emulate human intelligence and selection-making skills in particular fields of expertise. They play an important role in demand response techniques, providing guidelines for sports consisting of load discounts, demand control, and electricity conservation via utilizing pre-established pointers and expertise repositories. Those professional structures enable utilities to make precisely knowledgeable choices in real-time (C. Xu & Huang, 2023).

Figure 5. Classification of AI approaches in demand response

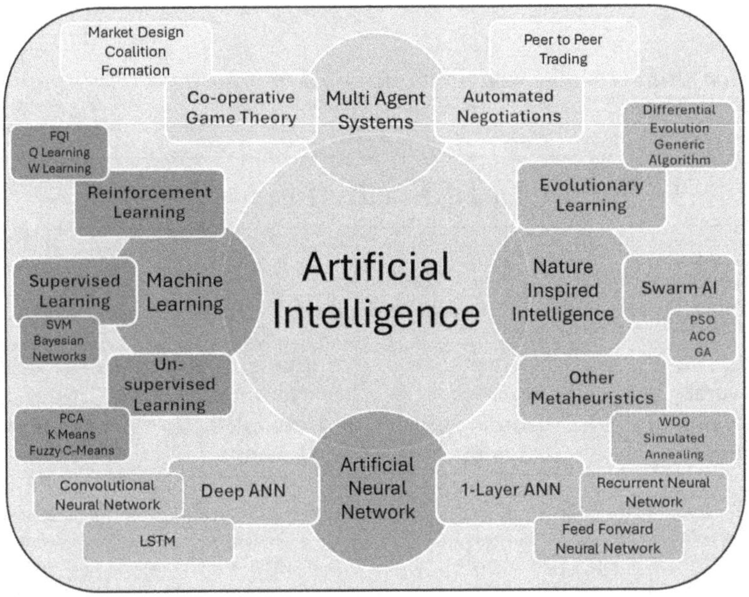

Various other AI techniques are used for energy consumption management as well. Figure 5. shows some of the AI methods used in demand response. These AI solutions are key for improving demand response, overall performance of the grid, power consumption, and smart grid infrastructure resilience (C. Xu & Huang, 2023). The interpretability of AI models is critical for utilities seeking insights into the thinking behind AI-powered suggestions and actions. When AI models are extremely complicated, such as deep learning neural networks, it might be difficult to understand the underlying mechanisms and decision-making processes. Transparency and interpretability of AI algorithms are critical for establishing confidence, gaining stakeholder acceptability, and aiding successful decision-making in demand response initiatives. By tackling the above-mentioned challenges and leveraging the opportunities provided by AI techniques, utilities can improve the effectualness of their DR initiatives, improve grid-resilience, and achieve greater energy efficiency in smart grid environments (M. A. Khan et al., 2022, C. Xu & Huang, 2023, Liu et al., 2022, Y. Xu et al., 2023).

6. AI APPROACH TOWARDS PREDICTION IN INTEGRATED RENEWABLE ENERGY SYSTEMS

Prediction methods in integrated renewable energy microgrids are critical for optimizing energy management, maintaining grid stability, while supporting the optimal usage of renewable energy resources. Several prediction procedures are used in integrated renewable energy microgrids. Renewable energy generation forecasting is one type of prediction methodology. One of the most important parts of planning grid operations is projecting the generation of renewable energy resources, like wind energy and solar power. Estimating renewable energy output involves using methods such as solar irradiance forecasts, wind speed prediction, and cloud tracking image analysis. Another type is Optimal energy management. Strategies for optimizing integrated renewable energy microgrids include energy storage, demand response programs, and renewable power generation.

Figure 6. Optimization strategies and load prediction – AI approaches

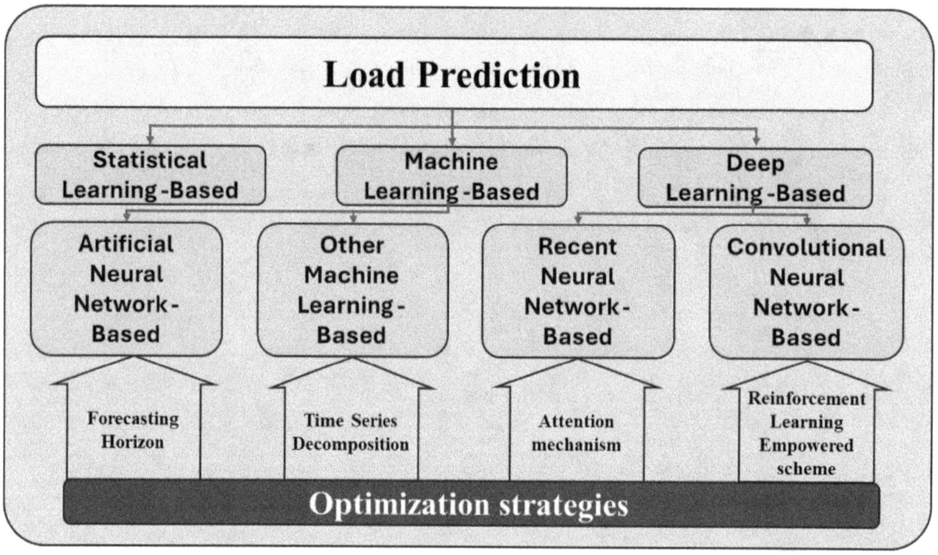

Unpredictable forecasting and adaptive approaches are also used in prediction methodologies. These solutions incorporate economic and environmental aspects to improve power scheduling and grid efficiency. By incorporating these prediction approaches into renewable energy systems, grid operators may achieve predictable and sustainable energy management, maximize resource usage, and increase grid resilience in the face of changing energy demands and fluctuating renewable energy generation patterns. Figure 6 illustrates improved load prediction systems based on artificial intelligence approach (Arumugham et al., 2023). By incorporating varied methodologies, hybrid models can deliver more trustworthy forecasts in microgrid integrated renewable energy systems.

Grid operators can improve grid stability, optimize energy planning, and encourage the effective integration of renewable energy sources for sustainable power generation by applying these AI-based prediction techniques in microgrid integrated renewable energy systems (D. Liu et al., 2018).

7. CATEGORIZATION OF ANOMALOUS DETECTION

Electricity utilities confront both technical and non-technical losses when seeking to meet demand. Technical losses originate from physical constraints in a power system, such as resistance, but non-technical losses are caused by theft and leakage, resulting in expensive costs and security breaches. Anomaly detection is critical in smart grids, especially on the demand side. Various data science techniques are utilized to train anomaly detection algorithms, which can be divided into general anomalies and other categories based on feature extraction methods. Anomaly detection offers feedback on energy usage, which aids in problem identification and aids both utilities and ecosystems (Ma et al., 2021).

Electrical load anomaly detection methods can be divided into two types: regression model-based and classification model-based.

Figure 7. Classification of anomaly detection in the demand side

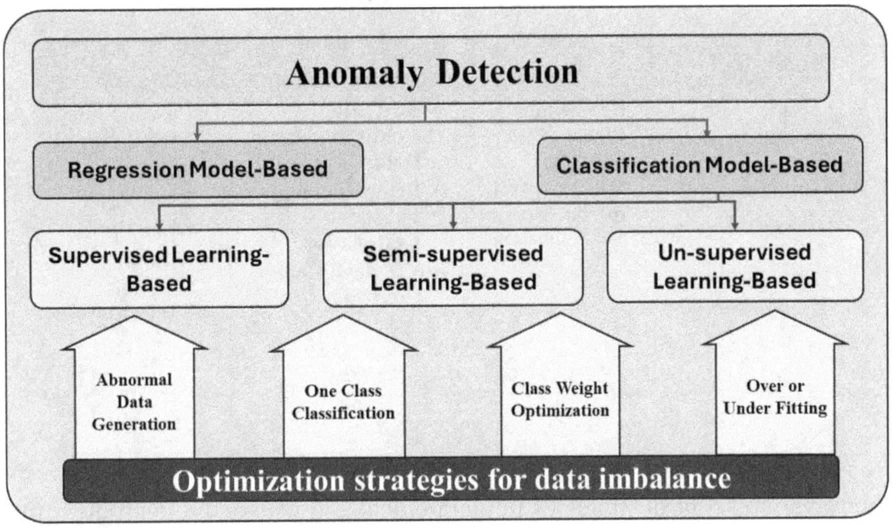

The regression-based strategy is useful for load forecasting and anomaly detection, but its identification accuracy is restricted because it depends on prediction results. Classification-based detection algorithms are intriguing, but each has drawbacks. Supervised learning-based approaches are highly accurate but incur data labeling expenses and raise customer privacy concerns. Unsupervised learning-based approaches provide a label-free methodology with significant computing costs and limited interpretability. Semi-supervised learning strikes a balance between these approaches; however, it may not be suitable for unknown datasets. Hyperspace Dimensional Computing (HDC) is a unique data-driven classification technique introduced in the energy domain that provides accurate detection results, reduces model optimization and pre-training costs, and saves training time (Ma et al., 2021). However, HDC relies on supervised learning, which may be hindered in some circumstances due to high data tagging monetary value. Figure 7 symbolizes a brief overview of cutting-edge techniques for anomaly detection mechanisms with artificial intelligence.

8. CONTROL SCHEMES AND ENERGY MANAGEMENT STRATEGIES

Two interrelated components of optimizing the performance of energy systems, such as direct current microgrids, are energy management and control. Microgrid control consists of a series comprising methodologies, algorithms, and technologies for managing and optimizing the operation of a microgrid, which is a small group of interconnected loads and dispersed energy supplies.

Figure 8. Microgrid control

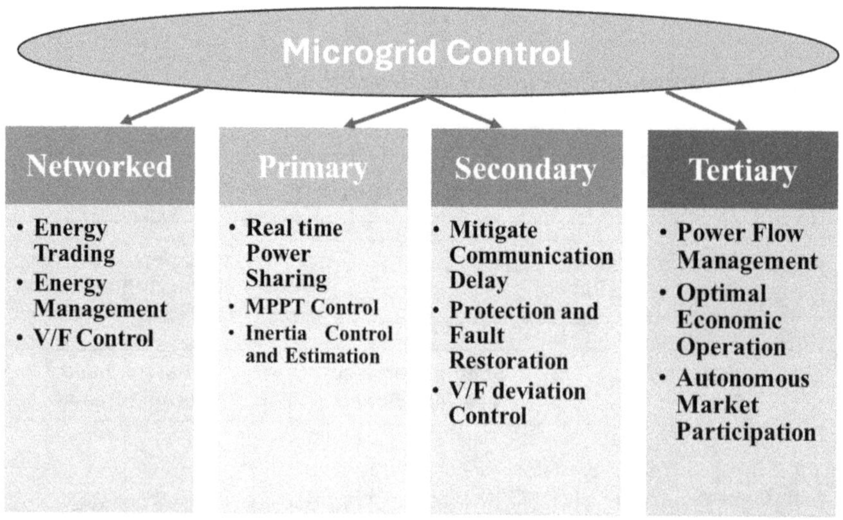

Figure 8 shows various control strategies in hierarchical and networked microgrids that use AI approaches and figure 9 represents different energy management strategies adopted in different microgrid systems (Trivedi & Khadem, 2022).

Figure 9. Different energy management schemes commonly adopted in microgrid systems

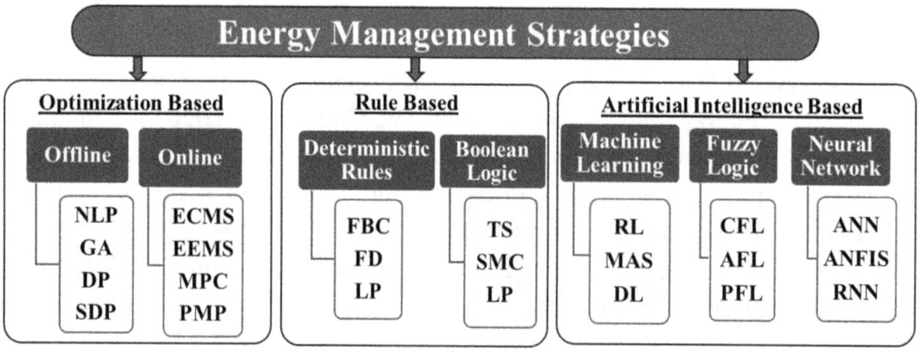

9. CONCLUSION

Intelligent energy providers and grid operators may use cutting-edge AI solutions to improve their demand management strategies and optimize smart grid operations. Data-driven forecasting, machine learning models, and expert-guided systems enable enterprises to evaluate large amounts of data, anticipate energy demand trends, and make enlightened decisions to successfully manage energy resources.

This proactive strategy enables utilities to forecast peak demand, apply load-shedding solutions, and educate customers on energy-efficient activities. By embracing these disruptive technologies, utilities may achieve a balance of energy supply and demand, reduce grid sarin, and improve grid dependability, resulting in a stable and efficient energy distribution network.

AI-driven technologies enable utility companies to reduce their energy costs. They accomplish this by improving energy procurement, power generation planning, and customer-focused management. Utilities may use AI-derived advice to identify energy-saving opportunities, implement energy-efficient initiatives, and reduce operating expenses. Embedding AI concepts and technologies into smart grid frameworks allows utilities and grid operators to improve energy efficiency, increase grid resilience, reduce costs, and ultimately enhance energy system performance to meet the changing demands of the modern energy landscape.

Advanced AI has enormous potential to transform the way smart grid systems handle catastrophe recovery. AI may significantly improve disaster recovery plans by improving predictive skills, allowing for faster decision-making, more effective customer engagement, and automated energy management. This chapter examined the potential of AI to transform integrated smart grid systems and looked at the ways AI may improve predictive analytics, control strategies, anomaly detection, demand response systems, and energy management strategies. Introducing AI into the energy distribution as well as reliability creates great opportunities. It would lead to new inventions, improving operational efficiency thereby enabling utilities to meet variable energy demands. Yet, additional research, industrial initiatives, and comprehensive testing are crucial ingredients for developing more sophisticated AI models and solutions capable of revolutionizing the industry.

REFERENCES

Abid, H., Thakur, J., Khatiwada, D., & Bauner, D. (2021). Energy storage integration with solar PV for increased electricity access: A case study of Burkina Faso. *Energy, 230*, 120656. DOI: 10.1016/j.energy.2021.120656

Al-Shetwi, A. Q., Hannan, M., Jern, K. P., Mansur, M., & Mahlia, T. (2019). Grid-connected renewable energy sources: Review of the recent integration requirements and control methods. *Journal of Cleaner Production, 253*, 119831. DOI: 10.1016/j.jclepro.2019.119831

Antonopoulos, I., Robu, V., Couraud, B., Kirli, D., Norbu, S., Kiprakis, A., Flynn, D., Elizondo-Gonzalez, S., & Wattam, S. (2020). Artificial intelligence and machine learning approaches to energy demand-side response: A systematic review. *Renewable & Sustainable Energy Reviews, 130*, 109899. DOI: 10.1016/j.rser.2020.109899

Arumugham, V., Ghanimi, H. M. A., Pustokhin, D. A., Pustokhina, I. V., Ponnam, V. S., Alharbi, M., Krishnamoorthy, P., & Sengan, S. (2023). An Artificial-Intelligence-Based renewable energy prediction program for Demand-Side management in smart grids. *Sustainability (Basel), 15*(6), 5453. DOI: 10.3390/su15065453

Assad, U., Hassan, M. S., Farooq, U., Kabir, A., Khan, M. Z., Bukhari, S. S. H., Jaffri, Z. U. A., Oláh, J., & Popp, J. (2022). Smart Grid, Demand Response and Optimization: A Critical Review of Computational methods. *Energies, 15*(6), 2003. DOI: 10.3390/en15062003

Bourhnane, S., Abid, M. R., Lghoul, R., Zine-Dine, K., Elkamoun, N., & Benhaddou, D. (2020). Machine learning for energy consumption prediction and scheduling in smart buildings. *SN Applied Sciences, 2*(2), 297. Advance online publication. DOI: 10.1007/s42452-020-2024-9

Cabrera, P., Carta, J. A., Lund, H., & Thellufsen, J. Z. (2021). Large-scale optimal integration of wind and solar photovoltaic power in water-energy systems on islands. *Energy Conversion and Management, 235*, 113982. DOI: 10.1016/j.enconman.2021.113982

Cavalca, D. L., Spavieri, G., & Fernandes, R. S. (2018). Comparative analysis between particle swarm optimization algorithms applied to Price-Based Demand response. In *Lecture notes in computer science* (pp. 323–332). DOI: 10.1007/978-3-319-91253-0_31

Chaianong, A., Bangviwat, A., Menke, C., Breitschopf, B., & Eichhammer, W. (2019). Customer economics of residential PV–battery systems in Thailand. *Renewable Energy, 146*, 297–308. DOI: 10.1016/j.renene.2019.06.159

Chen, C. J. (2024). *Physics of solar energy and energy Storage*. John Wiley & Sons. DOI: 10.1002/9781394203642

Chen, J., Qi, B., Rong, Z., Peng, K., Zhao, Y., & Zhang, X. (2020). Multi-energy coordinated microgrid scheduling with integrated demand response for flexibility improvement. *Energy, 217*, 119387. DOI: 10.1016/j.energy.2020.119387

Dinata, N. F. P., Ramli, M. M., Jambak, M. I., Sidik, M. B., & Alqahtani, M. M. (2024). Designing an optimal microgrid control system using deep reinforcement learning: A systematic review. *Engineering Science and Technology, an International Journal*, *51*, 101651. DOI: 10.1016/j.jestch.2024.101651

Du, R., Zou, P., & Ma, C. (2021). Multi-objective optimal sizing of hybrid energy storage systems for grid-connected wind farms using fuzzy control. *Journal of Renewable and Sustainable Energy*, *13*(1), 014103. Advance online publication. DOI: 10.1063/5.0031696

Duman, A. C., & Güler, Ö. (2018). Techno-economic analysis of off-grid PV/wind/fuel cell hybrid system combinations with a comparison of regularly and seasonally occupied households. *Sustainable Cities and Society*, *42*, 107–126. DOI: 10.1016/j.scs.2018.06.029

Efkarpidis, N., Imoscopi, S., Bratukhin, A., Brännvall, R., Franzl, G., Leopold, T., Bauer, V., Goranovic, A., Wilker, S., Yang, C., Gustafsson, J., Geidl, M., & Sauter, T. (2023). Proactive scheduling of mixed energy resources at different grid levels. *IEEE Transactions on Sustainable Energy*, *15*(2), 952–963. DOI: 10.1109/TSTE.2023.3320055

Falope, T., Lao, L., Hanak, D., & Huo, D. (2024). Hybrid energy system integration and management for solar energy: A review. *Energy Conversion and Management X*, *21*, 100527. DOI: 10.1016/j.ecmx.2024.100527

Figaj, R. (2024). Energy and economic sustainability of a Small-Scale hybrid renewable energy system powered by biogas, solar energy, and wind. *Energies*, *17*(3), 706. DOI: 10.3390/en17030706

Fraija, A., Henao, N., Agbossou, K., Kelouwani, S., Fournier, M., & Nagarsheth, S. H. (2024). Deep reinforcement learning based dynamic pricing for demand response considering market and supply constraints. *Smart Energy*, *14*, 100139. DOI: 10.1016/j.segy.2024.100139

Gamonwet, P., & Dhakal, S. (2023). The assessment of the value of electricity saving and economic benefit to residential solar rooftop PV customer: The case of Thailand. *Energy Strategy Reviews*, *50*, 101203. DOI: 10.1016/j.esr.2023.101203

Ghenai, C., & Bettayeb, M. (2019). Modelling and performance analysis of a stand-alone hybrid solar PV/Fuel Cell/Diesel Generator power system for university building. *Energy*, *171*, 180–189. DOI: 10.1016/j.energy.2019.01.019

Ghenai, C., Bettayeb, M., Brdjanin, B., & Hamid, A. K. (2019). Hybrid solar PV/PEM fuel Cell/Diesel Generator power system for cruise ship: A case study in Stockholm, Sweden. *Case Studies in Thermal Engineering*, *14*, 100497. DOI: 10.1016/j.csite.2019.100497

Ghorbani, N., Kasaeian, A., Toopshekan, A., Bahrami, L., & Maghami, A. (2017). Optimizing a hybrid wind-PV-battery system using GA-PSO and MOPSO for reducing cost and increasing reliability. *Energy*, *154*, 581–591. DOI: 10.1016/j.energy.2017.12.057

Goudarzi, A., Ghayoor, F., Waseem, M., Fahad, S., & Traore, I. (2022). A survey on IoT-Enabled Smart Grids: Emerging, applications, challenges, and outlook. *Energies*, *15*(19), 6984. DOI: 10.3390/en15196984

Gulay, E., Sen, M., & Akgun, O. B. (2023). Forecasting electricity production from various energy sources in Türkiye: A predictive analysis of time series, deep learning, and hybrid models. *Energy, 286*, 129566. DOI: 10.1016/j.energy.2023.129566

Hirsch, A., Parag, Y., & Guerrero, J. (2018). Microgrids: A review of technologies, key drivers, and outstanding issues. *Renewable & Sustainable Energy Reviews, 90*, 402–411. DOI: 10.1016/j.rser.2018.03.040

Jaramillo, M., Pavón, W., & Jaramillo, L. (2024). Adaptive Forecasting in Energy Consumption: A Bibliometric Analysis and review. *Data, 9*(1), 13. DOI: 10.3390/data9010013

Juma, D., Munda, J., & Kabiri, C. (2023). Power-System flexibility: A necessary complement to variable renewable energy optimal capacity configuration. *Energies, 16*(21), 7432. DOI: 10.3390/en16217432

Kadri, A., Marzougui, H., Aouiti, A., & Bacha, F. (2019). Energy management and control strategy for a DFIG wind turbine/fuel cell hybrid system with super capacitor storage system. *Energy, 192*, 116518. DOI: 10.1016/j.energy.2019.116518

Katche, M. L., Makokha, A. B., Zachary, S. O., & Adaramola, M. S. (2024). Techno-Economic Assessment of Solar–Grid–Battery hybrid energy systems for Grid-Connected university campuses in Kenya. *Electricity, 5*(1), 61–74. DOI: 10.3390/electricity5010004

Khan, M. A., Khan, T., Waseem, M., Saleh, A. M., Qamar, N., & Muqeet, H. A. (2024). Investigation and analysis of demand response approaches, bottlenecks, and future potential capabilities for IoT-enabled smart grid. *IET Renewable Power Generation, 18*(15), 3509–3535. DOI: 10.1049/rpg2.13011

Khan, M. A., Saleh, A. M., Waseem, M., & Sajjad, I. A. (2022). Artificial intelligence Enabled Demand Response: Prospects and challenges in smart grid environment. *IEEE Access : Practical Innovations, Open Solutions, 11*, 1477–1505. DOI: 10.1109/ACCESS.2022.3231444

Khan, M. W., & Wang, J. (2017). The research on multi-agent system for microgrid control and optimization. *Renewable & Sustainable Energy Reviews, 80*, 1399–1411. DOI: 10.1016/j.rser.2017.05.279

Liu, D., Sun, Y., Qu, Y., Li, B., & Xu, Y. (2018). Analysis and accurate prediction of user's response behavior in Incentive-Based Demand response. *IEEE Access : Practical Innovations, Open Solutions, 7*, 3170–3180. DOI: 10.1109/ACCESS.2018.2889500

Liu, Z., Sun, Y., Xing, C., Liu, J., He, Y., Zhou, Y., & Zhang, G. (2022). Artificial intelligence powered large-scale renewable integrations in multi-energy systems for carbon neutrality transition: Challenges and future perspectives. *Energy and AI, 10*, 100195. DOI: 10.1016/j.egyai.2022.100195

Luo, X., & Oyedele, L. O. (2022). A self-adaptive deep learning model for building electricity load prediction with moving horizon. *Machine Learning with Applications, 7*, 100257. DOI: 10.1016/j.mlwa.2022.100257

Luo, Y., Shi, L., & Tu, G. (2014). Optimal sizing and control strategy of isolated grid with wind power and energy storage system. *Energy Conversion and Management, 80*, 407–415. DOI: 10.1016/j.enconman.2014.01.061

Ma, X., Wu, J., Xue, S., Yang, J., Zhou, C., Sheng, Q. Z., Xiong, H., & Akoglu, L. (2021). A comprehensive survey on graph anomaly detection with deep learning. *IEEE Transactions on Knowledge and Data Engineering*, *35*(12), 12012–12038. DOI: 10.1109/TKDE.2021.3118815

Marrero, L., Sbarbaro, D., & Garcia-Santander, L. (2024). Online demand response characterization based on variability in customer behavior. *Journal of Modern Power Systems and Clean Energy*, *12*(3), 936–946. DOI: 10.35833/MPCE.2023.000516

Mehrabani, A., Aalami, H. A., & Neghab, A. P. (2023). Integrated planning framework for microgrid incorporating the flexibility provision capabilities of demand response programs. *International Journal of Electrical Power & Energy Systems*, *155*, 109521. DOI: 10.1016/j.ijepes.2023.109521

Memarzadeh, G., & Keynia, F. (2023). A new hybrid CBSA-GA optimization method and MRMI-LSTM forecasting algorithm for PV-ESS planning in distribution networks. *Journal of Energy Storage*, *72*, 108582. DOI: 10.1016/j.est.2023.108582

Mena, R., Godoy, D. R., Kristjanpoller, F., & Viveros, P. (2024). A multi-objective two-stage stochastic unit commitment model for wind and battery-integrated power systems. *Journal of Energy Storage*, *89*, 111723. DOI: 10.1016/j.est.2024.111723

Nair, S. S., Krishnia, L., Trukhanov, A., Thakur, P., & Thakur, A. (2022). Prospect of double perovskite over conventional perovskite in photovoltaic applications. *Ceramics International*, *48*(23), 34128–34147. DOI: 10.1016/j.ceramint.2022.08.184

Nair, S. S., Thakur, P., Wan, F., & Thakur, A. (2024). Structural modification and analysis of novel inverted perovskite photo-voltaic device by incorporating atomic layer deposition and surface passivation. *Solar Energy*, *269*, 112366. DOI: 10.1016/j.solener.2024.112366

Nair, S. S., Thakur, P., Wan, F., Trukhanov, A., Panina, L., & Thakur, A. (2023). Performance evaluation and the optimization of an inverted photo-voltaic cell with lead-free double perovskite material and inorganic transport layer materials. *Solar Energy*, *262*, 111823. DOI: 10.1016/j.solener.2023.111823

Pan, C., Tao, S., Fan, H., Shu, M., Zhang, Y., & Sun, Y. (2021). Multi-Objective optimization of a Battery-Supercapacitor hybrid energy storage system based on the concept of Cyber-Physical System. *Electronics (Basel)*, *10*(15), 1801. DOI: 10.3390/electronics10151801

Ponnuru, S., Kumar, R. A., & Swaroopan, N. J. (2020). Intelligent control and power management of wind-solar integration of renewable energy sources using microgrid. *Materials Today: Proceedings*, *45*, 2323–2328. DOI: 10.1016/j.matpr.2020.10.687

Quiles, E., Roldán-Blay, C., Escrivá-Escrivá, G., & Roldán-Porta, C. (2020). Accurate sizing of residential Stand-Alone photovoltaic systems considering system reliability. *Sustainability (Basel)*, *12*(3), 1274. DOI: 10.3390/su12031274

Rani, P., Parkash, V., & Sharma, N. K. (2023). Technological aspects, utilization and impact on power system for distributed generation: A comprehensive survey. *Renewable & Sustainable Energy Reviews*, *192*, 114257. DOI: 10.1016/j.rser.2023.114257

Sutton, R. S., & Barto, A. G. (2018). Reinforcement Learning, *second edition: An Introduction*. MIT Press.

Tabassum, T., Lim, S., & Khalghani, M. R. (2023). Artificial intelligence-based detection and mitigation of cyber disruptions in microgrid control. *Electric Power Systems Research, 226*, 109925. DOI: 10.1016/j.epsr.2023.109925

Talaat, M., Elgarhy, A., Elkholy, M., & Farahat, M. (2021). Integration of fuel cells into an off-grid hybrid system using wave and solar energy. *International Journal of Electrical Power & Energy Systems, 130*, 106939. DOI: 10.1016/j.ijepes.2021.106939

Talaat, M., Elkholy, M. H., Alblawi, A., & Said, T. (2023). Artificial intelligence applications for microgrids integration and management of hybrid renewable energy sources. *Artificial Intelligence Review, 56*(9), 10557–10611. DOI: 10.1007/s10462-023-10410-w

Tang, Y., Zhang, L., Huang, D., Yang, S., & Kuang, Y. (2024). Ultra-Short-Term photovoltaic power generation prediction based on Hunter–Prey optimized K-Nearest neighbors and simple recurrent unit. *Applied Sciences (Basel, Switzerland), 14*(5), 2159. DOI: 10.3390/app14052159

Thakur, A., Thakur, P., Alex, T., Krishnia, L., & Nair, S. S. (2022). Optimization based on computational analysis of lead-free inverted planar photo-voltaic device structure using halide based double perovskite material. *Materials Today: Proceedings, 71*, 312–316. DOI: 10.1016/j.matpr.2022.09.219

Tostado-Véliz, M., Escámez, A., Aguado, R., Sánchez-Lozano, D., Jurado, F., & Vera, D. (2024). Optimal sizing of hybrid PV–diesel–biomass gasification plants for electrification of off-grid communities: An efficient approach based on Benders' decomposition. *Journal of Cleaner Production, 458*, 142360. DOI: 10.1016/j.jclepro.2024.142360

Trivedi, R., & Khadem, S. (2022). Implementation of artificial intelligence techniques in microgrid control environment: Current progress and future scopes. *Energy and AI, 8*, 100147. DOI: 10.1016/j.egyai.2022.100147

Wang, B., Wang, X., Wang, Z., Wei, C., Zhang, X., Zhou, M., Gao, J., & Han, Z. (2024). A four-stage fast reliability assessment framework for renewables-dominated strong power systems with large-scale energy storage by temporal decoupling and contingencies filtering. *Applied Energy, 362*, 123035. DOI: 10.1016/j.apenergy.2024.123035

Wang, C., You, X., Gao, X., Zhu, X., Li, Z., Zhang, C., Wang, H., Huang, Y., Chen, Y., Haas, H., Thompson, J. S., Larsson, E. G., Di Renzo, M., Tong, W., Zhu, P., Shen, X., Poor, H. V., & Hanzo, L. (2023). On the Road to 6G: Visions, Requirements, Key Technologies, and Testbeds. *IEEE Communications Surveys and Tutorials, 25*(2), 905–974. DOI: 10.1109/COMST.2023.3249835

Xiao, Y., Zou, C., Dong, M., Chi, H., Yan, Y., & Jiang, S. (2024). Feasibility study: Economic and technical analysis of optimal configuration and operation of a hybrid CSP/PV/wind power cogeneration system with energy storage. *Renewable Energy, 225*, 120273. DOI: 10.1016/j.renene.2024.120273

Xu, C., & Huang, Y. (2023). Integrated Demand Response in Multi-Energy Microgrids: A Deep Reinforcement Learning-Based Approach. *Energies, 16*(12), 4769. DOI: 10.3390/en16124769

Xu, F., Liu, J., Lin, S., Dai, Q., & Li, C. (2018). A multi-objective optimization model of hybrid energy storage system for non-grid-connected wind power: A case study in China. *Energy, 163*, 585–603. DOI: 10.1016/j.energy.2018.08.152

Xu, Y., Li, J., Wang, H., & Du, P. (2023). A novel probabilistic forecasting system based on quantile combination in electricity price. *Computers & Industrial Engineering*, *187*, 109834. DOI: 10.1016/j.cie.2023.109834

Yang, X. (2020). *Nature-Inspired optimization algorithms*. Academic Press.

Zhang, X., Wu, H., Zhu, M., Dong, M., & Dong, S. (2024). Integrated demand response for Micro-Energy grid accounting for dispatchable loads. *Energies*, *17*(5), 1255. DOI: 10.3390/en17051255

Zhang, Z., Zhang, D., & Qiu, R. C. (2019). Deep reinforcement learning for power system: An overview. *CSEE Journal of Power and Energy Systems*. Advance online publication. DOI: 10.17775/CSEEJPES.2019.00920

Chapter 22
Public Education Campaigns on AI and IoT Applications in Wildfire Prevention and Management: Fostering Ethics, Responsibility, and Sustainability

Divya Mishra
https://orcid.org/0000-0002-9375-8720
Sharda University, India

Bhupinder Singh
https://orcid.org/0009-0006-4779-2553
Sharda University, India

ABSTRACT

This chapter explores the crucial role of public education campaigns in advancing the adoption of AI and IoT technologies for wildfire prevention and management. It introduces a robust communication model tailored for effective public education and an implementation framework designed to actively engage diverse stakeholders, demystify complex technologies, and cultivate trust. By emphasizing tailored messaging, interactive tools, real-world examples, and continuous feedback, the chapter addresses the challenges of technical complexity and public skepticism. The proposed strategies seek to deepen understanding and foster proactive community engagement, ultimately enhancing resilience and safety against wildfires. These insights offer valuable guidance for researchers, practitioners, and policymakers in developing impactful education campaigns to promote the integration of AI and IoT in wildfire management.

DOI: 10.4018/979-8-3373-0045-0.ch022

INTRODUCTION

Wildfires have turned out to be one of the most significant future environmental threats in the course of the twenty-first century, as the incidence of dangerous fires, their intensity, and duration have been on the rise globally. Climate change, land usage practices, increased human population/ settlement plus interference with climate in the specific fire-prone regions have aggravated and intensified the risk/ effect of wildfires making it a global issue (Moritz et al., 2012). Australia, California, Southern Europe, and many other countries have felt the wrath of unprecedented wildfires, and in the process, lost many people, properties, and biodiversity (Bowman et al., 2011). In addition to the direct impacts, socio-economic costs of wildfires are very high in terms of damages, health and social impact, changes in the economy, environmental impacts, the loss of ecosystem services (Johnston et al., 2020). Considering the fact that the threat of wildfire is on the rise, efforts need to be made in order to boost methods of managing with the intention of preserving human and natural environments alike.

It is clear that the use of enhanced technologies such as artificial intelligence and the internet of things has advanced in its capability in early perception and management of wildfire. AI variants are widely used to forecast the wildfire frequency by considering factors that are challenging to decipher by humans, such as the weather conditions, vegetation density, and prior fire events (Rein & Huang, 2021). Such models can be used to establish early warning mechanisms that improve readiness and better facilitate the distribution of resources in affected regions. Similarly, IoT technologies like sensor networks and remote monitoring are as valuable for detecting new wildfires as they are for monitoring environmental conditions that could lead to initiation (Vadrevu et al., 2020). For example, in California, there are IoT based systems that are used in monitoring the moisture in vegetation and even atmospheric conditions to help firefighters to note favorable conditions for fire outbreaks (Myoung et al., 2018). AI and IoT together form a synergy of technologies that can supplement the conventional wildfire management practices and decrease wildfire risks as and when they emerge.

However, the key to managing AI and IoT for controlling wild fire requires public sensitivity and knowledge. One must develop an understanding of how these technologies operate, what the advantages might be achieved, and what part people and groups can play in leveraging them (Kapucu & Garayev, 2011). Thus, the objective of this chapter is to elucidate the best approaches to the development and implementation of public awareness campaigns for ensuring the acceptance of AI and IoT solutions in fire safety. Hence, this chapter aims at enriching the process of developing effective assertions to the public through the identification of the target audience, choice of words and phrases to be used, and the application of simple graphics and engaging tools in a bid to ensure the creation of a well-informed society that can easily cope with emerging crises.

STRATEGIES FOR DESIGNING EFFECTIVE PUBLIC EDUCATION CAMPAIGNS

Approaches to promote awareness and acceptance of AI and IoT in wildfire fighting include developing proper marketing and educating the public on these programs and their applications. These campaigns, therefore, need to be effectiveness campaigns reaching the right stakeholders, explaining complicated information and showing real-life implications of such technologies. There are certain the key strategies involved in creating impactful public education campaigns such as focusing on audience identification, message crafting, the use of visual aids and interactive tools, and the incorporation of success stories and real-world examples.

Identifying and Targeting the Appropriate Audience

The first activity that needs to be done when designing any public education campaign is the identification of the right audience. There are a number of concerns that need to addressed with regard to wildfires and their management, and these concerns will vary from one stakeholder to another depending on their level of knowledge, and their perception of the role that AI and IoT will play in this context (Mishra, 2024b). Such stakeholders may involve residents of homes in the regions affected by wildfires, political leaders at the local level, fire brigades, rescue teams, organizations, and individuals that fund conservation, learning institutions, students, Firefighters and other emergency services and members of the public. These groups must all be engaged differently so that the information presented to them is useful and justifiably understandable to them (Steelman & McCaffrey, 2012).

For instance, owners of houses may be interested in how AI and IoT will help to secure their homes and families. When communicating with this group, the emphasis should be on the real use of such technologies as early warning systems, smart home systems that can identify and prevent fire hazards (McCaffrey, 2015). Emergency service providers, local government authorities, and fire departments may be of comparatively more interest in both the general possibilities of these technologies in the framework of safety threats and the distributions of government's stamina across the country than immediate advantageous to their organization. In this regard, messaging to this audience should capture the cost efficiency and enduring advantages of applying AI and IoT in current frameworks of wildfire management.

Demographic, geographic and psychographic targeting is necessary when invading a new market and addressing the specific audience. For instance, the campaign that aims at reaching communities in rural areas that often suffer from wild fires may require less perception of technology use and available digital media before changing its strategies. Conducting surveys, focus groups and or consulting with the communities would be greatly useful in describing the specific worries and expectations of particular target audiences to offer custom and accurate approaches (Morss et al., 2005).

Crafting Clear and Impactful Messages

Such information will be useful once the target audience has been identified and more focused messages need to be developed to the society to make them understand why AI and IoT are vital in suppressing wildfires. Because these are sophisticated technologies, it becomes incumbent upon the author to avoid complicated terminology and technical language. Its aim is to convey the information to the target audience in a clearly and easily digestible manner, while at the same time not losing the objectivity and truthfulness of the material (Kasperson, 2014).

Messages should focus on the practical advantage brought by AI and IoT, including safety, response readiness, and resource optimization. For instance, an early warning message might inform communities that the use of AI-based predictive analytics can result in a warning that provides extra time for the people to escape or another might explain how fire-detection IoT sensors reduce the chances of undetected fires (Wong et al., 2019). These benefits should be tailored in a way that resonates with the intended audience in relation to the interests of the community by covering areas such as safety of lives, property and the environment (Morss et al., 2015).

Further, information communicated to the audience should be uniform across platforms to guarantee that the target audience receives a clear, detailed facet of the same information. It enhances the trust and credibility within the audience hence increases chances of them to act on the information promoted. Campaigns should also target misconceptions/concerns about AI- IoT, clarifying how they will be harmless, efficient fairly utilized in practices. It also ensures that every argument necessary to overcome objections and skepticism that may accompany the offer is already put forward before such an occasion (Bostrom & Yudkowsky, 2014).

The Use of Visual Aids and Interactive Tools

To enhance the effectiveness of information dissemination about complicated technologies such as Artificial Intelligence and Internet of Things, free hand should be given to visual accoutrements and exhibiting instruments. Visual aids can help to clarify difficult to explain ideas and work to be studied; Information and communications Technologies can enhance student participation (Mayer, 2009).

For instance, infographics can be applied for illustrating how AI models work to estimate the risks of wildfire, or how IoT networks observe the environment in real time. These help in presentation of information, which is easier to understand especially by audiences that are not conversant with technological information. Classifications, such as diagrams and flowcharts, or animations can extend the account of the procedures linked to AI and IoT in relation to wildfire prevention and control. For example, creating realistic wildfire simulation like VFire has been given evidence to work in a way to educate firefighters, policy makers and public, on the reality of wildfire behavior and the impact of the preventative measures that they take; as a result letting users experiment with different suppression technique modes in the constructive space (Hoang et al., 2010). Interactive visualization tools also improve comprehension as these facilitate easy understanding of concepts normally demanding higher levels of comprehension among the audiences (Brusilovsky et al., 2010).

Use of learning by doing techniques like online simulations, use of mobile applications and the virtual world can be used to present the topic of AI and IoT in a more engaging perspective. For example, visual presentation of the part of how precisely an AI system predicts the spread of a wildfire based on collected data or how IoT sensors recognize changes in a particular environment and alert about it. Not only does it make learning fun, they also enable the target group to envision the use and importance of those technologies. Another aspect of the application of cognitive visualization to educational tools can be identified as the possibility of increasing the efficiency of the education process due to the provision of greater clarity of abstract concepts and turning them into interactive objects (Kalinichenko, 2017).

Further, visual aids and interactive tools can be particularly designed and developed for a specific group of users. For example, an app on IoT for homeowners will teach them how to safeguard their homes using these smart devices interconnectivity while a simulation that targets policymakers in the region will show how AI affects regional wildfire risks. The use of such visual aids to inform decision-making, especially in conditions such as the wildfires, has been proved, as some of types of the visualization enhance stress-induce decision making (Cheong et al., 2016). This way using these tools in the way that correspond to the targeted group's needs and preferences, campaigns will have more impact.

Incorporating Success Stories and Real-World Examples

Including success stories and best practices and practical examples into people's information campaigns can be very powerful in showing the benefits of AI and IoT approaches to wildfire management. Such real-life examples make these technologies easier to embrace by the audience as they can pinpoint to the effectiveness of process implementation.

For instance, using examples such as using artificial intelligence and predictive analysis to contain the loss during a fire outbreak in California can help in making the audience to see the full potential of these technologies (Jain et al., 2020). Likewise, demonstrating how IoT sensors aided in identifying and preventing a fire that occurred in a distant region of Australia can teach about the utility of IoT and the opportunity to eliminate dangers and shield assets (Thangavel et al., 2023). It could be in form of videos, use of testifiers, case studies and reports, and many more. Perhaps videos or testimonials may be loaded and impactful in as much as the audience is able to hear from people or groups that have been beneficiaries of AI and IoT. More information is available in case studies and reports, including quantitative data and metrics that reveal how well the technologies work in specific contexts (Thompson et al., 2019).

Suiting these instances does not only enhance the solidity of the message of the campaign but also raises trust and reliance on AI and IoT technologies. Since people often question the efficiency and feasibility of certain technologies, presenting evidences that they have been applied in other territories or conditions can help campaigns increase the use of such technologies.

COMMUNICATION CHANNELS FOR PUBLIC EDUCATION CAMPAIGNS

Using the appropriate channels in the dissemination of information about AI and IoT in managing wild fire is very crucial. Every channel has advantages and disadvantages for the given target audience hence the need to adopt a combination of channels in the communication effort to reach and influence the targeted subject audience. In general, currently, when popularizing public education, apart from work on the Internet and in the media space, it is necessary to use various types of offline events, mass media, and printed materials. A few most effective channels for public connect and enlightenment are identified, focusing on their parts in augmenting the understanding and acceptance of this sophisticated technologies.

Social Media Platforms: Engagement and Real-Time Updates

Social media platforms are the most helpful sources for public education, especially when it comes to involving the audience and making updates in real-time (Kaushal & Mishra, 2024). Some of these platforms include; Face book, Twitter, Intagram, You-tube that are important in that they can provide interactive communication through which information can be shared frequently and communicated to the public and embraced immediately. Since social media is readily used by people of different ages, gender and other aspects, it is an essential component of any public education regarding any issues especially regarding the youth as they are most active on social media platforms.

Social media's ability to reach a broad and diverse audience is complemented by its capacity to foster dialogue (Mishra, 2024a) and community building. For example, IoT sensors on wildfire risks provide real-time data, which could be Tweeted to the public as part of a marketing strategy to keep people informed in case of an outbreak, thereby allowing people to react in haste (Wang et al., 2016). They also afford sharing multimedia information like the video tutorials and infographics which could be used to pass information about AI and IoT in simple formats that would easier be understood by the general populace (Slavkovikj et al., 2014). However, with application of social media, people are able to engage in a constant dialogue, and be able to ask questions and share their experiences, as well as feedback, which enhances the creation of a community, thus enhancing trust in the technologies being promoted. Social media campaigns can also employ the use of advertising which can target people with specific messages, which is relevant to a particular group (Reuter et al., 2021).

Community Workshops, Seminars, and Demonstrations

Organizing community sessions for example, workshops, seminars or demonstrations goes a long way in establishing a comprehensive understanding of AI and IoT products within the public. These events are rather practical, during which participants get a firsthand experience about the technologies under discussion, may ask questions, and get answers on the spot. This is because workshops give hands-on experience on the utilisation of AI together with IoT devices including how to use smart home to prevent fires while seminars on the other hand may help a person have an overview of how AI and IoT can assist in suppressing wildfires. These events are particularly useful in popularizing and establishing confidence in new technologies on account of the fact that they provide an opportunity for direct and unmediated engagements with experts and real-life performance of these technologies (White, 2004).

This is especially important since demonstrations make the technology seem more familiar and tangible since they allow for exposure to how it works in practice. For instance, making a real-time data flow of an IoT sensor network designed to track environmental conditions in a fire-endangered region, can present simply and ad hoc practical demonstration of the IoT sensors in use (Hoang et al., 2010). It can also involve specifics in the needed information and concerns of the selected population, which makes the event more practical and interesting. This is especially true in rural areas where people may have poor technological literacy; these workshops will therefore concern itself with concrete ideas and where possible the fundamental ideas, to ensure that whoever attends the workshop will be in a position to know how to use AI and IoT to protect homes and communities from the ravages of wildfire (Parkinson et al., 2003). Furthermore, it provides people from the community with concerns being voiced, and the organizers who can set the record straight diminishing further doubts as to the safety of the technology.

Traditional Media: Newspapers, Television, and Radio

While there has been an increased emphasis on digital media many of the traditional media types like newspapers, Television and Radio are still essential means to target certain audiences particularly the elder and those in the rural areas. These channels are considered to be more creditable and therefore can be very useful in supporting the digital strategies by giving more detailed information and reaching larger audience. Newspapers can present extensive articles on the function of AI in predicting the wildfire spread, and since newspapers have had a longstanding history of being trusted sources of information for the public, readers learn how these technologies function (Baroutsis et al., 2018). Public awareness can be favoured through feature stories and, or op-eds on specific newspapers, which will expound on the benefits derived from applying AI and IoTs in the management of wild fires, as well as put into consideration anticipated limitations and drawbacks.

Television due to its visual impact can help the public see IoT through showcasing the technologies that support it thus creating easier understanding of esoteric issues. For example, the AI could be presented in the form of an advertisement where a news crew demonstrates the AI's capabilities for forecasting wildfire behavior, while interviewing professionals who could give a detailed insight into the AI algorithms. Such segment not only provides information but also gives the feel to the people by illustrating that how these technologies are useful in the real-world scenario (Kow, 2002). The utilization of radio to pass time-specific information to the communities is very relevant since it does not have issues with the internet in passing information as found in other electronic media. Through radio, people can act upon a message as quickly as it was delivered, especially important, for instance, to inform the population about the risks of wildfire that IoT sensors have spotted or describe the measures people should take in the case of a fire threat predicted by AI. The use of traditional media makes it easy to pass information to the public without leaving any social status unaddressed because not everybody uses the internet.

Distribution of Educational Materials: Brochures and Online Resources

Educational and informative printed and online materials are crucial to introduce people to elaborate information which they can further use in case of need. Flyers and booklets widely spread throughout centers for the population, schools, and other public grounds present brief and clear descriptions of AI and IoT with directions for using these technologies in the fight against wildfires. Such materials may include the procedures of installing IoT devices and using them, diagrams exhibiting how AI models estimate wildfire spread, and directories of the needed local aid contacts (McCaffrey, 2004).

Websites, blogs, and other downloadable PDFs provide additional information about these technologies and feature more detailed descriptions, examples, and information about other resources people can read or study. These are the resources can be used at any given time because it will foster continuous learning about techniques and approaches in the fight against wildfire incidents by AI and IoT (Robichaud & Ashmun, 2013). Wildfire related websites, particular those maintained by federal or state organizations, can directly or through links, offer the use of AI and/or Web based applications like wildfire risk assessment decision-support calculators, virtual or otherwise, essential in providing an application interface where individuals can input and compare the results of various alternatives with regards to the effectiveness of certain forms of fire management (Kalabokidis et al., 2015). Also, there is a possibility to hold online postings with expert interviews, success stories and the latest news on the growth of Artificial intelligence and IoT to keep the audiences interested in the project for long time. Printed as well as the electronic formats are utilized in the preparation of materials provided in education which makes it easier for all people to access educations materials irrespective of the type of media, and the level of technology available to them.

COLLABORATIVE EFFORTS IN PUBLIC EDUCATION CAMPAIGNS

People's cooperation is essential when promoting AI and IoT use in the fight against wildfires in public awareness campaigns. Campaigns need to pull together resources, experience, and contacts from multiple stakeholders in order to have wide promotion, applicability and effect. With the support of local governments, fire departments, educational institutions, environmental organizations and other entities within the community it becomes possible to create a well-informed and an active community within those campaigns. It was noted that there are certain works which could significantly improve the effectiveness of public education focusing on cooperation with local administrators and fire departments, involving educational establishments and environmental non-governmental organizations, including AI and IoT specialists, and using volunteer networks.

Partnering with Local Governments and Fire Departments

Local governments and fire departments are critical partners in the formulation and execution of public education about AI and IoT in wildfire issues. These sources are usually more credible during emergencies, and willingness by these organizations can greatly improve the effectiveness of public education campaigns. Engagement with the local government helps campaigns leverage these existing channels — workshop, newsletters, town hall, social media platforms which are critical in reaching the public with essential information (Wise, 2021).

Furthermore, fire departments themselves get exposure to the practicality and drawbacks of how to tackle wildfires and could prove useful in discovering the best way to incorporate AI and IoT into practice. One of the best ways of ensuring that campaigns create and disseminate proper content is to engage fire departments in the creation of the content that is to be disseminated. For instance, fire departments can also show how sensors in the IoT system can be employed to detect areas that are prone to fire incidence, as well as how AI based models can be used to predict fire rapidness thus enabling the public to understand and trust these technologies (Madsen et al., 2018).

Engaging Educational Institutions and Environmental Organizations

Governmental institutions, educational institutions and non – governmental organizations working in the field of environmental protection are most valuable in the framework of public campaigns, especially the ones targeted at youth and environmentally concerned people. The SYSTEM can be implemented at schools, college, and universities, where AI and IoT subjects can be included in the curriculum facilitating the knowledge of the role of these technologies in preventing and combating wildfires by students. Through the use of case studies, simulation and practical activities in stem education, the education facilities can nurture the new generation citizens who are informed citizens and be.

In addition, environmental organizations offer valuable expertise of the local environment and the problems related to the environment in different areas. These organizations usually have credibility in these communities, thus becoming good partners for any campaign targeting to sensitize communities on the ecological advantages of using AI and IoT in wildfire management. Engaging with environmental organizations may also assist in making sure the educational content reflects other large-scope objectives of environmental conservation including, conservation of habitats and mitigation of climate change (Moritz et al., 2014).

Involvement of AI and IoT Experts in Public Discussions

It is important to involve AI and IoT professionals in the public conversations so that information provided in educative campaigns is relevant and current with advances in technology. These specialists can give a detailed description regarding the functions of AI and IoT systems, the possibilities of their application, as well as the potential problems and concerns regarding AI and IoT systems. In reaching to the public through public forums, webinars, and similar community gatherings, AI and IoT experts can answer to questions and allay fears not only of the general public but at least of the concerned sectors (Kaur & Sood, 2020).

However, interaction of such experts with the general public and participation of the later in public awareness campaigns can close this gap. Scientists and researchers in these interconnected fields work at enhancing their communication styles and explaining principles and concepts in simpler language that is well understood by wider audiences, thus the need for popular or general explanations to the practicality and necessity of artificial intelligence and IoT in the eradication of wildfires. This also helps in the convinces the public since the information is from recognized authorities in the field of study (Williams et al., 2020).

Leveraging Partnerships and Volunteer Networks

Using partnerships and volunteers is a great way to increase the extension rationale of public education messages. Community activists and those concerned with environmental conservation and triage during wildfires may become strong allies for the use of AI and IoT techniques in managing these disasters. Such people can assist share and spread the information, coordinate community activities and create a base for mobilizing the population and advocacy in connection with the application of new technologies for preventing wildfires (Grant & Langer, 2021).

Other stakeholder engagement tools, such as cooperation with community-based organizations, non-governmental and non-profit organizations and local business can also improve the efficiency of public awareness raising activities. These partnerships can bring enhancement to other forms of support like funding, event space for activities and contacts to even more people who may be interested in supporting the cause. For instance, cooperation with a hardware shop may involve providing the shop with equipment and conducting demonstration of how to fit and operate the IoT devices for the purpose of preventing wildfire in a house, Cooperation with a non-profit organization may involve offering education especially to a disadvantaged society. Thus, campaigns can enhance the community by bringing together passionate partners and volunteers aware of the situation and ready to apply the AI and IoT in combating wildfires.

EVALUATING THE IMPACT OF PUBLIC EDUCATION CAMPAIGNS

It is important to assess the efficacy of public education campaigns as part of the strategy of using AI and IoT in wildfire management because without such an assessment the success, or lack thereof, of the campaigns in raising awareness about the technologies cannot be determined. It is not only beneficial for evaluating the achieved results at the end of the campaigns though important for providing suggestions of how the campaigns can be further improved. These include, client feedback, monitoring audience interaction with the content through statistics, and studying past similar projects and campaigns, public perception, and limited funds. There are specific approaches that may be employed to assess the effectiveness of public education intervention and as well confirm the appropriateness and relevance of the approach utilized in meeting the objective of the target population.

Methods for Gathering Feedback and Insights

Gathering feedback of the target audience is one of the most important components of assessing the effectiveness of public awareness campaigns. One of the biggest advantages of feedback is the ability to learn about the audience's comprehension, perception and interaction with AI and IoT technologies in combating wildfires. When this feedback is collected in a systematic manner, it gives the campaign organizers the best insight of how well the campaigns' messages went down, what went wrong and what could be done in order to improve in the future.

There are several ways of getting this feedback and these include; use of questionnaires, focus group discussion, interviews and social media analysis. Questionnaires are the most widespread research instruments for measuring the level of audience awareness and attitudes quantitatively. They can be released to the public through websites blogs, local events or offered in collaboration with organizations. Website-based surveys must be developed to contain questions whose answers indicate the respondent's level of awareness of AI and IoT, as well as the perceived risks and benefits of embracing those technologies. The survey results can ensure a first glance at the push campaign success, the aspects that were successful, and the areas that need improvement (Rodrigues & Riva, 2014).

While, focus groups and interviews provide method that can be defined as qualitative. These methods enable an exploration of issues and permit participants to voice their concerns, questions and detailed input. In general, focus groups involve the acquisition of limited numbers of people from each social group for discussion of certain aspects of the campaign. Focus groups are less structured than is the case with survey data and enable one to capture attitudes and beliefs not apparent in a survey. In the same regard, the views of formal and informal leaders within the community, from the local government, and those from environmental nongovernmental organizations can be sought to give an assessment of the efficiency of the campaign, as well as its relevance to local interests and values (Bot & Borges, 2022).

Another effective approach deemed to provide real time information is social media monitoring. Through the separately provided comments, shares and discussions on Twitter, Facebook and Instagram the campaign organizers may get an insight of how people and groups perceive particular messages of the campaign and how these messages spread. Social networking sites provide a constantly changing and real-time reaction to the message of the campaign proponents where the campaign organizers themselves can respond to misunderstandings or issues immediately. Furthermore, the demographics and geographic distribution of the audiences consuming content produced through social media can in fact be effectively analyzed, which can aid in better targeting subsequent campaign initiatives (Lovejoy & Saxton, 2012).

Using Data Analytics to Measure Engagement and Adjust Strategies

The information gained through data analytics is particularly useful when analyzing the success of public educational campaigns and their impact on the target group. The overall engagements act as a guide to evaluating how well the campaign is being understood by the target market as well as if messages being passed are understood well. Through web metrics, visitors as measured by website impressions, likes shared on social media platforms, view rate for videos, impressions to the webinars, and participants to the workshops can be examined to gauge the campaign coverage. These quantifiers assist the campaign organizers know the parts of the campaign that perform well and the parts that call for enhancement (Vytasek, Nosek, & Corley, 2019).

Google Analytics is the perfect tool for analyzing the website traffic and its users. It gives specifics of the manner in which visitors engage the available online materials concerning the campaign, which material type gets the most hits, the extent of time the visitors spend on the site, and the most visited pages. Such information can inspire the creators of campaigns to define what topics or messages are potentially interesting to the target audience and modify the content of web sites. Furthermore, Google Analytics can capture the location of the visitor of the website and therefore provide regional information on the campaign. In addition, tools like Smart Learning Analytics not only capture and report the user interactions but indeed show captive and futuristic insights by analyzing the tendency of actions and resulting effects (Uskov et al., 2020).

Some of the most used tools include Hootsuite, Sprout Social and the Twitter Analytics as they include data on engagement in the social media form of communication. Some of these tools can track engagements like; likes, shares, comments, new followers, and particular post engagement. From these metrics, the coordinators of the campaign can then identify which social networks and kind of posts best suits the targeted audience. For instance, if an infographic or video that was circulated got a lot of engagement then it might mean that the campaign's audience appreciates visuals and this will make the organizers to create more visuals in the future (Foster & Francis, 2020). Moreover, employing concept like educational data mining which currently is in downfall in enhance the identification of potential behaviors and interest of various audience in order to deliver efficient campaign (Slater et al., 2017).

The use of data analysis also ensures the campaigns' ongoing monitoring and customization. Through the analysis of engagement information, the initiators of the campaign can recognize the strategies and messages that have proved more or less effective. That is, if data reveals that performance decreases at specific times, social media managers may revise the schedule to achieve higher results. Likewise, if the individual messages that are promoted also garner little response the campaign may need to change those messages or try other message types. This means that the campaigns can now be optimized for the audience in real time, changes can be made where necessary to make the campaigns to meet the needs of the audience better than they are currently (Nasruddin et al., 2017). Also, it allows assessing the further level of possible engagement and applying this information to long-term promotion planning to achieve continuous results (Sghir et al., 2022).

Analyzing Case Studies: Australia, California, and Canada

Examining the examples of regions that have launched public awareness campaigns on AI and IoT use in combating wildfires offers insights and useful experiences. Three states that have experienced many challenges due to wild fires include Australia, California and Canada and some of the measures getting employed included making a public awareness to adopt improved technologies. These examples demonstrate what barriers lie in the successful implementation of such campaigns and what obstacles should be avoided.

Australia has been among the leading nations in incorporating AI technologies for purposes of combating wildfires more especially through the Bushfire and Natural Hazards Cooperative Research Centre (BNHCRC). Working to halt and mitigate wildfires, the BNHCRC has produced and shared AI technologies for predicting and improving management of the phenomena, and to ensure the public is aware and embracing the technology being employed. Studies on these campaigns have found out that what these campaigns need is consistent and clear messages alongside active community engagement. While the campaigns highlighted the factors which create communities' readiness for AI implementation, the targeting of multiple practical benefits of using the AI technologies made people more inclined and trusting in the latter (Neale et al., 2016).

In California for instance, there has been massive awareness being created on the use of IoT sensors for monitoring the prone areas to fires in real time. The state has engaged technology companies and environmentally conscious organizations, with the IoT networks that offer the early signs of wildfire probability. An analysis of these measures underscores the need to show that IoT solutions can make a difference in the real world to spur consumer acceptance. The same procedure favors California because campaigns require preparedness, and since the state has encouraged an early readiness plan toward disasters, people have become familiar with the concepts (Keivanpour, 2021).

Another successful example of the integration of AI and IoT into the promotion of Fire Smart program in Canada. The Fire Smart program has therefore embraced community engagement and artificial intelligence technologies on the prevention and control of extreme wildfires. The success of the program especially in the remote area and Indigenous communities has been attributed to the focus of addressing local issues and delivering solutions that would appeal to such areas. The reviews of Fire Smart program have demonstrated that people accept and support public education campaigns if the campaigns reflect the cultural believes and norms of the entire community as posited by McGee 2011.

These case studies demonstrate that context specific approaches are appropriate when using public education messages. Although the principles of efficient communication, involvement of the community and openness are the same everywhere, the methods and the messaged should be tuned to the specifics of each area. In this way, the achieved results and problems identified in these case studies can become the knowledge for the further campaign planning by the organizers.

Addressing Challenges: Public Skepticism and Resource Constraints

Some of the issues, which must be effectively managed, include; Challenge of skepticism among the public; and the challenge of limited funds to finance the campaign. Concerns about AI and IoT technologies come from ignorance, people's phobias, or just their common fear caused by emergent technologies of the lack of trust in the authorities and professionalism of companies and states. In order to counter this skepticism campaigns must promote clarity and accuracy of the campaign's message offering concrete data and dispelling myths. For example, public trust as a value was noted as being especially important in relation to new technologies for management of wildfires (Olsen et al., 2014).

A useful approach of combating skepticism is the inclusion of local authorities and influential persons in the campaign. These people can have considerable influence in their respective societies and can help support the messages being disseminated. The support of these people is crucial for establishing the trust of society and, in consequence, letting the people adopt new technologies more willingly. Also, the presentation of live models and findings of accomplishment backed by AI and IoT can explain the real-life significance of these technologies and reduce their fearsomeness in the eyes of the common populace (Gill et al., 2013).

The other main challenge is resource limitations especially financial and other support necessitating the public education campaigns. Some disadvantages of using this approach include While the lending and campaigns may be expensive, limited funding may reduce the range of the campaign and limit the range of the educational materials and activities places believes that logistical difficulties may also limit the transport of the educational materials and activities to remote and difficult areas. To solve these problems, the campaigns can either partner with companies, non-governmental organizations and other governmental agencies to share the burden of most of the costs associated with campaign activities. Such partnerships can bring extra funding, knowledge, and market, all of which will improve the campaign's efficacy.

Use of digital media and social media is another way of overcoming a challenge arising from limited resources. Campaigns are cheap through Internet based platforms and the targets an extensive reach compared to physical forms of media. In particular, social networks provide an opportunity to disseminate information, interact with the audience, and receive feedback quickly and, most importantly, for little or no money. Through adoption of advertisement specific targeting and contents that have been optimized for their reach, the campaigns can thus be effective irrespective of the amount of funds available.

Addressing the effects of public education campaigns along AI and IoT technologies in addressing wildfire calls for multifaceted and flexibility in assessment. Including feedback collection, data, case studies analysis, and issues like public skepticism reward and resource limitations whereby the campaign organizers can understand the effectiveness of efforts, the consideration of the responses and concerns of target groups, and the relevance of organized campaigns to target communities. Campaigns of this nature need constant monitoring within their targeted areas with the aim of fine-tuning them for the best intended target audience outreach with the end aim of preserving lives and property through better and sustainable prevention and combat of wildfires.

Based on the findings of this chapter, a comprehensive model (Figure 1) and an implementation framework (Figure 2) are proposed to improve the effectiveness of public education campaigns on AI and IoT applications in wildfire prevention and management. This type of model serves to ensure that every communication strategy formulated and developed is coherent, relevant and effective in dealing with the issue of these technologies with respect to different audiences who may be uncomfortable with such technologies. It consists of seven key components: audience identification, message crafting, use of visual aids and interactive tools, incorporation of success stories, multi-channel communication, partnership and collaboration, and feedback and evaluation. Table 1 provides a detailed overview of the key components, objectives, and strategies involved in the proposed strategic communication model. This table serves as a guide for effectively designing and executing public education campaigns, emphasizing the need for clarity, relevance, and adaptability in communication efforts. By following this model, public education campaigns can more effectively promote the understanding and adoption of AI and IoT technologies in wildfire management, ultimately enhancing community resilience and safety.

FUTURE DIRECTIONS AND RECOMMENDATIONS

To increase the effectiveness of using AI and IoT technologies for promoting public education in wildfire prevention and management several future directions should be taken. First of all, it is necessary to advance the adaptive communication approach based on potential AI data analysis. Such campaigns can utilize real time data on top of machine learning models to ensure the messages get to the different stakeholders when the information is still relevant depending on developments in technology and conditions in the physical world. To implement this form of learning, systematic processes must be developed that facilitate networking between technology developers, wildfire management professionals, educators and policymakers.

Secondly, the extension of utilizing some enhanced new media and interactive features in a plethora of the public educational campaigns can enhance the level of audience participation and understanding dramatically. Using AR and VR it is possible to stage learning scenarios that would provide the key stakeholders – community members, first responders, etc. – with a better sense about the behavior of wildfires as well as useful applications of AI and IoT. These tools can also help make complex information much easier to understand and be much more engaging which will then increase the citizens' trust and acceptance of the given information.

Thus, future studies should work on constructing AI models for communities that will be adapted to the features of a concrete region. Because these models are developed and tested with input from the local population in conjunction with public health campaigns regarding these AI tools, they can be even more effective. This application does not only improve the way wildfires are forecasted through technology but also ensures people embrace such interventions willingly.

It is also important to increase informational literacy going beyond explaining cases of AI and IoT usage for ethical purposes and precautionary measures of personal data leaks. This can be done in an ongoing effort to observe users' wants regarding data assembly, utilization, and safety, which help to build affection and coordination between the central and sub-state levels for the long term in the face of the rising public concern with surveillance and data protection. It is recommended that more specific directions on these concerns must be produced to contribute to the moral enrolment of AI and IoT in combating wildfire.

Finally, another direction for further development is the affiliation of renewable energy sources with IoT systems for use in such technologies in wildfire management. Such an integration may improve the reliability of such technological applications especially in areas which require power especially in the event of a wildfire calamity. As a result, it is crucial to continue emphasizing following areas as a potential for future work in raising public awareness.

CONCLUSION

The integration of AI and IoT technologies in wildfire prevention and management has the potential to revolutionize how communities prepare for and respond to wildfires. Such efforts in public education are imperative for the transformation as they help inform the populace and ensure their acceptance of the said sophisticated technologies. The success of these campaigns depends on how well sophisticated technology information is translated and delivered to the public, special audiences targeted, and how ethical issues in the uses of technology and data are tackled.

Moving forward, the ability to meet the goals of public education will depend on the extent to which these strategies remain relevant with the technological advances together with the environment existing today. We can improve their quality incorporating real-time data analysis, AI models tailored for communities, augmented and virtual reality. In addition, the encouragement of partnerships between technology developers, wildfire management workers, policy makers and the public as key stakeholders will be crucial for the development of a coordinated and appropriately normative approach to wildfire management.

Ultimately, a well-informed and engaged public is vital for the effective implementation of AI and IoT technologies in wildfire prevention and management. By continuing to advance public education campaigns and addressing emerging challenges, we can build a more resilient and proactive society capable of mitigating the devastating impacts of wildfires.

REFERENCES

Bostrom, N., & Yudkowsky, E. (2014). The ethics of artificial intelligence. In Frankish, K., & Ramsey, W. M. (Eds.), *The Cambridge Handbook of Artificial Intelligence* (pp. 316–334). Cambridge University Press. DOI: 10.1017/CBO9781139046855.020

Bot, K., & Borges, J. G. (2022). A Systematic review of applications of machine learning techniques for wildfire Management decision support. *Inventions (Basel, Switzerland)*, *7*(1), 15. DOI: 10.3390/inventions7010015

Bowman, D. M. J. S., Balch, J., Artaxo, P., Bond, W. J., Cochrane, M. A., D'Antonio, C. M., DeFries, R., Johnston, F. H., Keeley, J. E., Krawchuk, M. A., Kull, C. A., Mack, M., Moritz, M. A., Pyne, S., Roos, C. I., Scott, A. C., Sodhi, N. S., & Swetnam, T. W. (2011). The human dimension of fire regimes on Earth. *Journal of Biogeography*, *38*(12), 2223–2236. DOI: 10.1111/j.1365-2699.2011.02595.x PMID: 22279247

Brusilovsky, Peter & Ahn, Jae-wook & Rasmussen, Edie. (2010). Teaching Information Retrieval With Web-based Interactive Visualization. *Journal of Education for Library and Information Science*, *51*. Advance online publication. DOI: 10.2307/40732598

Butler, W. H., & Goldstein, B. E. (2010). The US Fire Learning Network: Springing a Rigidity Trap through Multiscalar Collaborative Networks. *Ecology and Society*, *15*(3), art21. Advance online publication. DOI: 10.5751/ES-03437-150321

Cheong, L., Bleisch, S., Kealy, A., Tolhurst, K., Wilkening, T., & Duckham, M. (2016). Evaluating the impact of visualization of wildfire hazard upon decision-making under uncertainty. *International Journal of Geographical Information Science*, *30*(7), 1377–1404. DOI: 10.1080/13658816.2015.1131829

Donovan, G., & Brown, T. (2007). Be Careful What You Wish for: The Legacy of Smokey Bear. Frontiers in Ecology and The Environment -. *Frontiers in Ecology and the Environment*, *5*(2), 73–79. DOI: 10.1890/1540-9295(2007)5[73:BCWYWF]2.0.CO;2

Foster, C., & Francis, P. (2019). A systematic review on the deployment and effectiveness of data analytics in higher education to improve student outcomes. *Assessment & Evaluation in Higher Education*, *45*(6), 822–841. DOI: 10.1080/02602938.2019.1696945

Gill, A. M., Stephens, S. L., & Cary, G. J. (2013). The worldwide "wildfire" problem. *Ecological Applications*, *23*(2), 438–454. DOI: 10.1890/10-2213.1 PMID: 23634593

Grant, A., & Langer, E. R. (2021). Wildfire volunteering and community disaster resilience in New Zealand: Institutional change in a dynamic rural social-ecological setting. *Ecology and Society*, *26*(3), art18. Advance online publication. DOI: 10.5751/ES-12474-260318

Hoang, R. V., Sgambati, M. R., Brown, T. J., Coming, D. S., & Harris, F. C.Jr. (2010). VFire: Immersive wildfire simulation and visualization. *Computers & Graphics*, *34*(6), 655–664. DOI: 10.1016/j.cag.2010.09.014

Hoang, R. V., Sgambati, M. R., Brown, T. J., Coming, D. S., & Harris, F. C.Jr. (2010b). VFire: Immersive wildfire simulation and visualization. *Computers & Graphics*, *34*(6), 655–664. DOI: 10.1016/j.cag.2010.09.014

Jain, P., Coogan, S. C., Subramanian, S. G., Crowley, M., Taylor, S. W., & Flannigan, M. D. (2020). A review of machine learning applications in wildfire science and management. *Environmental Reviews*, *28*(4), 478–505. DOI: 10.1139/er-2020-0019

Johnston, F. H., Borchers-Arriagada, N., Morgan, G. G., Jalaludin, B., Palmer, A. J., Williamson, G. J., & Bowman, D. M. J. S. (2020). Unprecedented health costs of smoke-related PM2.5 from the 2019–20 Australian megafires. *Nature Sustainability*, *4*(1), 42–47. DOI: 10.1038/s41893-020-00610-5

Kalabokidis, K., Ager, A., Finney, M., Athanasis, N., Palaiologou, P., & Vasilakos, C. (2016). AEGIS: A wildfire prevention and management information system. *Natural Hazards and Earth System Sciences*, *16*(3), 643–661. DOI: 10.5194/nhess-16-643-2016

Kalinichenko, A. (2017). Interactive electronic educational tools with cognitive visualization. *Scientific and Technical Journal of Information Technologies Mechanics and Optics*, 359–364. DOI: 10.17586/2226-1494-2017-17-2-359-364

Kapucu, N., & Garayev, V. (2011). Collaborative Decision-Making in emergency and disaster management. *International Journal of Public Administration*, *34*(6), 366–375. DOI: 10.1080/01900692.2011.561477

Kaur, H., & Sood, S. K. (2020). A smart Disaster management framework for wildfire detection and prediction. *The Computer Journal*, *63*(11), 1644–1657. DOI: 10.1093/comjnl/bxz091

Kaushal, S., & Mishra, D. (2024). Strategic implications of AI in contemporary business and society. 2024 *2nd International Conference on Disruptive Technologies (ICDT), Greater Noida, India*. DOI: 10.1109/ICDT61202.2024.10489027

Keivanpour, S. (2021). Internet of things in Optimizing Emergency logistics of Wildfire: A Review and Application Perspective in Canada. *2021 International Conference on Electrical, Computer and Energy Technologies (ICECET)*. https://doi.org/DOI: 10.1109/ICECET52533.2021.9698250

Kow, J. (2002). Television as a Tool of Environmental Adult Education: Limits and Possibilities. *Adult Learning*, *13*(2-3), 30–34. DOI: 10.1177/104515950201300207

Madsen, R., Haynes, H., & McCaffrey, S. (2018). Wildfire risk reduction in the United States: Leadership staff perceptions of local fire department roles and responsibilities. *International Journal of Disaster Risk Reduction*, *27*, 451–458. DOI: 10.1016/j.ijdrr.2017.11.009

Mandel, J., Beezley, J., Coen, J., & Kim, M. (2009). Data assimilation for wildland fires. *IEEE Control Systems*, *29*(3), 47–65. DOI: 10.1109/MCS.2009.932224

Marjani, M., Nasaruddin, F., Gani, A., Karim, A., Hashem, I. T., Siddiqa, A., & Yaqoob, I. (2017). Big IoT Data Analytics: Architecture, opportunities, and open research challenges. *IEEE Access : Practical Innovations, Open Solutions*, *5*, 5247–5261. DOI: 10.1109/ACCESS.2017.2689040

Mayer, R. E. (2009). *Multimedia learning* (2nd ed.). Cambridge University Press., DOI: 10.1017/CBO9780511811678

Mayer, R. E. (2009). *Multimedia learning*. HTTPs://DOI: 10.1017/CBO9780511811678

McCaffrey, S. (2015). Community Wildfire Preparedness: A Global State-of-the-Knowledge Summary of Social Science Research. *Current Forestry Reports*, *1*(2), 81–90. DOI: 10.1007/s40725-015-0015-7

McCaffrey, S. M. (2004). Fighting Fire with Education: What Is the Best Way to Reach Out to Homeowners? *Journal of Forestry*, *102*(5), 12–19. DOI: 10.1093/jof/102.5.12

McGee, T. (2011). Public engagement in neighbourhood level wildfire mitigation and preparedness: Case studies from Canada, the US and Australia. *Journal of Environmental Management*, *92*(10), 2524–2532. DOI: 10.1016/j.jenvman.2011.05.017 PMID: 21684061

Mishra, D. (2024a). Role of Social Media and AI in Transforming Healthcare Communication. In Vig, K., Kaunert, C., & Singh, B. (Eds.), *Intersections of Law and Computational Intelligence in Health Governance* (pp. 29–51). IGI Global., DOI: 10.4018/979-8-3693-5976-1.ch002

Mishra, D. (2024b). Strategic Communication for AI-Driven Sustainability Initiatives: Bridging Technology, Policy, and Public Engagement in Achieving SDGs. In Riswandi, B., Singh, B., Kaunert, C., & Vig, K. (Eds.), *AI Applications for Clean Energy and Sustainability* (pp. 187–211). IGI Global., DOI: 10.4018/979-8-3693-6567-0.ch010

Monroe, M. C., Ballard, H. L., Oxarart, A., Sturtevant, V. E., Jakes, P. J., & Evans, E. R. (2015). Agencies, educators, communities and wildfire: Partnerships to enhance environmental education for youth. *Environmental Education Research*, *22*(8), 1098–1114. DOI: 10.1080/13504622.2015.1057555

Moritz, M. A., Batllori, E., Bradstock, R. A., Gill, A. M., Handmer, J., Hessburg, P. F., Leonard, J., McCaffrey, S., Odion, D. C., Schoennagel, T., & Syphard, A. D. (2014). Learning to coexist with wildfire. *Nature*, *515*(7525), 58–66. DOI: 10.1038/nature13946 PMID: 25373675

Moritz, M. A., Parisien, M., Batllori, E., Krawchuk, M. A., Van Dorn, J., Ganz, D. J., & Hayhoe, K. (2012). Climate change and disruptions to global fire activity. *Ecosphere*, *3*(6), 1–22. DOI: 10.1890/ES11-00345.1

Morss, R. E., Wilhelmi, O. V., Downton, M. W., & Gruntfest, E. (2005). Flood Risk, Uncertainty, and Scientific Information for Decision Making: Lessons from an Interdisciplinary Project. *Bulletin of the American Meteorological Society*, *86*(11), 1593–1602. DOI: 10.1175/BAMS-86-11-1593

Myoung, B., Kim, S., Nghiem, S., Jia, S., Whitney, K., & Kafatos, M. (2018). Estimating Live Fuel Moisture from MODIS Satellite Data for Wildfire Danger Assessment in Southern California USA. *Remote Sensing (Basel)*, *10*(2), 87. DOI: 10.3390/rs10010087

Neale, T., Weir, J. K., & McGee, T. K. (2016). Knowing wildfire risk: Scientific interactions with risk mitigation policy and practice in Victoria, Australia. *Geoforum*, *72*, 16–25. DOI: 10.1016/j.geoforum.2016.03.008

Olsen, C. S., Mazzotta, D. K., Toman, E., & Fischer, A. P. (2014). Communicating About Smoke from Wildland Fire: Challenges and Opportunities for Managers. *Environmental Management*, *54*(3), 571–582. DOI: 10.1007/s00267-014-0312-0 PMID: 25034752

Parkinson, T., Force, J., & Smith, J. (2003). Hands-on Learning: Its Effectiveness in Teaching the Public about Wildland Fire. *Journal of Forestry*, *101*(7), 21–26. DOI: 10.1093/jof/101.7.21

Rein, G., & Huang, X. (2021). Smouldering wildfires in peatlands, forests and the arctic: Challenges and perspectives. *Current Opinion in Environmental Science & Health*, *24*, 100296. DOI: 10.1016/j.coesh.2021.100296 PMID: 34950823

Reuter, K., Wilson, M. L., Moran, M., Le, N., Angyan, P., Majmundar, A., Kaiser, E. M., & Unger, J. B. (2021). General audience engagement with antismoking public health messages across multiple social media sites: Comparative analysis. *JMIR Public Health and Surveillance*, *7*(2), e24429. DOI: 10.2196/24429 PMID: 33605890

Rhoades, E. B., Irani, T., & Telg, R. (2005). Assessing internet use in Florida newsrooms. *Journal of Applied Communications*, *89*(2). Advance online publication. DOI: 10.4148/1051-0834.1304

Robichaud, P. R., & Ashmun, L. E. (2013). Tools to aid post-wildfire assessment and erosion-mitigation treatment decisions. *International Journal of Wildland Fire*, *22*(1), 95. DOI: 10.1071/WF11162

Rodrigues, M., & De La Riva, J. (2014). An insight into machine-learning algorithms to model human-caused wildfire occurrence. *Environmental Modelling & Software*, *57*, 192–201. DOI: 10.1016/j.envsoft.2014.03.003

Sghir, N., Adadi, A., Mouden, Z. E., & Lahmer, M. (2022). Using learning analytics to improve students' enrollments in higher education. 2022 2nd International Conference on Innovative Research in Applied Science, Engineering and Technology (IRASET). DOI: 10.1109/IRASET52964.2022.9737993

Slater, S., Joksimović, S., Kovanovic, V., Baker, R. S., & Gasevic, D. (2016). Tools for educational data mining. *Journal of Educational and Behavioral Statistics*, *42*(1), 85–106. DOI: 10.3102/1076998616666808

Slavkovikj, V., Verstockt, S., Van Hoecke, S., & Van De Walle, R. (2014). Review of wildfire detection using social media. *Fire Safety Journal*, *68*, 109–118. DOI: 10.1016/j.firesaf.2014.05.021

Steelman, T. A., & McCaffrey, S. (2012). Best practices in risk and crisis communication: Implications for natural hazards management. *Natural Hazards*, *65*(1), 683–705. DOI: 10.1007/s11069-012-0386-z

Thangavel, K., Spiller, D., Sabatini, R., Amici, S., Sasidharan, S. T., Fayek, H., & Marzocca, P. (2023). Autonomous satellite wildfire detection using hyperspectral imagery and neural networks: A case study on Australian wildfire. *Remote Sensing (Basel)*, *15*(3), 720. DOI: 10.3390/rs15030720

Thompson, M. P., Wei, Y., Calkin, D. E., O'Connor, C. D., Dunn, C. J., Anderson, N. M., & Hogland, J. S. (2019). Risk management and analytics in wildfire response. *Current Forestry Reports*, *5*(4), 226–239. DOI: 10.1007/s40725-019-00101-7

Uskov, V. L., Bakken, J. P., Gayke, K., Fatima, J., Galloway, B., Ganapathi, K. S., & Jose, D. (2020). Smart Learning Analytics: student academic performance data representation, processing and prediction. In *Smart innovation, systems and technologies* (pp. 3–18). https://doi.org/DOI: 10.1007/978-981-15-5584-8_1

Vadrevu, K. P., Eaturu, A., Biswas, S., Lasko, K., Sahu, S., Garg, J. K., & Justice, C. (2020). Spatial and temporal variations of air pollution over 41 cities of India during the COVID-19 lockdown period. *Scientific Reports*, *10*(1), 16574. Advance online publication. DOI: 10.1038/s41598-020-72271-5 PMID: 33024128

Vytasek, J. M., Patzak, A., & Winne, P. H. (2019). Analytics for student engagement. In *Intelligent systems reference library* (pp. 23–48). DOI: 10.1007/978-3-030-13743-4_3

Wang, Z., Ye, X., & Tsou, M. (2016). Spatial, temporal, and content analysis of Twitter for wildfire hazards. *Natural Hazards*, *83*(1), 523–540. DOI: 10.1007/s11069-016-2329-6

White, S. M. (2004). *Bridging the worlds of fire managers and researchers: lessons and opportunities from the Wildland Fire Workshops.* DOI: 10.2737/PNW-GTR-599

Williams, K. J. H., Ford, R. M., & Rawluk, A. (2020). The role of collaborative research in learning to incorporate values of the public in social–ecological system governance: Case study of bushfire risk planning. *Ecology and Society*, *25*(4), art31. Advance online publication. DOI: 10.5751/ES-11987-250431

Wise, C. R. (2021). Accountability in Collaborative Federal Programs—Multidimensional and multi-level performance measures needed: The case of wildland fire prevention. *American Review of Public Administration*, *52*(2), 95–108. DOI: 10.1177/02750740211050367

Chapter 23
Scaling Green Hydrogen Production Through AI Innovations:
Review of Production Technologies and Global Policies Impact

Bhupinder Singh
https://orcid.org/0009-0006-4779-2553
Sharda University, India

Saquib Ahmed
https://orcid.org/0009-0008-1891-6910
Sharda University, India

Saurabh Chandra
https://orcid.org/0000-0003-4172-9968
Bennett University, Greater Noida, India

Anita Singh
Sharda University, India

ABSTRACT

Green hydrogen, which is produced via water electrolysis powered by renewable energy, has emerged as a favored solution to decarbonize sectors such as transportation, heavy industry and power generation. But things like high production cost, inefficiency, and a difficult operation still pose challenges for its widespread adoption. The transformative potential usually sits with artificial intelligence (AI). The analytics on data through AI algorithms means far better adjusted to live demands with solar and wind inputs for instance. Less energy is wasted and production costs fall. Predictive maintenance, backed with AI-driven algorithms, enables electrolysis systems to run continuously with less downtime, thereby reducing operational risks while ensuring better scalability.

DOI: 10.4018/979-8-3373-0045-0.ch023

1. INTRODUCTION

The world's energy landscape is poised for transformation as the pressure to address climate change rises. Such energy carrier that has emerged as a key enabler of this transition is green hydrogen, which represents an abundant and clean renewable option (Islam et al., 2024). Green hydrogen is a zero-emission replacement to fossil fuels created via water electrolysis running from renewable energy sources, providing a sustainable pathway for the decarbonisation of (sectors) hard to abate such as steel-making, chemicals production, aviation and heavy transportation (Sinha et al., 2021). Today, governments, industries and researchers acknowledge the potential of this technology, but scaling production to meet the skyrocketing demand on a global level remains an important challenge alongside technological or economical barriers. Predicting output fluctuations of renewable energy sources, machine learning models can help integrate intermittent renewables to the grid and produce hydrogen continuously.

Green hydrogen is unique in the energy pecking order due to its versatile and spatially diverse opportunities for sector integration. It can be kept for a long time, transferred through pipelines and buses, used in fuel cells to produce electricity or applied as an energy carrier in industrial activities. Even so, the scalability is severely constrained by the high price of production, scarcity of rare earth catalysts materials, low energy efficiency of existing production technologies and absence of storage and distribution system (Salam & Salam, 2020). However, while improvements in production techniques are being investigated, including water electrolysis, biomass-based production and photocatalytic methods, the progress is still too slow to match a decarbonized world by the 2050.

A new paradigm in the form of artificial intelligence (AI) can be the solution to this. With its ability to analyze large volumes of data, discover patterns and trends, and offer predictive analytics, AI has the potential to be a game changer when it comes to scaling up production for green hydrogen. The scope of its application ranges from production upside such as increased electrolysis efficiency to the discovery of new alternative catalysts and other materials. In addition, AI is instrumental in strengthening the coupling of variable renewable energy sources such as solar and wind with hydrogen production systems by guaranteeing that the intermittency associated with these resources does not limit accelerated deployment of green hydrogen. AI impact is not limited to production but spans the complete green hydrogen value chain. For instance, it can automate supply chains by predicting the demand in the market while also allocating resources and optimizing logistics such as storage and distribution. Digital twin and simulation models powered by AI provide stakeholders the unique ability to design and manage hydrogen production facilities through simulations providing a high-fidelity viewpoint of operational conduct (Bibri, 2018). They improve operational efficiencies while providing insights to cut costs, carbon footprints and make the system more reliable.

While there is a strong case for the potential of green hydrogen as a technology, its applicability and widespread adoption are indelibly linked to global policy. The role of governments and international organizations is crucial for its development by enabling through important frameworks, funding mechanisms and regulatory opportunities. Noteworthy sectoral developments spurring traction include the European Union Hydrogen Strategy, U. S. clean energy programs, and hydrogen initiatives in Asia-Pacific to mention a few. Clean Hydrogen Policies Industry-related measures are being implemented in the form of tax incentives for clean hydrogen production, research & development funding, infrastructure investments which have fueled innovation. Divergences in regulatory frameworks, standards and how to define "green" hydrogen differ from country to country presenting a challenge for global cooperation and the functioning of markets. At the same time, AI innovations can help raise matching creative and production practices in step with changing policy organizations. For example, industry can comply with lifecycle emission requirements and certification standards set by the regulators through machine learning algorithms. In addition, AI can pinpoint gaps in policies and provide implementable recommendations for accelerating adoption of green hydrogen worldwide.

The chapter explores the nexus of green hydrogen production technologies, AI-driven innovations and the global policy landscape. It summarizes available approaches to production, their merits and demerits. It discusses how AI will optimize these technologies, speed up material discovery and aid integration with renewable energy. In addition, the paper highlights the global policies driving green hydrogen production and the challenges and opportunities for scaling up its use. Besides, it also assesses the effects of global policies over the green hydrogen ecosystem through a financial point of view instead of technologic ones. It showcases top national and regional initiatives including the European Union's Green Deal, the United States' Inflation Reduction Act, and initiatives for hydrogen mobility in Japan. Among others, include the absence of standardized definitions and definition loopholes, as well as certification systems and fragmented legal regulation framework. The paper concludes with challenges to align policy and use AI for coherence in global action and innovation (Varjovi & Babaie, 2020).

The broader goal of this is to present a holistic framework for scaling green hydrogen production via AI innovations, where technology and policy interact in a mutually reinforcing manner. With the world moving toward a sustainable energy transition, developing holistic solutions to the complex challenges of green hydrogen adoption will be key. This provides bridging insights between technological developments and appropriate policy frameworks to set up a sustainable and fast-scaled green hydrogen ecosystem in line with the goals of global decarbonization. Green Hydrogen is not just an alternative energy resource rather it will play a vital role in the transition towards a sustainable and carbon-neutral global economy. But this potential will only be realised through concerted action to combine leading-edge technologies from AI to precision fermentation, and beyond with the policies and investments that enable them. It presents insights that contribute towards this effort and serve as a roadmap for effectively realizing the potential of green hydrogen as an essential component of the energy transition.

2. GREEN HYDROGEN PRODUCTION TECHNOLOGIES

The diversity of technologies available for green hydrogen production is a testament to the burgeoning urgency of scaling up its uptake across the globe. Although water electrolysis is a most applicable way, the development of PEM and SOEC technologies is really encouraged. The biomass-based approaches and solar driven processes such as photocatalysis and PEC cells further enrich the toolbox, responding to regional or sector-specific requirements. Innovation and investment will be key in overcoming technical, economic and material challenges to establishing green hydrogen as a linchpin of the energy transition (Ahmad & Zhang, 2021). Combined with AI-enabled optimizations, these technologies can help unlock the potential of green hydrogen to power a sustainable future. The simplest and most abundant element, hydrogen is now a key player in the energy transition to sustainability. However, all hydrogen are not the same. There are three primary hydrogen production pathways: grey, blue, and green, which are distinguished by the source used and environmental implications of the process.

2.1 Differences between Grey, Blue and Green Hydrogen

Grey Hydrogen: The most common way of producing grey hydrogen is from fossil fuels, chiefly via steam methane reforming (SMR). It is the most common and cheapest way to be done, but it emits a large amount of carbon dioxide (CO_2), which is not compatible with climate targets.

Blue Hydrogen: Blue hydrogen is similar to grey hydrogen, in that they both are based on fossil fuels, but rely on carbon capture and storage (CCS) technologies to avoid greenhouse gas emissions. However, it is not free from pollution due to dependence on nonrenewable resources and partial CO_2 capture.

Green Hydrogen: Green hydrogen uses sustainable energy to power up water electrolysis, which divides the water into hydrogen and oxygen. It is the solution that delivers zero emissions and therefore, the most sustainable choice (Bibri & Krogstie, 2020). The prohibitive production costs and technological inefficiencies have kept it out of mass adoption. Mirroring green hydrogen's transformative potential as a zero-emission solution for hard-to-abate sectors like heavy industry and aviation, there is international frenzy to scale production which as all the investigating into green H2 shows almost always comes at reducing costs.

2.2 Water Electrolysis: Technological Fundamentals and Improvements

The electrolysis of water is fundamental for the generation of green hydrogen. It uses electric current to separate hydrogen and oxygen from water molecules. Electrolyzers are key to making this process more efficient, scalable and cost effective.

2.2.1 Alkaline Electrolysis

Alkaline electrolysis is one of the most matured and long-established technologies for hydrogen production.

Mechanism: It uses a liquid alkaline electrolyte (usually potassium hydroxide) with non-precious metal electrodes.

Benefits: Alkaline electrolyzers can be made at a low cost, they are very durable and can work on a continous basis.

But their slower response times also make them less ideal for the variable output of renewable energy solar and wind. Their efficiency is moderate compared to newer technologies, which also make them less attractive for large scale green hydrogen projects.

2.2.2 Proton Exchange Membrane (PEM) Electrolysis

PEM electrolysis is a more advanced and efficient way of producing hydrogen.

Mechanism: Polymetric Electrolye Membrane (PEM) electrolyzers consist of a solid polymer membrane as the electrolyte and noble metal-based catalysts i.e platinum, iridium etc.

Advantages: Very efficient, relatively small and fast power systems that can rapidly respond to changing renewable generation inputs make them ideal partners for solar and wind.

Challenges of PEM: The primary challenge faced by PEM electrolyzers is their relatively high cost, largely due to the use of precious and uncommon materials. This restriction is being explored in researchers to generate inexpensive catalysts.

2.2.3 SOECs (Solid Oxide Electrolysis Cells)

SOECs are high-temperature electrolyzers that work at a temperature range greater than 700 °C.

SOECs (Solid Oxide Electrolysis Cells): SOECs use heat energy to greatly reduce the electrical energy associated with electrolysis which can improve efficiency.
Pros: They can fit in with existing industrial processes that naturally produce waste heat, making this technology a potential candidate for hydrogen production at an industrial scale.
Challenges: There are still issues with material durability at high operating temperatures, and SOECs are only in the experimental/pilot stage at present and need more development before they can be commercialised.

2.3 Hydrogen Production from Biomass

Hydrogen production through biomass utilizes organic materials including agricultural waste, forestry and algae and it proposes a carbon-neutral route.

Processes:
Gasification: it transforms the biomass to syngas a gas mixture consisting of hydrogen, carbon monoxide, and methane through partial combustion Once this is done, hydrogen can be collected through separation processes.
Fermentation: Uses microorganisms to convert organic matter into hydrogen without the presence of oxygen.
Benefits: It utilizes waste material, so it has a dual benefit of both managing waste and producing renewable hydrogen gas.
Limitations: Difficult to scale-up due to limited availability of feedstock, competing with food crops for available arable land and the energy intensiveness of some biomass conversion processes (Albreem et al., 2021).

2.4 Photo-Induced Process (Photocatalytic and Photoelectrochemical) Methods

Photocatalytic and photoelectrochemical (PEC) approaches are an exciting frontier in hydrogen generation, using sunlight directly to drive the water splitting process.

Photocatalysis:

Entails applying photocatalytic semiconductor materials to water splitting under sunlight. They are investigating nanostructured materials and advanced coatings for enhanced light absorption and improved catalytic efficiency.

Photoelectrochemical Cells:

PEC cells integrate photovoltaic with electrolytic functionality in one system, producing the electrical energy to drive the electrolysis process from sunlight. Titanium dioxide and perovskites lie at the core of this research for better efficiency and stability.

Advantages:

With such solutions, there is no external power which lowers operational costs. They match with this vision of producing hydrogen not in massive plants but directly on top of high demand areas potentially facilitating localized energy solutions.

Challenges:

At present, commercial viability is limited due to low conversion efficiencies and the cost of advanced materials. There will be a considerable amount of R and D to go from the lab scale up to an industrial scale.

3. ROLE OF ARTIFICIAL INTELLIGENCE IN HYDROGEN PRODUCTION

This has led to a race for green hydrogen throughout the world, forcing new ideas in order to tackle and solve existing challenges such as production, efficiency, and scalability. Artificial intelligence (AI) has become an invaluable tool in revolutionizing key components of hydrogen production and fueling its development through its ability to analyze large datasets, predict trends in the space, and optimize processes. The integration of AI across the hydrogen value chain from material discovery to system optimization is unlocking new efficiencies and encouraging the large-scale deployment of green hydrogen in the energy sector

Green hydrogen is largely produced through electrolysis, so the process involves catalysts and materials that efficiently break down water molecules. Historically, finding optimal and low-cost catalysts has been a time-consuming and expensive enterprise containing much trial-and-error experimentation. With the help of high-throughput computational screening combined with machine learning-based predictive models AI has transformed this process. In this approach, machine learning (ML) algorithms are used to analyze data from experiments, simulations, and literature in order to identify high catalytic activity, durability and low cost candidate materials from a larger pool of potential materials (Issa Zadeh et al., 2023). For example, AI models can predict how well non-precious metal catalysts will perform to replace expensive precious materials like platinum and iridium in proton exchange membrane (PEM) electrolyzers. Likewise, AI accelerates research and development of novel membranes and electrodes by simulating their behaviors under a wide range of conditions to cut material development time.

Optimization using AI is constantly changing the efficiency of the electrolyzers. Electrolysis is governed by many factors including temperature, pressure, electrolyte concentration and current density, so hydrogen yield and energy consumption are variable. Reinforcement Learning and Neural Networks based AI algorithms are capable to analyse these parameters in real-time, recommending adjustments to set for maximum efficiency. Dynamic optimization techniques driven by AI, for instance, make certain that electrolyzers always run at their highest levels of efficiency even when being connected to intermittent renewable energy sources such as solar and the wind. This can be accomplished by predicting deviations in energy input and using AI to adapt operating conditions of the systems for reduced losses during energy input fluctuations, while stabilizing hydrogen output. Similarly, predictive maintenance algorithms track the performance of electrolyzers, allowing them to predict failures before they occur—which can help reduce downtime and maintenance costs.

Energy management systems harnessing AI utilize real-time data from weather forecasts, energy grids and electrolysis units to maximize the distribution of renewable energy for hydrogen production. Such systems forecast the availability of renewable energy and regulate the operation of electrolyzers to correspond with supply, preventing energy loss. AI assists for demand-response approaches so that electrolysis systems operate when renewable energy sources are generating excess electricity during strong winds or significant solar beam radiation. Because green hydrogen production is dependent on renewable energy sources, the variability of those resources has a direct bearing on electrolysis efficiency and reliability. AI is key to this coordination over multiple systems, so the variability of renewables does not become a bottleneck for the scale-up of hydrogen production.

AI also adds value in other domains, such as grid balancing and storage. AI predicts patterns in energy supply and demand, enabling the necessary surplus of renewables to be directed towards hydrogen production whilst also coordinating with grid operators to avoid overloading. By using integrated AI in both smart grids and hydrogen production facilities, the Green Hydrogen ecosystems have an improved eco-friendly and polluting-free economy. After production, hydrogen needs to be stored and transported efficiently in order to supply end-users with the required amounts. AI improves storage and distribution systems by optimizing logistics, tracking safety, and cutting down costs.

Optimization tools powered by AI are used to determine the design and operation of hydrogen storage plants, allowing for minimal energy loss due to selection of best pressure and temperature conditions. By modeling how these materials will behave under various conditions researchers can also use similar tools to develop advanced storage MOFs and solid-state hydrogen carriers (Cantarero, 2020). During distribution, predictive analytics powered by AI simplify transportation routes that maintain a balance between cost efficiency and speed of delivery. Also, AI improves hydrogen system security by monitoring conditions of storage and transport systems, detecting potential leaks or failures and sending real-time warnings to reduce risks.

AI-based innovations are also making significant contributions to the development of emerging hydrogen production methods, including biomass gasification, photocatalysis, and photoelectrochemical (PEC) processes. Although these methods have potential, they are often complex reactions between complicated and delicate materials that need to be finely-tuned. AI expedites research and development in these domains, allowing a rapid simulation of reaction mechanisms, optimisation of process parameters, and performance predictive modelling. In photocatalysis, for instance, AI algorithms search for the most suitable semiconductor materials before simulating their light absorption properties and subsequently assessing the stability of these catalysts under comparable conditions to those in which they would be implemented. Likewise, AI in PEC systems helps researchers optimize the photoelectrode design and predict its overall performance over time to ensure scalability and reliability.

But as hydrogen is produced on a larger scale, sustainability of the entire value chain becomes ever more critical. AI-based lifecycle assessment (LCA) tools analyze how hydrogen production processes affect the environment, including resource extraction and enduse applications. These tools assess carbon footprints, water and energy consumption, enhancing data-driven decisions for sustainability improvements by stakeholders (Sodiq et al., 2019). It can assist industries in complying with changing regulatory frameworks and certification standards for green hydrogen. AI systems analyze global sustainability goals and compliance requirements to ensure that production practices conform accordingly, thereby reducing environmental footprints.

The high production cost of green hydrogen is a major obstacle to its adoption. AI sees extensive application in this challenge, aiding in economic analyses as well as market forecasting. These predictive analytics models evaluate areas along the value chain from where materials and components will be sourced through opportunities for cost reduction in their operations. AI also predicts what can be market trends, like which sector has a greater demand for hydrogen and at what price. This allows policymakers and investors to make informed decisions and create an adequate environment for scaling up green hydrogen production.

Digital twins or virtual coupling of physical systems, has emerged in the hydrogen domain due to advances stemming from AI technologies. Digital twins can test the behavior of hydrogen-generating facilities under different scenarios, helping stakeholders evaluate designs and optimize before manufacturing physical components. Digital twins integrate current data from sensors and AI algorithms that generate actionable insights to optimize operations and enhance system reliability for greater efficiency and less downtime. They also enable predictive maintenance to ensure hydrogen systems are always up and running with minimum downtime.

The potential of AI in hydrogen production should be mitigated through ethical considerations for its deployment. And that needs to be carefully balanced against questions of data privacy, algorithmic bias, and access to the benefits of the AI-fueled economy. Joint efforts from the government, industries, and research institutions are needed to fully harness in green hydrogen production. Public-private partnerships, international standardization efforts, and knowledge-sharing platforms can play a role in speeding up technological innovations while helping to solve the scaling challenges of hydrogen ecosystems. AI in hydrogen production is useful in diverse ways, ranging from material discovery to process optimisation and renewable energy integration to logistics tackling some of the most critical challenges associated with them. AI used in the energy sector can expedite transition to green hydrogen by enabling efficiencies, lowering costs and de-risking sustainability. The combination of technological advancements and global collaborations is about to make AR the backbone of green hydrogen revolution, thus helping us out in reaching a decarbonized energy future (Clark & Cooke, 2016).

4. GLOBAL POLICIES AND GREEN HYDROGEN MARKET DYNAMICS

Green hydrogen has established itself as a key element in the worldwide shift to clean energy, supporting ambitions to decarbonise some of the hardest-to-abate sectors – heavy industry, transport and energy storage. Given its strategic importance, nations worldwide are developing policies to stimulate domestic green hydrogen production, support market uptake, and build a global hydrogen economy. The interaction of these policies and market dynamics are contributing to the evolution of the green hydrogen space. Based on this, the governments are applying an all encompassing policy frameworks for scaling up green hydrogen production and consumption. These frameworks comprise monetary incentives, regulatory backing, strategic partnerships to spur innovation and minimize manufacturing costs (Tiwari et al., 2022). There are subsidies and tax breaks in many countries getting introduced to alleviate some of the expenses required for green hydrogen production as well as infrastructure development. As an example the EU is putting billions of euros towards renewable hydrogen projects with its Green Deal, whereas in the US tax credits for clean hydrogen production exist through the Inflation Reduction Act (IRA).

Many countries have drafted national hydrogen strategies that include targets for production, investments, and legislation. To position itself as an international leader in hydrogen, Germany announced its National Hydrogen Strategy (NWS) with a major emphasis on both domestic production and global partnerships. Japan, also a hydrogen pioneer, has made hydrogen development cross-sectoral, underpinned by decades of government funding. Carbon pricing schemes are indirectly encouraging the adoption of green hydrogen by rendering competing carbon-intensive options like grey hydrogen less economically attractive. Such a policy is especially successful in places like the EU, where the Emissions Trading System (ETS) already has carbon costs set high (Li et al., 2023).

Nations are banding together to establish trans-national green hydrogen trading networks. Projects like the Green Hydrogen Catapult are trying to cut production costs by building new projects and connecting as many countries as possible. In addition, bilateral deals between countries rich in renewable energy resources (e.g., Australia, Chile) and large consumer regions (e.g. Europe, Japan) are also making a global green hydrogen global supply chain possible. Produced, much affected by costs of production, upkeep devt., request maturity stage and geo political ecstasy in the green hydrogen market. These dynamics play an enormous role in how stakeholders recognize and seize emerging opportunities.

Although the price of renewable energy, electrolysis technology and other factors have affected the production cost of green hydrogen, hydrogen production from renewables still represents a limiting price barrier for market penetration. Yet, the cost of renewables is dropping thanks to technology dynamics, scale effects and more significant investments in renewable infrastructure. In many regions, green hydrogen is projected to be cost-neutral with grey hydrogen by the 2030s given further accelerating policies and innovation. This market cannot grow without a strong hydrogen infrastructure, including production facilities, storage systems, pipelines and refueling stations. This prevents large-scale adoption as things are not standardized through infrastructure. To mitigate this gap between production and consumption, both governments and private sector players are focusing on some serious infrastructure developments like hydrogen corridors in Europe or dedicated hydrogen hubs in North America (Elavarasan et al., 2021).

Green hydrogen is quickly becoming a priority for the industrial sector, as firms work to decarbonise steel and ammonia production. Another key hydrogen fuel cell market is the transportation sector, particularly heavy-duty and long-haul vehicles. Hydrogen-based solutions are also being looked into by the maritime and aviation industry to stay within tight emission reduction targets. Investments from the private sector are crucial for expediting green hydrogen expansion. Energy majors like Shell and Total Energies spend money on hydrogen projects, while startups and corporations try to innovate around creating something like artificial trees that produce as little as half of the current energy needed (Ahmed et al., 2017). These initiatives are supported by venture capital and private equity funding, which is stoking research into novel electrolyzer technologies and solid-state hydrogen storage options. Green hydrogen is geopolitically important because it fundamentally changes the geographical nature of energy markets, moving them away from fossil fuel abundant regions and towards areas with plentiful renewable resources. There are benefits and risks evolved in this shift for world energy. And countries rich in solar and wind resources, like Australia, Saudi Arabia and Chile are aiming to become leading green hydrogen exporters. Green hydrogen provides a way for importing nations to improve energy security by reducing their dependence on fossil fuel imports. But it still hard to make a stable and cheap route from hydrogen (and derivatives) for trade like ammonia

Generating green hydrogen entails huge amounts of renewable energy resources, which can result in conflicts with other RE applications (RE projects competing for limited land/energy resources). Decision-makers need to find a balance between these opposing requiremetns in order to achieve equity of resource allocation. There exists a substantial amount of renewable energy potential in many developing nations which cannot currently be exploited due partly to limitations on the infrastructure and financial capital necessary for financing such green hydrogen opportunities (Silva et al., 2023). These countries need international assistance, such as technology transfer and finance mechanisms, to be able to generate a share in the global hydrogen economy.

While much is being done, important gaps in global policies remain that need to be closed for unlocking the full potential of green hydrogen. Part of that would mean having standardized regulations, such as certification schemes for green hydrogen, so that there is a consistency among markets. This need more innovative finance tools from green bonds to climate finance to provide the financing necessary for infrastructure projects in developing countries.

Global coordination is vital to prevent a market split and also to foster an environment in which actors can innovate together. Institutions like the International Renewable Energy Agency (IRENA) as well as the Hydrogen Council are ideally suited to enable a dialogue and collaboration at global level. The evolution of the green hydrogen industry as a basis for climate transition is influenced by global policies shaping market dynamics. Supportive policies and increasing market demand are accelerating progress, but challenges such as cost, infrastructure and resource allocation remain (Cortese et al., 2022). However, if these challenges can be tackled through specific actions and support mechanisms, which also requires international cooperation, the world can make strides towards an increased deployment of green hydrogen as a key pillar of energy systems around the globe.

5. INTEGRATING AI INTO GREEN HYDROGEN ECOSYSTEMS

AI is transforming industries around the world, and integrating its capabilities into green hydrogen ecosystems could help in improving efficiencies, scalability, and sustainability. AI can help stakeholders along the entire hydrogen value chain from production, to storage, and from distribution to end-use to optimize operations, lower costs and address barriers. Combining the forces of AI and green hydrogen ecosystems holds immense potential for revolutionary impact on the energy sector, smoothing our transition toward a carbon-neutral future. The main challenge is to enhance the efficiency of electrolysis processes, which has been recognized as one of the bottlenecks in green hydrogen production. AI powered predictive analytics and machine learning models help enhance water electrolysis systems performance including alkaline electrolysis, proton exchange membrane (PEM) system, solid oxide electrolysis cell (SOEC) etc (Thornbush & Golubchikov, 2019).

Using real-time data from electrolysis units, AI algorithms can optimize operating conditions including temperature, pressure, and current density. To achieve high hydrogen production with lower energy consumption. Machine learning models predict possible faults in the electrolysis gadget and assist in predictive maintenance. This reduces downtime and prolongs the life of mission-critical components, resulting in improved bottoms-up cost efficiency. AI enables intermittent renewable energy sources, like solar and wind, to be smoothly integrated into hydrogen production. AI predicts when renewable energy will be available based on weather prediction models, ensuring electrolysis units are designed to operate during the maximum availability of renewables and minimizing dependence on grid electricity.

Hydrogen storage and transportation is an essential part of the green hydrogen value chain. AI technologies are being utilized to mitigate the issues of storage efficiency, safety and distribution logistics. With the help of AI, smart storage systems are being developed that will observe pressure, temperature, and containment integrity for hydrogen. Such systems enable the optimization of storage conditions to reduce energy losses and maintain safety.

Hydrogen delivery routes can be optimised on AI-powered logistics platforms, substantially cutting down on fuel consumption and emissions. AI, in turn, can analyze traffic patterns as well as fuel costs and weather conditions and recommend the most efficient distribution pathways. Machine learning models that reserve hydrogen storage and transport for possible safety threats, like leaks or pressure failures, to take action prior to risks. Supply of hydrogen, which has traditionally been periodic in character and driven by industrial requirements hypothesizing prospective demand forecasting (Zhukovskiy et al., 2021). AI is able to analyse significant data sets to identify the hydrogen demand across various sectors, including industrial processes, transport and energy storage components.

AI shares insight on sectors via historical data and market trends which inform production scaling and infrastructure development. This guarantees the match between hydrogen supply and demand whilst preventing oversupply or shortfall of output. AI algorithms can also model market dynamics, such as policy changes, renewable energy prices and consumer behaviour. That makes it a key aspect to the development of competitive pricing strategies and market stability. AI can also assist policymakers and regulatory bodies in the design of effective templates geared towards facilitating green hydrogen development. AI aids in designing growth-oriented, sustainable regulations through analysis of the impacts of potential policies and what-if simulations on policy performance (Jayachandran et al., 2022). By using AI tools, it is possible to understand how the current and even potential future policies for hydrogen ecosystems are performing and where they can do better. This guarantees that regulatory frameworks are dynamic and respond to the needs of sectors (Jele et al., 2024).

Green hydrogen is a new trend that has not yet been certified, and AI can assist in automating this certification process by ensuring compliance with environmental and safety standards. This also enables international commerce and increases user confidence in hydrogen technologies. Green hydrogen markets will be global, requiring close collaboration between nations and industries (Patel et al., 2024). With AI-based platforms, they will be able to share knowledge, transfer technology and create collaborative research initiatives. With the aid of AI, digital twins represent virtual copies of hydrogen ecosystems; stakeholders can create them to assess different scenarios and performance or test new technologies in real-world situations without producing a single molecule. Such AI-powered platforms make communication and collaboration between academia, policymakers as well as industry actors seamless (SaberiKamarposhti et Al., 2024). These platforms help to spur innovation and enable quicker deployment of best practices along the hydrogen value chain. The rise of green hydrogen ecosystems where AI integration is introducing game-changer opportunities that can tackle many technical, economic and logistical challenges (Bade et al., 2024). AI is going to be a pillar of the hydrogen economy whether improving production or storage safety, and market dynamics or policy innovation. The world is investing in green hydrogen to mitigate climate change, and the AI-hydrogen tech synergies will be critical towards realizing a decarbonized energy future.

6. CONCLUSION

The scaling of green hydrogen production lies at the crossroads of technology, policy and global collaboration. Green hydrogen, which is an essential building block of the decarbonization agenda, has a great potential to change energy systems, curb climate change and boost the economy. The review illustrates the way Artificial Intelligence (AI) is transforming hydrogen production technologies, with process optimization and closing existing gaps for second generation and beyond modes of H2 production which are indispensable for scalability and sustainability. Meanwhile, forward-looking global policies are laying out a solid framework to encourage the use of green hydrogen, paving the way for market growth and innovation.

Green hydrogen ecosystems face critical challenges, such as energy-intensive production and infrastructure inefficiencies, alongside fluctuating renewable energy availability; AI has been instrumental in meeting these needs. Through the application of machine learning, predictive analytics and smart automation AI optimizes water electrolysis, renewables usage as well as hydrogen storage and transport. Not only does it decrease the cost of production, but also increases safety, reliability and operational efficiency. Simultaneously, worldwide policies from financial incentives to collaborative trade agreements are speeding green hydrogen to market, creating a fertile ground for innovation and investment.

7. FUTURE SCOPE

In the future, there are even more possibilities for integrating AI into green hydrogen production. Innovations in areas like digital twins, robotics and other new decentralized energy systems could make them even more efficient and scalable through AI. AI-driven hydrogen production hubs, for example, can quantify real-time energy inputs and respond dynamically to maximize the output when renewable power fluctuates from surplus to scarcity. Additionally, utilization of the AI algorithms mentioned above have potential to be translated into novel hydrogen production pathways e.g., kinetics-related breakthroughs in photocatalytic and photoelectrochemical processes.

Achieving an integrated global hydrogen market also requires aligned standards and certification of hydrogen activities on a policy level. By bringing different governments, private sector stakeholders and research institutions together we can enable shared infrastructure to speed up developments for technology transfer and narrow the resource gap between developed nations and developing countries. Also, the ethical and social considerations associated with AI deployment such as ensuring equal access to green hydrogen technologies and limiting environmental effects will be crucial to enable truly sustainable development. AI innovations in scaling green hydrogen production are a pivotal opportunity to create a robust, regenerative, and equitable energy future. In this way, the potential of green hydrogen to set the groundwork not only for a carbon free future but also for real economic and environmental progress can be unlocked by making use of what AI has on offer in tandem with global policies aligned with long-term climate action.

REFERENCES

Ahmad, T., & Zhang, D. (2021). Using the internet of things in smart energy systems and networks. *Sustainable Cities and Society*, *68*, 102783. DOI: 10.1016/j.scs.2021.102783

Ahmed, F., Naeem, M., & Iqbal, M. (2017). ICT and renewable energy: A way forward to the next generation telecom base stations. *Telecommunication Systems*, *64*(1), 43–56. DOI: 10.1007/s11235-016-0156-4

Albreem, M. A., Sheikh, A. M., Alsharif, M. H., Jusoh, M., & Yasin, M. N. M. (2021). Green Internet of Things (GIoT): Applications, practices, awareness, and challenges. *IEEE Access : Practical Innovations, Open Solutions*, *9*, 38833–38858. DOI: 10.1109/ACCESS.2021.3061697

Bade, S. O., Tomomewo, O. S., Meenakshisundaram, A., Ferron, P., & Oni, B. A. (2024). Economic, social, and regulatory challenges of green hydrogen production and utilization in the US: A review. *International Journal of Hydrogen Energy*, *49*, 314–335.

Bibri, S. E. (2018). The IoT for smart sustainable cities of the future: An analytical framework for sensor-based big data applications for environmental sustainability. *Sustainable Cities and Society*, *38*, 230–253.

Bibri, S. E., & Krogstie, J. (2020). Environmentally data-driven smart sustainable cities: Applied innovative solutions for energy efficiency, pollution reduction, and urban metabolism. *Energy Informatics*, *3*(1), 1–59. DOI: 10.1186/s42162-020-00130-8

Cantarero, M. M. V. (2020). Of renewable energy, energy democracy, and sustainable development: A roadmap to accelerate the energy transition in developing countries. *Energy Research & Social Science*, *70*, 101716. DOI: 10.1016/j.erss.2020.101716

Clark, I. I. W., & Cooke, G. (2016). *Smart green cities: toward a carbon neutral world*. Routledge. DOI: 10.4324/9781315609386

Cortese, T. T. P., Almeida, J. F. S. D., Batista, G. Q., Storopoli, J. E., Liu, A., & Yigitcanlar, T. (2022). Understanding sustainable energy in the context of smart cities: A PRISMA review. *Energies*, *15*(7), 2382. DOI: 10.3390/en15072382

Elavarasan, R. M., Pugazhendhi, R., Jamal, T., Dyduch, J., Arif, M. T., Kumar, N. M., & Nadarajah, M. (2021). Envisioning the UN Sustainable Development Goals (SDGs) through the lens of energy sustainability (SDG 7) in the post-COVID-19 world. *Applied Energy*, *292*, 116665. DOI: 10.1016/j.apenergy.2021.116665

Islam, A., Islam, T., Mahmud, H., Raihan, O., Islam, M. S., Marwani, H. M., & Awual, M. R. (2024). Accelerating the green hydrogen revolution: A comprehensive analysis of technological advancements and policy interventions. *International Journal of Hydrogen Energy*, *67*, 458–486.

Issa Zadeh, S. B., Esteban Perez, M. D., López-Gutiérrez, J. S., & Fernández-Sánchez, G. (2023). Optimizing Smart Energy Infrastructure in Smart Ports: A Systematic Scoping Review of Carbon Footprint Reduction. *Journal of Marine Science and Engineering*, *11*(10), 1921. DOI: 10.3390/jmse11101921

Jayachandran, M., Gatla, R. K., Rao, K. P., Rao, G. S., Mohammed, S., Milyani, A. H., & Geetha, S. (2022). Challenges in achieving sustainable development goal 7: Affordable and clean energy in light of nascent technologies. *Sustainable Energy Technologies and Assessments*, *53*, 102692.

Jeje, S. O., Marazani, T., Obiko, J. O., & Shongwe, M. B. (2024). Advancing the hydrogen production economy: A comprehensive review of technologies, sustainability, and future prospects. *International Journal of Hydrogen Energy*, *78*, 642–661. DOI: 10.1016/j.ijhydene.2024.06.344

Li, D., Guan, X., Tang, T., Zhao, L., Tong, W., & Wang, Z. (2023). The clean energy development path and sustainable development of the ecological environment driven by big data for mining projects. *Journal of Environmental Management*, *348*, 119426. DOI: 10.1016/j.jenvman.2023.119426 PMID: 37879178

Patel, J., Patel, A. R., & Tyagi, H. (2024). Advances in the Solar Thermal Systems Assisted Production of Green Hydrogen: Its Analysis, Scaling-Up Techniques, and Economics Aspects as Applied to Tropical Regions. In *Challenges and Opportunities in Green Hydrogen Production* (pp. 239–267). Springer Nature Singapore. DOI: 10.1007/978-981-97-1339-4_10

SaberiKamarposhti, M., Kamyab, H., Krishnan, S., Yusuf, M., Rezania, S., Chelliapan, S., & Khorami, M. (2024). A comprehensive review of AI-enhanced smart grid integration for hydrogen energy: Advances, challenges, and future prospects. *International Journal of Hydrogen Energy*.

Salam, A., & Salam, A. (2020). Internet of things in sustainable energy systems. *Internet of Things for Sustainable Community Development: Wireless Communications. Sensory Systems*, 183–216.

Silva, B. V., Holm-Nielsen, J. B., Sadrizadeh, S., Teles, M. P., Kiani-Moghaddam, M., & Arabkoohsar, A. (2023). Sustainable, green, or smart? Pathways for energy-efficient healthcare buildings. *Sustainable Cities and Society*, 105013.

Sinha, M., Chacko, E., Makhija, P., & Pramanik, S. (2021). Energy-Efficient smart cities with green internet of things. *Green Technological Innovation for Sustainable Smart Societies: Post Pandemic Era*, 345-361.

Sodiq, A., Baloch, A. A., Khan, S. A., Sezer, N., Mahmoud, S., Jama, M., & Abdelaal, A. (2019). Towards modern sustainable cities: Review of sustainability principles and trends. *Journal of Cleaner Production*, *227*, 972–1001. DOI: 10.1016/j.jclepro.2019.04.106

Thornbush, M. J., & Golubchikov, O. (2019). *Sustainable urbanism in digital transitions: from low carbon to smart sustainable cities*. Springer.

Tiwari, S., Rosak-Szyrocka, J., & Żywiołek, J. (2022). Internet of things as a sustainable energy management solution at tourism destinations in India. *Energies*, *15*(7), 2433. DOI: 10.3390/en15072433

Varjovi, A. E., & Babaie, S. (2020). Green Internet of Things (GIoT): Vision, applications and research challenges. *Sustainable Computing : Informatics and Systems*, *28*, 100448. DOI: 10.1016/j.suscom.2020.100448

Zhukovskiy, Y. L., Batueva, D. E., Buldysko, A. D., Gil, B., & Starshaia, V. V. (2021). Fossil energy in the framework of sustainable development: Analysis of prospects and development of forecast scenarios. *Energies*, *14*(17), 5268.

Chapter 24
Unfolding Artificial Intelligence (AI) in Green Hydrogen Production:
Towards an Intelligent Energy Transition Fostering Sustainability

Bhupinder Singh
https://orcid.org/0009-0006-4779-2553
Sharda University, India

Saurabh Chandra
https://orcid.org/0000-0003-4172-9968
Bennett University, Greater Noida, India

ABSTRACT

Artificial intelligence (AI) is increasingly being integrated with green hydrogen production, so as to enable a seamless transition from conventionally suffocating energy sources. Water electrolysis, using renewable energies such as solar and wind energy as a power source, is the key technology of green hydrogen with considerable advantages towards decarbonizing energy-intensive industries and promoting global sustainability. Innovative application of AI is changing the way green hydrogen operates for efficiency and scalability by improving production processes saving energy in electrolysis making it cost effective. The AI and green hydrogen production benefits not only energy efficiency, but also serves to further the cause of climate action by weaning society off fossil fuels. And this clever, foresighted investment clears a path for a cleaner future which also adds one more way out of many possible ways that the world can reach its sustainability goals and move off carbon-neutral energy.

1. INTRODUCTION

AI-driven applications like machine learning models are now beginning to change the way electrolysis systems function, scheduling maintenance operations before it's too late, improving local conditions for proper hydrogen generation and much more (Boretti, 2024). AI is also heavily relied on to help manage

DOI: 10.4018/979-8-3373-0045-0.ch024

the storage and transportation of hydrogen to make the supply chain process efficient and less energy wasteful (Singh & Kaunert, 2024). This way industries are able to effectively manage and integrate renewable energy sources within their hydrogen production process using predictive forecasting, artificial intelligence based data analysis and real-time optimization towards a more modestly cleaner energy ecosystem (Singh et al., 2024).

With the world facing a two-front challenge fighting climate change and moving to renewable resources, the importance of renewable energy sources has grown phenomenally (Singh & Kaunert, 2024). One of the most exciting renewable energy replacements is green hydrogen, which has been hailed as a game-changer in our quest to be carbon neutral and promote energy independence. Made from water electrolysis powered by 100% renewable energies such as solar, wind and hydroelectric power, green hydrogen is an easy and efficient solution for industries and sectors where fossil fuels are traditionally used (Singh et al., 2024). The positioning as a solution for decarbonizing heavier-polluting, higher-growth sectors like transportation and manufacturing (also known as hard-to-abate industrial carbon sourcing) or heavy industry makes blue hydrogen sound like Corri-doors but also ultimately vital.

The large-scale deployment and manufacture of green hydrogen encounter numerous technological and economic difficulties. Hydrogen production via electrolysis is inefficient and energy-intensive, and it also faces challenges with regard to storage and distribution that must be overcome if green hydrogen is ever to become a viable global energy source (Singh et al., 2024). And this is where the role of Artificial Intelligence (AI) com into picture. The AI includes machine learning, deep learning, and data analytics that is helping sectors like healthcare to finance. The electricity sector uses Artificial Intelligence for renewable energy production, grid management, and demand predictions (Singh & Kaunert, 2024). The use of AI in green hydrogen production is still early, but holds potential to eliminate key bottlenecks, decrease costs, and offer scalability. AI-based solutions can help to optimize hydrogen production pathways -including improving the energy efficiency of less efficient modes like SMR or easing the deployment increase in renewable energy into the value chain. In addition, A.I. can be instrumental in storing, transferring, and utilizing green hydrogen to help take it from theoretical potential to real-world application (Raghav et al., 2024).

It undertake this study to investigate the potential of AI for green hydrogen production, assess and predict how machine learning can accelerate an intelligent energy transition, particularly in light of sustainability efforts. It, provides a look into the massive unlocked potential in green hydrogen by prospecting four powerful facets of AI and their role within this sustainability domain: an example being AI-powered electrolysis enhancement where researchers are exploring efficiency improvements to leverage further savings (Kaushik et al., 2024).

1.1 Significance of the Study

The importance of this study of the confluence, AI and green hydrogen creation is set in the context of nations around the world being on a war-footing to address the rapid advance of climate change. Green hydrogen is set to be important for decarbonizing hard-to-electrify sectors, as the world transitions to a net-zero emissions economy. Green hydrogen can offer an alternative for industries like steel manufacturing, aviation and shipping that may find direct electrification technologically challenging or economically infeasible but are facing pressure to meet decarbonization targets. But to realize this opportunity, other bottlenecks must be overcome production inefficiencies, energy consumption that is high, cost limitations.

Solving Production Inefficiencies with AI: The high-energy intensity of electrolysis-the process by which water is broken down into hydrogen and oxygen using electricity-is a key hurdle in the production of green hydrogen. Even "green" electrolysis is largely inefficient hundreds of megajoules per kilogram ultimately put into the process, leading to over half a terajoule of energy in waste heat for every tonne of hydrogen produced (Singh et al., 2024). This in turn leads to higher production costs, making green hydrogen significantly less price-competitive than grey hydrogen from natural gas or coal which is almost always environmentally dirtier (Singh et al., 2024). This is where AI comes into play as the technology provides a path to turbocharge the electrolysis process. AI can optimize operational parameters temperature, pressure, and electrical input of electrolyzers using machine learning algorithms and real-time data analytics (Singh et al., 2024). These refinements enable a more efficient conversion of water to hydrogen, therefore lowering energy waste and production costs. Also, by cross-referencing renewable energy supply data, AI can anticipate when the best time to produce hydrogen; then electrolysis will be carried out at times where peak solar or wind generation is occurring. In practice, it means that the green hydrogen is extracted as energy-efficiently as possible from this process, saving an even more considerable portion of carbon emissions (Song et al., 2024).

Predictive Maintenance & Operations Efficiency: The other key space AI can help in the area of green hydrogen production is AI-driven predictive maintenance models. Industrial machinery experiences mechanical wear and tear over time, which results in operational inefficiencies and may cause downtime, including those installed to provide electrolysis. With traditional maintenance strategies come the inevitable breakdowns and disruptions, leading to high repair costs (Singh & Kaunert, 2024). This AI is capable of analyzing continuous sensor data, detecting patterns that might indicate a problem and notifying relevant staff of those imminent problems long before they have the opportunity to lead to wasted time or unrecoverable health. By proactively identifying possible issues, operators are able to take action before the equipment fails, significantly increasing the reliability and lifespan of hydrogen production infrastructure (Wang et al., 2024).

Application of AI for Better Storage and Distribution: The major challenges for green hydrogen value chain have to do with storage and distribution. It is because hydrogen is the lightest element and therefore needs expensive storage solutions such as high-pressure tanks or cryogenic systems which are energy intensive to build and very much expensive. Transportation also poses challenges, since hydrogen must be compressed or liquefied to more efficiently. transported over long distances, adding even further to the energy requirements (Singh & Kaunert, 2024). Artificial intelligence (AI) could be used to build predictive models of energy demand and patterns of hydrogen consumption and distribution logistics to optimize the network for such storage and distribution. It can be used to help design hydrogen storage systems that are more lenient, forecast energy demand patterns or determine delivery pathways in order to deliver the hydrogen where it is most needed using as little energy as possible (Singh & Kaunert, 2024). Additionally, AI is being used to support the integration of hydrogen storage into the wider grid and in so doing balancing hydrogen supply (electrolysers) and demand (fuel cell applications) with renewable energy source sielding (Dutta et al., 2024).

AI Can Drive Sustainability: Green hydrogen production is another case where AI plays a role that supports global sustainability objectives, not only matching but potentially exceeding the UN Sustainable Development Goals (SDGs) (Singh et al., 2024). Optimising energy use, reducing waste and facilitating the widespread adoption of renewable hydrogen are ways in which AI-driven solutions make those figures add up by contributing to SDG 7 Affordable and Clean Energy and SDG 13 Climate Action (Singh & Kaunert, 2024). Moreover, AI-supported production of green hydrogen can help make the economy more circular and enable us to use renewable energy for hydrogen generation/storage/distribution, thus limiting the consumption of fossil fuels and emissions this way (Qiu & Zhao, 2024).

The intersection of Artificial Intelligence and green hydrogen production is a game-changer in accelerating sustainable energy transition. This means that while green hydrogen is already the cleanest source of energy available for various uses, AI applications to optimize manufacturing processes, improve storage and distribution or reduce operational inefficiencies can make green hydrogen economically viable on a global scale. In order to meet increasingly aggressive carbon neutrality goals around the world, and within individual industries, advances in green hydrogen production led by improvements in AI technology are more important than ever for creating a greener and cleaner energy future. It helps demonstrate the growing role of AI in this crucial aspect and emphasizes its importance in powering up a green switch worldwide (Singh, 2024).

2. AI IN OPTIMIZING GREEN HYDROGEN PRODUCTION

Green hydrogen, produced from the electrolysis of water using renewable energy sources such as wind and solar, is touted to be pivotal in the world's clean energy future. Nonetheless, conventional hydrogen production remains uncompetitive with obstacles involving energy consumption, price, and expandability (Singh & Kaunert, 2024). These challenges can be mitigated by Artificial Intelligence (AI) to enable optimization of green hydrogen production, thus helping increase the efficiency of the process, and decreasing costs to improve its coordination when integrated with renewable energy systems (Singh & Kaunert, 2024).

The use of AI is upending the electrolysis process which takes water and splits it into hydrogen and oxygen with an electric current. The process is efficient or not depending on several factors such as temperature, pressure, electricity quality and etc. These variables can be monitored and optimized in real time by using AI algorithms, mostly machine learning (ML) models to assure the highest efficiency operation of the electrolyzers (Singh & Kaunert, 2024). This system helps control the operations of an electrolyzer more effectively and helps it maximize its yield in terms of hydrogen production using minimum amount of energy by constantly adjusting operational parameters as per input energy available at all times & condition of the Electrolyzer.

AI applications could predict how well electrolyzers would perform in various conditions, which operators could use to optimize the process and maximize efficiency. They use sensors to monitor temperature, pressure, and electricity input as data sources for AI systems that self-optimize the electrolysis process. Not only does this improve efficiency but also prolongs the operational life of electrolyzers by protecting them from wear and tear due to suboptimal operating conditions. Energy prices are a big hurdle to overcome for scaling production of green hydrogen. The electrolysis process to generate hydrogen is a high energy cost process made more complicated with renewable resources such as wind and solar, that can be intermittent. From energy savings and the use of renewable options in producing hydrogen, AI has the answer (Asif, 2024).

By studying past data of renewable energy production, the AI algorithms can predict when there will be large amounts of energy available. AI can combine this predictive ability with hydrogen production systems, so that the electrolysis processes are only powered when surplus renewable energy is being generated (instead of drawing upon non-renewable sources). AI could also enable hydrogen production to be scheduled so that demand and supply of energy can have a more balanced power, perhaps activating the hydrogen production during off-peak hours when electrical energy is cheaper (Singh & Kaunert, 2024). But the approach could see significant savings on the operational costs of hydrogen production facilities, and also ensure that a larger percentage of green energy ends up being used. AI is also essential in maintaining the hydrogen production infrastructure. Then, there are regular maintenance activities for each electrolyzers and other components to ensure the system works effectively and avoid downtime. AI-based predictive maintenance systems can process sensor data in real-time and identify the earliest symptoms of equipment wear, degradation or failure. It enables real-time intervention in case of any significant problem that help to avoid the heavy damages, leading to less operational disruption and lower maintenance cost (Kumar et al., 2024).

Machine learning models can also spot anomalies in the production process that may be an indication of equipment issues. If detected early, these anomalies can notify the AI system of when maintenance is required and they can schedule maintenance during times when hydrogen production impact is minimal. It provides not just improved operational efficiency and longer-life critical equipment in hydrogen plants -but also reduced costs. Because renewable energy sources such as solar and wind power are intermittent, they cause the amount of hydrogen gas that can be produced at facilities to fluctuate (Singh & Kaunert, 2024). The sun and wind generate power intermittently, resulting in fluctuations in the energy supply. The incorporation of hydrogen production to renewable energy grids can be optimized by AI. AI can predict energy generation from clean solar / wind sources in real-time and forecast generation and thus modify hydrogen production. For example, when there is more renewable energy than the grid needs AI can ramp up hydrogen production so that more of that excess can be converted to hydrogen and stored (as hydrogen). Later as power demand picks back up or whenever there are not enough renewables to meet demand. This rapidly responding dynamic energy management capability will significantly increase the stability and efficiency of operation in green hydrogen utility, so that renewable power is best used with no waste (Singh et al., 2024).

Artificial intelligence (AI) has the potential to revolutionize green hydrogen production andclear many of the hurdles tied to efficiency, cost and scalability. Through better electrolysis, less energy usage, predictive maintenance, and greater integration with renewable energy sources AI is shaping a more efficient and sustainable hydrogen economy. In a world increasingly moving towards carbon neutrality, AI will play a big role in facilitating the true potential of green hydrogen as an energy source (Singh & Kaunert, 2024).

3. AI-DRIVEN PREDICTIVE MAINTENANCE IN HYDROGEN INFRASTRUCTURE

As the green hydrogen production landscape takes shape in one of the quickly transforming sectors, keeping up the infrastructure is crucial to achieve peak performance and minimize downtime. In short, the hydrogen infrastructure electrolyzers, storage systems, and distribution networks — are very complex bits of equipment that must operate with absolute reliability if hydrogen production is to be economically viable (Singh et al., 2024). Artificial Intelligence (AI) based predictive maintenance is changing the landscape of monitoring and maintenance practices in hydrogen production facilities, saving these high-value assets from expensive failures and improving operational efficiency (Singh et al., 2024). That is because most traditional ways to approach maintenance, like reactive and scheduled preventive maintenance, can eat up quite scarce resources for not a lot of value in return. Reactive maintenance can be very expensive and generate unexpected downtimes, where preventive maintenance may cause undue interventions in equipment in good operation. By contrast, predictive maintenance takes a more proactive approach and uses AI models to predict the time when machines will fail or need servicing using real-time data feeding into historical behavior patterns (Singh et al., 2024).

Predictive maintenance is crucial when it comes to the hydrogen production, as electrolyzers, compressors, storage tanks, etc., are extremely susceptible to many parameters including temperature and pressure or the quality of input electricity. All deviations will limit the efficiency, increase wear or lead to a total failure. Over time, this will help in managing the above risks as predictive maintenance through AI will constantly monitor equipment health and alert if something goes wrong, such that we can overcome the issue before it can trigger a failure. AI provides the power necessary to process huge data from a variety of sensors embedded in hydrogen infrastructure and orchestrate large hydrogen infrastructures across various domains. These sensors generate massive real-time data like temperature, pressure, energy consumption, vibration along with other KPIs. These data are analyzed by AI algorithms, mostly to machine learning (ML) models, to detect patterns and correlations missed by human operators (Singh et al., 2024).

The applications is a system that uses machine learning to identify components on the electrolyzer showing early signs of wear, for example variations in temperature or pressure that indicate declining performance. Through ongoing education from a growing pool of data, an increasingly robust AI system will have an ever better idea of when machines are going to fail, which in turn will tell the operators when they need to repair or replace parts before one of them finally plunges to its demise (Lal et al., 2024). The result is an improvement from higher accuracy and dependability that hydrogen production centers which also reduces operational costs. AI-based predictive maintenance also shines in anomaly detection and when the system can recognise deviations from normal operating behavior (Arora et al., 2024). This means that as soon as a piece of equipment begins to behave atypically, AI models can identify these aberrations quickly, minimizing downtime by alerting operators when something may be wrong. A small departure from that vibration pattern in, say, a compressor's operating signature could indicate a mechanical problem that might cause big trouble down the road if it isn't dealt with (Kaushik et al., 2024).

AI can also use the operational history of equipment to predict failures. Based on historical failures, AI algorithms predict the time when a specific component is likely to fail by comparing the current data. This enables scheduled maintenance during non-critical periods to minimise interruptions to hydrogen production and avoid costly emergency repairs (Raghav et al., 2024). There are several strategic pluses from the standpoint of hydrogen infrastructure: The predictive maintenance element that AI provides carries benefits for the following reasons:

Minimize downtime: The early failure detection that AI offers allows for faster identification of potential faults and decreased unplanned outages, which increases hydrogen production system availability.

Less operation and maintenance costs: AI eliminates the need for rote preventive maintenance since machines can dispatch warnings when they are about to break down, saving time and money undefined.

Extend the life of electrolysers, compressors, and storage systems by preventing them from experiencing operating conditions that are not optimal.

Improves safety: Keeping a check on the conditions of the equipment guarantees regular maintenance preventing any downtimes converting into dangerous situations, also leading to a safe environment (Singh, 2024).

Hydrogen infrastructure benefits from AI-driven predictive maintenance which yields improved efficiency, cost-effectiveness, and reliability of hydrogen production systems (Singh & Kaunert, 2024). And with the increasing relevance of hydrogen in the wider global energy transition, this will also represent a necessary infrastructure component for scalable, sustainable hydrogen production on an industrial scale; requiring new solutions are called for, as we head into playing our part in clean green engineering challenges and opportunities (Singh & Kaunert, 2024).

4. AI IN HYDROGEN STORAGE AND DISTRIBUTION

High-capacity storage and efficient distribution of hydrogen are essential for the viability of hydrogen as a sustainable energy carrier. Not only is hydrogen very combustable and low density lIt makes it a huge challenge for to transport and store it in safe, energy efficient methods (Singh & Kaunert, 2024). It is as mentioned above AI can be a game-changer in the case of hydrogen storage and distribution system at all levels of supply chain by enhancing safety, efficiency and cost-effectiveness.

For storage, hydrogen is usually stored in one of three types: compressed gas (gaseous), liquid or absorbed to solid chemical compounds like metal hydrides. There are drawbacks to each of these methods, like keeping pumps powerful enough to work with high pressures, or engines cold without complex chemical reactions etc (Singh & Kaunert, 2024). If a hydrogen storage is monitored by a sensor to report parameters such as pressure, temperature or containment integrity, AI can be used to classify the mean to different actions how the system behavior should work optimally. So, at present several AI algorithms can predict the best conditions to store hydrogen which would minimize energy lose and safety standards. Examples include using artificial intelligence to monitor pressure levels in compressed hydrogen storage systems and predicting impending failures or leaks where changes are required prior failure. Artificial intelligence can be used to optimize refrigeration systems for cryogenic temperatures necessary to liquefy hydrogen thus saving a lot of energy and reducing operating costs in liquid hydrogen storage procedure. AI can also help determine the optimal storage technology for different use cases, thereby creating storage solutions that are more versatile and scalable to varying energy demands (Dutta et al., 2024).

Because of its small molecular size, hydrogen can escape easily and lead to safety risks during distribution, whether through pipelines or vehicles (Singh & Kaunert, 2024). Artificial Intelligence may greatly increase the safety in supply chains for hydrogen with the help of real-time tracking and predictive maintenance. An example of the use case with AI systems is using data from sensors installed along the pipeline or inside the transport vessel to capture any anomaly that may indicate there is something happening, typically a drop in pressure, changing some temperature or even altered chemical composition what implies there might be leak or valve failure (Singh & Kaunert, 2024). In addition to monitoring, models can predict when equipment is likely to fail based on its historical performance so that operators can preemptively schedule maintenance before a leak or failure. This would not only make them safer, but also save the time and expense of dealing with emergency repairs. In addition to routing the transport of hydrogen more intelligently, AI algorithms can find the most fuel- and emission-efficient routes for transportation vehicles that help in reducing environmental footprints in the entire biomass supply chain (Singh et al., 2024).

With hydrogen increasingly part of the energy system, balancing its distribution with other renewables is set to become more important. AI can further extend hydrogen storage and distribution into a larger smart grid, to maintain real-time balancing of supply and demand. For example, AI solutions can predict energy demand and optimize the distribution of hydrogen to regions with high consumption (above) while storing surplus hydrogen during low periods of demand. This flexible H2 energy management strategy ensures that hydrogen arrives where and when it is most needed, making the delivery of Hydrogen as an energy carrier more reliable and efficient (Singh et al., 2024). It is really revolutionizing the way hydrogen storage and distribution is done by making it more safe, efficient as well close to 30% cost savings. By giving stakeholders access to real-time monitoring and predictive analytics edged with the AI based smart grid integration, we would ensure hydrogen could be stored and transported as fuel for overall scalability and sustainability of the hydrogen economy (Singh et al., 2024). In view of the imperative for hydrogen to support the global energy transition, the use of AI solutions will be crucial in tackling these issues with storage and transport (Singh & Kaunert, 2024).

5. AI INTEGRATION WITH RENEWABLE ENERGY SOURCES

Artificial Intelligence (AI) is increasingly being integrated with renewable energy sources to improve the world´s supply of energy by allowing more intelligent management of production, distribution, and consumption. This variability is inherent in renewable energy sources for example solar, wind and hydropower because these are dependent on certain environmental conditions: weather and amount of daylight (Singh & Kaunert, 2024). To overcome these hurdles, AI is crucial in improving the prediction, efficiency as well as optimization of renewable energy systems. The biggest thing AI brought is forecasting how much energy will be generated from wind turbines at any given time. Sophisticated AI-powered systems can predict solar irradiance, wind speeds, and hydrological conditions to a great degree, which is paramount to design energy systems. By providing predictability these forecasts enable grid operators to further optimise the scheduling and dispatch of energy from renewable sources and help better balance electricity supply with demand. Proper forecasting lowers the amount of energy you need to over-produce, reduces the need for backup using fossil fuels and increases marginally the stability of the grid (Singh & Kaunert, 2024).

Renewable energy is commonly subject to curtailment when excess generation occurs, such as during sunny or windy days. E.g. it can control when to store/release energy in an energy storage system like batteries or green hydrogen etc since an AI model can identify through historical big data analysis, and predict the optimal state by optimization techniques rigesimal circuits, and the old classical methods are still used for that kind of execution (Singh et al., 2024). AI can, for example, analyse the up-to-the-minute demand on a grid and autonomously decide to store any excess solar power in batteries or turn that extra wind power into clean green hydrogen. On the distribution side, AI coordinates the distribution of renewable energy across a grid to deliver power where it is most needed in real-time limiting transmission losses (Singh et al., 2024).

The integration of AI can turn these into more dynamic, responsive energy grids known as “smart grids”. The usage of AI in these grids include managing energy flows, demand forecasting, balancing and other operations for efficient operation in real time. As a second example, AI can make demand response systems possible—these are systems that automatically change the pattern of consumption based on power grid conditions (Singh & Kaunert, 2024). For example, AI-defined algorithms can alert users to save power when demand is high by sourcing peak time consumption from other areas (Singh, 2024). The importance of having AI integrated with the renewable source of energy can have a long way in making our future life better in terms of sustainable, more reliable and efficient energy. AI is enabling the integration of renewables, and thereby advancing the world-wide transition to 100% clean energy (Singh, 2024).

6. SUSTAINABILITY IMPLICATIONS OF AI IN GREEN HYDROGEN PRODUCTION

Artificial Intelligence (AI) integrated green hydrogen production holds the promise to cater with global energy requirements and environmental friendly approach. Green hydrogen, which is produced by the electrolysis of water powered by renewable energy sources, is a clean energy carrier capable of decarbonizing difficult to electrify and transport sectors, such as heavy industries or aviation and shipping sector (Singh & Kaunert, 2024). But green hydrogen is energy-intensive and comes with its own technical and economic challenges. The ability of AI to offer innovative solutions that boost production efficiency, cost-effectiveness and scalability in the case of green hydrogen can have wide-ranging positive sustainability impacts (Singh, 2024).

A major advantage of AI-enabled green hydrogen production is the improvement in energy efficiency which underpins virtually all sustainability claims. The technology itself requires a great deal of electricity, and through AI optimization th is electricity can be used as efficiently as possible. AI can be used to detect wide-ranging data from electrolyzers such as voltage,current, temperature and water quality (Singh & Kaunert, 2024). Using AI to optimize energy consumption is one of these methods and, when applied to key variables such as temperature, time and current density, the AI can redefine performance without loss in hydrogen output—a way of increasing the overall efficiency of the system (Singh & Kaunert, 2024). AI can also help merge sources of renewable energies e.g., solar or wind with hydrogen production, such that the transition is smoother and carried out at a larger scale. For instance, AI algorithms can predict opportunities for large quantities of renewable power generation on account of weather events and operate electrolysis at a time when there is plenty renewable electricity. This could become less reliant on grid electricity from non-renewable sources, partially decarbonizing the production of hydrogen (Singh, 2024).

Green hydrogen is too expensive, to say the least. Because of this, AI is able to reduce production costs via optimization on multiple tiers. It lowers the operational costs associated with electricity consumption, which is one of the most important factors while calculating green hydrogen's total cost. In addition to this, AI-supported predictive maintenance can also guarantee that your machines never go out of business and stop costly failures by recognizing equipment issues before they result in large scale damage. For green hydrogen production to be deployed at the global scale, AI will be needed for operating systems at these scales. With the support of AI we can reduce human intervention and automate crucial processes which would thereby enable a large production plant to run 24×7 with least downtime. This scalability is important to green hydrogen filling a substantial role in decarbonizing our global economy especially for those hard-to-abate sectors of the economy like heavy industry and transportation, that are difficult to decarbonize (Boretti, 2024).

In any case, water is important for producing green hydrogen and its sustainable usage must be ensured to make the use of hydrogen as an energy carrier a long-lasting solution. On one hand, for green hydrogen production, water management can be made sustainable through AI and recycle of excess or waste water coming out during electrolysis process. Water quality: An AI algorithm will be able to monitor the water that is being used in the tenant suite for electrolysis and ensure that it does not contain impurities that can hinder the efficiency of electrolyzers or damage long term equipment. In addition, water recycling within the production system can be boosted by AI, which will help to decrease water loss. Especially for water stressed regions, availability of big-data led AI based water management can possibly play a substantial role to ensure that the green hydrogen production is not worsening local scarcity of water (Song et al., 2024).

In industries like steelmaking, cement production and heavy transport, using temperature heat or long-distance fuel to produce power is still a big challenge. These industries can be switched from fossil fuels to green hydrogen, but the production of this new hydrogen must be as globally scalable and cost-effective as possible in order to do this. The ability of AI to streamline production, bring costs down, and integrate seamlessly into the renewable market is essential for green hydrogen to supplant fossil fuels. The use of AI in green hydrogen will have important sustainability consequences (Qiu & Zhao, 2024). Green hydrogen could become a cornerstone of the global energy transition, thanks to AI that can drive efficiencies in energy use and cost savings for manufacturing, as well as enable large-scale adoption and sustainable water management. With industry and policy increasingly moving towards sustainability and carbon neutrality, AI in the production of green hydrogen will have a significant role to play in defining an energy landscape expected to be cleaner and more robust (Rasheed et al., 2024).

7. CONCLUSION AND FUTURE SCOPE

This means that one of the biggest moves in transitioning global energy to a more sustainable state has occurred from the infusion of AI technology into green hydrogen production. From improving the efficiency of electrolysis for hydrogen production to predictive maintenance, smart grid integration and efficient water management among others can be augmented with AI which helps in optimization at almost every stage of hydrogen production. AI can unlock the potential of green hydrogen as a clean, scalable energy source by increasing energy efficiency, lowering production costs, and facilitating wide-scale use. By combining AI with green hydrogen on αlist tech, difficult-to-decarbonize sectors like heavy industry and transport can be seen to become carbon neutral, while the environmental issues due to CO2 emissions or resource depletion are related to our pressing need.

The future that AI in green hydrogen production holds is enormous. Further efficiencies, cost reductions and system scalability will come with continued advances in AI algorithms, machine learning and predictive analytics. AI can improve hydrogen integration in the global energy systems assisting as a key energy carrier for a decarbonized economy. So, advances in AI technology will unlock a newer and greenerwave of innovation in hydrogen storage, distribution and hybrid energy systems, leading to an intelligent, sustainable energy ecosystem that facilitates global climate goals while ensuring a long-term environmental stewardship.

REFERENCES

Arora, M. K., Lal, S., Singh, B., & Olyaee, M. (2024). Through an Energy Lens: Examining SDGs From a Policy Perspective. In Riswandi, B., Singh, B., Kaunert, C., & Vig, K. (Eds.), *AI Applications for Clean Energy and Sustainability* (pp. 320–343). IGI Global. DOI: 10.4018/979-8-3693-6567-0.ch016

Asif, M. (n.d.). Sustainable Energy Transition: Technological Perspective. In *Handbook of Energy and Environment in the 21st Century* (pp. 1-21). CRC Press.

Boretti, A. (2024). Artificial intelligence driven evolution of perovskite-based solar thermochemical systems for hydrogen production: A narrative review. *Emergent Materials*, 1–15.

Boretti, A. (2024). Artificial intelligence driven evolution of perovskite-based solar thermochemical systems for hydrogen production: A narrative review. *Emergent Materials*, 1–15. DOI: 10.1007/s42247-024-00823-w

Dutta, P. K., Singh, B., Towfeek, A.-S. K., Adamopoulou, J. P., Bardavouras, A. N., Bamwerinde, W., Turyasingura, B., & Ayiga, N. (2024). IoT Revolutionizes Humidity Measurement and Management in Smart Cities to Enhance Health and Wellness. *Mesopotamian Journal of Artificial Intelligence in Healthcare*, *2024*, 110–117. DOI: 10.58496/MJAIH/2024/013

Dutta, P. K., Singh, B., Towfeek, A.-S. K., Adamopoulou, J. P., Bardavouras, A. N., Bamwerinde, W., Turyasingura, B., & Ayiga, N. (2024). IoT Revolutionizes Humidity Measurement and Management in Smart Cities to Enhance Health and Wellness. *Mesopotamian Journal of Artificial Intelligence in Healthcare*, *2024*, 110–117. DOI: 10.58496/MJAIH/2024/013

Kaushik, T. K., Singh, A., Raghav, A., Singh, B., Arora, M. K., & Lal, S. (2024). Untapping Artificial Intelligence in Environmental and Sustainable Progression: Projecting Legal and Ethical Considerations. In B. Singh, C. Kaunert, K. Vig, & S. Dutta (Eds.), *Maintaining a Sustainable World in the Nexus of Environmental Science and AI* (pp. 109-130). IGI Global. DOI: 10.4018/979-8-3693-6336-2.ch005

Kaushik, T. K., Singh, A., Raghav, A., & Singh, B. (2024). Potential of Artificial Intelligence in the Agricultural Industry: A Comprehensive Review. In B. Riswandi, B. Singh, C. Kaunert, & K. Vig (Eds.), *AI Applications for Clean Energy and Sustainability* (pp. 344-365). IGI Global. DOI: 10.4018/979-8-3693-6567-0.ch017

Kumar, S., Malviya, R., Uniyal, P., Srivastava, S., Sridhar, S. B., Obaidur Rab, S., & Singh, B. (2024, August 23). (2024). Plant Fiber Composites for Biomedical Applications: Advances and Prospective. *Current Pharmaceutical Biotechnology*, *25*. Advance online publication. DOI: 10.2174/0113892010314833240815113622 PMID: 39185642

Lal, S., Arora, M. K., Riswandi, B. A., Singh, N., & Singh, B. (2024). Melding Artificial Intelligence and Remote Sensing Data for Real-Time Monitoring of Marine Biodiversity. In Riswandi, B., Singh, B., Kaunert, C., & Vig, K. (Eds.), *AI Applications for Clean Energy and Sustainability* (pp. 299–319). IGI Global. DOI: 10.4018/979-8-3693-6567-0.ch015

Qiu, K., & Zhao, K. (2024). The integration of green energy and artificial intelligence in next-generation energy supply chain: An analysis of economic, social, and environmental impacts. *Sustainable Energy Technologies and Assessments, 64*, 103660.

Qiu, K., & Zhao, K. (2024). The integration of green energy and artificial intelligence in next-generation energy supply chain: An analysis of economic, social, and environmental impacts. *Sustainable Energy Technologies and Assessments, 64*, 103660. DOI: 10.1016/j.seta.2024.103660

Raghav, A., Singh, B., & Jermsittiparsert, K. (2024). Role of Artificial Intelligence and Machine Learning in Clean and Blue Energy for Enhancing Efficiency and Sustainability. In Riswandi, B., Singh, B., Kaunert, C., & Vig, K. (Eds.), *AI Applications for Clean Energy and Sustainability* (pp. 366–386). IGI Global. DOI: 10.4018/979-8-3693-6567-0.ch018

Raghav, A., Singh, B., Jermsittiparsert, K., Raghav, R., & Yadav, U. (2024). Artificial Intelligence in Environmental and Climate Changes: A Sustainable Future. In Singh, B., Kaunert, C., Vig, K., & Dutta, S. (Eds.), *Maintaining a Sustainable World in the Nexus of Environmental Science and AI* (pp. 485–506). IGI Global. DOI: 10.4018/979-8-3693-6336-2.ch019

Rasheed, M. Q., Yuhuan, Z., Haseeb, A., Ahmed, Z., & Saud, S. (2024). Asymmetric relationship between competitive industrial performance, renewable energy, industrialization, and carbon footprint: Does artificial intelligence matter for environmental sustainability? *Applied Energy, 367*, 123346. DOI: 10.1016/j.apenergy.2024.123346

Singh, B. (2023). Blockchain Technology in Renovating Healthcare: Legal and Future Perspectives. In Kaushik, K., Dahiya, S., Aggarwal, S., & Dwivedi, A. (Eds.), *Revolutionizing Healthcare Through Artificial Intelligence and Internet of Things Applications* (pp. 177–186). IGI Global. DOI: 10.4018/978-1-6684-5422-0.ch012

Singh, B. (2024). Cherish Data Privacy and Human Rights in the Digital Age: Harmonizing Innovation and Individual Autonomy. In Pucelj, M., & Bohinc, R. (Eds.), *Balancing Human Rights, Social Responsibility, and Digital Ethics* (pp. 199–226). IGI Global. DOI: 10.4018/979-8-3693-3334-1.ch007

Singh, B. (2024). Lensing Legal Dynamics for Examining Responsibility and Deliberation of Generative AI-Tethered Technological Privacy Concerns: Infringements and Use of Personal Data by Nefarious Actors. In Ara, A., & Ara, A. (Eds.), *Exploring the Ethical Implications of Generative AI* (pp. 146–167). IGI Global. DOI: 10.4018/979-8-3693-1565-1.ch009

Singh, B. (2024). Social Cognition of Incarcerated Women and Children: Addressing Exposure to Infectious Diseases and Legal Outcomes. In Reddy, K. (Ed.), *Principles and Clinical Interventions in Social Cognition* (pp. 236–251). IGI Global. DOI: 10.4018/979-8-3693-1265-0.ch014

Singh, B. (2024). Evolutionary Global Neuroscience for Cognition and Brain Health: Strengthening Innovation in Brain Science. In Prabhakar, P. (Ed.), *Biomedical Research Developments for Improved Healthcare* (pp. 246–272). IGI Global. DOI: 10.4018/979-8-3693-1922-2.ch012

Singh, B. (n.d.). Digital Twins as New Paradigm for Bridging the Gap from Personalized Medicine to Specific Rural Public Health: Exploring Foundations, Legal Angels, and Technologies for Health 5.0—Future Research Directions. In *Artificial Intelligence Solutions for Cyber-Physical Systems* (pp. 286-309). Auerbach Publications. https://www.taylorfrancis.com/chapters/edit/10.1201/9781032694375-16/digital-twins-new-paradigm-bridging-gap-personalized-medicine-specific-rural-public-health-bhupinder-singh

Singh, B., Dutta, P. K., Gautam, R., & Kaunert, C. (2024). Uncapping the Potential of Quantum Computing Towards Manufacturing Optimization: Routing Supply Chain Projecting Sustainability. In Hassan, A., Bhattacharya, P., Dutta, P., Verma, J., & Kundu, N. (Eds.), *Quantum Computing and Supply Chain Management: A New Era of Optimization* (pp. 395–419). IGI Global. DOI: 10.4018/979-8-3693-4107-0.ch024

Singh, B., Dutta, P. K., & Kaunert, C. (2024). Boosting Quantum Computing to Redefine Supply Chain in the Hospital Industry: Pitch the Vision for Medical and Healthcare Needs. In Hassan, A., Bhattacharya, P., Dutta, P., Verma, J., & Kundu, N. (Eds.), *Quantum Computing and Supply Chain Management: A New Era of Optimization* (pp. 196–219). IGI Global. DOI: 10.4018/979-8-3693-4107-0.ch014

Singh, B., Dutta, P. K., & Kaunert, C. (2024). Replenish Artificial Intelligence in Renewable Energy for Sustainable Development: Lensing SDG 7 Affordable and Clean Energy and SDG 13 Climate Actions With Legal-Financial Advisory. In Derbali, A. (Ed.), *Social and Ethical Implications of AI in Finance for Sustainability* (pp. 198–227). IGI Global. DOI: 10.4018/979-8-3693-2881-1.ch009

Singh, B., Jain, V., Kaunert, C., Dutta, P. K., & Singh, G. (2024). Privacy Matters: Espousing Blockchain and Artificial Intelligence (AI) for Consumer Data Protection on E-Commerce Platforms in Ethical Marketing. In Saluja, S., Nayyar, V., Rojhe, K., & Sharma, S. (Eds.), *Ethical Marketing Through Data Governance Standards and Effective Technology* (pp. 167–184). IGI Global. DOI: 10.4018/979-8-3693-2215-4.ch015

Singh, B., & Kaunert, C. (2024). Adventure in High Altitude of Mountainous Topographies and Health Impacts: Lensing Tourism Sustainability via Reducing Ecological and Sociocultural Footprint and Health Emergency and Medical Assistance Management. In Meraj, G., Hashimoto, S., & Kumar, P. (Eds.), *Navigating Natural Hazards in Mountainous Topographies. Disaster Risk Reduction*. Springer. DOI: 10.1007/978-3-031-65862-4_15

Singh, B., & Kaunert, C. (2024). Transforming Smart City and Smart Sustainable Tourism Projecting Artificial Intelligence and IoT: Immersive Potential for Revolutionizing Urban Living and Enhancing Travel Experience. In Jermsittiparsert, K., & Suanpang, P. (Eds.), *Technological and Managerial Approaches to Fostering Sustainable Travel* (pp. 55–80). IGI Global. DOI: 10.4018/979-8-3693-8115-1.ch003

Singh, B., & Kaunert, C. (2024). AI-Driven Strategies for Customer Engagement, Market Segmentation, and Resource Optimization: Projecting End User Satisfaction and Futuristic Growth of Business. In Nadda, V., Tyagi, P., Singh, A., & Singh, V. (Eds.), *AI Innovations in Service and Tourism Marketing* (pp. 104–128). IGI Global. DOI: 10.4018/979-8-3693-7909-7.ch006

Singh, B., & Kaunert, C. (2024). Blue Ocean and Machine Learning Trajectories SDG 14: Life Below Water for Handling Ocean Pollution: Metaverse Conserve Ocean Health Sustainability Through the Lens of Transboundary Legal-Policy Regulations as Articulating Space for Futuristic Changes. In *Artificial Intelligence and Edge Computing for Sustainable Ocean Health*. The Springer Series in Applied Machine Learning. Springer. DOI: 10.1007/978-3-031-64642-3_4

Singh, B., & Kaunert, C. (2024). Hitching Artificial Intelligence and IoT and for Livestock Management Competence: Line Up Nature and Technology for Smart Agriculture. In Zia Ul Haq, M., & Ali, I. (Eds.), *Revolutionizing Pest Management for Sustainable Agriculture* (pp. 499–520). IGI Global. DOI: 10.4018/979-8-3693-3061-6.ch020

Singh, B., & Kaunert, C. (2024). Redefining the Nodes of Metaverse and Luxury Hospitality: Unveiling AI and Tech-Driven Solutions in Hospitality and Tourism Industry. In Talukder, M., Kumar, S., & Tyagi, P. (Eds.), *Impact of AI and Tech-Driven Solutions in Hospitality and Tourism* (pp. 175–202). IGI Global. DOI: 10.4018/979-8-3693-6755-1.ch010

Singh, B., & Kaunert, C. (2024). Trans-Disciplinary Collaborative Disciplines Generating Cohesive Curriculum: Runway for Innovation, Cooperation, and Intentionality in Higher Education. In R. Kumar, E. Ong, S. Anggoro, T. Toh, & M. Fukui (Eds.), *Transdisciplinary Teaching and Technological Integration for Improved Learning: Case Studies and Practical Approaches* (pp. 152-176). IGI Global. DOI: 10.4018/979-8-3693-8217-2.ch008

Singh, B., & Kaunert, C. (2024). Human-Machine Nexus for Digital Rebound Fostering Futuristic Energy-Efficiency: Ecological Footprint of Technology for Smart-Sustainable Urban Mobility. In González-Lezcano, R., & Sansaniwal, S. (Eds.), *Design Strategies for Efficient and Sustainable Building Facilities* (pp. 222–244). IGI Global. DOI: 10.4018/979-8-3693-3200-9.ch010

Singh, B., & Kaunert, C. (2024). Hidden Gems Breakthrough Dynamic Landscape of Adaptive Learning Technologies for Higher Education: Bridging the Gap Between Theoretical and Practical Knowledge Projecting Student Learning Outcomes. In Minh Tung, T. (Ed.), *Adaptive Learning Technologies for Higher Education* (pp. 222–247). IGI Global. DOI: 10.4018/979-8-3693-3641-0.ch010

Singh, B., & Kaunert, C. (2024). Embryonic Machine-Deep Learning, Smart Healthcare and Privacy Deliberations in Hospital Industry: Lensing Confidentiality of Patient's Information and Personal Data in Legal-Ethical Landscapes Projecting Futuristic Dimensions. In Kumar, P., Singh, P., Diwakar, M., & Garg, D. (Eds.), *Healthcare Industry Assessment: Analyzing Risks, Security, and Reliability. Engineering Cyber-Physical Systems and Critical Infrastructures* (Vol. 11). Springer. DOI: 10.1007/978-3-031-65434-3_7

Singh, B., & Kaunert, C. (2024). AI-Driven Strategies for Customer Engagement, Market Segmentation, and Resource Optimization: Projecting End User Satisfaction and Futuristic Growth of Business. In Nadda, V., Tyagi, P., Singh, A., & Singh, V. (Eds.), *AI Innovations in Service and Tourism Marketing* (pp. 104–128). IGI Global. DOI: 10.4018/979-8-3693-7909-7.ch006

Singh, B., & Kaunert, C. (2024). Vertical Assimilation of Artificial Intelligence and Machine Learning in Safeguarding Financial Data. In Naz, F., & Karim, S. (Eds.), *Safeguarding Financial Data in the Digital Age* (pp. 169–197). IGI Global. DOI: 10.4018/979-8-3693-3633-5.ch010

Singh, B., & Kaunert, C. (2024). Computational Thinking for Innovative Solutions and Problem-Solving Techniques: Transforming Conventional Education to Futuristic Interdisciplinary Higher Education. In Fonkam, M., & Vajjhala, N. (Eds.), *Revolutionizing Curricula Through Computational Thinking, Logic, and Problem Solving* (pp. 60–82). IGI Global. DOI: 10.4018/979-8-3693-1974-1.ch004

Singh, B., & Kaunert, C. (2024). Wind and Solar Energy as Renewable Energy for Sustainable Global Future: Projecting Future Multi-Sector Sustainable Policies and Innovation. In Ara, A., & Thakore, R. (Eds.), *Promoting Multi-Sector Sustainability With Policy and Innovation* (pp. 210–245). IGI Global. DOI: 10.4018/979-8-3693-2113-3.ch009

Singh, B., & Kaunert, C. (2024). Aroma of Highly Smart Internet of Medical Things (IoMT) and Lightweight EdgeTrust Expansion Medical Care Facilities for Electronic Healthcare Systems: Fortified-Chain Architecture for Remote Patient Monitoring and Privacy Protection Beyond Imagination. In Hassan, A., Bhattacharya, P., Tikadar, S., Dutta, P., & Sagayam, M. (Eds.), *Lightweight Digital Trust Architectures in the Internet of Medical Things (IoMT)* (pp. 196–212). IGI Global. DOI: 10.4018/979-8-3693-2109-6.ch011

Singh, B., & Kaunert, C. (2024). Augmented Reality and Virtual Reality Modules for Mindfulness: Boosting Emotional Intelligence and Mental Wellness. In Hiran, K., Doshi, R., & Patel, M. (Eds.), *Applications of Virtual and Augmented Reality for Health and Wellbeing* (pp. 111–128). IGI Global. DOI: 10.4018/979-8-3693-1123-3.ch007

Singh, B., & Kaunert, C. (2024). Salvaging Responsible Consumption and Production of Food in the Hospitality Industry: Harnessing Machine Learning and Deep Learning for Zero Food Waste. In Singh, A., Tyagi, P., & Garg, A. (Eds.), *Sustainable Disposal Methods of Food Wastes in Hospitality Operations* (pp. 176–192). IGI Global. DOI: 10.4018/979-8-3693-2181-2.ch012

Singh, B., & Kaunert, C. (2024). Revealing Green Finance Mobilization: Harnessing FinTech and Blockchain Innovations to Surmount Barriers and Foster New Investment Avenues. In Jafar, S., Rodriguez, R., Kannan, H., Akhtar, S., & Plugmann, P. (Eds.), *Harnessing Blockchain-Digital Twin Fusion for Sustainable Investments* (pp. 265–286). IGI Global. DOI: 10.4018/979-8-3693-1878-2.ch011

Singh, B., & Kaunert, C. (2024). Harnessing Sustainable Agriculture Through Climate-Smart Technologies: Artificial Intelligence for Climate Preservation and Futuristic Trends. In Kannan, H., Rodriguez, R., Paprika, Z., & Ade-Ibijola, A. (Eds.), *Exploring Ethical Dimensions of Environmental Sustainability and Use of AI* (pp. 214–239). IGI Global. DOI: 10.4018/979-8-3693-0892-9.ch011

Singh, B., & Kaunert, C. (2025). Innovative Business Models in Digital and Virtual Reality for the Gig Economy: Space-Age Sustainable Development Projecting Smart Business Practices. In Taherdoost, H., Drazenovic, G., Madanchian, M., Khan, I., & Arshi, O. (Eds.), *Business Transformation in the Era of Digital Disruption* (pp. 127–156). IGI Global. DOI: 10.4018/979-8-3693-7056-8.ch005

Singh, B., & Kaunert, C. (2025). Dynamic Landscape of Artificial General Intelligence (AGI) for Advancing Renewable Energy in Urban Environments: Synergies with SDG 11—Sustainable Cities and Communities Lensing Policy and Governance. In El Hajjami, S., Kaushik, K., & Khan, I. U. (Eds.), *Artificial General Intelligence (AGI) Security. Advanced Technologies and Societal Change*. Springer. DOI: 10.1007/978-981-97-3222-7_12

Singh, B., & Kaunert, C. (2025). Blockchain Framework for Precision Medicine, Clinical Trials, and Genomic Biomarkers Discovery: Focuses on Monitoring and Managing Data in Health Dimensions. In Jermsittiparsert, K., & Marzo, R. (Eds.), *AI Technologies and Advancements for Psychological Well-Being and Healthcare* (pp. 255–288). IGI Global. DOI: 10.4018/979-8-3693-9158-7.ch010

Singh, B., & Kaunert, C. FedHealth in Wearable Healthcare, Orchestrated Federated Deep Learning for Smart Healthcare: Health Monitoring and Healthcare Informatics Lensing Challenges and Future Directions. In *Federated Deep Learning for Healthcare* (pp. 207–223). CRC Press. https://www.taylorfrancis.com/chapters/edit/10.1201/9781032694870-16/fedhealth-wearable-healthcare-orchestrated-federated-deep-learning-smart-healthcare-bhupinder-singh-christian-kaunert

Singh, B., & Kaunert, C. (n.d.). Tagging Blockchain Technology Fostering Panacea for Data Privacy, Cloud Computing, and Integrity: Legal Framework in India and at a Globe. In *Artificial Intelligence Solutions for Cyber-Physical Systems* (pp. 206-213). Auerbach Publications. https://www.taylorfrancis.com/chapters/edit/10.1201/9781032694375-11/tagging-blockchain-technology-fostering-panacea-data-privacy-cloud-computing-integrity-bhupinder-singh-christian-kaunert

Singh, B., & Kaunert, C. Marvel of Biosensors in Smart Healthcare and Application of Internet of Medical Things for Diagnosis: Unveiling Benefits, Challenges and Futuristic Approach. In *Smart Healthcare Systems* (pp. 187–198). CRC Press. https://www.taylorfrancis.com/chapters/edit/10.1201/9781032698519-14/marvel-biosensors-smart-healthcare-application-internet-medical-things-diagnosis-bhupinder-singh-christian-kaunert

Singh, B., Kaunert, C., & Gautam, R. (2024). Harnessing Machine Learning Intelligence Against Cyber Threats: Navigating Legal and Ethical Challenges in Global Cyber Investigations. In Özsungur, F. (Ed.), *Empowering Entrepreneurial Mindsets With AI* (pp. 319–352). IGI Global. DOI: 10.4018/979-8-3693-7658-4.ch016

Singh, B., Kaunert, C., & Jermsittiparsert, K. (2024). Environment-Biodiversity Protection and SDG 15 (Life on Land): Legal Landscapes and Probable Resolutions. In Singh, B., Kaunert, C., Vig, K., & Dutta, S. (Eds.), *Maintaining a Sustainable World in the Nexus of Environmental Science and AI* (pp. 259–282). IGI Global. DOI: 10.4018/979-8-3693-6336-2.ch010

Singh, B., Kaunert, C., & Jermsittiparsert, K. (2025). Advocating Medical Tourism and Healthcare in Global Health Market Places: Promoting Sustainability and Technological Advancements. In Jermsittiparsert, K., & Marzo, R. (Eds.), *AI Technologies and Advancements for Psychological Well-Being and Healthcare* (pp. 327–350). IGI Global. DOI: 10.4018/979-8-3693-9158-7.ch012

Singh, B., Kaunert, C., Jermsittiparsert, K., Lal, S., Arora, M. K., & Raghav, A. (2024). Ultimate Deep Learning Acclimatization With Artificial Intelligence Drones for Environment Shield and Climate Change: Lifting Sustainability Towards Fostering SDG 13. In Singh, B., Kaunert, C., Vig, K., & Dutta, S. (Eds.), *Maintaining a Sustainable World in the Nexus of Environmental Science and AI* (pp. 59–80). IGI Global. DOI: 10.4018/979-8-3693-6336-2.ch003

Singh, B., Kaunert, C., & Kaushik, T. K. (2024). Unscrambling Financial Fraud With AI and Machine Learning in E-Commerce Transactions: Airing Into Ad Clicks, Credit Card Management. In Pandow, B., Masoodi, F., Iqbal, J., & Hussain, G. (Eds.), *Navigating the Future of Finance in the Age of AI* (pp. 253–271). IGI Global. DOI: 10.4018/979-8-3693-4382-1.ch013

Singh, B., Kaunert, C., Lal, S., Arora, M. K., & Singh, G. (2024). Abundance Ocean Wave Energy to Electricity With Artificial Intelligence and IoT Solutions: Future of Clean Energy and Sustainability in Smart Cities. In Riswandi, B., Singh, B., Kaunert, C., & Vig, K. (Eds.), *AI Applications for Clean Energy and Sustainability* (pp. 274–298). IGI Global. DOI: 10.4018/979-8-3693-6567-0.ch014

Singh, B., Kaunert, C., & Malviya, R. (2024). Featuring Machine Learning in Municipal Solid Waste Management: Technological Advancements in Waste Sorting and Processing Ropeway for SDG 13 (Climate Action). In Mehra, R., Kumar, P., & Hung, B. (Eds.), *Municipal Solid Waste Management and Recycling Technologies* (pp. 343–366). IGI Global. DOI: 10.4018/979-8-3693-4054-7.ch013

Singh, B., Kaunert, C., & Malviya, R. (2024). Coupling Human-Computer Interface Lensing Artificial Emotional Intelligence: Transforming for Futuristic Cognitive Recognition for Decision-Marking in Industry 5.0. In Kumar, N., Pal, S., Agarwal, P., Rosak-Szyrocka, J., & Jain, V. (Eds.), *Human-Machine Collaboration and Emotional Intelligence in Industry 5.0* (pp. 233–250). IGI Global. DOI: 10.4018/979-8-3693-6806-0.ch013

Singh, B., Kaunert, C., Malviya, R., Gautam, R., & Singh, G. (2024). Generative Artificial Intelligence in the Tourism-Hospitality Industry: Strengthening Customer Relationships, Delivering Better Experiences. In Talukder, M., Kumar, S., & Tyagi, P. (Eds.), *Hotel and Travel Management in the AI Era* (pp. 275–300). IGI Global. DOI: 10.4018/979-8-3693-7898-4.ch013

Singh, B., Kaunert, C., Malviya, R., Lal, S., & Arora, M. K. (2024). Scrutinizing Consumer Sentiment on Social Media and Data-Driven Decisions for Business Insights: Fusion of Artificial Intelligence (AI) and Business Intelligence (BI) Foster Sustainable Growth. In Natarajan, A., Galety, M., Iwendi, C., Das, D., & Shankar, A. (Eds.), *Intersection of AI and Business Intelligence in Data-Driven Decision-Making* (pp. 183–210). IGI Global. DOI: 10.4018/979-8-3693-5288-5.ch007

Singh, B., Malviya, R., & Kaunert, C. (2024). Elevating Workplace Sustainability for Employees Lensing Mental Health Advancements: Runway for Future Ready Healthcare Services Projecting SDG 3 (Good Health and Well-Being). In Shaikh, E. (Ed.), *Impact of Corporate Social Responsibility on Employee Wellbeing* (pp. 285–310). IGI Global. DOI: 10.4018/979-8-3693-3470-6.ch014

Singh, B., Singh, A., Kaunert, C., Arora, M. K., Lal, S., & Ravesangar, K. (2025). Unleashing the Ethical and Legal Implications of E-Business Revolution: Consumer Privacy and Security Concerns in Phase of Digital Disruption. In Taherdoost, H., Drazenovic, G., Madanchian, M., Khan, I., & Arshi, O. (Eds.), *Business Transformation in the Era of Digital Disruption* (pp. 157–180). IGI Global. DOI: 10.4018/979-8-3693-7056-8.ch006

Singh, N., Lal, S., Raghav, A., Arora, M. K., Singh, B., & Kaunert, C. (2024). Power of ML and DL in Precision Modeling and Climate Predictions: NextGen Ocean Insights. In Riswandi, B., Singh, B., Kaunert, C., & Vig, K. (Eds.), *AI Applications for Clean Energy and Sustainability* (pp. 254–273). IGI Global. DOI: 10.4018/979-8-3693-6567-0.ch013

Song, G., Ifaei, P., Ha, J., Kang, D., Won, W., Liu, J. J., & Na, J. (2024). The AI circular hydrogen economist: Hydrogen supply chain design via hierarchical deep multi-agent reinforcement learning. *Chemical Engineering Journal*, *497*, 154464.

Song, G., Ifaei, P., Ha, J., Kang, D., Won, W., Liu, J. J., & Na, J. (2024). The AI circular hydrogen economist: Hydrogen supply chain design via hierarchical deep multi-agent reinforcement learning. *Chemical Engineering Journal*, *497*, 154464. DOI: 10.1016/j.cej.2024.154464

Wang, H., Wen, C., Duan, L., Li, X., Liu, D., & Guo, W. (2024). Sustainable energy transition in cities: A deep statistical prediction model for renewable energy sources management for low-carbon urban development. *Sustainable Cities and Society*, *107*, 105434.

Chapter 25
Unfolding Artificial Intelligence (AI) in Green Hydrogen Production:
Towards an Intelligent Energy Transition Fostering Sustainability

Bhupinder Singh
https://orcid.org/0009-0006-4779-2553
Sharda University, India

Saurabh Chandra
https://orcid.org/0000-0003-4172-9968
Bennett University, Greater Noida, India

ABSTRACT

Artificial intelligence (AI) is increasingly being integrated with green hydrogen production, so as to enable a seamless transition from conventionally suffocating energy sources. Water electrolysis, using renewable energies such as solar and wind energy as a power source, is the key technology of green hydrogen with considerable advantages towards decarbonizing energy-intensive industries and promoting global sustainability. Innovative application of AI is changing the way green hydrogen operates for efficiency and scalability by improving production processes saving energy in electrolysis making it cost effective. The AI and green hydrogen production benefits not only energy efficiency, but also serves to further the cause of climate action by weaning society off fossil fuels. And this clever, foresighted investment clears a path for a cleaner future which also adds one more way out of many possible ways that the world can reach its sustainability goals and move off carbon-neutral energy.

1. INTRODUCTION

AI-driven applications like machine learning models are now beginning to change the way electrolysis systems function, scheduling maintenance operations before it's too late, improving local conditions for proper hydrogen generation and much more (Boretti, 2024). AI is also heavily relied on to help manage

DOI: 10.4018/979-8-3373-0045-0.ch025

the storage and transportation of hydrogen to make the supply chain process efficient and less energy wasteful (Singh & Kaunert, 2024). This way industries are able to effectively manage and integrate renewable energy sources within their hydrogen production process using predictive forecasting, artificial intelligence based data analysis and real-time optimization towards a more modestly cleaner energy ecosystem (Singh et al., 2024).

With the world facing a two-front challenge fighting climate change and moving to renewable resources, the importance of renewable energy sources has grown phenomenally (Singh & Kaunert, 2024). One of the most exciting renewable energy replacements is green hydrogen, which has been hailed as a game-changer in our quest to be carbon neutral and promote energy independence. Made from water electrolysis powered by 100% renewable energies such as solar, wind and hydroelectric power, green hydrogen is an easy and efficient solution for industries and sectors where fossil fuels are traditionally used (Singh et al., 2024). The positioning as a solution for decarbonizing heavier-polluting, higher-growth sectors like transportation and manufacturing (also known as hard-to-abate industrial carbon sourcing) or heavy industry makes blue hydrogen sound like Corri-doors but also ultimately vital.

The large-scale deployment and manufacture of green hydrogen encounter numerous technological and economic difficulties. Hydrogen production via electrolysis is inefficient and energy-intensive, and it also faces challenges with regard to storage and distribution that must be overcome if green hydrogen is ever to become a viable global energy source (Singh et al., 2024). And this is where the role of Artificial Intelligence (AI) com into picture. The AI includes machine learning, deep learning, and data analytics that is helping sectors like healthcare to finance. The electricity sector uses Artificial Intelligence for renewable energy production, grid management, and demand predictions (Singh & Kaunert, 2024). The use of AI in green hydrogen production is still early, but holds potential to eliminate key bottlenecks, decrease costs, and offer scalability. AI-based solutions can help to optimize hydrogen production pathways -including improving the energy efficiency of less efficient modes like SMR or easing the deployment increase in renewable energy into the value chain. In addition, A.I. can be instrumental in storing, transferring, and utilizing green hydrogen to help take it from theoretical potential to real-world application (Raghav et al., 2024).

It undertake this study to investigate the potential of AI for green hydrogen production, assess and predict how machine learning can accelerate an intelligent energy transition, particularly in light of sustainability efforts. It, provides a look into the massive unlocked potential in green hydrogen by prospecting four powerful facets of AI and their role within this sustainability domain: an example being AI-powered electrolysis enhancement where researchers are exploring efficiency improvements to leverage further savings (Kaushik et al., 2024).

1.1 Significance of the Study

The importance of this study of the confluence, AI and green hydrogen creation is set in the context of nations around the world being on a war-footing to address the rapid advance of climate change. Green hydrogen is set to be important for decarbonizing hard-to-electrify sectors, as the world transitions to a net-zero emissions economy. Green hydrogen can offer an alternative for industries like steel manufacturing, aviation and shipping that may find direct electrification technologically challenging or economically infeasible but are facing pressure to meet decarbonization targets. But to realize this opportunity, other bottlenecks must be overcome production inefficiencies, energy consumption that is high, cost limitations.

Solving Production Inefficiencies with AI: The high-energy intensity of electrolysis-the process by which water is broken down into hydrogen and oxygen using electricity-is a key hurdle in the production of green hydrogen. Even "green" electrolysis is largely inefficient hundreds of megajoules per kilogram ultimately put into the process, leading to over half a terajoule of energy in waste heat for every tonne of hydrogen produced (Singh et al., 2024). This in turn leads to higher production costs, making green hydrogen significantly less price-competitive than grey hydrogen from natural gas or coal which is almost always environmentally dirtier (Singh et al., 2024). This is where AI comes into play as the technology provides a path to turbocharge the electrolysis process. AI can optimize operational parameters temperature, pressure, and electrical input of electrolyzers using machine learning algorithms and real-time data analytics (Singh et al., 2024). These refinements enable a more efficient conversion of water to hydrogen, therefore lowering energy waste and production costs. Also, by cross-referencing renewable energy supply data, AI can anticipate when the best time to produce hydrogen; then electrolysis will be carried out at times where peak solar or wind generation is occurring. In practice, it means that the green hydrogen is extracted as energy-efficiently as possible from this process, saving an even more considerable portion of carbon emissions (Song et al., 2024).

Predictive Maintenance & Operations Efficiency: The other key space AI can help in the area of green hydrogen production is AI-driven predictive maintenance models. Industrial machinery experiences mechanical wear and tear over time, which results in operational inefficiencies and may cause downtime, including those installed to provide electrolysis. With traditional maintenance strategies come the inevitable breakdowns and disruptions, leading to high repair costs (Singh & Kaunert, 2024). This AI is capable of analyzing continuous sensor data, detecting patterns that might indicate a problem and notifying relevant staff of those imminent problems long before they have the opportunity to lead to wasted time or unrecoverable health. By proactively identifying possible issues, operators are able to take action before the equipment fails, significantly increasing the reliability and lifespan of hydrogen production infrastructure (Wang et al., 2024).

Application of AI for Better Storage and Distribution: The major challenges for green hydrogen value chain have to do with storage and distribution. It is because hydrogen is the lightest element and therefore needs expensive storage solutions such as high-pressure tanks or cryogenic systems which are energy intensive to build and very much expensive. Transportation also poses challenges, since hydrogen must be compressed or liquefied to more efficiently. transported over long distances, adding even further to the energy requirements (Singh & Kaunert, 2024). Artificial intelligence (AI) could be used to build predictive models of energy demand and patterns of hydrogen consumption and distribution logistics to optimize the network for such storage and distribution. It can be used to help design hydrogen storage systems that are more lenient, forecast energy demand patterns or determine delivery pathways in order to deliver the hydrogen where it is most needed using as little energy as possible (Singh & Kaunert, 2024). Additionally, AI is being used to support the integration of hydrogen storage into the wider grid and in so doing balancing hydrogen supply (electrolysers) and demand (fuel cell applications) with renewable energy source sielding (Dutta et al., 2024).

AI Can Drive Sustainability: Green hydrogen production is another case where AI plays a role that supports global sustainability objectives, not only matching but potentially exceeding the UN Sustainable Development Goals (SDGs) (Singh et al., 2024). Optimising energy use, reducing waste and facilitating the widespread adoption of renewable hydrogen are ways in which AI-driven solutions make those figures add up by contributing to SDG 7 Affordable and Clean Energy and SDG 13 Climate Action (Singh & Kaunert, 2024). Moreover, AI-supported production of green hydrogen can help make the economy more circular and enable us to use renewable energy for hydrogen generation/storage/distribution, thus limiting the consumption of fossil fuels and emissions this way (Qiu & Zhao, 2024).

The intersection of Artificial Intelligence and green hydrogen production is a game-changer in accelerating sustainable energy transition. This means that while green hydrogen is already the cleanest source of energy available for various uses, AI applications to optimize manufacturing processes, improve storage and distribution or reduce operational inefficiencies can make green hydrogen economically viable on a global scale. In order to meet increasingly aggressive carbon neutrality goals around the world, and within individual industries, advances in green hydrogen production led by improvements in AI technology are more important than ever for creating a greener and cleaner energy future. It helps demonstrate the growing role of AI in this crucial aspect and emphasizes its importance in powering up a green switch worldwide (Singh, 2024).

2. AI IN OPTIMIZING GREEN HYDROGEN PRODUCTION

Green hydrogen, produced from the electrolysis of water using renewable energy sources such as wind and solar, is touted to be pivotal in the world's clean energy future. Nonetheless, conventional hydrogen production remains uncompetitive with obstacles involving energy consumption, price, and expandability (Singh & Kaunert, 2024). These challenges can be mitigated by Artificial Intelligence (AI) to enable optimization of green hydrogen production, thus helping increase the efficiency of the process, and decreasing costs to improve its coordination when integrated with renewable energy systems (Singh & Kaunert, 2024).

The use of AI is upending the electrolysis process which takes water and splits it into hydrogen and oxygen with an electric current. The process is efficient or not depending on several factors such as temperature, pressure, electricity quality and etc. These variables can be monitored and optimized in real time by using AI algorithms, mostly machine learning (ML) models to assure the highest efficiency operation of the electrolyzers (Singh & Kaunert, 2024). This system helps control the operations of an electrolyzer more effectively and helps it maximize its yield in terms of hydrogen production using minimum amount of energy by constantly adjusting operational parameters as per input energy available at all times & condition of the Electrolyzer.

AI applications could predict how well electrolyzers would perform in various conditions, which operators could use to optimize the process and maximize efficiency. They use sensors to monitor temperature, pressure, and electricity input as data sources for AI systems that self-optimize the electrolysis process. Not only does this improve efficiency but also prolongs the operational life of electrolyzers by protecting them from wear and tear due to suboptimal operating conditions. Energy prices are a big hurdle to overcome for scaling production of green hydrogen. The electrolysis process to generate hydrogen is a high energy cost process made more complicated with renewable resources such as wind and solar, that can be intermittent. From energy savings and the use of renewable options in producing hydrogen, AI has the answer (Asif, 2024).

By studying past data of renewable energy production, the AI algorithms can predict when there will be large amounts of energy available. AI can combine this predictive ability with hydrogen production systems, so that the electrolysis processes are only powered when surplus renewable energy is being generated (instead of drawing upon non-renewable sources). AI could also enable hydrogen production to be scheduled so that demand and supply of energy can have a more balanced power, perhaps activating the hydrogen production during off-peak hours when electrical energy is cheaper (Singh & Kaunert, 2024). But the approach could see significant savings on the operational costs of hydrogen production facilities, and also ensure that a larger percentage of green energy ends up being used. AI is also essential in maintaining the hydrogen production infrastructure. Then, there are regular maintenance activities for each electrolyzers and other components to ensure the system works effectively and avoid downtime. AI-based predictive maintenance systems can process sensor data in real-time and identify the earliest symptoms of equipment wear, degradation or failure. It enables real-time intervention in case of any significant problem that help to avoid the heavy damages, leading to less operational disruption and lower maintenance cost (Kumar et al., 2024).

Machine learning models can also spot anomalies in the production process that may be an indication of equipment issues. If detected early, these anomalies can notify the AI system of when maintenance is required and they can schedule maintenance during times when hydrogen production impact is minimal. It provides not just improved operational efficiency and longer-life critical equipment in hydrogen plants -but also reduced costs. Because renewable energy sources such as solar and wind power are intermittent, they cause the amount of hydrogen gas that can be produced at facilities to fluctuate (Singh & Kaunert, 2024). The sun and wind generate power intermittently, resulting in fluctuations in the energy supply. The incorporation of hydrogen production to renewable energy grids can be optimized by AI. AI can predict energy generation from clean solar / wind sources in real-time and forecast generation and thus modify hydrogen production. For example, when there is more renewable energy than the grid needs AI can ramp up hydrogen production so that more of that excess can be converted to hydrogen and stored (as hydrogen). Later as power demand picks back up or whenever there are not enough renewables to meet demand. This rapidly responding dynamic energy management capability will significantly increase the stability and efficiency of operation in green hydrogen utility, so that renewable power is best used with no waste (Singh et al., 2024).

Artificial intelligence (AI) has the potential to revolutionize green hydrogen production andclear many of the hurdles tied to efficiency, cost and scalability. Through better electrolysis, less energy usage, predictive maintenance, and greater integration with renewable energy sources AI is shaping a more efficient and sustainable hydrogen economy. In a world increasingly moving towards carbon neutrality, AI will play a big role in facilitating the true potential of green hydrogen as an energy source (Singh & Kaunert, 2024).

3. AI-DRIVEN PREDICTIVE MAINTENANCE IN HYDROGEN INFRASTRUCTURE

As the green hydrogen production landscape takes shape in one of the quickly transforming sectors, keeping up the infrastructure is crucial to achieve peak performance and minimize downtime. In short, the hydrogen infrastructure electrolyzers, storage systems, and distribution networks — are very complex bits of equipment that must operate with absolute reliability if hydrogen production is to be economically viable (Singh et al., 2024). Artificial Intelligence (AI) based predictive maintenance is changing the landscape of monitoring and maintenance practices in hydrogen production facilities, saving these high-value assets from expensive failures and improving operational efficiency (Singh et al., 2024). That is because most traditional ways to approach maintenance, like reactive and scheduled preventive maintenance, can eat up quite scarce resources for not a lot of value in return. Reactive maintenance can be very expensive and generate unexpected downtimes, where preventive maintenance may cause undue interventions in equipment in good operation. By contrast, predictive maintenance takes a more proactive approach and uses AI models to predict the time when machines will fail or need servicing using real-time data feeding into historical behavior patterns (Singh et al., 2024).

Predictive maintenance is crucial when it comes to the hydrogen production, as electrolyzers, compressors, storage tanks, etc., are extremely susceptible to many parameters including temperature and pressure or the quality of input electricity. All deviations will limit the efficiency, increase wear or lead to a total failure. Over time, this will help in managing the above risks as predictive maintenance through AI will constantly monitor equipment health and alert if something goes wrong, such that we can overcome the issue before it can trigger a failure. AI provides the power necessary to process huge data from a variety of sensors embedded in hydrogen infrastructure and orchestrate large hydrogen infrastructures across various domains. These sensors generate massive real-time data like temperature, pressure, energy consumption, vibration along with other KPIs. These data are analyzed by AI algorithms, mostly to machine learning (ML) models, to detect patterns and correlations missed by human operators (Singh et al., 2024).

The applications is a system that uses machine learning to identify components on the electrolyzer showing early signs of wear, for example variations in temperature or pressure that indicate declining performance. Through ongoing education from a growing pool of data, an increasingly robust AI system will have an ever better idea of when machines are going to fail, which in turn will tell the operators when they need to repair or replace parts before one of them finally plunges to its demise (Lal et al., 2024). The result is an improvement from higher accuracy and dependability that hydrogen production centers which also reduces operational costs. AI-based predictive maintenance also shines in anomaly detection and when the system can recognise deviations from normal operating behavior (Arora et al., 2024). This means that as soon as a piece of equipment begins to behave atypically, AI models can identify these aberrations quickly, minimizing downtime by alerting operators when something may be wrong. A small departure from that vibration pattern in, say, a compressor's operating signature could indicate a mechanical problem that might cause big trouble down the road if it isn't dealt with (Kaushik et al., 2024).

AI can also use the operational history of equipment to predict failures. Based on historical failures, AI algorithms predict the time when a specific component is likely to fail by comparing the current data. This enables scheduled maintenance during non-critical periods to minimise interruptions to hydrogen production and avoid costly emergency repairs (Raghav et al., 2024). There are several strategic pluses from the standpoint of hydrogen infrastructure: The predictive maintenance element that AI provides carries benefits for the following reasons:

Minimize downtime: The early failure detection that AI offers allows for faster identification of potential faults and decreased unplanned outages, which increases hydrogen production system availability.

Less operation and maintenance costs: AI eliminates the need for rote preventive maintenance since machines can dispatch warnings when they are about to break down, saving time and money undefined.

Extend the life of electrolysers, compressors, and storage systems by preventing them from experiencing operating conditions that are not optimal.

Improves safety: Keeping a check on the conditions of the equipment guarantees regular maintenance preventing any downtimes converting into dangerous situations, also leading to a safe environment (Singh, 2024).

Hydrogen infrastructure benefits from AI-driven predictive maintenance which yields improved efficiency, cost-effectiveness, and reliability of hydrogen production systems (Singh & Kaunert, 2024). And with the increasing relevance of hydrogen in the wider global energy transition, this will also represent a necessary infrastructure component for scalable, sustainable hydrogen production on an industrial scale; requiring new solutions are called for, as we head into playing our part in clean green engineering challenges and opportunities (Singh & Kaunert, 2024).

4. AI IN HYDROGEN STORAGE AND DISTRIBUTION

High-capacity storage and efficient distribution of hydrogen are essential for the viability of hydrogen as a sustainable energy carrier. Not only is hydrogen very combustable and low density lIt makes it a huge challenge for to transport and store it in safe, energy efficient methods (Singh & Kaunert, 2024). It is as mentioned above AI can be a game-changer in the case of hydrogen storage and distribution system at all levels of supply chain by enhancing safety, efficiency and cost-effectiveness.

For storage, hydrogen is usually stored in one of three types: compressed gas (gaseous), liquid or absorbed to solid chemical compounds like metal hydrides. There are drawbacks to each of these methods, like keeping pumps powerful enough to work with high pressures, or engines cold without complex chemical reactions etc (Singh & Kaunert, 2024). If a hydrogen storage is monitored by a sensor to report parameters such as pressure, temperature or containment integrity, AI can be used to classify the mean to different actions how the system behavior should work optimally. So, at present several AI algorithms can predict the best conditions to store hydrogen which would minimize energy lose and safety standards. Examples include using artificial intelligence to monitor pressure levels in compressed hydrogen storage systems and predicting impending failures or leaks where changes are required prior failure. Artificial intelligence can be used to optimize refrigeration systems for cryogenic temperatures necessary to liquefy hydrogen thus saving a lot of energy and reducing operating costs in liquid hydrogen storage procedure. AI can also help determine the optimal storage technology for different use cases, thereby creating storage solutions that are more versatile and scalable to varying energy demands (Dutta et al., 2024).

Because of its small molecular size, hydrogen can escape easily and lead to safety risks during distribution, whether through pipelines or vehicles (Singh & Kaunert, 2024). Artificial Intelligence may greatly increase the safety in supply chains for hydrogen with the help of real-time tracking and predictive maintenance. An example of the use case with AI systems is using data from sensors installed along the pipeline or inside the transport vessel to capture any anomaly that may indicate there is something happening, typically a drop in pressure, changing some temperature or even altered chemical composition what implies there might be leak or valve failure (Singh & Kaunert, 2024). In addition to monitoring, models can predict when equipment is likely to fail based on its historical performance so that operators can preemptively schedule maintenance before a leak or failure. This would not only make them safer, but also save the time and expense of dealing with emergency repairs. In addition to routing the transport of hydrogen more intelligently, AI algorithms can find the most fuel- and emission-efficient routes for transportation vehicles that help in reducing environmental footprints in the entire biomass supply chain (Singh et al., 2024).

With hydrogen increasingly part of the energy system, balancing its distribution with other renewables is set to become more important. AI can further extend hydrogen storage and distribution into a larger smart grid, to maintain real-time balancing of supply and demand. For example, AI solutions can predict energy demand and optimize the distribution of hydrogen to regions with high consumption (above) while storing surplus hydrogen during low periods of demand. This flexible H2 energy management strategy ensures that hydrogen arrives where and when it is most needed, making the delivery of Hydrogen as an energy carrier more reliable and efficient (Singh et al., 2024). It is really revolutionizing the way hydrogen storage and distribution is done by making it more safe, efficient as well close to 30% cost savings. By giving stakeholders access to real-time monitoring and predictive analytics edged with the AI based smart grid integration, we would ensure hydrogen could be stored and transported as fuel for overall scalability and sustainability of the hydrogen economy (Singh et al., 2024). In view of the imperative for hydrogen to support the global energy transition, the use of AI solutions will be crucial in tackling these issues with storage and transport (Singh & Kaunert, 2024).

5. AI INTEGRATION WITH RENEWABLE ENERGY SOURCES

Artificial Intelligence (AI) is increasingly being integrated with renewable energy sources to improve the world´s supply of energy by allowing more intelligent management of production, distribution, and consumption. This variability is inherent in renewable energy sources for example solar, wind and hydropower because these are dependent on certain environmental conditions: weather and amount of daylight (Singh & Kaunert, 2024). To overcome these hurdles, AI is crucial in improving the prediction, efficiency as well as optimization of renewable energy systems. The biggest thing AI brought is forecasting how much energy will be generated from wind turbines at any given time. Sophisticated AI-powered systems can predict solar irradiance, wind speeds, and hydrological conditions to a great degree, which is paramount to design energy systems. By providing predictability these forecasts enable grid operators to further optimise the scheduling and dispatch of energy from renewable sources and help better balance electricity supply with demand. Proper forecasting lowers the amount of energy you need to over-produce, reduces the need for backup using fossil fuels and increases marginally the stability of the grid (Singh & Kaunert, 2024).

Renewable energy is commonly subject to curtailment when excess generation occurs, such as during sunny or windy days. E.g. it can control when to store/release energy in an energy storage system like batteries or green hydrogen etc since an AI model can identify through historical big data analysis, and predict the optimal state by optimization techniques rigesimal circuits, and the old classical methods are still used for that kind of execution (Singh et al., 2024). AI can, for example, analyse the up-to-the-minute demand on a grid and autonomously decide to store any excess solar power in batteries or turn that extra wind power into clean green hydrogen. On the distribution side, AI coordinates the distribution of renewable energy across a grid to deliver power where it is most needed in real-time limiting transmission losses (Singh et al., 2024).

The integration of AI can turn these into more dynamic, responsive energy grids known as “smart grids”. The usage of AI in these grids include managing energy flows, demand forecasting, balancing and other operations for efficient operation in real time. As a second example, AI can make demand response systems possible—these are systems that automatically change the pattern of consumption based on power grid conditions (Singh & Kaunert, 2024). For example, AI-defined algorithms can alert users to save power when demand is high by sourcing peak time consumption from other areas (Singh, 2024). The importance of having AI integrated with the renewable source of energy can have a long way in making our future life better in terms of sustainable, more reliable and efficient energy. AI is enabling the integration of renewables, and thereby advancing the world-wide transition to 100% clean energy (Singh, 2024).

6. SUSTAINABILITY IMPLICATIONS OF AI IN GREEN HYDROGEN PRODUCTION

Artificial Intelligence (AI) integrated green hydrogen production holds the promise to cater with global energy requirements and environmental friendly approach. Green hydrogen, which is produced by the electrolysis of water powered by renewable energy sources, is a clean energy carrier capable of decarbonizing difficult to electrify and transport sectors, such as heavy industries or aviation and shipping sector (Singh & Kaunert, 2024). But green hydrogen is energy-intensive and comes with its own technical and economic challenges. The ability of AI to offer innovative solutions that boost production efficiency, cost-effectiveness and scalability in the case of green hydrogen can have wide-ranging positive sustainability impacts (Singh, 2024).

A major advantage of AI-enabled green hydrogen production is the improvement in energy efficiency which underpins virtually all sustainability claims. The technology itself requires a great deal of electricity, and through AI optimization th is electricity can be used as efficiently as possible. AI can be used to detect wide-ranging data from electrolyzers such as voltage,current, temperature and water quality (Singh & Kaunert, 2024). Using AI to optimize energy consumption is one of these methods and, when applied to key variables such as temperature, time and current density, the AI can redefine performance without loss in hydrogen output—a way of increasing the overall efficiency of the system (Singh & Kaunert, 2024). AI can also help merge sources of renewable energies e.g., solar or wind with hydrogen production, such that the transition is smoother and carried out at a larger scale. For instance, AI algorithms can predict opportunities for large quantities of renewable power generation on account of weather events and operate electrolysis at a time when there is plenty renewable electricity. This could become less reliant on grid electricity from non-renewable sources, partially decarbonizing the production of hydrogen (Singh, 2024).

Green hydrogen is too expensive, to say the least. Because of this, AI is able to reduce production costs via optimization on multiple tiers. It lowers the operational costs associated with electricity consumption, which is one of the most important factors while calculating green hydrogen's total cost. In addition to this, AI-supported predictive maintenance can also guarantee that your machines never go out of business and stop costly failures by recognizing equipment issues before they result in large scale damage. For green hydrogen production to be deployed at the global scale, AI will be needed for operating systems at these scales. With the support of AI we can reduce human intervention and automate crucial processes which would thereby enable a large production plant to run 24×7 with least downtime. This scalability is important to green hydrogen filling a substantial role in decarbonizing our global economy especially for those hard-to-abate sectors of the economy like heavy industry and transportation, that are difficult to decarbonize (Boretti, 2024).

In any case, water is important for producing green hydrogen and its sustainable usage must be ensured to make the use of hydrogen as an energy carrier a long-lasting solution. On one hand, for green hydrogen production, water management can be made sustainable through AI and recycle of excess or waste water coming out during electrolysis process. Water quality: An AI algorithm will be able to monitor the water that is being used in the tenant suite for electrolysis and ensure that it does not contain impurities that can hinder the efficiency of electrolyzers or damage long term equipment. In addition, water recycling within the production system can be boosted by AI, which will help to decrease water loss. Especially for water stressed regions, availability of big-data led AI based water management can possibly play a substantial role to ensure that the green hydrogen production is not worsening local scarcity of water (Song et al., 2024).

In industries like steelmaking, cement production and heavy transport, using temperature heat or long-distance fuel to produce power is still a big challenge. These industries can be switched from fossil fuels to green hydrogen, but the production of this new hydrogen must be as globally scalable and cost-effective as possible in order to do this. The ability of AI to streamline production, bring costs down, and integrate seamlessly into the renewable market is essential for green hydrogen to supplant fossil fuels. The use of AI in green hydrogen will have important sustainability consequences (Qiu & Zhao, 2024). Green hydrogen could become a cornerstone of the global energy transition, thanks to AI that can drive efficiencies in energy use and cost savings for manufacturing, as well as enable large-scale adoption and sustainable water management. With industry and policy increasingly moving towards sustainability and carbon neutrality, AI in the production of green hydrogen will have a significant role to play in defining an energy landscape expected to be cleaner and more robust (Rasheed et al., 2024).

7. CONCLUSION AND FUTURE SCOPE

This means that one of the biggest moves in transitioning global energy to a more sustainable state has occurred from the infusion of AI technology into green hydrogen production. From improving the efficiency of electrolysis for hydrogen production to predictive maintenance, smart grid integration and efficient water management among others can be augmented with AI which helps in optimization at almost every stage of hydrogen production. AI can unlock the potential of green hydrogen as a clean, scalable energy source by increasing energy efficiency, lowering production costs, and facilitating wide-scale use. By combining AI with green hydrogen on αlist tech, difficult-to-decarbonize sectors like heavy industry and transport can be seen to become carbon neutral, while the environmental issues due to CO2 emissions or resource depletion are related to our pressing need.

The future that AI in green hydrogen production holds is enormous. Further efficiencies, cost reductions and system scalability will come with continued advances in AI algorithms, machine learning and predictive analytics. AI can improve hydrogen integration in the global energy systems assisting as a key energy carrier for a decarbonized economy. So, advances in AI technology will unlock a newer and greenerwave of innovation in hydrogen storage, distribution and hybrid energy systems, leading to an intelligent, sustainable energy ecosystem that facilitates global climate goals while ensuring a long-term environmental stewardship.

REFERENCES

Arora, M. K., Lal, S., Singh, B., & Olyaee, M. (2024). Through an Energy Lens: Examining SDGs From a Policy Perspective. In Riswandi, B., Singh, B., Kaunert, C., & Vig, K. (Eds.), *AI Applications for Clean Energy and Sustainability* (pp. 320–343). IGI Global. DOI: 10.4018/979-8-3693-6567-0.ch016

Asif, M. (n.d.). Sustainable Energy Transition: Technological Perspective. In *Handbook of Energy and Environment in the 21st Century* (pp. 1-21). CRC Press.

Boretti, A. (2024). Artificial intelligence driven evolution of perovskite-based solar thermochemical systems for hydrogen production: A narrative review. *Emergent Materials*, 1–15.

Boretti, A. (2024). Artificial intelligence driven evolution of perovskite-based solar thermochemical systems for hydrogen production: A narrative review. *Emergent Materials*, 1–15. DOI: 10.1007/s42247-024-00823-w

Dutta, P. K., Singh, B., Towfeek, A.-S. K., Adamopoulou, J. P., Bardavouras, A. N., Bamwerinde, W., Turyasingura, B., & Ayiga, N. (2024). IoT Revolutionizes Humidity Measurement and Management in Smart Cities to Enhance Health and Wellness. *Mesopotamian Journal of Artificial Intelligence in Healthcare, 2024*, 110–117. DOI: 10.58496/MJAIH/2024/013

Dutta, P. K., Singh, B., Towfeek, A.-S. K., Adamopoulou, J. P., Bardavouras, A. N., Bamwerinde, W., Turyasingura, B., & Ayiga, N. (2024). IoT Revolutionizes Humidity Measurement and Management in Smart Cities to Enhance Health and Wellness. *Mesopotamian Journal of Artificial Intelligence in Healthcare, 2024*, 110–117. DOI: 10.58496/MJAIH/2024/013

Kaushik, T. K., Singh, A., Raghav, A., Singh, B., Arora, M. K., & Lal, S. (2024). Untapping Artificial Intelligence in Environmental and Sustainable Progression: Projecting Legal and Ethical Considerations. In B. Singh, C. Kaunert, K. Vig, & S. Dutta (Eds.), *Maintaining a Sustainable World in the Nexus of Environmental Science and AI* (pp. 109-130). IGI Global. DOI: 10.4018/979-8-3693-6336-2.ch005

Kaushik, T. K., Singh, A., Raghav, A., & Singh, B. (2024). Potential of Artificial Intelligence in the Agricultural Industry: A Comprehensive Review. In B. Riswandi, B. Singh, C. Kaunert, & K. Vig (Eds.), *AI Applications for Clean Energy and Sustainability* (pp. 344-365). IGI Global. DOI: 10.4018/979-8-3693-6567-0.ch017

Kumar, S., Malviya, R., Uniyal, P., Srivastava, S., Sridhar, S. B., Obaidur Rab, S., & Singh, B. (2024, August 23). (2024). Plant Fiber Composites for Biomedical Applications: Advances and Prospective. *Current Pharmaceutical Biotechnology, 25*. Advance online publication. DOI: 10.2174/0113892010314833240815113622 PMID: 39185642

Lal, S., Arora, M. K., Riswandi, B. A., Singh, N., & Singh, B. (2024). Melding Artificial Intelligence and Remote Sensing Data for Real-Time Monitoring of Marine Biodiversity. In Riswandi, B., Singh, B., Kaunert, C., & Vig, K. (Eds.), *AI Applications for Clean Energy and Sustainability* (pp. 299–319). IGI Global. DOI: 10.4018/979-8-3693-6567-0.ch015

Qiu, K., & Zhao, K. (2024). The integration of green energy and artificial intelligence in next-generation energy supply chain: An analysis of economic, social, and environmental impacts. *Sustainable Energy Technologies and Assessments*, *64*, 103660.

Qiu, K., & Zhao, K. (2024). The integration of green energy and artificial intelligence in next-generation energy supply chain: An analysis of economic, social, and environmental impacts. *Sustainable Energy Technologies and Assessments*, *64*, 103660. DOI: 10.1016/j.seta.2024.103660

Raghav, A., Singh, B., & Jermsittiparsert, K. (2024). Role of Artificial Intelligence and Machine Learning in Clean and Blue Energy for Enhancing Efficiency and Sustainability. In Riswandi, B., Singh, B., Kaunert, C., & Vig, K. (Eds.), *AI Applications for Clean Energy and Sustainability* (pp. 366–386). IGI Global. DOI: 10.4018/979-8-3693-6567-0.ch018

Raghav, A., Singh, B., Jermsittiparsert, K., Raghav, R., & Yadav, U. (2024). Artificial Intelligence in Environmental and Climate Changes: A Sustainable Future. In Singh, B., Kaunert, C., Vig, K., & Dutta, S. (Eds.), *Maintaining a Sustainable World in the Nexus of Environmental Science and AI* (pp. 485–506). IGI Global. DOI: 10.4018/979-8-3693-6336-2.ch019

Rasheed, M. Q., Yuhuan, Z., Haseeb, A., Ahmed, Z., & Saud, S. (2024). Asymmetric relationship between competitive industrial performance, renewable energy, industrialization, and carbon footprint: Does artificial intelligence matter for environmental sustainability? *Applied Energy*, *367*, 123346. DOI: 10.1016/j.apenergy.2024.123346

Singh, B. (2023). Blockchain Technology in Renovating Healthcare: Legal and Future Perspectives. In Kaushik, K., Dahiya, S., Aggarwal, S., & Dwivedi, A. (Eds.), *Revolutionizing Healthcare Through Artificial Intelligence and Internet of Things Applications* (pp. 177–186). IGI Global. DOI: 10.4018/978-1-6684-5422-0.ch012

Singh, B. (2024). Cherish Data Privacy and Human Rights in the Digital Age: Harmonizing Innovation and Individual Autonomy. In Pucelj, M., & Bohinc, R. (Eds.), *Balancing Human Rights, Social Responsibility, and Digital Ethics* (pp. 199–226). IGI Global. DOI: 10.4018/979-8-3693-3334-1.ch007

Singh, B. (2024). Lensing Legal Dynamics for Examining Responsibility and Deliberation of Generative AI-Tethered Technological Privacy Concerns: Infringements and Use of Personal Data by Nefarious Actors. In Ara, A., & Ara, A. (Eds.), *Exploring the Ethical Implications of Generative AI* (pp. 146–167). IGI Global. DOI: 10.4018/979-8-3693-1565-1.ch009

Singh, B. (2024). Social Cognition of Incarcerated Women and Children: Addressing Exposure to Infectious Diseases and Legal Outcomes. In Reddy, K. (Ed.), *Principles and Clinical Interventions in Social Cognition* (pp. 236–251). IGI Global. DOI: 10.4018/979-8-3693-1265-0.ch014

Singh, B. (2024). Evolutionary Global Neuroscience for Cognition and Brain Health: Strengthening Innovation in Brain Science. In Prabhakar, P. (Ed.), *Biomedical Research Developments for Improved Healthcare* (pp. 246–272). IGI Global. DOI: 10.4018/979-8-3693-1922-2.ch012

Singh, B. (n.d.). Digital Twins as New Paradigm for Bridging the Gap from Personalized Medicine to Specific Rural Public Health: Exploring Foundations, Legal Angels, and Technologies for Health 5.0—Future Research Directions. In *Artificial Intelligence Solutions for Cyber-Physical Systems* (pp. 286-309). Auerbach Publications. https://www.taylorfrancis.com/chapters/edit/10.1201/9781032694375-16/digital-twins-new-paradigm-bridging-gap-personalized-medicine-specific-rural-public-health-bhupinder-singh

Singh, B., Dutta, P. K., Gautam, R., & Kaunert, C. (2024). Uncapping the Potential of Quantum Computing Towards Manufacturing Optimization: Routing Supply Chain Projecting Sustainability. In Hassan, A., Bhattacharya, P., Dutta, P., Verma, J., & Kundu, N. (Eds.), *Quantum Computing and Supply Chain Management: A New Era of Optimization* (pp. 395–419). IGI Global. DOI: 10.4018/979-8-3693-4107-0.ch024

Singh, B., Dutta, P. K., & Kaunert, C. (2024). Boosting Quantum Computing to Redefine Supply Chain in the Hospital Industry: Pitch the Vision for Medical and Healthcare Needs. In Hassan, A., Bhattacharya, P., Dutta, P., Verma, J., & Kundu, N. (Eds.), *Quantum Computing and Supply Chain Management: A New Era of Optimization* (pp. 196–219). IGI Global. DOI: 10.4018/979-8-3693-4107-0.ch014

Singh, B., Dutta, P. K., & Kaunert, C. (2024). Replenish Artificial Intelligence in Renewable Energy for Sustainable Development: Lensing SDG 7 Affordable and Clean Energy and SDG 13 Climate Actions With Legal-Financial Advisory. In Derbali, A. (Ed.), *Social and Ethical Implications of AI in Finance for Sustainability* (pp. 198–227). IGI Global. DOI: 10.4018/979-8-3693-2881-1.ch009

Singh, B., Jain, V., Kaunert, C., Dutta, P. K., & Singh, G. (2024). Privacy Matters: Espousing Blockchain and Artificial Intelligence (AI) for Consumer Data Protection on E-Commerce Platforms in Ethical Marketing. In Saluja, S., Nayyar, V., Rojhe, K., & Sharma, S. (Eds.), *Ethical Marketing Through Data Governance Standards and Effective Technology* (pp. 167–184). IGI Global. DOI: 10.4018/979-8-3693-2215-4.ch015

Singh, B., & Kaunert, C. (2024). Adventure in High Altitude of Mountainous Topographies and Health Impacts: Lensing Tourism Sustainability via Reducing Ecological and Sociocultural Footprint and Health Emergency and Medical Assistance Management. In Meraj, G., Hashimoto, S., & Kumar, P. (Eds.), *Navigating Natural Hazards in Mountainous Topographies. Disaster Risk Reduction*. Springer. DOI: 10.1007/978-3-031-65862-4_15

Singh, B., & Kaunert, C. (2024). Transforming Smart City and Smart Sustainable Tourism Projecting Artificial Intelligence and IoT: Immersive Potential for Revolutionizing Urban Living and Enhancing Travel Experience. In Jermsittiparsert, K., & Suanpang, P. (Eds.), *Technological and Managerial Approaches to Fostering Sustainable Travel* (pp. 55–80). IGI Global. DOI: 10.4018/979-8-3693-8115-1.ch003

Singh, B., & Kaunert, C. (2024). AI-Driven Strategies for Customer Engagement, Market Segmentation, and Resource Optimization: Projecting End User Satisfaction and Futuristic Growth of Business. In Nadda, V., Tyagi, P., Singh, A., & Singh, V. (Eds.), *AI Innovations in Service and Tourism Marketing* (pp. 104–128). IGI Global. DOI: 10.4018/979-8-3693-7909-7.ch006

Singh, B., & Kaunert, C. (2024). Blue Ocean and Machine Learning Trajectories SDG 14: Life Below Water for Handling Ocean Pollution: Metaverse Conserve Ocean Health Sustainability Through the Lens of Transboundary Legal-Policy Regulations as Articulating Space for Futuristic Changes. In *Artificial Intelligence and Edge Computing for Sustainable Ocean Health*. The Springer Series in Applied Machine Learning. Springer. DOI: 10.1007/978-3-031-64642-3_4

Singh, B., & Kaunert, C. (2024). Hitching Artificial Intelligence and IoT and for Livestock Management Competence: Line Up Nature and Technology for Smart Agriculture. In Zia Ul Haq, M., & Ali, I. (Eds.), *Revolutionizing Pest Management for Sustainable Agriculture* (pp. 499–520). IGI Global. DOI: 10.4018/979-8-3693-3061-6.ch020

Singh, B., & Kaunert, C. (2024). Redefining the Nodes of Metaverse and Luxury Hospitality: Unveiling AI and Tech-Driven Solutions in Hospitality and Tourism Industry. In Talukder, M., Kumar, S., & Tyagi, P. (Eds.), *Impact of AI and Tech-Driven Solutions in Hospitality and Tourism* (pp. 175–202). IGI Global. DOI: 10.4018/979-8-3693-6755-1.ch010

Singh, B., & Kaunert, C. (2024). Trans-Disciplinary Collaborative Disciplines Generating Cohesive Curriculum: Runway for Innovation, Cooperation, and Intentionality in Higher Education. In R. Kumar, E. Ong, S. Anggoro, T. Toh, & M. Fukui (Eds.), *Transdisciplinary Teaching and Technological Integration for Improved Learning: Case Studies and Practical Approaches* (pp. 152-176). IGI Global. DOI: 10.4018/979-8-3693-8217-2.ch008

Singh, B., & Kaunert, C. (2024). Human-Machine Nexus for Digital Rebound Fostering Futuristic Energy-Efficiency: Ecological Footprint of Technology for Smart-Sustainable Urban Mobility. In González-Lezcano, R., & Sansaniwal, S. (Eds.), *Design Strategies for Efficient and Sustainable Building Facilities* (pp. 222–244). IGI Global. DOI: 10.4018/979-8-3693-3200-9.ch010

Singh, B., & Kaunert, C. (2024). Hidden Gems Breakthrough Dynamic Landscape of Adaptive Learning Technologies for Higher Education: Bridging the Gap Between Theoretical and Practical Knowledge Projecting Student Learning Outcomes. In Minh Tung, T. (Ed.), *Adaptive Learning Technologies for Higher Education* (pp. 222–247). IGI Global. DOI: 10.4018/979-8-3693-3641-0.ch010

Singh, B., & Kaunert, C. (2024). Embryonic Machine-Deep Learning, Smart Healthcare and Privacy Deliberations in Hospital Industry: Lensing Confidentiality of Patient's Information and Personal Data in Legal-Ethical Landscapes Projecting Futuristic Dimensions. In Kumar, P., Singh, P., Diwakar, M., & Garg, D. (Eds.), *Healthcare Industry Assessment: Analyzing Risks, Security, and Reliability. Engineering Cyber-Physical Systems and Critical Infrastructures* (Vol. 11). Springer. DOI: 10.1007/978-3-031-65434-3_7

Singh, B., & Kaunert, C. (2024). AI-Driven Strategies for Customer Engagement, Market Segmentation, and Resource Optimization: Projecting End User Satisfaction and Futuristic Growth of Business. In Nadda, V., Tyagi, P., Singh, A., & Singh, V. (Eds.), *AI Innovations in Service and Tourism Marketing* (pp. 104–128). IGI Global. DOI: 10.4018/979-8-3693-7909-7.ch006

Singh, B., & Kaunert, C. (2024). Vertical Assimilation of Artificial Intelligence and Machine Learning in Safeguarding Financial Data. In Naz, F., & Karim, S. (Eds.), *Safeguarding Financial Data in the Digital Age* (pp. 169–197). IGI Global. DOI: 10.4018/979-8-3693-3633-5.ch010

Singh, B., & Kaunert, C. (2024). Computational Thinking for Innovative Solutions and Problem-Solving Techniques: Transforming Conventional Education to Futuristic Interdisciplinary Higher Education. In Fonkam, M., & Vajjhala, N. (Eds.), *Revolutionizing Curricula Through Computational Thinking, Logic, and Problem Solving* (pp. 60–82). IGI Global. DOI: 10.4018/979-8-3693-1974-1.ch004

Singh, B., & Kaunert, C. (2024). Wind and Solar Energy as Renewable Energy for Sustainable Global Future: Projecting Future Multi-Sector Sustainable Policies and Innovation. In Ara, A., & Thakore, R. (Eds.), *Promoting Multi-Sector Sustainability With Policy and Innovation* (pp. 210–245). IGI Global. DOI: 10.4018/979-8-3693-2113-3.ch009

Singh, B., & Kaunert, C. (2024). Aroma of Highly Smart Internet of Medical Things (IoMT) and Lightweight EdgeTrust Expansion Medical Care Facilities for Electronic Healthcare Systems: Fortified-Chain Architecture for Remote Patient Monitoring and Privacy Protection Beyond Imagination. In Hassan, A., Bhattacharya, P., Tikadar, S., Dutta, P., & Sagayam, M. (Eds.), *Lightweight Digital Trust Architectures in the Internet of Medical Things (IoMT)* (pp. 196–212). IGI Global. DOI: 10.4018/979-8-3693-2109-6.ch011

Singh, B., & Kaunert, C. (2024). Augmented Reality and Virtual Reality Modules for Mindfulness: Boosting Emotional Intelligence and Mental Wellness. In Hiran, K., Doshi, R., & Patel, M. (Eds.), *Applications of Virtual and Augmented Reality for Health and Wellbeing* (pp. 111–128). IGI Global. DOI: 10.4018/979-8-3693-1123-3.ch007

Singh, B., & Kaunert, C. (2024). Salvaging Responsible Consumption and Production of Food in the Hospitality Industry: Harnessing Machine Learning and Deep Learning for Zero Food Waste. In Singh, A., Tyagi, P., & Garg, A. (Eds.), *Sustainable Disposal Methods of Food Wastes in Hospitality Operations* (pp. 176–192). IGI Global. DOI: 10.4018/979-8-3693-2181-2.ch012

Singh, B., & Kaunert, C. (2024). Revealing Green Finance Mobilization: Harnessing FinTech and Blockchain Innovations to Surmount Barriers and Foster New Investment Avenues. In Jafar, S., Rodriguez, R., Kannan, H., Akhtar, S., & Plugmann, P. (Eds.), *Harnessing Blockchain-Digital Twin Fusion for Sustainable Investments* (pp. 265–286). IGI Global. DOI: 10.4018/979-8-3693-1878-2.ch011

Singh, B., & Kaunert, C. (2024). Harnessing Sustainable Agriculture Through Climate-Smart Technologies: Artificial Intelligence for Climate Preservation and Futuristic Trends. In Kannan, H., Rodriguez, R., Paprika, Z., & Ade-Ibijola, A. (Eds.), *Exploring Ethical Dimensions of Environmental Sustainability and Use of AI* (pp. 214–239). IGI Global. DOI: 10.4018/979-8-3693-0892-9.ch011

Singh, B., & Kaunert, C. (2025). Innovative Business Models in Digital and Virtual Reality for the Gig Economy: Space-Age Sustainable Development Projecting Smart Business Practices. In Taherdoost, H., Drazenovic, G., Madanchian, M., Khan, I., & Arshi, O. (Eds.), *Business Transformation in the Era of Digital Disruption* (pp. 127–156). IGI Global. DOI: 10.4018/979-8-3693-7056-8.ch005

Singh, B., & Kaunert, C. (2025). Dynamic Landscape of Artificial General Intelligence (AGI) for Advancing Renewable Energy in Urban Environments: Synergies with SDG 11—Sustainable Cities and Communities Lensing Policy and Governance. In El Hajjami, S., Kaushik, K., & Khan, I. U. (Eds.), *Artificial General Intelligence (AGI) Security. Advanced Technologies and Societal Change*. Springer. DOI: 10.1007/978-981-97-3222-7_12

Singh, B., & Kaunert, C. (2025). Blockchain Framework for Precision Medicine, Clinical Trials, and Genomic Biomarkers Discovery: Focuses on Monitoring and Managing Data in Health Dimensions. In Jermsittiparsert, K., & Marzo, R. (Eds.), *AI Technologies and Advancements for Psychological Well-Being and Healthcare* (pp. 255–288). IGI Global. DOI: 10.4018/979-8-3693-9158-7.ch010

Singh, B., & Kaunert, C. FedHealth in Wearable Healthcare, Orchestrated Federated Deep Learning for Smart Healthcare: Health Monitoring and Healthcare Informatics Lensing Challenges and Future Directions. In *Federated Deep Learning for Healthcare* (pp. 207–223). CRC Press. https://www.taylorfrancis.com/chapters/edit/10.1201/9781032694870-16/fedhealth-wearable-healthcare-orchestrated-federated-deep-learning-smart-healthcare-bhupinder-singh-christian-kaunert

Singh, B., & Kaunert, C. (n.d.). Tagging Blockchain Technology Fostering Panacea for Data Privacy, Cloud Computing, and Integrity: Legal Framework in India and at a Globe. In *Artificial Intelligence Solutions for Cyber-Physical Systems* (pp. 206-213). Auerbach Publications. https://www.taylorfrancis.com/chapters/edit/10.1201/9781032694375-11/tagging-blockchain-technology-fostering-panacea-data-privacy-cloud-computing-integrity-bhupinder-singh-christian-kaunert

Singh, B., & Kaunert, C. Marvel of Biosensors in Smart Healthcare and Application of Internet of Medical Things for Diagnosis: Unveiling Benefits, Challenges and Futuristic Approach. In *Smart Healthcare Systems* (pp. 187–198). CRC Press. https://www.taylorfrancis.com/chapters/edit/10.1201/9781032698519-14/marvel-biosensors-smart-healthcare-application-internet-medical-things-diagnosis-bhupinder-singh-christian-kaunert

Singh, B., Kaunert, C., & Gautam, R. (2024). Harnessing Machine Learning Intelligence Against Cyber Threats: Navigating Legal and Ethical Challenges in Global Cyber Investigations. In Özsungur, F. (Ed.), *Empowering Entrepreneurial Mindsets With AI* (pp. 319–352). IGI Global. DOI: 10.4018/979-8-3693-7658-4.ch016

Singh, B., Kaunert, C., & Jermsittiparsert, K. (2024). Environment-Biodiversity Protection and SDG 15 (Life on Land): Legal Landscapes and Probable Resolutions. In Singh, B., Kaunert, C., Vig, K., & Dutta, S. (Eds.), *Maintaining a Sustainable World in the Nexus of Environmental Science and AI* (pp. 259–282). IGI Global. DOI: 10.4018/979-8-3693-6336-2.ch010

Singh, B., Kaunert, C., & Jermsittiparsert, K. (2025). Advocating Medical Tourism and Healthcare in Global Health Market Places: Promoting Sustainability and Technological Advancements. In Jermsittiparsert, K., & Marzo, R. (Eds.), *AI Technologies and Advancements for Psychological Well-Being and Healthcare* (pp. 327–350). IGI Global. DOI: 10.4018/979-8-3693-9158-7.ch012

Singh, B., Kaunert, C., Jermsittiparsert, K., Lal, S., Arora, M. K., & Raghav, A. (2024). Ultimate Deep Learning Acclimatization With Artificial Intelligence Drones for Environment Shield and Climate Change: Lifting Sustainability Towards Fostering SDG 13. In Singh, B., Kaunert, C., Vig, K., & Dutta, S. (Eds.), *Maintaining a Sustainable World in the Nexus of Environmental Science and AI* (pp. 59–80). IGI Global. DOI: 10.4018/979-8-3693-6336-2.ch003

Singh, B., Kaunert, C., & Kaushik, T. K. (2024). Unscrambling Financial Fraud With AI and Machine Learning in E-Commerce Transactions: Airing Into Ad Clicks, Credit Card Management. In Pandow, B., Masoodi, F., Iqbal, J., & Hussain, G. (Eds.), *Navigating the Future of Finance in the Age of AI* (pp. 253–271). IGI Global. DOI: 10.4018/979-8-3693-4382-1.ch013

Singh, B., Kaunert, C., Lal, S., Arora, M. K., & Singh, G. (2024). Abundance Ocean Wave Energy to Electricity With Artificial Intelligence and IoT Solutions: Future of Clean Energy and Sustainability in Smart Cities. In Riswandi, B., Singh, B., Kaunert, C., & Vig, K. (Eds.), *AI Applications for Clean Energy and Sustainability* (pp. 274–298). IGI Global. DOI: 10.4018/979-8-3693-6567-0.ch014

Singh, B., Kaunert, C., & Malviya, R. (2024). Featuring Machine Learning in Municipal Solid Waste Management: Technological Advancements in Waste Sorting and Processing Ropeway for SDG 13 (Climate Action). In Mehra, R., Kumar, P., & Hung, B. (Eds.), *Municipal Solid Waste Management and Recycling Technologies* (pp. 343–366). IGI Global. DOI: 10.4018/979-8-3693-4054-7.ch013

Singh, B., Kaunert, C., & Malviya, R. (2024). Coupling Human-Computer Interface Lensing Artificial Emotional Intelligence: Transforming for Futuristic Cognitive Recognition for Decision-Marking in Industry 5.0. In Kumar, N., Pal, S., Agarwal, P., Rosak-Szyrocka, J., & Jain, V. (Eds.), *Human-Machine Collaboration and Emotional Intelligence in Industry 5.0* (pp. 233–250). IGI Global. DOI: 10.4018/979-8-3693-6806-0.ch013

Singh, B., Kaunert, C., Malviya, R., Gautam, R., & Singh, G. (2024). Generative Artificial Intelligence in the Tourism-Hospitality Industry: Strengthening Customer Relationships, Delivering Better Experiences. In Talukder, M., Kumar, S., & Tyagi, P. (Eds.), *Hotel and Travel Management in the AI Era* (pp. 275–300). IGI Global. DOI: 10.4018/979-8-3693-7898-4.ch013

Singh, B., Kaunert, C., Malviya, R., Lal, S., & Arora, M. K. (2024). Scrutinizing Consumer Sentiment on Social Media and Data-Driven Decisions for Business Insights: Fusion of Artificial Intelligence (AI) and Business Intelligence (BI) Foster Sustainable Growth. In Natarajan, A., Galety, M., Iwendi, C., Das, D., & Shankar, A. (Eds.), *Intersection of AI and Business Intelligence in Data-Driven Decision-Making* (pp. 183–210). IGI Global. DOI: 10.4018/979-8-3693-5288-5.ch007

Singh, B., Malviya, R., & Kaunert, C. (2024). Elevating Workplace Sustainability for Employees Lensing Mental Health Advancements: Runway for Future Ready Healthcare Services Projecting SDG 3 (Good Health and Well-Being). In Shaikh, E. (Ed.), *Impact of Corporate Social Responsibility on Employee Wellbeing* (pp. 285–310). IGI Global. DOI: 10.4018/979-8-3693-3470-6.ch014

Singh, B., Singh, A., Kaunert, C., Arora, M. K., Lal, S., & Ravesangar, K. (2025). Unleashing the Ethical and Legal Implications of E-Business Revolution: Consumer Privacy and Security Concerns in Phase of Digital Disruption. In Taherdoost, H., Drazenovic, G., Madanchian, M., Khan, I., & Arshi, O. (Eds.), *Business Transformation in the Era of Digital Disruption* (pp. 157–180). IGI Global. DOI: 10.4018/979-8-3693-7056-8.ch006

Singh, N., Lal, S., Raghav, A., Arora, M. K., Singh, B., & Kaunert, C. (2024). Power of ML and DL in Precision Modeling and Climate Predictions: NextGen Ocean Insights. In Riswandi, B., Singh, B., Kaunert, C., & Vig, K. (Eds.), *AI Applications for Clean Energy and Sustainability* (pp. 254–273). IGI Global. DOI: 10.4018/979-8-3693-6567-0.ch013

Song, G., Ifaei, P., Ha, J., Kang, D., Won, W., Liu, J. J., & Na, J. (2024). The AI circular hydrogen economist: Hydrogen supply chain design via hierarchical deep multi-agent reinforcement learning. *Chemical Engineering Journal*, *497*, 154464.

Song, G., Ifaei, P., Ha, J., Kang, D., Won, W., Liu, J. J., & Na, J. (2024). The AI circular hydrogen economist: Hydrogen supply chain design via hierarchical deep multi-agent reinforcement learning. *Chemical Engineering Journal*, *497*, 154464. DOI: 10.1016/j.cej.2024.154464

Wang, H., Wen, C., Duan, L., Li, X., Liu, D., & Guo, W. (2024). Sustainable energy transition in cities: A deep statistical prediction model for renewable energy sources management for low-carbon urban development. *Sustainable Cities and Society*, *107*, 105434.

Compilation of References

. Verma, T. L. (2019). Role of micro, small and medium enterprises (MSMES) in achieving sustainable development goals. *Small And Medium Enterprises (MSMEs) In Achieving Sustainable Development Goals (April 1, 2019).*

Abbe, G., & Smith, H. (2016). Technological development trends in Solar-powered Aircraft Systems. *Renewable & Sustainable Energy Reviews*, *60*, 770–783. DOI: 10.1016/j.rser.2016.01.053

Abd Aziz, A. J., Baharuddin, N. A., Khalid, R. M., & Kamarudin, S. K. (2024). Review of the policies and development programs for renewable energy in Malaysia: Progress, achievements and challenges. *Energy Exploration & Exploitation*, *42*(4), 1472–1501. DOI: 10.1177/01445987241227509

Abdelghany, M., Al-Durra, A., & Gao, F. (2024). A Coordinated Optimal Operation of a Grid-Connected Wind-Solar Microgrid Incorporating Hybrid Energy Storage Management Systems. *IEEE Transactions on Sustainable Energy*, *15*(1), 39–51. DOI: 10.1109/TSTE.2023.3263540

Abdella, G. M., Kucukvar, M., Cihat Onat, N., Al-Yafay, H. M., & Enis Bulak, M. (2020). Sustainability assessment and modeling based on supervised machine learning techniques: The case for food consumption. *Journal of Cleaner Production*, *251*, 1–16. DOI: 10.1016/j.jclepro.2019.119661

Abid, H., Thakur, J., Khatiwada, D., & Bauner, D. (2021). Energy storage integration with solar PV for increased electricity access: A case study of Burkina Faso. *Energy*, *230*, 120656. DOI: 10.1016/j.energy.2021.120656

Abualigah, L., Zitar, R. A., Almotairi, K. H., Hussein, A. M., Abd Elaziz, M., Nikoo, M. R., & Gandomi, A. H. (2022). Wind, solar, and photovoltaic renewable energy systems with and without energy storage optimization: A survey of advanced machine learning and deep learning techniques. *Energies*, *15*(2), 578. DOI: 10.3390/en15020578

Abubakar, A., Almeida, C., & Gemignani, M. (2021). Review of Artificial Intelligence-Based Failure Detection and Diagnosis Methods for Solar Photovoltaic Systems. *Machines (Basel)*, *9*(12), 328. Advance online publication. DOI: 10.3390/machines9120328

Adebiyi, A. A., & Moloi, K. (2024). Renewable energy source utilization progress in South Africa: A review. *Energies*, *17*(14), 3487. DOI: 10.3390/en17143487

Adewale, B. A., Ene, V. O., Ogunbayo, B. F., & Aigbavboa, C. O. (2024). A systematic review of the applications of AI in a sustainable building's lifecycle. *Buildings (Basel, Switzerland)*, *14*(7), 2137. DOI: 10.3390/buildings14072137

Agência Portuguesa do Ambiente. (2015). *Estratégia Nacional de Desenvolvimento Sustentável*. JMF – Agência Portuguesa do Ambiente.

Aghabiglou, A., Chu, C. S., Dabbech, A., & Wiaux, Y. (2024). The R2D2 deep neural network series paradigm for fast precision imaging in radio astronomy. *arXiv preprint arXiv:2403.05452*.

Agostinelli, S. (2024). Optimization and management of microgrids in the built environment based on intelligent digital twins.

Ahamed, N. N., Karthikeyan, P., Velliangiri, S., & Gunjan, V. K. (2023). Blockchain Security for the Internet of Things. In *Modern Approaches in IoT and Machine Learning for Cyber Security: Latest Trends in AI* (pp. 75–92). Springer International Publishing.

Ahmad, S., Miskon, S., Alabdan, R., & Tlili, I. (2020). Towards sustainable textile and apparel industry: Exploring the role of business intelligence systems in the era of industry 4.0. *Sustainability (Basel), 12*(7), 2632. DOI: 10.3390/su12072632

Ahmad, T., Madonski, R., Zhang, D., Huang, C., & Mujeeb, A. (2022). Data-driven probabilistic machine learning in sustainable smart energy/smart energy systems: Key developments, challenges, and future research opportunities in the context of smart grid paradigm. *Renewable & Sustainable Energy Reviews, 160*, 112128. DOI: 10.1016/j.rser.2022.112128

Ahmad, T., & Zhang, D. (2021). Using the internet of things in smart energy systems and networks. *Sustainable Cities and Society, 68*, 102783. DOI: 10.1016/j.scs.2021.102783

Ahmad, T., Zhang, D., Huang, C., Zhang, H., Dai, N., Song, Y., & Chen, H. (2021). Artificial intelligence in sustainable energy industry: Status Quo, challenges and opportunities. *Journal of Cleaner Production, 289*, 125834. Advance online publication. DOI: 10.1016/j.jclepro.2021.125834

Ahmad, T., Zhang, D., Huang, C., Zhang, H., Dai, N., Song, Y., & Chen, H. (2021). Artificial intelligence in the sustainable energy industry: Status quo, challenges and opportunities. [Google Scholar] [CrossRef]. *Journal of Cleaner Production, 289*, 125834.

Ahmed, A. (2022). Marketing 4.0: The Unseen Potential of AI in Consumer Relations. *International Journal of New Media Studies: International Peer Reviewed Scholarly Indexed Journal, 9*(1), 5–12.

Ahmed, F., Naeem, M., & Iqbal, M. (2017). ICT and renewable energy: A way forward to the next generation telecom base stations. *Telecommunication Systems, 64*(1), 43–56. DOI: 10.1007/s11235-016-0156-4

Ahmed, S., Ali, A., & D'angola, A. (2024). A review of renewable energy communities: Concepts, scope, progress, challenges, and recommendations. *Sustainability (Basel), 16*(5), 1749. DOI: 10.3390/su16051749

Aitkazinov, A. (2023). The role of artificial intelligence in auditing: Opportunities and challenges. International Journal of Research in Engineering. *Science and Management, 6*(6), 117–119.

Aizpurietis, J. (2024). Developing an AI-powered chatbot for CO_2 emissions estimations utilizing Botpress framework. Theseus Repository. Retrieved from https://www.theseus.fi/bitstream/handle/10024/861047/Aizpurietis_Janis.pdf?sequence=2&isAllowed=y

Ajayi, N. F., Ademola, N. O. M., Amuda, N. K. F., & Alade, N. B. (2024). AI-driven decarbonization of buildings: Leveraging predictive analytics and automation for sustainable energy management. World Journal of Advanced Research and Reviews, 24(1), 061–079. DOI: 10.30574/wjarr.2024.24.1.2997

Akhavan, P., Azizi, N., Akhtari, S., Haass, O., Jan, T., & Sajeev, S. (2023). Understanding critical success factors for implementing medical tourism in a multi-case analysis. *Knowledge Management & E-Learning*, *15*(1), 43.

Akhtar, M. M., Rizvi, D. R., Ahad, M. A., Kanhere, S. S., Amjad, M., & Coviello, G. (2021). Efficient data communication using distributed ledger technology and iota-enabled internet of things for a future machine-to-machine economy. *Sensors (Basel)*, *21*(13), 4354. PMID: 34202173

Akhter, M. N., Mekhilef, S., Mokhlis, H., & Mohamed Shah, N. (2019). Review on forecasting of photovoltaic power generation based on machine learning and metaheuristic techniques. *IET Renewable Power Generation*, *13*(7), 1009–1023.

Akram, S. V., Malik, P. K., Singh, R., Anita, G., & Tanwar, S. (2020). Adoption of blockchain technology in various realms: Opportunities and challenges. *Security and Privacy*, *3*(5), e109. DOI: 10.1002/spy2.109

Akram, S. V., Malik, P. K., Singh, R., Gehlot, A., Juyal, A., Ghafoor, K. Z., & Shrestha, S. (2022). Implementation of digitalized technologies for fashion industry 4.0: Opportunities and challenges. *Scientific Programming*, *2022*, 2022. DOI: 10.1155/2022/7523246

Alam, M., Alshahrani, T., Khan, F., Hakami, J., Shinde, S., & Azim, R. (2023). AI-based efficiency analysis technique for photovoltaic renewable energy system. *Physica Scripta*, *98*(12), 126006. Advance online publication. DOI: 10.1088/1402-4896/ad0bb4

Albreem, M. A., Sheikh, A. M., Alsharif, M. H., Jusoh, M., & Yasin, M. N. M. (2021). Green Internet of Things (GIoT): Applications, practices, awareness, and challenges. *IEEE Access : Practical Innovations, Open Solutions*, *9*, 38833–38858. DOI: 10.1109/ACCESS.2021.3061697

Aldieri, L., Grafström, J., & Vinci, C. P. (2021). The effect of Marshallian and Jacobian knowledge spillovers on jobs in the solar, wind and energy efficiency sector. *Energies*, *14*(14), 4269. DOI: 10.3390/en14144269

Aljohani, A. (2023). Predictive Analytics and Machine Learning for Real-Time Supply Chain Risk Mitigation and Agility. *Sustainability (Basel)*, *15*(20), 15088. DOI: 10.3390/su152015088

Allal-Cherif, O., Simon-Moya, V., & Ballester, A. C. C. (2021). Intelligent purchasing: How artificial intelligence can redefine the purchasing function. *Journal of Business Research*, *124*, 69–76. DOI: 10.1016/j.jbusres.2020.11.050

Al-Masri, R. A., Chenoweth, J., & Murphy, R. J. (2019). Exploring the status quo of water-energy nexus policies and governance in Jordan. [Google Scholar] [CrossRef]. *Environmental Science & Policy*, *100*, 192–204. DOI: 10.1016/j.envsci.2019.06.012

Al-Orabi, A. M., Osman, M. G., & Sedhom, B. E. (2023). Evaluation of green hydrogen production using solar, wind, and hybrid technologies under various technical and financial scenarios for multi-sites in Egypt. *International Journal of Hydrogen Energy, 48*(98), 38535–38556. Advance online publication. DOI: 10.1016/j.ijhydene.2023.06.218

Alruwaill, M., Bapatla, A. K., Mohanty, S. P., & Kougianos, E. (2023, October). FarmIns: Blockchain Leveraged Secure and Reliable Crop Insurance Management System. In *IFIP International Internet of Things Conference* (pp. 381-389). Cham: Springer Nature Switzerland.

Al-Shareeda, M., Ali, M., & Manickam, S. (2023). The blockchain internet of things: Review, opportunities, challenges, and recommendations. *Indonesian Journal of Electrical Engineering and Computer Science, 31*(3), 1673–1683. DOI: 10.11591/ijeecs.v31.i3.pp1673-1683

Al-Shetwi, A. Q., Hannan, M., Jern, K. P., Mansur, M., & Mahlia, T. (2019). Grid-connected renewable energy sources: Review of the recent integration requirements and control methods. *Journal of Cleaner Production, 253*, 119831. DOI: 10.1016/j.jclepro.2019.119831

Al-tabatabaie, K. F., Hossain, M. B., Islam, M. K., Awual, M. R., TowfiqulIslam, A. R. M., Hossain, M. A., Esraz-Ul-Zannat, M., & Islam, A. (2022). Taking strides towards decarbonization: The viewpoint of Bangladesh. *Energy Strategy Reviews, 44*, 100948. DOI: 10.1016/j.esr.2022.100948

Alves, A. F., Costa Curta, L., & da Silva Dias, L. (2017). Analysis of correlation among HDI (Human Development Index), violence and corruption perceptions index. *Proceedings of the 5th CR3+ Conference - Making Corporate Responsibility Useful*, Helsinki, Finland.

Alzoubi, Y. I., Al-Ahmad, A., Kahtan, H., & Jaradat, A. (2022). Internet of things and blockchain integration: Security, privacy, technical, and design challenges. *Future Internet, 14*(7), 216. DOI: 10.3390/fi14070216

Amoah, E., & Oh, J. Y. (2020). BLOCKCHAIN IN IoT AND PROJECT MANAGEMENT. *Issues in Information Systems, 21*(3).

Amran, Y. A., Amran, Y. M., Alyousef, R., & Alabduljabbar, H. (2020). Renewable and sustainable energy production in Saudi Arabia according to Saudi Vision 2030; Current status and future prospects. *Journal of Cleaner Production, 247*, 119602. DOI: 10.1016/j.jclepro.2019.119602

Analytics, M. (2016). *The age of analytics: competing in a data-driven world.* McKinsey Global Institute Research.

Ancillai, C., Terho, H., Cardinali, S., & Pascucci, F. (2019). Advancing social media driven sales research: Establishing conceptual foundations for B-to-B social selling. *Industrial Marketing Management, 82*, 293–308. DOI: 10.1016/j.indmarman.2019.01.002

Andrews, R. N. (1997). National environmental policies: The United States. In Jaenicke, M., & Weidner, H. J. (Eds.), *National environmental policies: A comparative study of capacity building* (pp. 25–43). Springer Verlag. DOI: 10.1007/978-3-642-60507-9_2

Andronie, M., Lazaroiu, G., Iatagan, M., U ă, C., tefănescu, R., & Coco atu, M. (2021). Artificial intelligence-based decision-making algorithms, internet of things sensing networks, and deep learning-assisted smart process management in cyber-physical production systems. *Electronics (Basel), 10*(20), 2497. DOI: 10.3390/electronics10202497

Ansarullah, S. I., Kirmani, M. M., Alshmrany, S., & Firdous, A. (2024). Ethical issues around artificial intelligence. In *A Biologist s Guide to Artificial Intelligence* (pp. 301–314). Academic Press.

Anthopoulos, L., & Kazantzi, V. (2022). Urban energy efficiency assessment models from an AI and big data perspective: Tools for policymakers. [Google Scholar] [CrossRef]. *Sustainable Cities and Society, 76*, 103492. DOI: 10.1016/j.scs.2021.103492

Antonopoulos, I., Robu, V., Couraud, B., Kirli, D., Norbu, S., Kiprakis, A., Flynn, D., Elizondo-Gonzalez, S., & Wattam, S. (2020). Artificial intelligence and machine learning approaches to energy demand-side response: A systematic review. *Renewable & Sustainable Energy Reviews, 130*, 109899. DOI: 10.1016/j.rser.2020.109899

Anuşlu, M. D., & Firat, S. Ü. (2019). Clustering analysis application on Industry 4.0-driven global indexes. *Procedia Computer Science, 158*, 145–152. DOI: 10.1016/j.procs.2019.09.037

Arbabzadeh, M., Gençer, E., Morris, J. F., Paltsev, S., & Armstrong, R. C. (2019). *Plausible Energy Futures: A Framework for Evaluating Options*. Impacts, and National Energy Choices.

Arent, D. J., Green, P., Abdullah, Z., Barnes, T., Bauer, S., Bernstein, A., & Turchi, C. (2022). Challenges and opportunities in decarbonizing the US energy system. *Renewable & Sustainable Energy Reviews, 169*, 112939. DOI: 10.1016/j.rser.2022.112939

Arridge, S., Maass, P., Öktem, O., & Schönlieb, C.-B. (2019). Solving inverse problems using data-driven models. [Google Scholar] [CrossRef]. *Acta Numerica, 28*, 1–174. DOI: 10.1017/S0962492919000059

Arshad, J., Siddique, M. A. B., Zulfiqar, Z., Khokhar, A., Salim, S., Younas, T., & Asad, A. (2020, February). A novel remote user authentication scheme by using private blockchain-based secure access control for agriculture monitoring. In *2020 International Conference on Engineering and Emerging Technologies (ICEET)* (pp. 1-9). IEEE. DOI: 10.1109/ICEET48479.2020.9048218

Arumugham, V., Ghanimi, H. M. A., Pustokhin, D. A., Pustokhina, I. V., Ponnam, V. S., Alharbi, M., Krishnamoorthy, P., & Sengan, S. (2023). An Artificial-Intelligence-Based renewable energy prediction program for Demand-Side management in smart grids. *Sustainability (Basel), 15*(6), 5453. DOI: 10.3390/su15065453

Arya, A., Dixit, S., Sahoo, A. K., Babu, Y. R., Purohit, P., & Abdullah, I. (2024). Investigation into designing VLSI of a flexible architecture for a deep neural network accelerator. *Journal of Electrical Systems, 20*(6s), 1024–1031.

Asif, M. (n.d.). Sustainable Energy Transition: Technological Perspective. In *Handbook of Energy and Environment in the 21st Century* (pp. 1-21). CRC Press.

Asif, M., Searcy, C., & Castka, P. (2023). ESG and Industry 5.0: The role of technologies in enhancing ESG disclosure. *Technological Forecasting and Social Change*, *195*, 122806. DOI: 10.1016/j.techfore.2023.122806

Aslam, S., Herodotou, H., Mohsin, S. M., Javaid, N., Ashraf, N., & Aslam, S. (2021). A survey on deep learning methods for power load and renewable energy forecasting in smart microgrids. *Renewable & Sustainable Energy Reviews*, *144*, 110992.

Assad, U., Hassan, M. S., Farooq, U., Kabir, A., Khan, M. Z., Bukhari, S. S. H., Jaffri, Z. U. A., Oláh, J., & Popp, J. (2022). Smart Grid, Demand Response and Optimization: A Critical Review of Computational methods. *Energies*, *15*(6), 2003. DOI: 10.3390/en15062003

Atalay, Y., Biermann, F., & Kalfagianni, A. (2016). Adoption of renewable energy technologies in oil-rich countries: Explaining policy variation in the Gulf Cooperation Council states. *Renewable Energy*, *85*, 206–214. DOI: 10.1016/j.renene.2015.06.045

Auttri, B., Chaitanya, K., Daida, S., & Jain, S. K. (2023). Digital Transformation in Customer Relationship Management: Enhancing Engagement and Loyalty. [EEL]. *European Economic Letters*, *13*(3), 1140–1149.

Awan, K. A., Ud Din, I., Almogren, A., & Almajed, H. (2020). AgriTrust—A trust management approach for smart agriculture in cloud-based internet of agriculture things. *Sensors (Basel)*, *20*(21), 6174. DOI: 10.3390/s20216174 PMID: 33138295

Babet, A. (2020). Utilization of personalization in marketing automation and email marketing.

Bade, S. O., Tomomewo, O. S., Meenakshisundaram, A., Ferron, P., & Oni, B. A. (2024). Economic, social, and regulatory challenges of green hydrogen production and utilization in the US: A review. *International Journal of Hydrogen Energy*, *49*, 314–335.

Badgett, A., Ruth, M., James, B., & Pivovar, B. (2021). Methods identifying cost reduction potential for water electrolysis systems. *Current Opinion in Chemical Engineering*, *33*, 100714. DOI: 10.1016/j.coche.2021.100714

Bagdi, T., Ghosh, S., Sarkar, A., Hazra, A. K., Balachandran, S., & Chaudhury, S. (2023). Evaluation of research progress and trends on gender and renewable energy: A bibliometric analysis. *Journal of Cleaner Production*, *423*, 138654. DOI: 10.1016/j.jclepro.2023.138654

Bahrami, A., Mohammadnejad, S., & Soleimaninezhad, S. (2013). Photovoltaic cells technology: Principles and recent developments. *Optical and Quantum Electronics*, *45*(2), 161–197. DOI: 10.1007/s11082-012-9613-9

Baják, I. (2013). A helyi fenntartható fejlődési stratégiák helyzete Magyarországon. *Acta Carolus Robertus*, *3*(2), 9–22. DOI: 10.22004/ag.econ.171428

Bajgoric, N. (2023). Always-on business: integrating business continuity and business resilience. *Continuity & Resilience Review*, (ahead-of-print).

Bala, K., & Kaur, P. D. (2022, January). Changing trends of blockchain in IoT: Benefits and challenges. In *2022 12th International Conference on Cloud Computing, Data Science & Engineering (Confluence)* (pp. 324-329). IEEE.

Bapatla, A. K., Mohanty, S. P., & Kougianos, E. (2021, November). sFarm: A distributed ledger based remote crop monitoring system for smart farming. In *IFIP International Internet of Things Conference* (pp. 13-31). Cham: Springer International Publishing.

Baran, Z., & Karaca, Ş. (2023). Next-Generation Technologies in Health Tourism. In *Global perspectives on the opportunities and future directions of health tourism* (pp. 138–164). IGI Global.

Bartoluci, M., Hendija, Z., & Petracic, M. (2015). Implementing Principles of Sustainable Development in Rural Tourism in Continental Croatia. Economy of eastern Croatia yesterday, today, tomorrow. *Josip Juraj Strossmayer University of Osijek. Faculty of Economics*, *4*, 26–37.

Basheer, M., Nechifor, V., Calzadilla, A., Ringler, C., Hulme, D., & Harou, J. J. (2022). Balancing national economic policy outcomes for sustainable development. *Nature Communications*, *13*(1), 5041. DOI: 10.1038/s41467-022-32415-9 PMID: 36028486

Basil-Jones, W. (2024). ESG in fashion: Your guide to navigating reporting regulations, https://plana.earth/academy/sustainability-fashion-guide-esg-regulations

Bassey, K. E., & Ibegbulam, C. (2023). Machine learning for green hydrogen production. *Computer Science & IT Research Journal*, *4*(3), 368–385.

Bassey, K. E., Juliet, A. R., & Stephen, A. O. (2024). AI-Enhanced lifecycle assessment of renewable energy systems. *Engineering Science & Technology Journal*, *5*(7), 2082–2099.

Bathalapalli, V. K., Mohanty, S. P., Kougianos, E., Yanambaka, V. P., Baniya, B. K., & Rout, B. (2021, December). A puf-based approach for sustainable cybersecurity in smart agriculture. In *2021 19th OITS International Conference on Information Technology (OCIT)* (pp. 375-380). IEEE. DOI: 10.1109/OCIT53463.2021.00080

Bayih, A. Z., Morales, J., Assabie, Y., & de By, R. A. (2022). Utilization of internet of things and wireless sensor networks for sustainable smallholder agriculture. *Sensors (Basel)*, *22*(9), 3273. DOI: 10.3390/s22093273 PMID: 35590963

Beatley, T., & Manning, K. (1998). *The ecology of place: Planning for environment, economy and community*. Island Press.

Bečić, E., Mulej, E. M., & Svarc, J. (2012). Measuring social progress by sustainable development indicators: Cases of Croatia and Slovenia. *Procedia: Social and Behavioral Sciences*, *37*, 458–465. DOI: 10.1016/j.sbspro.2012.03.311

Behera, U. S., Sangwai, J. S., & Byun, H. S. (2025). A comprehensive review on the recent advances in applications of nanofluids for effective utilization of renewable energy. *Renewable & Sustainable Energy Reviews*, *207*, 114901. DOI: 10.1016/j.rser.2024.114901

Behura, A., Sahu, S., & Kabat, M. R. (2021). Advancement of Machine Learning and Cloud Computing in the Field of Smart Health Care. *Machine Learning Approach for Cloud Data Analytics in IoT*, 273-306.

Bello, N. S. F., Wada, N. I. U., Ige, N. O. B., Chianumba, N. E. C., & Adebayo, N. S. A. (2024). AI-driven predictive maintenance and optimization of renewable energy systems for enhanced operational efficiency and longevity. *International Journal of Science and Research Archive, 13*(1), 2823–2837. DOI: 10.30574/ijsra.2024.13.1.1992

Benchikh, S., Jarou, T., & Lamrani, R. (2024). Enhancing renewable energy systems with advanced artificial intelligence solutions. *SCT Proceedings in Interdisciplinary Insights and Innovations, 2*, 319-319.

Benda, V., & Černá, L. (2020). PV cells and modules–State of the art, limits and trends. *Heliyon, 6*(12), e05666. DOI: 10.1016/j.heliyon.2020.e05666 PMID: 33364478

Benedetti, M., Leonardi, F., Messina, F., Santoro, C., & Vasilakos, A. (2018). Anomaly detection and predictive maintenance for photovoltaic systems. *Neurocomputing, 310*, 59–68. DOI: 10.1016/j.neucom.2018.05.017

Bennagi, A., AlHousrya, O., Cotfas, D. T., & Cotfas, P. A. (2024). Comprehensive study of the artificial intelligence applied in renewable energy. *Energy Strategy Reviews, 54*, 101446. DOI: 10.1016/j.esr.2024.101446

Bera, S., Dey, T., Mukherjee, A., & Buyya, R. (2023). E-CropReco: A dew-edge-based multi-parametric crop recommendation framework for internet of agricultural things. *The Journal of Supercomputing, 79*(11), 1–35. DOI: 10.1007/s11227-023-05131-4

Bertola, P., & Teunissen, J. (2018). Fashion 4.0. Innovating fashion industry through digital transformation. *Research Journal of Textile and Apparel, 22*(4), 352–369. DOI: 10.1108/RJTA-03-2018-0023

Bhatnagar, S. (2022). Digital Disruptions and Transformation of Bank Marketing.

Bhattacharjee, S., & Nandi, C. (2019). Implementation of industrial internet of things in the renewable energy sector. *The Internet of Things in the industrial sector: security and device connectivity, smart environments, and industry 4.0*, 223-259.

Bhushan, B., Sahoo, C., Sinha, P., & Khamparia, A. (2021). Unification of Blockchain and Internet of Things (BIoT): Requirements, working model, challenges and future directions. *Wireless Networks, 27*(1), 55–90. DOI: 10.1007/s11276-020-02445-6

Bhutta, M. S., Li, Y., Abubakar, M., Almasoudi, F. M., Alatawi, K. S. S., Altimania, M. R., & Al-Barashi, M. (2024). Optimizing solar power efficiency in smart grids using hybrid machine learning models for accurate energy generation prediction. *Scientific Reports, 14*(1), 17101. PMID: 39048605

Bibri, S. E. (2018). The IoT for smart sustainable cities of the future: An analytical framework for sensor-based big data applications for environmental sustainability. *Sustainable Cities and Society, 38*, 230–253. DOI: 10.1016/j.scs.2017.12.034

Bibri, S. E., & Krogstie, J. (2020). Environmentally data-driven smart sustainable cities: Applied innovative solutions for energy efficiency, pollution reduction, and urban metabolism. *Energy Informatics, 3*(1), 1–59. DOI: 10.1186/s42162-020-00130-8

Birol, F. (2022). *World energy outlook 2022*. International Energy Agency. (Google Scholar)

BlackRock. (2023). *Advancing ESG integration in investment decisions*. BlackRock Sustainability Insights.

Bodkhe, U., Tanwar, S., Parekh, K., Khanpara, P., Tyagi, S., Kumar, N., & Alazab, M. (2020). Blockchain for industry 4.0: A comprehensive review. *IEEE Access : Practical Innovations, Open Solutions*, *8*, 79764–79800. DOI: 10.1109/ACCESS.2020.2988579

Bogdanov, D., Ram, M., Aghahosseini, A., Gulagi, A., Oyewo, A. S., Child, M., & Breyer, C. (2021). Low-cost renewable electricity as the key driver of the global energy transition towards sustainability. *Energy*, *227*, 120467. DOI: 10.1016/j.energy.2021.120467

Böll Stiftung, H. (2012). *We need to change. Mapping Croatia's Potential for Sustainable Development*. Kerschoffset.

Boretti, A. (2024). Artificial intelligence driven evolution of perovskite-based solar thermochemical systems for hydrogen production: A narrative review. *Emergent Materials*, 1–15.

Bostrom, N., & Yudkowsky, E. (2014). The ethics of artificial intelligence. In Frankish, K., & Ramsey, W. M. (Eds.), *The Cambridge Handbook of Artificial Intelligence* (pp. 316–334). Cambridge University Press. DOI: 10.1017/CBO9781139046855.020

Bot, K., & Borges, J. G. (2022). A Systematic review of applications of machine learning techniques for wildfire Management decision support. *Inventions (Basel, Switzerland)*, *7*(1), 15. DOI: 10.3390/inventions7010015

Bourhnane, S., Abid, M. R., Lghoul, R., Zine-Dine, K., Elkamoun, N., & Benhaddou, D. (2020). Machine learning for energy consumption prediction and scheduling in smart buildings. *SN Applied Sciences*, *2*(2), 297. Advance online publication. DOI: 10.1007/s42452-020-2024-9

Bowman, D. M. J. S., Balch, J., Artaxo, P., Bond, W. J., Cochrane, M. A., D'Antonio, C. M., DeFries, R., Johnston, F. H., Keeley, J. E., Krawchuk, M. A., Kull, C. A., Mack, M., Moritz, M. A., Pyne, S., Roos, C. I., Scott, A. C., Sodhi, N. S., & Swetnam, T. W. (2011). The human dimension of fire regimes on Earth. *Journal of Biogeography*, *38*(12), 2223–2236. DOI: 10.1111/j.1365-2699.2011.02595.x PMID: 22279247

Boza, P., & Evgeniou, T. (2021). Artificial intelligence to support the integration of variable renewable energy sources to the power system. *Applied Energy*, *290*, 116754. DOI: 10.1016/j.apenergy.2021.116754

BP. (2024). Net Zero by 2050: Transforming Energy. BP Sustainability Report.

Braganza, A., Chen, W., Canhoto, A., & Sap, S. (2020). Productive employment and decent work: The impact of AI adoption on psychological contracts, job engagement and employee trust. *Journal of Business Research*, *131*, 485–494. DOI: 10.1016/j.jbusres.2020.08.018 PMID: 32836565

Brandful Cobbinach, P., Erdiaw-Kwasie, M. O., & Amoateng, P. (2014). Rethinking sustainable development within the framework of poverty and urbanisation in developing countries. *Environmental Development*, *13*, 18–32. DOI: 10.1016/j.envdev.2014.11.001

Braun, R. (2015). *Vállalati társadalmi felelősségvállalás*. Akadémiai Könyvkiadó.

Brennan, L. J., Byrne, M. T., Bari, M., & Gun'ko, Y. K. (2011). Carbon nanomaterials for dye-sensitized solar cell applications: A bright future. *Advanced Energy Materials*, *1*(4), 472–485. DOI: 10.1002/aenm.201100136

Briška, I., & Rungule, R. (2011). Sustainable development: Changes in the evaluation of environmental risks in Latvia. *Management Theory and Studies for Rural Business and Infrastructure Development*, *1*(25), 36–43. DOI: 10.25143/prom-rsu_2013-23_dts

Brizga, J. (2012). How Well Sustainable Development Is Integrated into Environmental Policies: Case study: Latvia. *Safety of Technogenic Environment*, *2*, 24–33.

Brown, L. R. (1982). *Building a Sustainable Society*. Norton for the Worldwatch Institute. DOI: 10.1007/BF02712913

Brusilovsky, Peter & Ahn, Jae-wook & Rasmussen, Edie. (2010). Teaching Information Retrieval With Web-based Interactive Visualization. *Journal of Education for Library and Information Science*, *51*. Advance online publication. DOI: 10.2307/40732598

Buah, E., Linnanen, L., Wu, H., & Kesse, M. A. (2020). Can artificial intelligence assist project developers in Long-Term management of energy projects? The case of CO2 capture and storage. *Energies*, *13*(23), 6259. DOI: 10.3390/en13236259

Budiman, I. (2018). Enabling community participation for social innovation in the energy sector. *Indonesian Journal of Energy*, *1*(2), 21–31. DOI: 10.33116/ije.v1i2.23

Bujang, A. S., Bern, C. J., & Brumm, T. J. (2016). Summary of energy demand and renewable energy policies in Malaysia. *Renewable & Sustainable Energy Reviews*, *53*, 1459–1467. DOI: 10.1016/j.rser.2015.09.047

Burja, C., & Burja, V. (2009). Some aspects of the Sustainable Development in Romania. *International Journal of Social, Behavioral, Educational, Economic. Business and Industrial Engineering*, *3*(6), 802–808.

Busari, G. A., & Lim, D. H. (2021). Crude oil price prediction: A comparison between AdaBoost-LSTM and AdaBoost-GRU for improving forecasting performance. [Google Scholar] [CrossRef]. *Computers & Chemical Engineering*, *155*, 107513. DOI: 10.1016/j.compchemeng.2021.107513

Butler, W. H., & Goldstein, B. E. (2010). The US Fire Learning Network: Springing a Rigidity Trap through Multiscalar Collaborative Networks. *Ecology and Society*, *15*(3), art21. Advance online publication. DOI: 10.5751/ES-03437-150321

Cabrera, P., Carta, J. A., Lund, H., & Thellufsen, J. Z. (2021). Large-scale optimal integration of wind and solar photovoltaic power in water-energy systems on islands. *Energy Conversion and Management*, *235*, 113982. DOI: 10.1016/j.enconman.2021.113982

Cantarero, M. M. V. (2020). Of renewable energy, energy democracy, and sustainable development: A roadmap to accelerate the energy transition in developing countries. *Energy Research & Social Science*, *70*, 101716. DOI: 10.1016/j.erss.2020.101716

Carney, M. (2019). Fifty shades of green: The world needs a new sustainable financial system to stop runaway climate change. IMF, The Economics of Climate. *Finance & Development*, •••, 12–15.

Casciani, D., Chkanikova, O., & Pal, R. (2022). Exploring the nature of digital transformation in the fashion industry: opportunities for supply chains, business models, and sustainability-oriented innovations. *Sustainability: Science. Sustainability, 18*(1), 773–795. DOI: 10.1080/15487733.2022.2125640

Cavalca, D. L., Spavieri, G., & Fernandes, R. S. (2018). Comparative analysis between particle swarm optimization algorithms applied to Price-Based Demand response. In *Lecture notes in computer science* (pp. 323–332). DOI: 10.1007/978-3-319-91253-0_31

Central Statistical Office. (2011). *Sustainable Development Indicators in Poland*. Central Statistical Office.

Chaianong, A., Bangviwat, A., Menke, C., Breitschopf, B., & Eichhammer, W. (2019). Customer economics of residential PV–battery systems in Thailand. *Renewable Energy, 146*, 297–308. DOI: 10.1016/j.renene.2019.06.159

Chattopadhyay, P., Patel, H. P., & Parmar, V. (2022, August). Internet of Things (IoT) in Smart Agriculture. In *2022 3rd International Conference on Electronics and Sustainable Communication Systems (ICESC)* (pp. 536-540). IEEE.

Chauhan, N., & Dwivedi, R. K. (2022, April). Designing a secure smart healthcare system with blockchain: a review. In *2022 1st International Conference on Informatics (ICI)* (pp. 214-219). IEEE. DOI: 10.1109/ICI53355.2022.9786895

Chawla, Y., Shimpo, F., & Sokołowski, M. M. (2022). Artificial intelligence and information management in the energy transition of India: Lessons from the global IT heart. *Digital Policy* [Google Scholar] [CrossRef]. *Regulation & Governance, 24*(1), 17–29. DOI: 10.1108/DPRG-05-2021-0062

Cheema, S. M., Ali, M., Pires, I. M., Gonçalves, N. J., Naqvi, M. H., & Hassan, M. (2022). IoAT Enabled smart farming: Urdu language-based solution for low-literate farmers. *Agriculture, 12*(8), 1277. DOI: 10.3390/agriculture12081277

Chen, C. P., & Zhang, C. Y. (2014). Data-intensive applications, challenges, techniques and technologies: A survey on Big Data. *Information sciences, 275*, 314-347. Mayer-Schönberger, V., & Cukier, K. (2013). *Big data: A revolution that will transform how we live, work, and think*. Houghton Mifflin Harcourt.

Chen, C. J. (2024). *Physics of solar energy and energy Storage*. John Wiley & Sons. DOI: 10.1002/9781394203642

Chen, F., Xiao, Z., Cui, L., Lin, Q., Li, J., & Yu, S. (2020). Blockchain for Internet of Things applications: A review and open issues. *Journal of Network and Computer Applications, 172*, 102839. DOI: 10.1016/j.jnca.2020.102839

Chen, J., Qi, B., Rong, Z., Peng, K., Zhao, Y., & Zhang, X. (2020). Multi-energy coordinated microgrid scheduling with integrated demand response for flexibility improvement. *Energy, 217*, 119387. DOI: 10.1016/j.energy.2020.119387

Cheong, L., Bleisch, S., Kealy, A., Tolhurst, K., Wilkening, T., & Duckham, M. (2016). Evaluating the impact of visualization of wildfire hazard upon decision-making under uncertainty. *International Journal of Geographical Information Science, 30*(7), 1377–1404. DOI: 10.1080/13658816.2015.1131829

Cho, J., & Yang, S. (2022). Data-driven Wasserstein distributionally robust unit commitment for power systems under uncertainty of renewable generation. *arXiv preprint arXiv:2203.15333.* (arXiv)

Choudhary, P., & Srivastava, R. K. (2019). Sustainability perspectives-a review for solar photovoltaic trends and growth opportunities. *Journal of Cleaner Production, 227,* 589–612. DOI: 10.1016/j.jclepro.2019.04.107

Christopoulos, S., Horvath, B., & Kull, M. (2012). Advancing the governance of cross-sectoral policies for sustainable development: A metagovernance perspective. *Public Administration and Development, 32*(3), 305–323. DOI: 10.1002/pad.1629

Clark, I. I. W., & Cooke, G. (2016). *Smart green cities: toward a carbon neutral world.* Routledge. DOI: 10.4324/9781315609386

Climate Finance, I. M. F. (2018). Carbon footprint of bank loans. Retrieved from https://climatedata.imf.org/pages/climate-finance

Cohen, S. (2011). *Sustainability Management.* Columbia University Press.

Cohen, S. M., Iyer, G. C., Brown, M., Macknick, J., Wise, M., Binsted, M., & Hejazi, M. (2021). How structural differences influence cross-model consistency: An electric sector case study. *Renewable & Sustainable Energy Reviews, 144,* 111009. DOI: 10.1016/j.rser.2021.111009

Connolly, D., Lund, H., Mathiesen, B. V., & Leahy, M. (2010). A review of computer tools for analysing the integration of renewable energy into various energy systems. *Applied Energy, 87*(4), 1059–1082. DOI: 10.1016/j.apenergy.2009.09.026

Cortese, T. T. P., Almeida, J. F. S. D., Batista, G. Q., Storopoli, J. E., Liu, A., & Yigitcanlar, T. (2022). Understanding sustainable energy in the context of smart cities: A PRISMA review. *Energies, 15*(7), 2382. DOI: 10.3390/en15072382

Costanza, R., Daly, H. E., & Bartholomew, J. A. (1991). Goals, agenda and policy recommendations for ecological economics. *Ecological Economics: The Science and Management of Sustainability,* 1-20.

Cruz-Rojas, T., Franco, J. A., Hernandez-Escobedo, Q., Ruiz-Robles, D., & Juarez-Lopez, J. M. (2023). A novel comparison of image semantic segmentation techniques for detecting dust in photovoltaic panels using machine learning and deep learning. *Renewable Energy, 217,* 119126. DOI: 10.1016/j.renene.2023.119126

Dai, H., Xie, X., Xie, Y., Liu, J., & Masui, T. (2016). Green growth: The economic impacts of large-scale renewable energy development in China. *Applied Energy, 162,* 435–449. DOI: 10.1016/j.apenergy.2015.10.049

Dalkıran, G. B. (2023). Health Tourism Components and Intermediary Institutions as Supplier Businesses in the Context of Health 4.0. In *Health 4.0 and Medical Supply Chain* (pp. 147–162). Springer Nature Singapore.

Daly, H. E. (2002). *Sustainable Development: Definitions, Principles, Policies.* World Bank.

Damman, S., Sandberg, E., Rosenberg, E., Pisciella, P., & Graabak, I. (2021). A hybrid perspective on energy transition pathways: Is hydrogen the key for Norway? *Energy Research & Social Science, 78*, 102116. DOI: 10.1016/j.erss.2021.102116

Danish, M. S. S. (2023). AI in Energy: Overcoming Unforeseen Obstacles. *AI, 4*(2), 406–425. DOI: 10.3390/ai4020022

Danish, M. S. S., Elsayed, M. E. L., Ahmadi, M., Senjyu, T., Karimy, H., & Zaheb, H. (2020). A strategic-integrated approach for sustainable energy deployment. [Google Scholar] [CrossRef]. *Energy Reports, 6*, 40–44. DOI: 10.1016/j.egyr.2019.11.039

Danish, M. S. S., & Senjyu, T. (2023). AI-Enabled Energy Policy for a Sustainable Future. *Sustainability (Basel), 15*(9), 7643. DOI: 10.3390/su15097643

Danish, M. S. S., & Senjyu, T. (2023). AI-Enabled Energy Policy for a Sustainable Future. *Sustainability, 15*(9), 7643.

Danish, M. S. S., Senjyu, T., Ibrahimi, A. M., Ahmadi, M., & Howlader, A. M. (2019). A managed framework for energy-efficient building. [Google Scholar] [CrossRef]. *Journal of Building Engineering, 21*, 120–128. DOI: 10.1016/j.jobe.2018.10.013

Danish, M. S. S., Senjyu, T., Zaheb, H., Sabory, N. R., Ibrahimi, A. M., & Matayoshi, H. (2019). A novel transdisciplinary paradigm for municipal solid waste to energy. [Google Scholar] [CrossRef]. *Journal of Cleaner Production, 233*, 880–892. DOI: 10.1016/j.jclepro.2019.05.402

Dannish Ministry of the Environment. (2002). *A Shared Future - Balanced Development. Denmark's National Strategy for Sustainable Development.* Dannish Ministry of the Environment.

Dash, R., McMurtrey, M., Rebman, C., & Kar, U. K. (2019). Application of artificial intelligence in automation of supply chain management. *Journal of Strategic Innovation and Sustainability, 14*(3), 43–53.

Das, O., Zafar, M. H., Sanfilippo, F., Rudra, S., & Kolhe, M. L. (2024). Advancements in digital twin technology and machine learning for energy systems: A comprehensive review of applications in smart grids, renewable energy, and electric vehicle optimisation. *Energy Conversion and Management, X*, 100715. DOI: 10.1016/j.ecmx.2024.100715

De Silva, P., Gunarathne, N., & Kumar, S. (2024). *Exploring the impact of digital knowledge, integration and performance on sustainable accounting, reporting and assurance.* Meditari Accountancy Research., DOI: 10.1108/MEDAR-02-2024-2383

Debisso, K. (2012). Az Európai Unió fejlesztési politikája. In Szabó, M., Láncos, P. L., & Gyeney, L. (Eds.), *Uniós szakpolitikák* (pp. 1–20). Szent István Társulat.

Del Ser, J., Casillas-Perez, D., Cornejo-Bueno, L., Prieto-Godino, L., Sanz-Justo, J., Casanova-Mateo, C., & Salcedo-Sanz, S. (2022). Randomization-based machine learning in renewable energy prediction problems: Critical literature review, new results and perspectives. *Applied Soft Computing, 118*, 108526. DOI: 10.1016/j.asoc.2022.108526

Dell'Angelo, J., D'Odorico, P., & Rulli, M. C. (2017). Threats to sustainable development posed by land and water grabbing. *Current Opinion in Environmental Sustainability, 26-27*, 120–128. DOI: 10.1016/j.cosust.2017.07.007

Demekas, D. G., & Grippa, P. (2021). *Financial regulation, climate change, and the transition to a low-carbon economy: A survey of the issues*. International Monetary Fund., DOI: 10.5089/9781616356521.001

Dent, C. M. (2012). Renewable energy and East Asia's new developmentalism: Towards a low carbon future? *The Pacific Review, 25*(5), 561–587. DOI: 10.1080/09512748.2012.728240

Deshmukh, M. K. G., Sameeroddin, M., Abdul, D., & Sattar, M. A. (2023). Renewable energy in the 21st century: A review. *Materials Today: Proceedings, 80*, 1756–1759. DOI: 10.1016/j.matpr.2021.05.501

Devabhaktuni, V., Alam, M., Depuru, S. S. S. R., Green, R. C.II, Nims, D., & Near, C. (2013). Solar energy: Trends and enabling technologies. *Renewable & Sustainable Energy Reviews, 19*, 555–564.

Dietterich, T. G. (2017). Steps toward robust artificial intelligence. *AI Magazine, 38*(3), 3–24.

Dinata, N. F. P., Ramli, M. M., Jambak, M. I., Sidik, M. B., & Alqahtani, M. M. (2024). Designing an optimal microgrid control system using deep reinforcement learning: A systematic review. *Engineering Science and Technology, an International Journal, 51*, 101651. DOI: 10.1016/j.jestch.2024.101651

Dinc, E., Kuscu, M., Bilgin, B. A., & Akan, O. B. (2019). Internet of everything: A unifying framework beyond internet of things. In *Harnessing the Internet of Everything (IoE) for Accelerated Innovation Opportunities* (pp. 1–30). IGI Global. DOI: 10.4018/978-1-5225-7332-6.ch001

Dobos, E. (2024): Álhírek és tévhitek közt élünk: a divatipar nem is annyira szennyező, mint hisszük. KRTK Blog. Portfolio. https://www.portfolio.hu/krtk/20240608/alhirek-es-tevhitek-kozt-elunk-a-divatipar-nem-is-annyira-szennyezo-mint-hisszuk-690857

Donovan, G., & Brown, T. (2007). Be Careful What You Wish for: The Legacy of Smokey Bear. Frontiers in Ecology and The Environment -. *Frontiers in Ecology and the Environment, 5*(2), 73–79. DOI: 10.1890/1540-9295(2007)5[73:BCWYWF]2.0.CO;2

Doost Mohammadian, H., & Rezaie, F. (2020). Blue-green smart mobility technologies as readiness for facing tomorrow's urban shock toward the world as a better place for living (Case studies: Songdo and Copenhagen). *Technologies, 8*(3), 39. DOI: 10.3390/technologies8030039

dos Santos, C. H., Gabriel, G. T., do Amaral, J. V. S., Montevechi, J. A. B., & de Queiroz, J. A. (2021). Decision-making in a fast fashion company in the Industry 4.0 era: A Digital Twin proposal to support operational planning. *International Journal of Advanced Manufacturing Technology, 116*(5-6), 1653–1666. DOI: 10.1007/s00170-021-07543-z

Drikakis, D., & Dbouk, T. (2022). The Role of Computational Science in Wind and Solar Energy: A Critical Review. *Energies, 15*(24), 9609. Advance online publication. DOI: 10.3390/en15249609

Dubravská, M. (2015). The analysis of chosen development indicators of the sustainable development of the Slovak Republic. *eXclusive e-JOURNAL, 3*, 1-3. DOI: 10.1016/S2212-5671(15)00569-9

Dudek, T., Dzhuguryan, T., & Lemke, J. (2019). Sustainable production network design for city multi-floor manufacturing cluster. *Procedia Computer Science, 159*, 2081–2090. DOI: 10.1016/j.procs.2019.09.381

Duman, A. C., & Güler, Ö. (2018). Techno-economic analysis of off-grid PV/wind/fuel cell hybrid system combinations with a comparison of regularly and seasonally occupied households. *Sustainable Cities and Society*, *42*, 107–126. DOI: 10.1016/j.scs.2018.06.029

Du, P., Wang, J., Yang, W., & Niu, T. (2019). A novel hybrid model for short-term wind power forecasting. *Applied Soft Computing*, *80*, 93–106. DOI: 10.1016/j.asoc.2019.03.035

Du, R., Zou, P., & Ma, C. (2021). Multi-objective optimal sizing of hybrid energy storage systems for grid-connected wind farms using fuzzy control. *Journal of Renewable and Sustainable Energy*, *13*(1), 014103. Advance online publication. DOI: 10.1063/5.0031696

Dutta, M. J., & Elers, S. (2019). Public relations, indigeneity and colonization: Indigenous resistance as dialogic anchor. *Public Relations Review*, *46*(1), 101852. DOI: 10.1016/j.pubrev.2019.101852

Dutta, P. K., Singh, B., Towfeek, A.-S. K., Adamopoulou, J. P., Bardavouras, A. N., Bamwerinde, W., Turyasingura, B., & Ayiga, N. (2024). IoT Revolutionizes Humidity Measurement and Management in Smart Cities to Enhance Health and Wellness. *Mesopotamian Journal of Artificial Intelligence in Healthcare*, *2024*, 110–117. DOI: 10.58496/MJAIH/2024/013

Dwivedi, S. K., Roy, P., Karda, C., Agrawal, S., & Amin, R. (2021). Blockchain-based internet of things and industrial IoT: A comprehensive survey. *Security and Communication Networks*, *2021*, 1–21. DOI: 10.1155/2021/7142048

Dwyer, J., & Bidwell, D. (2018). Chains of trust: Energy justice, public engagement, and the first offshore wind farm in the United States. *Energy Research & Social Science*, *47*, 166–176. DOI: 10.1016/j.erss.2018.08.019

Eboli, F. 2017. SDGs in Italy: current status and the process towards the National Strategy on Sustainable Development. *IAERE Fifth Annual Conference*. Rome.

Efkarpidis, N., Imoscopi, S., Bratukhin, A., Brännvall, R., Franzl, G., Leopold, T., Bauer, V., Goranovic, A., Wilker, S., Yang, C., Gustafsson, J., Geidl, M., & Sauter, T. (2023). Proactive scheduling of mixed energy resources at different grid levels. *IEEE Transactions on Sustainable Energy*, *15*(2), 952–963. DOI: 10.1109/TSTE.2023.3320055

Ege Jorgensen, C. (2008). *Strategy plans for sustainable development: case studies from Denmark*. Nordic Council of Minister's Office in Latvia.

Ehring, B., Minutella, D., & Shah, N. (2024)Taken as a whole, today's global fashion industry still struggles to work in a circular manner, https://www.kearney.com/industry/consumer-retail/article/the-kearney-cfx-2024-report-navigating-material-and-product-choices-as-key-unlocks-to-circularity?utm_campaign=climate-tech-monday&utm_medium=email&utm_source=newsletter&mkt_tok=MjExLU5KWS0xNjUAAAGUOjNgC3w6DC2s-OBHkF9WSPo5md9IAdb4CtgbjIwlK4yw veauFYpZEBZw5imZE4HVEQnn8Ddtak-xdEXbagHlni3e8C_SfS4zPpA0350boyNd_g

Ehteshami, S., & Chan, S. (2014). The role of hydrogen and fuel cells to store renewable energy in the future energy network – potentials and challenges. *Energy Policy*, *73*, 103–109. DOI: 10.1016/j.enpol.2014.04.046

El Jaadi, M., Haidi, T., & Belfqih, A. (2024). Advancements in wind farm layout optimization: A comprehensive review of artificial intelligence approaches. *Telkomnika*, *22*(3), 763–772.

Elavarasan, R. M., Pugazhendhi, R., Jamal, T., Dyduch, J., Arif, M. T., Kumar, N. M., & Nadarajah, M. (2021). Envisioning the UN Sustainable Development Goals (SDGs) through the lens of energy sustainability (SDG 7) in the post-COVID-19 world. *Applied Energy*, *292*, 116665. DOI: 10.1016/j.apenergy.2021.116665

Elavarasan, R. M., Shafiullah, G. M., Padmanaban, S., Kumar, N. M., Annam, A., Vetrichelvan, A. M., & Holm-Nielsen, J. B. (2020). A comprehensive review on renewable energy development, challenges, and policies of leading Indian states with an international perspective. *IEEE Access : Practical Innovations, Open Solutions*, *8*, 74432–74457. DOI: 10.1109/ACCESS.2020.2988011

ElFar, O. A., Chang, C. K., Leong, H. Y., Peter, A. P., Chew, K. W., & Show, P. L. (2021). Prospects of Industry 5.0 in algae: Customization of production and new advance technology for clean bioenergy generation. *Energy Conversion and Management: X*, *10*, 100048.

Elham, M. N., Sabeghi, M. S., Al-Rasbi, F. S., Daud, S. M., Noor, N. Q., Sahak, R., & Mazlan, A. A. (2020, April). A preliminary study on poultry farm environmental monitoring using internet of things and blockchain technology. In *2020 IEEE 10th Symposium on Computer Applications & Industrial Electronics (ISCAIE)* (pp. 273-276). IEEE. DOI: 10.1109/ISCAIE47305.2020.9108820

Eljack, F., & Kazi, M. K. (2021). Prospects and challenges of green hydrogen economy via multi-sector global symbiosis in Qatar. *Frontiers in Sustainability*, *1*, 612762. DOI: 10.3389/frsus.2020.612762

Elmalt, D., Igan, D., & Kirti, D. (2021). Limits to private climate change mitigation. IMF Working Paper. DOI: 10.5089/9781513582504.001

Elrahmani, A., Hannun, J., Eljack, F., & Kazi, M. K. (2021). Status of renewable energy in the GCC region and future opportunities. *Current Opinion in Chemical Engineering*, *31*, 100664. DOI: 10.1016/j.coche.2020.100664

Entezari, A., Aslani, A., Zahedi, R., & Noorollahi, Y. (2023). Artificial intelligence and machine learning in energy systems: A bibliographic perspective. [Google Scholar] [CrossRef]. *Energy Strategy Reviews*, *45*, 101017. DOI: 10.1016/j.esr.2022.101017

Estonian Ministry of the Environment. (2005). *Estonian National Strategy on Sustainable Development. Sustainable Estonia 21*. Estonian Ministry of the Environment.

European Commission. (2021). *European Commission | Choose your language | Choisir une langue | Wählen Sie eine Sprache*. Europa.eu; European Union. https://ec.europa.eu

European Commission. (2024). The Green Deal and Farm to Fork Strategy: Sustainable Food Systems in Europe. Retrieved from https://ec.europa.eu/

European Commission. (2024a). *What is the EU emissions trading system (EU ETS)?* European Commission. Retrieved December 20, 2024, from https://climate.ec.europa.eu/eu-action/eu-emissions-trading-system-eu-ets/what-eu-ets_en

European Commission. (2024b). *A European Green Deal*. European Commission. Retrieved December 20, 2024, from https://commission.europa.eu/strategy-and-policy/priorities-2019-2024/european-green-deal_en

European Commission. (2024c). *Corporate sustainability reporting*. European Commission. Retrieved December 20, 2024, from https://finance.ec.europa.eu/capital-markets-union-and-financial-markets/company-reporting-and-auditing/company-reporting/corporate-sustainability-reporting_en

European Environment Agency. (2002). *Environmental Signals 2002. Benchmarking the millennium*. Office for Official Publications of the European Communities.

ExxonMobil. (2024). Workplace safety protocols and risk management in the energy sector. ExxonMobil Corporate Report.

Ezeh, N. M. O., Ogbu, N. D., Ikevuje, N. H., & George, N. E. P. (2024). Stakeholder engagement and influence: Strategies for successful energy projects. *International Journal of Management & Entrepreneurship Research*, *6*(7), 2375–2395. DOI: 10.51594/ijmer.v6i7.1330

Faaique, M. (2024). Overview of Big Data Analytics in Modern Astronomy. *International Journal of Mathematics, Statistics, and Computer Science*, *2*, 96–113.

Fair Trade International. (2023). Annual Report 2023: Impact and Progress. Retrieved from https://www.fairtrade.net/

Falope, T., Lao, L., Hanak, D., & Huo, D. (2024). Hybrid energy system integration and management for solar energy: A review. *Energy Conversion and Management X*, *21*, 100527. DOI: 10.1016/j.ecmx.2024.100527

Fan, Z., Yan, Z., & Wen, S. (2023). Deep Learning and Artificial Intelligence in Sustainability: A Review of SDGs, Renewable Energy, and Environmental Health. *Sustainability (Basel)*, *15*(18), 13493. Advance online publication. DOI: 10.3390/su151813493

Farooq, M. S., Javid, R., Riaz, S., & Atal, Z. (2022). Internet of Things (IoT) Technologies in Greenhouse Farming: A Systematic Literature Review. *IEEE Access : Practical Innovations, Open Solutions*.

Federal Planning Bureau. (2011). *Twenty Years of Political Commitment to Sustainable Development. Federal Report on sustainable development 2011*. Henri Bogaert.

Feng, Y., Cui, N., Chen, Y., Gong, D., & Hu, X. (2019). Development of data-driven models for prediction of daily global horizontal irradiance in Northwest China. [Google Scholar] [CrossRef]. *Journal of Cleaner Production*, *223*, 136–146. DOI: 10.1016/j.jclepro.2019.03.091

Ferdaus, M. M., Dam, T., Anavatti, S., & Das, S. (2024). Digital technologies for a net-zero energy future: A comprehensive review. *Renewable & Sustainable Energy Reviews*, *202*, 114681. DOI: 10.1016/j.rser.2024.114681

Fernández-Caramés, T. M., & Fraga-Lamas, P. (2018). A Review on the Use of Blockchain for the Internet of Things. *IEEE Access : Practical Innovations, Open Solutions*, *6*, 32979–33001. DOI: 10.1109/ACCESS.2018.2842685

Ferrag, M. A., & Shu, L. (2021). The performance evaluation of blockchain-based security and privacy systems for the Internet of Things: A tutorial. *IEEE Internet of Things Journal*, *8*(24), 17236–17260. DOI: 10.1109/JIOT.2021.3078072

Ferrag, M. A., Shu, L., Yang, X., Derhab, A., & Maglaras, L. (2020). Security and privacy for green IoT-based agriculture: Review, blockchain solutions, and challenges. *IEEE Access : Practical Innovations, Open Solutions*, *8*, 32031–32053. DOI: 10.1109/ACCESS.2020.2973178

Figaj, R. (2024). Energy and economic sustainability of a Small-Scale hybrid renewable energy system powered by biogas, solar energy, and wind. *Energies*, *17*(3), 706. DOI: 10.3390/en17030706

Finnish National Commission on Sustainable Development. (2008). *Finnish National Commission on Sustainable Development. Towards a globally and nationally sustainable Finland.* Ministry of the Environment.

Fischer, C., & Preonas, L. (2010). Combining policies for renewable energy: Is the whole less than the sum of its parts?. *Resource for the future discussion chapter*, (10-19).

Fischer, J., Helman, K., Kramulová, J., Petkovová, L., & Zeman, J. (2013). Sustainable Development Indicators at the Regional Level in the Czech Republic. *Statistika*, *93*(1), 5–18.

Fleischer, T. (2007). Fenntartható fejlődés: környezeti, társadalmi és gazdasági tényezők. In Farkas, P., & Fóti, G. (Eds.), *Magyarország globális környezete 2020-ig. Háttértanulmányok a magyar külstratégiához I* (pp. 192–202). MTA Világgazdasági Kutatóintézet.

Fleming, A., Wise, R. M., Hansen, H., & Sams, L. (2017). The sustainable development goals: A case study. *Marine Policy*, *86*, 94–103. DOI: 10.1016/j.marpol.2017.09.019

Flores, A., Morales, A., Campos, G., & Gelso, J. (2021, December). Energy Efficiency Using IOTA Tangle for Greenhouse Agriculture. In *Annual International Conference on Information Management and Big Data* (pp. 122-138). Cham: Springer International Publishing.

Food and Agriculture Organization of the United Nations (FAO). (2024). The State of Food and Agriculture 2024: Value-driven transformation of agrifood systems. Rome. DOI: 10.4060/cd2616en

Foster, C., & Francis, P. (2019). A systematic review on the deployment and effectiveness of data analytics in higher education to improve student outcomes. *Assessment & Evaluation in Higher Education*, *45*(6), 822–841. DOI: 10.1080/02602938.2019.1696945

Foster, K., Blakstad, S., Bos, M., & Gazi, S. (2021). *BigFintechs and their impacts on sustainable development*. The Dialogue on Global Finance Digital Finance Governance Project. DOI: 10.2139/ssrn.3871385

Fraija, A., Henao, N., Agbossou, K., Kelouwani, S., Fournier, M., & Nagarsheth, S. H. (2024). Deep reinforcement learning based dynamic pricing for demand response considering market and supply constraints. *Smart Energy*, *14*, 100139. DOI: 10.1016/j.segy.2024.100139

Fried, S., Novan, K., & Peterman, W. B. (2022). Climate policy transition risk and the macroeconomy. *European Economic Review, 147*, 104174. DOI: 10.1016/j.euroecorev.2022.104174

Gaba, S., Khan, H., Almalki, K. J., Jabbari, A., Budhiraja, I., Kumar, V., Singh, A., Singh, K. K., Askar, S. S., & Abouhawwash, M. (2023). Holochain: An agent-centric distributed hash table security in smart iot applications. *IEEE Access : Practical Innovations, Open Solutions, 11*, 81205–81223. DOI: 10.1109/ACCESS.2023.3300220

Gabr, A. Z., Helal, A. A., & Abbasy, N. H. (2020). Economic evaluation of rooftop grid-connected photovoltaic systems for residential buildings in Egypt. [Google Scholar] [CrossRef]. *International Transactions on Electrical Energy Systems, 30*(12), e12379.

Gajić, T., Minasyan, L. A., Petrović, M. D., Bakhtin, V. A., Kaneeva, A. V., & Wiegel, N. L. (2023). Travelers' (in) Resilience to environmental risks emphasized in the media and their redirecting to medical destinations: Enhancing sustainability. *Sustainability, 15*(21), 15297.

Gamonwet, P., & Dhakal, S. (2023). The assessment of the value of electricity saving and economic benefit to residential solar rooftop PV customer: The case of Thailand. *Energy Strategy Reviews, 50*, 101203. DOI: 10.1016/j.esr.2023.101203

Gao, Y., & Liu, H. (2022). Artificial intelligence-enabled personalization in interactive marketing: a customer journey perspective. *Journal of Research in Interactive Marketing*, (ahead-of-print), 1-18.

Gao, Z., Zhao, Y., Li, L., & Hao, Y. (2024). Economic effects of sustainable energy technology progress under carbon reduction targets: An analysis based on a dynamic multi-regional CGE model. *Applied Energy, 363*, 123071. DOI: 10.1016/j.apenergy.2024.123071

Garan, M., Tidriri, K., & Kovalenko, I. (2022). A Data-Centric Machine Learning Methodology: Application on Predictive Maintenance of Wind Turbines. *Energies, 15*(3), 826. Advance online publication. DOI: 10.3390/en15030826

Garrido, S., Sequeira, T., & Santos, M. (2020). Renewable energy and sustainability from the supply side: A critical review and analysis. *Applied Sciences (Basel, Switzerland), 10*(17), 5755. DOI: 10.3390/app10175755

Gatti, A., Barbierato, E., & Pozzi, A. (2024). Toward Greener Smart Cities: A Critical Review of Classic and Machine-Learning-Based Algorithms for Smart Bin Collection. *Electronics (Basel), 13*(5), 836.

Gatto, A., & Drago, C. (2021). When renewable energy, empowerment, and entrepreneurship connect: Measuring energy policy effectiveness in 230 countries. *Energy Research & Social Science, 78*, 101977. DOI: 10.1016/j.erss.2021.101977

Ge, S., Isah, H., Zulkernine, F., & Khan, S. (2019). A scalable framework for multilevel streaming data analytics using deep learning. arXiv preprint arXiv:1907.06690. DOI: 10.1109/COMPSAC.2019.10205

George, A. S. (2023). Future Economic Implications of Artificial Intelligence. *Partners Universal International Research Journal, 2*(3), 20–39.

George, G. M., & Jayashree, L. S. (2022). Fusion of Blockchain-IoT network to improve supply chain traceability using Ethermint Smart chain: A Review. *KSII Transactions on Internet and Information Systems*, *16*, 3694–3722.

Georgiou, M. (2019). *The role of ai technology in improving the renewable energy sector*. Insider.

Ghadikolaei, S. S. C. (2021). Solar photovoltaic cells performance improvement by cooling technology: An overall review. *International Journal of Hydrogen Energy*, *46*(18), 10939–10972. DOI: 10.1016/j.ijhydene.2020.12.164

Ghai, D., Tripathi, S. L., Saxena, S., Chanda, M., & Alazab, M. (Eds.). (2022). *Machine Learning Algorithms for Signal and Image Processing*. John Wiley & Sons. DOI: 10.1002/9781119861850

Ghenai, C., & Bettayeb, M. (2019). Modelling and performance analysis of a stand-alone hybrid solar PV/Fuel Cell/Diesel Generator power system for university building. *Energy*, *171*, 180–189. DOI: 10.1016/j.energy.2019.01.019

Ghenai, C., Bettayeb, M., Brdjanin, B., & Hamid, A. K. (2019). Hybrid solar PV/PEM fuel Cell/Diesel Generator power system for cruise ship: A case study in Stockholm, Sweden. *Case Studies in Thermal Engineering*, *14*, 100497. DOI: 10.1016/j.csite.2019.100497

Ghorbani, N., Kasaeian, A., Toopshekan, A., Bahrami, L., & Maghami, A. (2017). Optimizing a hybrid wind-PV-battery system using GA-PSO and MOPSO for reducing cost and increasing reliability. *Energy*, *154*, 581–591. DOI: 10.1016/j.energy.2017.12.057

Gielen, D., & Boshell, F. (2021, April 30). *Renewable energy is the cornerstone of the energy transition*. World Economic Forum; World Economic Forum. https://www.weforum.org/stories/2021/04/why-renewables-are-the-cornerstone-of-the-energy-transition/

Gill, A. M., Stephens, S. L., & Cary, G. J. (2013). The worldwide "wildfire" problem. *Ecological Applications*, *23*(2), 438–454. DOI: 10.1890/10-2213.1 PMID: 23634593

Global Water Partnership. (2024). Sustainable Water Management in Agriculture. Retrieved from https://www.gwp.org/

Goel, A., Goel, A. K., & Kumar, A. (2023). The role of artificial neural network and machine learning in utilizing spatial information. *Spatial Information Research*, *31*(3), 275–285.

Golchha, A. (2024). *Artificial Intelligence: A Catalyst to Transform Energy Sector for Sustainable Future*. Csis.org; Center for Strategic and International Studies. https://www.csis.org/analysis/artificial-intelligence-catalyst-transform-energy-sector-sustainable-future

Gopi, A., Sharma, P., Sudhakar, K., Ngui, W., Kirpichnikova, I., & Cuce, E. (2022). Weather Impact on Solar Farm Performance: A Comparative Analysis of Machine Learning Techniques. *Sustainability (Basel)*, *15*(1), 439. Advance online publication. DOI: 10.3390/su15010439

Goti, A., Querejeta-Lomas, L., Almeida, A., de la Puerta, J. G., & López-de-Ipiña, D. (2023). Artificial Intelligence in Business-to-Customer Fashion Retail: A Literature Review. *Mathematics*, *11*(13), 2943. DOI: 10.3390/math11132943

Goudarzi, A., Ghayoor, F., Waseem, M., Fahad, S., & Traore, I. (2022). A survey on IoT-Enabled Smart Grids: Emerging, applications, challenges, and outlook. *Energies*, *15*(19), 6984. DOI: 10.3390/en15196984

Grant, A., & Langer, E. R. (2021). Wildfire volunteering and community disaster resilience in New Zealand: Institutional change in a dynamic rural social-ecological setting. *Ecology and Society*, *26*(3), art18. Advance online publication. DOI: 10.5751/ES-12474-260318

Griggs, D., Stafford Smith, M., Rockström, J., Öhman, M. C., Gaffney, O., Glaser, G., Kanie, N., Noble, J., Steffen, W., & Shyamsundar, P. (2014). An integrated framework for sustainable development goals. *Ecology and Society*, *19*(4), 1–24. DOI: 10.5751/ES-07082-190449

Group, D. G. B. (2023, January 23). *The history and future of renewable energy*. Www.green.earth; DGB Group. https://www.green.earth/blog/the-history-and-future-of-renewable-energy

Grubb, M., Drummond, P., Poncia, A., McDowall, W., Popp, D., Samadi, S., & Pavan, G. (2021). Induced innovation in energy technologies and systems: A review of evidence and potential implications for CO2 mitigation. *Environmental Research Letters*, *16*(4), 043007. DOI: 10.1088/1748-9326/abde07

Guchhait, R., & Sarkar, B. (2023). Increasing Growth of Renewable Energy: A State of Art. *Energies*, *16*(6), 2665. DOI: 10.3390/en16062665

Gulay, E., Sen, M., & Akgun, O. B. (2023). Forecasting electricity production from various energy sources in Türkiye: A predictive analysis of time series, deep learning, and hybrid models. *Energy*, *286*, 129566. DOI: 10.1016/j.energy.2023.129566

Guo, W., Yang, B., Ji, J., & Liu, X. (2023). Green finance development drives renewable energy development: Mechanism analysis and empirical research. *Renewable Energy*, *215*, 118982. DOI: 10.1016/j.renene.2023.118982

Gupta, D., Bhatt, P., & Bhatt, S. (2020, December). A game theoretic analysis for cooperative smart farming. In *2020 IEEE International Conference on Big Data (Big Data)* (pp. 2350-2359). IEEE. DOI: 10.1109/BigData50022.2020.9377935

Guven, I., & Parlak, M. (2022, November). Blockchain, AI and IoT Empowered Swarm Drones for Precision Agriculture Applications. In *2022 IEEE 1st Global Emerging Technology Blockchain Forum: Blockchain & Beyond (iGETblockchain)* (pp. 1-6). IEEE.

Hainsch, K., Löffler, K., Burandt, T., Auer, H., del Granado, P. C., Pisciella, P., & Zwickl-Bernhard, S. (2022). Energy transition scenarios: What policies, societal attitudes, and technology developments will realize the EU Green Deal? *Energy*, *239*, 122067. DOI: 10.1016/j.energy.2021.122067

Halmai, P. (Ed.). (2023). *A jegybank és a pénzügyi szektor portfólióinak zöldülése - Fenntarthatóság a közgazdaság-tudományban: Elméleti alapok, alkalmazások*. MERSZ., Retrieved from https://mersz.hu/dokumentum/m1101fak__33/

Hanif, M. F., & Mi, J. (2024). Harnessing AI for solar energy: Emergence of transformer models. *Applied Energy*, *369*, 123541.

Hannan, M. A., Al-Shetwi, A. Q., Ker, P. J., Begum, R. A., Mansor, M., Rahman, S. A., Dong, Z. Y., Tiong, S. K., Mahlia, T. M. I., & Muttaqi, K. M. (2021). Impact of renewable energy utilization and artificial intelligence in achieving sustainable development goals. *Energy Reports*, *7*, 5359–5373. DOI: 10.1016/j.egyr.2021.08.172

Haque, M. A., Haque, S., Zeba, S., Kumar, K., Ahmad, S., Rahman, M., & Ahmed, L. (2023). Sustainable and efficient E-learning internet of things system through blockchain technology. *E-Learning and Digital Media*, ●●●, 20427530231156711.

Harvard Business School. (2024). *Competing in the Age of AI—Virtual*. Harvard Business School; Harvard Business School. https://www.exed.hbs.edu/competing-age-ai-virtual?utm_source=google&utm_medium=paid-search&utm_campaign=brand-program-aiv-global-none-phrase-cross-device-all&utm_id=core&gad_source=1&gclid=Cj0KCQiAgJa6BhCOARIsAMiL7V9B4zcM_s7dxWNHJm6GVmK2acLqeLq4t0r2WKeizsoxz59HW0HG1lsaAmHKEALw_wcB&gclsrc=aw.ds

HarvestPlus. (2024). Biofortification Progress and Impact Report. Retrieved from https://www.harvestplus.org/

Hassan, Q., Sameen, A. Z., Olapade, O., Alghoul, M., Salman, H. M., & Jaszczur, M. (2023). Hydrogen fuel as an important element of the energy storage needs for future smart cities. *International Journal of Hydrogen Energy*, *48*(78), 30247–30262. DOI: 10.1016/j.ijhydene.2023.03.413

Hassija, V., Zeadally, S., Jain, I., Tahiliani, A., Chamola, V., & Gupta, S. (2021). Framework for determining the suitability of blockchain: Criteria and issues to consider. *Transactions on Emerging Telecommunications Technologies*, *32*(10), e4334. DOI: 10.1002/ett.4334

HB, M., & Ahammed, , ASM, U. (2024). Optimized Efficiency of IoT-Based Next Generation Smart Wireless Sensor Networks Using a Machine Learning Algorithm. *International Journal of Computing and Digital Systems*, *17*(1), 1–13.

Henke, N., & Jacques Bughin, L. (2016). The age of analytics: Competing in a data-driven world.

Herremans, D. (2021). aiSTROM—A roadmap for developing a successful AI strategy. *arXiv preprint arXiv:2107.06071*. (arXiv)

Hirij, Z., & Pashankar, S. (2024) Why It's So Hard to Track the Fashion Industry's Emissions, A growing number of fashion companies are talking about substantially cutting their greenhouse gas emissions. But evaluating those efforts is tricky. https://www.bloomberg.com/news/features/2024-05-16/is-your-favorite-fashion-brand-cutting-emissions-it-s-tricky

Hirsch, A., Parag, Y., & Guerrero, J. (2018). Microgrids: A review of technologies, key drivers, and outstanding issues. *Renewable & Sustainable Energy Reviews*, *90*, 402–411. DOI: 10.1016/j.rser.2018.03.040

Hoang, R. V., Sgambati, M. R., Brown, T. J., Coming, D. S., & Harris, F. C.Jr. (2010). VFire: Immersive wildfire simulation and visualization. *Computers & Graphics*, *34*(6), 655–664. DOI: 10.1016/j.cag.2010.09.014

Hoffmann, M. W., Wildermuth, S., Gitzel, R., Boyaci, A., Gebhardt, J., Kaul, H., & Tornede, T. (2020). Integration of novel sensors and machine learning for predictive maintenance in medium voltage switchgear to enable the energy and mobility revolutions. *Sensors (Basel)*, *20*(7), 2099. DOI: 10.3390/s20072099 PMID: 32276442

Horton, P., & Horton, B. P. (2019). Re-defining Sustainability: Living in Harmony with Life on Earth. *One Earth*, *1*, 86–94.

IBM. (2023). *Artificial Intelligence (AI) Solutions | IBM*. Ibm.com; International Business Machines Corporation. https://www.ibm.com/artificial-intelligence?utm_content=SRCWW&p1=Search&p4=43700077827237093&p5=p&p9=58700008530962751&gad_source=1&gclid=Cj0KCQiAgJa6BhCOARIsAMiL7V8neyLIiAdc8eOnwhXADP78b9vQOC_vPudPXEQbUeZV5eDfH-zLsDUaAqaHEALw_wcB&gclsrc=aw.ds

IEA. (2021, May). *Net Zero by 2050 – Analysis*. International Energy Agency; International Energy Agency. https://www.iea.org/reports/net-zero-by-2050

Ilyas, A., Parkinson, S., Vinca, A., Byers, E., Manzoor, T., Riahi, K., & Muhammad, A. (2022). Balancing smart irrigation and hydropower investments for sustainable water conservation in the Indus basin. *Environmental Science & Policy*, *135*, 147–161. DOI: 10.1016/j.envsci.2022.04.012

IMAD (Institute of Macroeconomic Analysis and Development). (2007). *Development Report 2007*. SOLOS.

IMAD (Institute of Macroeconomic Analysis and Development). (2012). *Development Report 2012*. IMAD.

Intergovernmental Panel on Climate Change (IPCC). (2014). Mitigation of Climate Change: Working Group III Contribution to the Fifth Assessment Report of the IPCC. Retrieved from https://www.ipcc.ch/site/assets/uploads/2018/02/ipcc_wg3_ar5_chapter7.pdf

International Energy Agency – IEA. (2024). *Smart grids*. IEA. Retrieved December 20, 2024, from https://www.iea.org/energy-system/electricity/smart-grids

International Energy Agency. (2023). *CO2 emissions in 2023: Executive summary*. Retrieved from https://www.iea.org/reports/co2-emissions-in-2023/executive-summary

International Federation of Organic Agriculture Movements (IFOAM). (2024). The Role of Organic Farming in Climate Change Mitigation. Retrieved from https://www.ifoam.bio/

International Renewable Energy Agency (IRENA). (2024). Renewable energy and global sustainability. IRENA Annual Report. Retrieved from https://www.irena.org

IPCC. (2023). [Mitigation of Climate Change. Intergovernmental Panel on Climate Change.]. *Climatic Change*, •••, 2023.

Islam, A., Islam, T., Mahmud, H., Raihan, O., Islam, M. S., Marwani, H. M., & Awual, M. R. (2024). Accelerating the green hydrogen revolution: A comprehensive analysis of technological advancements and policy interventions. *International Journal of Hydrogen Energy*, *67*, 458–486.

Issa Zadeh, S. B., Esteban Perez, M. D., López-Gutiérrez, J. S., & Fernández-Sánchez, G. (2023). Optimizing Smart Energy Infrastructure in Smart Ports: A Systematic Scoping Review of Carbon Footprint Reduction. *Journal of Marine Science and Engineering*, *11*(10), 1921. DOI: 10.3390/jmse11101921

Issa Zadeh, S. B., López Gutiérrez, J. S., Esteban, M. D., Fernández-Sánchez, G., & Garay-Rondero, C. L. (2023). Scope of the Literature on Efforts to Reduce the Carbon Footprint of Seaports. *Sustainability (Basel)*, *15*(11), 8558. DOI: 10.3390/su15118558

Izam, N. S. M. N., Itam, Z., Sing, W. L., & Syamsir, A. (2022). Sustainable development perspectives of solar energy technologies with focus on solar Photovoltaic—A review. *Energies*, *15*(8), 2790. DOI: 10.3390/en15082790

Jabareen, Y. (2008). A new conceptual framework for sustainable development. *Environment, Development and Sustainability*, *10*(2), 179–192. DOI: 10.1007/s10668-006-9058-z

Jain, A., Sharma, A., Jately, V., & Azzopardi, B. (Eds.). (2023). *Sustainable Energy Solutions with Artificial Intelligence, Blockchain Technology, and Internet of Things*. CRC Press. DOI: 10.1201/9781003356639

Jain, P., Coogan, S. C., Subramanian, S. G., Crowley, M., Taylor, S. W., & Flannigan, M. D. (2020). A review of machine learning applications in wildfire science and management. *Environmental Reviews*, *28*(4), 478–505. DOI: 10.1139/er-2020-0019

Jaramillo, M., Pavón, W., & Jaramillo, L. (2024). Adaptive Forecasting in Energy Consumption: A Bibliometric Analysis and review. *Data*, *9*(1), 13. DOI: 10.3390/data9010013

Järvelä, M., & Juhola, S. (2011). *Energy, policy, and the environment: Modeling sustainable development for the North*. Springer Science & Business Media. (Google Scholar)

Javaid, U., Usman, R., & Javaid, A. (2023). Investigating the Energy Production through Sustainable Sources by Incorporating Multifarious Machine Learning Methodologies. 2023 3rd International Conference on Artificial Intelligence (ICAI), 233-237. https://doi.org/DOI: 10.1109/ICAI58407.2023.10136677

Javaid, S., Ioannou, I., Pitsillides, A., & Tan, Y. (2023). Motivating an Agent-based Distributed AI Framework for Renewable Integration: Power Balancing. *2023 International Conference on Consumer Electronics - Taiwan (ICCE-Taiwan)*, 189-190. https://doi.org/DOI: 10.1109/ICCE-Taiwan58799.2023.10226739

Jayachandran, M., Gatla, R. K., Rao, K. P., Rao, G. S., Mohammed, S., Milyani, A. H., Azhari, A. A., Kalaiarasy, C., & Geetha, S. (2022). Challenges in achieving sustainable development goal 7: Affordable and clean energy in light of nascent technologies. *Sustainable Energy Technologies and Assessments*, *53*, 102692. DOI: 10.1016/j.seta.2022.102692

Jeje, S. O., Marazani, T., Obiko, J. O., & Shongwe, M. B. (2024). Advancing the hydrogen production economy: A comprehensive review of technologies, sustainability, and future prospects. *International Journal of Hydrogen Energy*, *78*, 642–661. DOI: 10.1016/j.ijhydene.2024.06.344

Jiang, H., Tjandra, R., Soh, C. B., Cao, S., Soh, D. C. L., Tan, K. T., Tseng, K. J., & Krishnan, S. B. (2024). Digital twin of microgrid for predictive power control to buildings. *Sustainability (Basel)*, *16*(2), 482. DOI: 10.3390/su16020482

Johnston, F. H., Borchers-Arriagada, N., Morgan, G. G., Jalaludin, B., Palmer, A. J., Williamson, G. J., & Bowman, D. M. J. S. (2020). Unprecedented health costs of smoke-related PM2.5 from the 2019–20 Australian megafires. *Nature Sustainability*, *4*(1), 42–47. DOI: 10.1038/s41893-020-00610-5

Juma, D., Munda, J., & Kabiri, C. (2023). Power-System flexibility: A necessary complement to variable renewable energy optimal capacity configuration. *Energies*, *16*(21), 7432. DOI: 10.3390/en16217432

Kadaba, D. M. K., & Aithal, P. S., & KRS, S. (2023). Government Initiatives and Digital Innovation for Atma Nirbhar MSMEs/SMEs: To Achieve Sustainable and Inclusive Economic Growth. [IJMTS]. *International Journal of Management, Technology, and Social Sciences*, *8*(1), 68–82.

Kadhem, A. A., Wahab, N. A., Aris, I., Jasni, J., Abdalla, A. N., & Matsukawa, Y. (2017). Reliability assessment of generating systems with wind power penetration via BPSO. Journal of International Journal on Advanced Science. *Engineering and Information Technology*, *7*(4), 1248–1254.

Kadlubek, M. (2015). Examples of Sustainable Development in the Area of Transport. *Procedia Economics and Finance*, *27*, 494–500. DOI: 10.1016/S2212-5671(15)01026-6

Kadri, A., Marzougui, H., Aouiti, A., & Bacha, F. (2019). Energy management and control strategy for a DFIG wind turbine/fuel cell hybrid system with super capacitor storage system. *Energy*, *192*, 116518. DOI: 10.1016/j.energy.2019.116518

Kahloot, K. M., & Ekler, P. (2021). Algorithmic splitting: A method for dataset preparation. *IEEE Access : Practical Innovations, Open Solutions*, *9*, 125229–125237. DOI: 10.1109/ACCESS.2021.3110745

Kalabokidis, K., Ager, A., Finney, M., Athanasis, N., Palaiologou, P., & Vasilakos, C. (2016). AEGIS: A wildfire prevention and management information system. *Natural Hazards and Earth System Sciences*, *16*(3), 643–661. DOI: 10.5194/nhess-16-643-2016

Kalinichenko, A. (2017). Interactive electronic educational tools with cognitive visualization. *Scientific and Technical Journal of Information Technologies Mechanics and Optics*, 359–364. DOI: 10.17586/2226-1494-2017-17-2-359-364

Kamal, M., & Himel, A. S. (2023). Redefining Modern Marketing: An Analysis of AI and NLP's Influence on Consumer Engagement, Strategy, and Beyond. *Eigenpub Review of Science and Technology*, *7*(1), 203–223.

Kamarposhti, M. S., Kamyab, H., Krishnan, S., Yusuf, M., Rezania, S., Chelliapan, S., & Khorami, M. (2024). A comprehensive review of AI-enhanced smart grid integration for hydrogen energy: Advances, challenges, and future prospects. *International Journal of Hydrogen Energy*, *67*, 1009–1025. Advance online publication. DOI: 10.1016/j.ijhydene.2024.01.129

Kannakumar, K., Manikandan, P., Ravivarman, G., Sivalingam, A., Moorthy, A. A., Karthik, M., & Girimurugan, R. (2023). Enhanced behavior of solar stills with modified absorber surfaces: Comparative analysis of floating wick designs. In *E3S Web of Conferences,* 455, 02002.

Kannan, N., & Vakeesan, D. (2016). Solar energy for future world:-A review. *Renewable & Sustainable Energy Reviews*, *62*, 1092–1105. DOI: 10.1016/j.rser.2016.05.022

Kapucu, N., & Garayev, V. (2011). Collaborative Decision-Making in emergency and disaster management. *International Journal of Public Administration*, *34*(6), 366–375. DOI: 10.1080/01900692.2011.561477

Karduri, R. K. R. (2019, February 1). *The Role of Artificial Intelligence in Optimizing Energy Systems*. Research Gate; Research Gate. https://www.researchgate.net/publication/375910911_The_Role_of_Artificial_Intelligence_in_Optimizing_Energy_Systems/citations

Karim, T. F., Lipu, M. H., & Mahmud, M. S. (2017). Electricity access improvement using renewable energy and energy efficiency: A case of urban poor area of dhaka, bangladesh. [IJRER]. *International Journal of Renewable Energy Research*, *7*(3), 1296–1306.

Karlsson, K., Jiang, W., Wicker, S., Adams, D., Ma, E., van Renesse, R., & Weatherspoon, H. (2018, July). Vegvisir: A partition-tolerant blockchain for the internet-of-things. In *2018 IEEE 38th International Conference on Distributed Computing Systems (ICDCS)* (pp. 1150-1158). IEEE. DOI: 10.1109/ICDCS.2018.00114

Kasinathan, P., Pugazhendhi, R., Elavarasan, R. M., Ramachandaramurthy, V. K., Ramanathan, V., Subramanian, S., & Alsharif, M. H. (2022). Realization of Sustainable Development Goals with Disruptive Technologies by Integrating Industry 5.0, Society 5.0, Smart Cities and Villages. *Sustainability (Basel)*, *14*(22), 15258. DOI: 10.3390/su142215258

Katamoura, S., Aksoy, M. S., & Alkhamees, B. F. (2024). Privacy and Security in Artificial Intelligence and Machine Learning Systems for Renewable Energy Big Data. 2024 21st Learning and Technology Conference (L&T), 209-214.

Katche, M. L., Makokha, A. B., Zachary, S. O., & Adaramola, M. S. (2024). Techno-Economic Assessment of Solar–Grid–Battery hybrid energy systems for Grid-Connected university campuses in Kenya. *Electricity*, *5*(1), 61–74. DOI: 10.3390/electricity5010004

Kathar, S., Hardel, H., Alam, Z., Phansalkar, S., & Sajjan, R. (2021, December). Auction System for Agricultural Trade Using Blockchain Technology: A Survey with Proposed Framework. In *2021 IEEE Pune Section International Conference (PuneCon)* (pp. 1-10). IEEE. DOI: 10.1109/PuneCon52575.2021.9686543

Katterbauer, K., Qasim, A., Marsala, A., & Yousef, A. (2021). A Data Driven Artificial Intelligence Framework for Hydrogen Production Optimization in Water flooded Hydrocarbon Reservoir. Day 4 Thu, November 18, 2021. .DOI: 10.2118/207847-MS

Katterbauer, K., Qasim, A., Shehri, A., & Yousif, A. (2022). A Novel Artificial Intelligence Framework for the Optimal Control of Wireless Temperature Sensors for Optimizing Oxygen Injection in Subsurface Reservoirs. Day 2 Wed, March 23, 2022. .DOI: 10.4043/31558-MS

Kaur, U., Kaur, M., & Khang, A. (2023). Futuristic Technologies in Agriculture Challenges and Future Prospects. *Handbook of Research on AI-Equipped IoT Applications in High-Tech Agriculture*, 347-364.

Kaur, H., & Sood, S. K. (2020). A smart Disaster management framework for wildfire detection and prediction. *The Computer Journal*, *63*(11), 1644–1657. DOI: 10.1093/comjnl/bxz091

Kaushal, S., & Mishra, D. (2024). Strategic implications of AI in contemporary business and society. 2024 2nd International Conference on Disruptive Technologies (ICDT), Greater Noida, India. DOI: 10.1109/ICDT61202.2024.10489027

Kaushal, V., & Yadav, R. (2023). Learning successful implementation of Chatbots in businesses from B2B customer experience perspective. *Concurrency and Computation*, *35*(1), e7450. DOI: 10.1002/cpe.7450

Kaushik, T. K. Ravish, Singh, A., Raghav, A., & Singh, B. (2024). Potential of Artificial Intelligence in the Agricultural Industry: A Comprehensive Review. In B. Riswandi, B. Singh, C. Kaunert, & K. Vig (Eds.), *AI Applications for Clean Energy and Sustainability* (pp. 344-365). IGI Global. https://doi.org/ DOI: 10.4018/979-8-3693-6567-0.ch017

Kaushik, T. K. Ravish, Singh, A., Raghav, A., Singh, B., Arora, M. K., & Lal, S. (2024). Untapping Artificial Intelligence in Environmental and Sustainable Progression: Projecting Legal and Ethical Considerations. In B. Singh, C. Kaunert, K. Vig, & S. Dutta (Eds.), *Maintaining a Sustainable World in the Nexus of Environmental Science and AI* (pp. 109-130). IGI Global. https://doi.org/DOI: 10.4018/979-8-3693-6336-2.ch005

Keivanpour, S. (2021). Internet of things in Optimizing Emergency logistics of Wildfire: A Review and Application Perspective in Canada. *2021 International Conference on Electrical, Computer and Energy Technologies (ICECET)*. https://doi.org/DOI: 10.1109/ICECET52533.2021.9698250

Kelly, R., & Moles, R. (2000). Towards sustainable development in the mid-west region of Ireland. *Environmental Management and Health*, *11*(5), 422–432. DOI: 10.1108/09566160010351881

Khan, M. A., Khan, T., Waseem, M., Saleh, A. M., Qamar, N., & Muqeet, H. A. (2024). Investigation and analysis of demand response approaches, bottlenecks, and future potential capabilities for IoT-enabled smart grid. *IET Renewable Power Generation*, *18*(15), 3509–3535. DOI: 10.1049/rpg2.13011

Khan, M. A., Saleh, A. M., Waseem, M., & Sajjad, I. A. (2022). Artificial intelligence Enabled Demand Response: Prospects and challenges in smart grid environment. *IEEE Access : Practical Innovations, Open Solutions*, *11*, 1477–1505. DOI: 10.1109/ACCESS.2022.3231444

Khan, M. K., Raza, M., Shahbaz, M., Farooq, U., & Akram, M. U. (2024). Recent advancement in energy storage technologies and their applications. *Journal of Energy Storage*, *92*, 112112. DOI: 10.1016/j.est.2024.112112

Khan, M. W., & Wang, J. (2017). The research on multi-agent system for microgrid control and optimization. *Renewable & Sustainable Energy Reviews*, *80*, 1399–1411. DOI: 10.1016/j.rser.2017.05.279

Kim, J., & Jang, H. W. (2023). Can Artificial Intelligence Boost developing electro catalysts for efficient water splitting to Produce Green Hydrogen? *Korean Journal of Materials Research*, *33*(5), 175–188.

King, K. (2022). *AI Strategy for Sales and Marketing: Connecting Marketing, Sales and Customer Experience*. Kogan Page Publishers.

Kis-Orloczki, M. (2014). *National Sustainable Development Strategies in the Visegrad Four.* Electronic International Interdisciplinary Conference, 1-5. September, 2014. (pp. 160-163).

Kitchin, R. (2013). Big data and human geography: Opportunities, challenges and risks. *Dialogues in Human Geography*, *3*(3), 262–267.

Klaiber, J., & Dinther, C. (2023). Deep Learning for Variable Renewable Energy: A Systematic Review. *ACM Computing Surveys*, *56*(1), 1–37. DOI: 10.1145/3586006

Kolozsi, P. P., Ladányi, S., & Straubinger, A. (2022). Pénzügyi eszközök klímakockázatának mérése: Módszertani kihívások és jegybanki gyakorlat. *Hitelintézeti Szemle*, *21*(1), 113–140. DOI: 10.25201/HSZ.21.1.113

Komala, C. R., Basha, M. M., Farook, S., Niranchana, R., Rajendiran, M., & Subhi, B. (2024). Smart Energy Systems-Integrated Machine Learning, IoT, and AI Tools. In Reshaping Environmental Science Through Machine Learning and IoT (pp. 201-229). IGIGlobal.

Kong, W., You, Z., Lyu, S., & Lv, X. (2024). Multi-dimensional stereo face reconstruction for psychological assistant diagnosis in medical meta-universe. *Information Sciences*, *654*, 119831.

Kose, U., Prasath, V. S., Mondal, M. R. H., Podder, P., & Bharati, S. (Eds.). (2022). *Artificial Intelligence and Smart Agriculture Technology*. CRC Press. DOI: 10.1201/9781003299059

Kothari, D. P., Ranjan, R., & Singal, K. C. (2021). Renewable energy sources and emerging technologies.

Kow, J. (2002). Television as a Tool of Environmental Adult Education: Limits and Possibilities. *Adult Learning*, *13*(2-3), 30–34. DOI: 10.1177/104515950201300207

Kralj, D., & Markič, M. (2008). Processes Innovation and Sustainable Development. *WSEAS Transactions on Environment and Development*, *4*(2), 99–108.

Krarti, M. (2019). Evaluation of energy efficiency potential for the building sector in the Arab region. [Google Scholar] [CrossRef]. *Energies*, *12*(22), 4279. DOI: 10.3390/en12224279

Kuhlman, T., & Farrington, J. (2010). What is Sustainability? *Sustainability (Basel)*, *2*(11), 3436–3448. DOI: 10.3390/su2113436

Kumar, N. R., Seenivasan, S., Hemalatha, P., Prakash, K. J., Pitchandi, P., Sathiya, R. D., & Girimurugan, R. (2023). Evaluating the influence of coir fibres on solar still efficiency and economic viability. In *E3S Web of Conferences,* 455, 02004. DOI: 10.1051/e3sconf/202345502004

Kumar, N., & Katiyar, A. (2024). Artificial Intelligence and the Future of Renewable Energy: Solar PV Power Forecasting. In New Innovations in AI, Aviation, and Air Traffic Technology (pp. 1-28). IGI Global.

Kumar, P. V., Kulkarni, A., Mendhe, D., Keshar, D. K., Babu, S. B. T., & Rajesh, N. (2024, April). AI-Optimized Hardware Design for Internet of Things (IoT) Devices. In *2024 5th International Conference on Recent Trends in Computer Science and Technology (ICRTCST)* (pp. 21-26). IEEE.

Kumar, A., Alaraj, M., Rizwan, M., & Nangia, U. (2021). Novel AI Based Energy Management System for Smart Grid With RES Integration. *IEEE Access : Practical Innovations, Open Solutions*, *9*, 162530–162542. DOI: 10.1109/ACCESS.2021.3131502

Kumar, A., Nagar, S., & Anand, S. (2021). Climate change and existential threats. In *Global Climate Change* (pp. 1–31). Elsevier.

Kumari, A., & Sharma, I. (2023, October). Augmentation of Internet of Things Security and Privacy with IOTA Tangle Network. In *2023 4th IEEE Global Conference for Advancement in Technology (GCAT)* (pp. 1-5). IEEE. DOI: 10.1109/GCAT59970.2023.10353394

Kumari, S., & Muthulakshmi, P. (2023). Artificial Intelligence—Blockchain Enabled Technology for Internet of Things: Research Statements, Open Issues, and Possible Applications in the Near Future. *Privacy Preservation of Genomic and Medical Data*, 433-480.

Kumar, N., Chand, A., Malvoni, M., Prasad, K., Mamun, K., Islam, F., & Chopra, S. (2020). Distributed Energy Resources and the Application of AI, IoT, and Blockchain in Smart Grids. *Energies*, *13*(21), 5739. Advance online publication. DOI: 10.3390/en13215739

Kumar, S., Lim, W. M., Sivarajah, U., & Kaur, J. (2022). Artificial Intelligence and Blockchain Integration in Business: Trends from a Bibliometric-Content Analysis. *Information Systems Frontiers*, *25*, 871–896. DOI: 10.1007/s10796-022-10279-0 PMID: 35431617

Kumar, S., Malviya, R., Uniyal, P., Srivastava, S., Sridhar, S. B., Obaidur Rab, S., & Singh, B. (2024, August 23). (2024). Plant Fiber Composites for Biomedical Applications: Advances and Prospective. *Current Pharmaceutical Biotechnology*, *25*. Advance online publication. DOI: 10.2174/0113892010314833240815113622 PMID: 39185642

Kumar, S., & Rathore, K. (2023). Renewable energy for sustainable development goal of clean and affordable energy. *International Journal of Materials Manufacturing and Sustainable Technologies*, *2*(1), 1–15.

Kumar, V. R., Bali, S. K., & Devarapalli, R. (2023). Application of AI-Based Algorithms for Industrial Photovoltaic Module Parameter Extraction. *SN Computer Science*, *4*(5), 525.

Kurtz, J., Bradley, T., Winkler, E., & Gearhart, C. (2020). Predicting demand for hydrogen station fueling. *International Journal of Hydrogen Energy*, *45*(56), 32298–32310. DOI: 10.1016/j.ijhydene.2019.10.014

Kuusinen, M. (2019). Scenarios for digital marketing: a Delphi-based analysis for 2028.

Ladoneczki, G., & Kósi, K. (2014). Makroszintű nemzeti stratégiák konzisztenciavizsgálata a biodiverzitás tükrében. *Gazdálkodás*, *58*(4), 331–340. DOI: 10.22004/ag.econ.206091

Lahmadi, K. (2023). *Artificial intelligence can improve renewable energy production, report shows.* Morocco World News.

Laimon, M., & Yusaf, T. (2024). Towards energy freedom: Exploring sustainable solutions for energy independence and self-sufficiency using integrated renewable energy-driven hydrogen system. *Renewable Energy*, *222*, 119948. DOI: 10.1016/j.renene.2024.119948

Lamtzidis, O., & Gialelis, J. (2018, December). An IOTA based distributed sensor node system. In *2018 IEEE Globecom Workshops (GC Wkshps)* (pp. 1-6). IEEE. DOI: 10.1109/GLOCOMW.2018.8644153

Lamtzidis, O., Pettas, D., & Gialelis, J. (2019). A novel combination of distributed ledger technologies on internet of things: Use case on precision agriculture. *Applied System Innovation*, *2*(3), 30. DOI: 10.3390/asi2030030

Latif, S., Idrees, Z., Huma, Z., & Ahmad, J. (2021). Blockchain technology for the industrial Internet of Things: A comprehensive survey on security challenges, architectures, applications, and future research directions. *Transactions on Emerging Telecommunications Technologies*, *32*(11), e4337. DOI: 10.1002/ett.4337

Lawrenz, L., Xiong, B., Lorenz, L., Krumm, A., Hosenfeld, H., Burandt, T., & Von Hirschhausen, C. (2018). Exploring energy pathways for the low-carbon transformation in India—A model-based analysis. *Energies*, *11*(11), 3001. DOI: 10.3390/en11113001

Lazzarin, R. M., & Noro, M. (2018). Past, present, future of solar cooling: Technical and economical considerations. *Solar Energy*, *172*, 2–13. DOI: 10.1016/j.solener.2017.12.055

Leduc, G., Kubler, S., & Georges, J. P. U,"Innovative Blockchain-based farming marketplace and smart contract performance evaluation,". *J of C. Production, Elsevier.*

Leduc, G., Kubler, S., & Georges, J. P. (2021). Innovative blockchain-based farming marketplace and smart contract performance evaluation. *Journal of Cleaner Production*, *306*, 127055. DOI: 10.1016/j.jclepro.2021.127055

Lee, C.-C., Fang, Y., Quan, S., & Li, X. (2024). Leveraging the power of artificial intelligence toward the energy transition: The key role of the digital economy. *Energy Economics*, *135*, 107654. DOI: 10.1016/j.eneco.2024.107654

Lee, Y. K. (2021). Transformation of the innovative and sustainable supply chain with upcoming real-time fashion systems. *Sustainability (Basel)*, *13*(3), 1081. DOI: 10.3390/su13031081

Leipute, B. (2014). Lithuania in the context of sustainable development. *Lithuanian Journal of Statistics*, *53*(1), 24–33. DOI: 10.15388/LJS.2014.13891

Leogrande, A. (2024). Integrating ESG principles into smart logistics: Toward sustainable supply chains. HAL Sciences. DOI: 10.20944/preprints202411.1432.v1

Lepik, E. (2018). *SDG Coordination and Implementation in Estonia.* Republic of Estonia Government Office.

Leslie, D., Rincon, C., Briggs, M., Perini, A., Jayadeva, S., Borda, A., Bennett, S., Burr, C., Aitken, M., Katell, M., Fischer, C., Wong, J., & Ismael, K. G. (2024). AI Sustainability in Practice Part One: Foundations for Sustainable AI projects. SSRN Electronic Journal. https://doi.org/DOI: 10.2139/ssrn.4731832

Leung, P. H., Chui, K. T., Lo, K., & de Pablos, P. O. (2021). A support vector machine–based voice disorders detection using human voice signal. In *Artificial Intelligence and Big Data Analytics for Smart Healthcare* (pp. 197–208). Academic Press.

Liao, M., & Yao, Y. (2021). Applications of artificial intelligence-based modeling for bioenergy systems: A review. *Global Change Biology. Bioenergy*, *13*(5), 774–802.

Li, D., Guan, X., Tang, T., Zhao, L., Tong, W., & Wang, Z. (2023). The clean energy development path and sustainable development of the ecological environment driven by big data for mining projects. *Journal of Environmental Management*, *348*, 119426. DOI: 10.1016/j.jenvman.2023.119426 PMID: 37879178

Li, J., & Xu, G. (2022). Circular economy towards zero waste and decarbonization. [Google Scholar] [CrossRef]. *Circular Economy*, *1*(1), 100002. DOI: 10.1016/j.cec.2022.100002

Lima, G. M., Belchior, F. N., Villena, J. E. N., Domingos, J. L., Freitas, M. A. V., & Hunt, J. D. (2024). Hybrid electrical energy generation from hydropower, solar photovoltaic and hydrogen. *International Journal of Hydrogen Energy*, *53*, 602–612. DOI: 10.1016/j.ijhydene.2023.12.092

Linardatos, P., Papastefanopoulos, V., & Kotsiantis, S. (2021). Explainable AI: A review of machine learning interpretability methods. *Entropy (Basel, Switzerland)*, *23*(1), 18. DOI: 10.3390/e23010018 PMID: 33375658

Lin, Y., Chen, Y., Hsieh, S., Liu, H., Chiang, C., & Yang, H. (2023). Reinforcement Learning-based Energy Management System for Microgrids with High Renewable Energy Penetration. *2023 IEEE International Conference on Energy Technologies for Future Grids (ETFG)*, 1-6. DOI: 10.1109/ETFG55873.2023.10407803

Lin, Y., & Fu, L. (2022). A Study for a Hybrid Wind-Solar-Battery System for Hydrogen Production in an Islanded MVDC Network. *IEEE Access : Practical Innovations, Open Solutions*, *10*, 85355–85367. DOI: 10.1109/ACCESS.2022.3193683

Li, Q., Xue, Z., Wu, Y., & Zeng, X. (2022). The status quo and prospect of sustainable development of smart clothing. *Sustainability (Basel)*, *14*(2), 990. DOI: 10.3390/su14020990

Liu, D., Sun, Y., Qu, Y., Li, B., & Xu, Y. (2018). Analysis and accurate prediction of user's response behavior in Incentive-Based Demand response. *IEEE Access : Practical Innovations, Open Solutions*, *7*, 3170–3180. DOI: 10.1109/ACCESS.2018.2889500

Liu, K. S., & Lin, M. H. (2021). Performance Assessment on the Application of Artificial Intelligence to Sustainable Supply Chain Management in the Construction Material Industry. *Sustainability (Basel)*, *13*(22), 12767. DOI: 10.3390/su132212767

Liu, L., Bouman, T., Perlaviciute, G., & Steg, L. (2019). Effects of trust and public participation on acceptability of renewable energy projects in the Netherlands and China. *Energy Research & Social Science*, *53*, 137–144. DOI: 10.1016/j.erss.2019.03.006

Liu, X., Yang, Y., Jiang, Y., Fu, Y., Zhong, R. Y., Li, M., & Huang, G. Q. (2023). Data-driven ESG assessment for blockchain services: A comparative study in textiles and apparel industry. *Resources, Conservation and Recycling*, *190*, 106837. DOI: 10.1016/j.resconrec.2022.106837

Liu, Z., Sun, Y., Xing, C., Liu, J., He, Y., Zhou, Y., & Zhang, G. (2022). Artificial intelligence powered large-scale renewable integrations in multi-energy systems for carbon neutrality transition: Challenges and future perspectives. *Energy and AI*, *10*, 100195. DOI: 10.1016/j.egyai.2022.100195

Li, W., Lu, C., & Zhang, Y. W. (2019). Prospective exploration of future renewable portfolio standard schemes in China via a multi-sector CGE model. *Energy Policy*, *128*, 45–56. DOI: 10.1016/j.enpol.2018.12.054

Li, X., Raorane, C. J., Xia, C., Wu, Y., Tran, T. K. N., & Khademi, T. (2023). Latest approaches on green hydrogen as a potential source of renewable energy towards sustainable energy: Spotlighting of recent innovations, challenges, and future insights. *Fuel, 334*, 126684. DOI: 10.1016/j.fuel.2022.126684

Lobo, G., Costa, S., Nogueira, R., Antunes, P., & Brito, A. (2005). *A Scenario Building Methodology to Support the Definition of Sustainable Development Strategies: the Case of the Azores Region. 11th Annual International Sustainable Development Research Conference*, 6-8 June, 2005. (pp. 1-13). Helsinki: Finland.

Lodhi, S. K., Gill, A. Y., & Hussain, I. (2024). Renewable energy technologies: Present patterns and upcoming paths in ecological power production. *Global Journal of Universal Studies, 1*(1), 108–131. DOI: 10.70445/gjus.1.1.10

Lucas, P., Ludwig, K., Kok, M., & Kruitwagen, S. (2016). *Sustainable Development Goals in the Netherlands. Building blocks for environmental policy for 2030.* PBL Netherlands Environmental Assessment Agency.

Luo, X., & Oyedele, L. O. (2022). A self-adaptive deep learning model for building electricity load prediction with moving horizon. *Machine Learning with Applications, 7*, 100257. DOI: 10.1016/j.mlwa.2022.100257

Luo, Y., Shi, L., & Tu, G. (2014). Optimal sizing and control strategy of isolated grid with wind power and energy storage system. *Energy Conversion and Management, 80*, 407–415. DOI: 10.1016/j.enconman.2014.01.061

Luo, Z., Wang, C., Tang, Q., & Tian, W. (2024). Renewable energy technology innovation effect on the economics growth. *Chemistry and Technology of Fuels and Oils, 59*(6), 1271–1278. DOI: 10.1007/s10553-024-01644-7

Lyytimäki, J., Rinne, J., & Kautto, P. (2011). Using indicators to assess sustainable development in the European Union, Finland, Malta and Slovakia. The Finnish Environment 4. Finnish Environment Institute, Environmental Policy Centre.

Madessa, H. B., Shakerin, M., Reinskau, E. H., & Rabani, M. (2024). Recent progress in the application of energy technologies in Large-Scale building Blocks: A State-of-the-Art review. *Energy Conversion and Management, 305*, 118210. DOI: 10.1016/j.enconman.2024.118210

Madhavan, V. M., Rahul, S., Vijayakumar, P., Dhal, P. K., Girimurugan, R., Ravivarman, G., & Joseph, J. (2023). Optimizing solar energy utilization and energy efficiency through thermal energy storage with phase change materials in a solar water heating system. In *E3S Web of Conferences,* 455, 02005. DOI: 10.1051/e3sconf/202345502005

Madsen, R., Haynes, H., & McCaffrey, S. (2018). Wildfire risk reduction in the United States: Leadership staff perceptions of local fire department roles and responsibilities. *International Journal of Disaster Risk Reduction, 27*, 451–458. DOI: 10.1016/j.ijdrr.2017.11.009

Maes, P. (2018). *Disruptive Selling: A New Strategic Approach to Sales, Marketing and Customer Service.* Kogan Page Publishers.

Magyar Nemzeti Bank. (2022). Körkép a jegybankok és nemzetközi szervezetek zöld programjairól. Retrieved from https://www.ngfs.net/sites/default/files/medias/documents/climate_related_litigation.pdf

Mahapatra, S., Kumar, D., Singh, B., & Sachan, P. K. (2021). Biofuels and their sources of production: A review on cleaner sustainable alternative against conventional fuel, in the framework of the food and energy nexus. *Energy Nexus, 4*, 100036.

Majeed, A., Ahmad, M., Rasheed, M. F., Khan, M. K., Popp, J., & Oláh, J. (2022). The dynamic impact of financial globalization, environmental innovations, and energy productivity on renewable energy consumption: Evidence from advanced panel techniques. *Frontiers in Environmental Science, 10*, 894857. DOI: 10.3389/fenvs.2022.894857

Makatjane, K., Moroke, N., & Ncube, B. (2021). Detecting Financial Fraud in South Africa: A Comparison of Logistic Model Tree and Gradient Boosting Decision Tree.

Makatjane, K. (2022). Deep Learning for Sentiment Analysis to Predict the Probability of Bank Loan Default. *American Journal of Data Mining and Knowledge Discovery, 7*(2), 5–12.

Maksimović, M., Jokić, S., & Bošković, M. Č. (2024, March). Digital Twins and 5G: Unlocking the Potential in the Energy Sector. In *2024 23rd International Symposium INFOTEH-JAHORINA (INFOTEH)* (pp. 1-6). IEEE.

Malik, P., Malik, P. K., Singh, R., & Gehlot, A. (2023). Sustainable Development, Renewable Energy and Environment. In *Micro-Electronics and Telecommunication Engineering: Proceedings of 6th ICMETE 2022* (pp. 455–463). Springer Nature Singapore.

Managi, S., Shimamura, T., & Tanaka, Y. (2024). Evaluating the impact of report readability on ESG scores: A generative AI approach. *SSRN*. DOI: 10.2139/ssrn.4888339

Mandel, J., Beezley, J., Coen, J., & Kim, M. (2009). Data assimilation for wildland fires. *IEEE Control Systems, 29*(3), 47–65. DOI: 10.1109/MCS.2009.932224

Mangalam, H. SS, K. S., & Pavithran, P. (2024, July). Hybrid Modeling System for Photovoltaic Monitoring of Solar Panels. In 2024 2nd International Conference on Sustainable Computing and Smart Systems (ICSCSS) (pp. 38-42). IEEE.

Manikandan, R., Gopalakrishnan, K., Ashokkumar, P., Maheskumar, P., Girimurugan, R., Ravivarman, G., & Anand, R. (2023). Experimental study of minimum-temperature hydrated salt latent heat thermal energy storage with sodium acetate trihydrate as phase change materials. In *E3S Web of Conferences*, 455, 02008. DOI: 10.1051/e3sconf/202345502008

Manoharan, K. G., Nehru, J. A., & Balasubramanian, S. (2021). *Artificial Intelligence and Fuzzy Logics*. Springer.

Manoharan, K. G., Nehru, J. A., & Balasubramanian, S. (2021). *Artificial Intelligence and IoT*. Springer. DOI: 10.1007/978-981-33-6400-4

Marjani, M., Nasaruddin, F., Gani, A., Karim, A., Hashem, I. A. T., Siddiqa, A., & Yaqoob, I. (2017). Big IoT data analytics: architecture, opportunities, and open research challenges. *ieee access, 5*, 5247-5261.

Marjani, M., Nasaruddin, F., Gani, A., Karim, A., Hashem, I. T., Siddiqa, A., & Yaqoob, I. (2017). Big IoT Data Analytics: Architecture, opportunities, and open research challenges. *IEEE Access : Practical Innovations, Open Solutions*, *5*, 5247–5261. DOI: 10.1109/ACCESS.2017.2689040

Marrero, L., Sbarbaro, D., & Garcia-Santander, L. (2024). Online demand response characterization based on variability in customer behavior. *Journal of Modern Power Systems and Clean Energy*, *12*(3), 936–946. DOI: 10.35833/MPCE.2023.000516

Marwa, B., Ali, Z. M., Hanen, B., & Faouzi, M. M. (2024, July). AI Empowered Solar Energy: Reinforcement Learning and Comparative Analysis for Grid-Connected Photovoltaic Systems Optimization. In 2024 IEEE 28th International Conference on Intelligent Engineering Systems (INES) (pp. 000251-000256). IEEE.

Mashood, K., Kayani, H. U. R., Malik, A. A., & Tahir, A. (2023). ARTIFICIAL INTELLIGENCE RECENT TRENDS AND APPLICATIONS IN INDUSTRIES. *Pakistan Journal of Science*, *75*(02). Advance online publication. DOI: 10.57041/pjs.v75i02.855

Mathews, I., Kantareddy, S. N., Buonassisi, T., & Peters, I. M. (2019). Technology and market perspective for indoor photovoltaic cells. *Joule*, *3*(6), 1415–1426. DOI: 10.1016/j.joule.2019.03.026

Maurya, N. P. K. (2024). Artificial intelligence to enhance energy management and distribution in smart grid communication networks. *Tuijin Jishu/Journal of Propulsion Technology*, *45*(02), 4366–4378. DOI: 10.52783/tjjpt.v45.i02.6648

Ma, X., Wu, J., Xue, S., Yang, J., Zhou, C., Sheng, Q. Z., Xiong, H., & Akoglu, L. (2021). A comprehensive survey on graph anomaly detection with deep learning. *IEEE Transactions on Knowledge and Data Engineering*, *35*(12), 12012–12038. DOI: 10.1109/TKDE.2021.3118815

Mayer, R. E. (2009). *Multimedia learning* (2nd ed.). Cambridge University Press., DOI: 10.1017/CBO9780511811678

Mazzeo, D., Herdem, M. S., Matera, N., Bonini, M., Wen, J. Z., Nathwani, J., & Oliveti, G. (2021). Artificial intelligence application for the performance prediction of a clean energy community. *Energy*, *232*, 120999.

McCaffrey, S. (2015). Community Wildfire Preparedness: A Global State-of-the-Knowledge Summary of Social Science Research. *Current Forestry Reports*, *1*(2), 81–90. DOI: 10.1007/s40725-015-0015-7

McCaffrey, S. M. (2004). Fighting Fire with Education: What Is the Best Way to Reach Out to Homeowners? *Journal of Forestry*, *102*(5), 12–19. DOI: 10.1093/jof/102.5.12

McGee, T. (2011). Public engagement in neighbourhood level wildfire mitigation and preparedness: Case studies from Canada, the US and Australia. *Journal of Environmental Management*, *92*(10), 2524–2532. DOI: 10.1016/j.jenvman.2011.05.017 PMID: 21684061

Mchirgui, N., Quadar, N., Kraiem, H., & Lakhssassi, A. (2024). The Applications and Challenges of Digital Twin Technology in Smart Grids: A Comprehensive Review. *Applied Sciences (Basel, Switzerland)*, *14*(23), 10933. DOI: 10.3390/app142310933

Mebratu, D. (1998). Sustainability and Sustainable Development: Historical and Conceptual Review. *Environmental Impact Assessment Review*, *18*(6), 493–520. DOI: 10.1016/S0195-9255(98)00019-5

Mehrabani, A., Aalami, H. A., & Neghab, A. P. (2023). Integrated planning framework for microgrid incorporating the flexibility provision capabilities of demand response programs. *International Journal of Electrical Power & Energy Systems*, *155*, 109521. DOI: 10.1016/j.ijepes.2023.109521

Memarzadeh, G., & Keynia, F. (2023). A new hybrid CBSA-GA optimization method and MRMI-LSTM forecasting algorithm for PV-ESS planning in distribution networks. *Journal of Energy Storage*, *72*, 108582. DOI: 10.1016/j.est.2023.108582

Mena, R., Godoy, D. R., Kristjanpoller, F., & Viveros, P. (2024). A multi-objective two-stage stochastic unit commitment model for wind and battery-integrated power systems. *Journal of Energy Storage*, *89*, 111723. DOI: 10.1016/j.est.2024.111723

Mennella, C., Maniscalco, U., Pietro, G. D., & Esposito, M. (2024). Ethical and regulatory challenges of AI technologies in healthcare: A narrative review. *Heliyon*, *10*(4), e26297–e26297. DOI: 10.1016/j.heliyon.2024.e26297 PMID: 38384518

Mikami, N., Yagi, E., & Ikebe, Y. (2014). Recent Trends and New Approaches of Public Engagement in Japan

Ministère de l'Écologie, du Développement durable et de l'Énergie (2015). *National Strategy of ecological transition towards sustainable development 2015-2020.* La Défense Cedex.

Ministry for Sustainable Development, the Environment and Climate Change (2016). *Greening Our Economy – Achieving a sustainable future.* Ministry for Sustainable Development, the Environment and Climate Change.

Ministry of Environmental Protection, Physical Planning and Construction (2009). *Strategy for Sustainable Development of the Republic of Croatia.* Ministry of Environmental protection, Physical planning and construction.

Ministry of the Environment of the Czech Republic. (2010, May 20). *The Strategic Framework for Sustainable Development in the Czech Republic.*https://www.vlada.cz/assets/ppov/udrzitelny-rozvoj/The-Strategic-Framework-for-SD_2011.pdf

Mishra, D. (2024a). Role of Social Media and AI in Transforming Healthcare Communication. In Vig, K., Kaunert, C., & Singh, B. (Eds.), *Intersections of Law and Computational Intelligence in Health Governance* (pp. 29–51). IGI Global., DOI: 10.4018/979-8-3693-5976-1.ch002

Mishra, D. (2024b). Strategic Communication for AI-Driven Sustainability Initiatives: Bridging Technology, Policy, and Public Engagement in Achieving SDGs. In Riswandi, B., Singh, B., Kaunert, C., & Vig, K. (Eds.), *AI Applications for Clean Energy and Sustainability* (pp. 187–211). IGI Global., DOI: 10.4018/979-8-3693-6567-0.ch010

Mistry, I., Tanwar, S., Tyagi, S., & Kumar, N. (2020). Blockchain for 5G-enabled IoT for industrial automation: A systematic review, solutions, and challenges. *Mechanical Systems and Signal Processing*, *135*, 106382. DOI: 10.1016/j.ymssp.2019.106382

MIT. (2024, September 19). *The MIT Energy Initiative's Future Energy Systems Center funds ten new energy research projects.* MIT Energy Initiative; MIT Energy Initiative. https://energy.mit.edu/news/the-mit-energy-initiatives-future-energy-systems-center-funds-ten-new-energy-research-projects/

Mitra, A., Vangipuram, S. L., Bapatla, A. K., Bathalapalli, V. K., Mohanty, S. P., Kougianos, E., & Ray, C. (2022). Everything you wanted to know about smart agriculture. *arXiv preprint arXiv:2201.04754.*

Mitra, A., Singhal, A., Mohanty, S. P., Kougianos, E., & Ray, C. (2022). eCrop: A novel framework for automatic crop damage estimation in smart agriculture. *SN Computer Science*, *3*(4), 319.

Mkrttchian, V. (2021). Artificial and natural intelligence techniques as IoP-and IoT-based technologies for sustainable farming and smart agriculture. In *Artificial Intelligence and IoT-Based Technologies for Sustainable Farming and Smart Agriculture* (pp. 40–53). IGI Global. DOI: 10.4018/978-1-7998-1722-2.ch003

Moemen, Y. S., Alshater, H., & El-Sayed, I. E. T. (2023). Clean Energy Management Based on Internet of Things and Sensor Networks for Climate Change Problems. In *The Power of Data: Driving Climate Change with Data Science and Artificial Intelligence Innovations* (pp. 117–136). Springer Nature Switzerland.

Mohammadian, H. D. (2022). Mapping the Future Global SMEs Growth via Hybrid SMEs/SME 5.0/ Tomorrow's SMEs Concept through the 5th Wave, i-Sustainability Plus and DCT Theories. In *Management and Information Technology in the Digital Era* (Vol. 29, pp. 173–206). Emerald Publishing Limited. DOI: 10.1108/S1877-636120220000029011

Mohanty, S. (2021). Internet-of-agro-things (ioat) makes smart agriculture. *IEEE Consumer Electronics Magazine*, *10*(4), 4–5. DOI: 10.1109/MCE.2021.3074775

Mohiuddin Babu, M., Akter, S., Rahman, M., Billah, M. M., & Hack-Polay, D. (2022). The role of artificial intelligence in shaping the future of Agile fashion industry. *Production Planning and Control*, •••, 1–15.

Molnár, T. (2015). *Empirikus területi kutatások.* Akadémiai Kiadó.

Monroe, M. C., Ballard, H. L., Oxarart, A., Sturtevant, V. E., Jakes, P. J., & Evans, E. R. (2015). Agencies, educators, communities and wildfire: Partnerships to enhance environmental education for youth. *Environmental Education Research*, *22*(8), 1098–1114. DOI: 10.1080/13504622.2015.1057555

Monteiro, A. C. B., França, R. P., Arthur, R., & Iano, Y. (2021). An overview of medical Internet of Things, artificial intelligence, and cloud computing employed in health care from a modern panorama. *The Fusion of Internet of Things, Artificial Intelligence, and Cloud Computing in Health Care*, 3-23.

Monteiro, E. S., da Rosa Righi, R., Barbosa, J. L. V., & Alberti, A. M. (2021). APTM: A model for pervasive traceability of agrochemicals. *Applied Sciences (Basel, Switzerland)*, *11*(17), 8149. DOI: 10.3390/app11178149

Moraga, J., Duzgun, H. S., Cavur, M., & Soydan, H. (2022). The geothermal artificial intelligence for geothermal exploration. *Renewable Energy*, *192*, 134–149.

Moritz, M. A., Batllori, E., Bradstock, R. A., Gill, A. M., Handmer, J., Hessburg, P. F., Leonard, J., McCaffrey, S., Odion, D. C., Schoennagel, T., & Syphard, A. D. (2014). Learning to coexist with wildfire. *Nature*, *515*(7525), 58–66. DOI: 10.1038/nature13946 PMID: 25373675

Moritz, M. A., Parisien, M., Batllori, E., Krawchuk, M. A., Van Dorn, J., Ganz, D. J., & Hayhoe, K. (2012). Climate change and disruptions to global fire activity. *Ecosphere*, *3*(6), 1–22. DOI: 10.1890/ES11-00345.1

Moroke, N. D., & Makatjane, K. (2022). Predictive Modelling for Financial Fraud Detection Using Data Analytics: A Gradient-Boosting Decision Tree. In Applications of Machine Learning and Deep Learning for Privacy and Cybersecurity (pp. 25-45). IGI Global. DOI: 10.4018/978-1-7998-9430-8.ch002

Morss, R. E., Wilhelmi, O. V., Downton, M. W., & Gruntfest, E. (2005). Flood Risk, Uncertainty, and Scientific Information for Decision Making: Lessons from an Interdisciplinary Project. *Bulletin of the American Meteorological Society*, *86*(11), 1593–1602. DOI: 10.1175/BAMS-86-11-1593

MSCI. (2024). *BP PLC ESG ratings and climate search tool.* MSCI. Retrieved December 20, 2024, from https://www.msci.com/our-solutions/esg-investing/esg-ratings-climate-search-tool/issuer/bp-plc/IID000000002140371

Mur, A., Dormido, R., Duro, N., Dormido-Canto, S., & Vega, J. (2016). Determination of the optimal number of clusters using a spectral clustering optimization. *Expert Systems with Applications*, *63*, 304–314. DOI: 10.1016/j.eswa.2016.08.059

Myoung, B., Kim, S., Nghiem, S., Jia, S., Whitney, K., & Kafatos, M. (2018). Estimating Live Fuel Moisture from MODIS Satellite Data for Wildfire Danger Assessment in Southern California USA. *Remote Sensing (Basel)*, *10*(2), 87. DOI: 10.3390/rs10010087

Nair, S. S., Krishnia, L., Trukhanov, A., Thakur, P., & Thakur, A. (2022). Prospect of double perovskite over conventional perovskite in photovoltaic applications. *Ceramics International*, *48*(23), 34128–34147. DOI: 10.1016/j.ceramint.2022.08.184

Nair, S. S., Thakur, P., Wan, F., & Thakur, A. (2024). Structural modification and analysis of novel inverted perovskite photo-voltaic device by incorporating atomic layer deposition and surface passivation. *Solar Energy*, *269*, 112366. DOI: 10.1016/j.solener.2024.112366

Nair, S. S., Thakur, P., Wan, F., Trukhanov, A., Panina, L., & Thakur, A. (2023). Performance evaluation and the optimization of an inverted photo-voltaic cell with lead-free double perovskite material and inorganic transport layer materials. *Solar Energy*, *262*, 111823. DOI: 10.1016/j.solener.2023.111823

Nakata, T. (2004). Energy-economic models and the environment. *Progress in Energy and Combustion Science*, *30*(4), 417–475. DOI: 10.1016/j.pecs.2004.03.001

Nam, K., Hwangbo, S., & Yoo, C. (2020). A deep learning-based forecasting model for renewable energy scenarios to guide sustainable energy policy: A case study of korea. *Renewable & Sustainable Energy Reviews*, *122*, 109725.

Naser, N., & Abdelbari, A. (2020, June). Estimation of Global Solar Radiation using Back Propagation Neural Network: A case study Tripoli, Libya. In *2020 International Conference on Electrical, Communication, and Computer Engineering (ICECCE)* (pp. 1-5). IEEE.

Nassiry, D. (2018). *The role of fintech in unlocking green finance: Policy insights for developing countries* (No. 883). ADBI working paper.

Nastasi, B., Markovska, N., Puksec, T., Duić, N., & Foley, A. (2022). Renewable and sustainable energy challenges to face for the achievement of Sustainable Development Goals. *Renewable & Sustainable Energy Reviews*, *157*, 112071. DOI: 10.1016/j.rser.2022.112071

Nazir, T., & Banday, M. T. (2018, October). Green Internet of Things: A survey of enabling techniques. In *2018 International Conference on Automation and Computational Engineering (ICACE)* (pp. 197-202). IEEE. DOI: 10.1109/ICACE.2018.8686974

Neale, T., Weir, J. K., & McGee, T. K. (2016). Knowing wildfire risk: Scientific interactions with risk mitigation policy and practice in Victoria, Australia. *Geoforum*, *72*, 16–25. DOI: 10.1016/j.geoforum.2016.03.008

Necula, S. (2023). Assessing the Potential of Artificial Intelligence in Advancing Clean Energy Technologies in Europe: A Systematic Review. *Energies*, *16*(22), 7633. Advance online publication. DOI: 10.3390/en16227633

Neuwirth, R. J. (2023). Prohibited artificial intelligence practices in the proposed EU artificial intelligence act (AIA). *Computer Law & Security Review*, *48*, 105798. DOI: 10.1016/j.clsr.2023.105798

News, M. I. T. (2023, November 27). *A green hydrogen innovation for clean energy*. MIT News | Massachusetts Institute of Technology; Massachusetts Institute of Technology. https://news.mit.edu/2023/green-hydrogen-innovation-clean-energy-sossina-haile-1127

Nguyen, T. Q., Hoang, T., Zhang, L., Dobre, O. A., & Duong, T. Q. (2024). A Survey on Smart Optimisation Techniques for 6G-oriented Integrated Circuits Design. *Mobile Networks and Applications*, •••, 1–18.

Nilashi, M., Rupani, P. F., Mobin Rupani, M., Kamyab, H., Shao, W., Ahmadi, H., Rashid, T. A., & Aljojo, N. (2019). Measuring sustainability through ecological sustainability and human sustainability: A machine learning approach. *Journal of Cleaner Production*, *240*, 1–10. DOI: 10.1016/j.jclepro.2019.118162

Noor, A., Saeed, M. A., Ullah, T., Uddin, Z., & Ullah Khan, R. M. W. (2022). A review of artificial intelligence applications in apparel industry. *Journal of the Textile Institute*, *113*(3), 505–514. DOI: 10.1080/00405000.2021.1880088

Noor, N. M., Razali, N. A. M., Malizan, N. A., Ishak, K. K., Wook, M., & Hasbullah, N. A. (2022). Decentralized Access Control using Blockchain Technology for Application in Smart Farming. *International Journal of Advanced Computer Science and Applications*, *13*(9). Advance online publication. DOI: 10.14569/IJACSA.2022.0130993

Noor, N. M., Razali, N. A. M., Sham, S. N. S. A., Ishak, K. K., Wook, M., & Hasbullah, N. A. (2023). Decentralised Access Control Framework using Blockchain: Smart Farming Case. *International Journal of Advanced Computer Science and Applications, 14*(5). Advance online publication. DOI: 10.14569/IJACSA.2023.0140560

Nouinou, H., Asadollahi-Yazdi, E., Baret, I., Nguyen, N. Q., Terzi, M., Ouazene, Y., Yalaoui, F., & Kelly, R. (2023). Decision-making in the context of Industry 4.0: Evidence from the textile and clothing industry. *Journal of Cleaner Production, 391*, 136184. DOI: 10.1016/j.jclepro.2023.136184

Nour El-Din, M., Poças Martins, J., Ramos, N. M., & Pereira, P. F. (2024). The Role of Blockchain-Secured Digital Twins in Promoting Smart Energy Performance-Based Contracts for Buildings. *Energies, 17*(14), 3392. DOI: 10.3390/en17143392

Nwokediegwu, Z. Q. S., Ibekwe, K. I., Ilojianya, V. I., Etukudoh, E. A., & Ayorinde, O. B. (2024). Renewable energy technologies in engineering: A review of current developments and future prospects. *Engineering science & technology journal,* 5, 367-384.

Obeidat, M., Ameryeen, B., Mansour, A., Salem, H., & Awwad, A. (2022). Wind Power Forecasting using Artificial Neural Network. *WSEAS Transactions on Power Systems, 17*, 269–279. Advance online publication. DOI: 10.37394/232016.2022.17.28

Oberer, B. J., & Erkollar, A. (2011). Monitoring of Sustainable Development: on the way to a sustainable Austria. *Proceedings of the 3rd International Conference on Governance, Fraud, Ethics & Social Responsibility*, 1-13.

Office of the Government of the Czech Republic. (2006). *Interim Report on the Czech Republic Strategy for Sustainable Development.* Office of the Government of the Czech Republic.

Ohalete, N. C., Aderibigbe, A. O., Ani, E. C., Ohenhen, P. E., Daraojimba, D. O., & Odulaja, B. A. (2023). AI-driven solutions in renewable energy: A review of data science applications in solar and wind energy optimization. *World Journal of Advanced Research and Reviews, 20*(3), 401–417. DOI: 10.30574/wjarr.2023.20.3.2433

Ohueri, C. C., Masrom, M. A. N., & Seghier, T. E. (2024). Digital twin for decarbonizing operating buildings: A systematic review and implementation framework development. *Energy and Building, 320*, 114567. DOI: 10.1016/j.enbuild.2024.114567

Oliveira, F., Costa, D. G., Assis, F., & Silva, I. (2024). Internet of Intelligent Things: A convergence of embedded systems, edge computing and machine learning. *Internet of Things : Engineering Cyber Physical Human Systems, 26*, 101153. DOI: 10.1016/j.iot.2024.101153

Olsen, C. S., Mazzotta, D. K., Toman, E., & Fischer, A. P. (2014). Communicating About Smoke from Wildland Fire: Challenges and Opportunities for Managers. *Environmental Management, 54*(3), 571–582. DOI: 10.1007/s00267-014-0312-0 PMID: 25034752

Olshannikova, E., Ometov, A., Koucheryavy, Y., & Olsson, T. (2015). Visualizing Big Data with augmented and virtual reality: Challenges and research agenda. *Journal of Big Data, 2*(1), 1–27.

Omitaomu, O. A., & Niu, H. (2021). Artificial intelligence techniques in smart grid: A survey. [Google Scholar] [CrossRef]. *Smart Cities*, *4*(2), 548–568. DOI: 10.3390/smartcities4020029

Ontiveros, A. (2024). Development and evaluation of an IT tool for ESG compliance consulting for companies using LLMs. University of Hamburg. Retrieved from https: https://www.inf.uni-hamburg .de/en/inst/ab/lt/teaching/theses/completed-theses/2024-ma-ontiveros.pdf

Onwusinkwue, S., Osasona, F., Ahmad, I. A. I., Anyanwu, A. C., Dawodu, S. O., Obi, O. C., & Hamdan, A. (2024). Artificial intelligence (AI) in renewable energy: A review of predictive maintenance and energy optimization. *World Journal of Advanced Research and Reviews*, *21*(1), 2487–2499.

Ordonez García, B. J., & Martos Nunez, M. V. (2023). Blockchain in Agriculture: A PESTELS Analysis JAVIER.

Ordóñez, J., Alexopoulos, A., Koutras, K., Kalogeras, A., Stefanidis, K., & Martos, V. (2023). Blockchain in Agriculture: A PESTELS Analysis. *IEEE Access : Practical Innovations, Open Solutions.*

Organisation for Economic Co-operation and Development. (2016). *Better Policies for 2030. An OECD Action Plan on the Sustainable Development Goals*. OECD.

Østergaard, P. A., Duic, N., Noorollahi, Y., & Kalogirou, S. (2023). Advances in renewable energy for sustainable development. *Renewable Energy*, *219*, 119377. DOI: 10.1016/j.renene.2023.119377

Otoum, S., & Mouftah, H. T. (2021, December). Enabling Trustworthiness in Sustainable Energy Infrastructure Through Blockchain and AI-Assisted Solutions. *IEEE Wireless Communications*, *28*(6), 19–25. DOI: 10.1109/MWC.018.2100194

Ozcanli, A. K., Yaprakdal, F., & Baysal, M. (2020). Deep learning methods and applications for electrical power systems: A comprehensive review. *International Journal of Energy Research*, *44*(9), 7136–7157. DOI: 10.1002/er.5331

Padmaja, M., Shitharth, S., Prasuna, K., Chaturvedi, A., Kshirsagar, P. R., & Vani, A. (2022). Grow of artificial intelligence to challenge security in IoT application. *Wireless Personal Communications*, *127*(3), 1829–1845.

Pai, P.-F., & Chen, T.-C. (2009). Rough set theory with discriminant analysis in analyzing electricity loads. [Google Scholar] [CrossRef]. *Expert Systems with Applications*, *36*(3), 8799–8806. DOI: 10.1016/j.eswa.2008.11.012

Pakulska, T. (2021). Green energy in central and eastern European (CEE) countries: New challenges on the path to sustainable development. *Energies*, *14*(4), 884. DOI: 10.3390/en14040884

Pan, C., Tao, S., Fan, H., Shu, M., Zhang, Y., & Sun, Y. (2021). Multi-Objective optimization of a Battery-Supercapacitor hybrid energy storage system based on the concept of Cyber-Physical System. *Electronics (Basel)*, *10*(15), 1801. DOI: 10.3390/electronics10151801

Panda, S. S., Mohanta, B. K., Dey, M. R., Satapathy, U., & Jena, D. (2020, July). Distributed ledger technology for securing IoT. In *2020 11th international conference on computing, communication and networking technologies (ICCCNT)* (pp. 1-6). IEEE. DOI: 10.1109/ICCCNT49239.2020.9225333

Panda, A., Dauda, A. K., Chua, H., Tan, R. R., & Aviso, K. B. (2023). Recent advances in the integration of renewable energy sources and storage facilities with hybrid power systems. *Cleaner Engineering and Technology*, *12*, 100598. DOI: 10.1016/j.clet.2023.100598

Pandey, V., Sircar, A., Bist, N., Solanki, K., & Yadav, K. (2023). Accelerating the renewable energy sector through Industry 4.0: Optimization opportunities in the digital revolution. *International Journal of Innovation Studies*, *7*(2), 171–188. DOI: 10.1016/j.ijis.2023.03.003

Papoutsi, M., Piazzesi, M., & Schneider, M. (2021). How unconventional is green monetary policy? Retrieved from https://web.stanford.edu/~piazzesi/How_unconventional_is_green_monetary_policy.pdf

Park, C., & Kim, M. (2024). Utilization and challenges of artificial intelligence in the energy sector. *Energy & Environment*, 0958305X241258795. Advance online publication. DOI: 10.1177/0958305X241258795

Parkinson, T., Force, J., & Smith, J. (2003). Hands-on Learning: Its Effectiveness in Teaching the Public about Wildland Fire. *Journal of Forestry*, *101*(7), 21–26. DOI: 10.1093/jof/101.7.21

Park, J., & Kim, H. (2023). Federated learning-based optimization for smart grid energy management. *IEEE Transactions on Energy Conversion*, *37*(4), 723–736. DOI: 10.1109/TEC.2023.0274571

Parry, I. (2019). Putting a Price on Pollution: Carbon pricing strategies could hold the key to meeting the world's climate stabilization goals. IMF, The Economics of Climate. *Finance & Development*, •••, 16–19.

Pascaris, A. S. (2021). Examining existing policy to inform a comprehensive legal framework for agrivoltaics in the US. *Energy Policy*, *159*, 112620. DOI: 10.1016/j.enpol.2021.112620

Patagonia. (2024). Worn Wear initiative: Driving circular economy in fashion. Patagonia Sustainability Report.

Patel, J., Patel, A. R., & Tyagi, H. (2024). Advances in the Solar Thermal Systems Assisted Production of Green Hydrogen: Its Analysis, Scaling-Up Techniques, and Economics Aspects as Applied to Tropical Regions. In *Challenges and Opportunities in Green Hydrogen Production* (pp. 239–267). Springer Nature Singapore. DOI: 10.1007/978-981-97-1339-4_10

Patel, R., & Mehta, P. (2024). Fuzzy logic controllers for optimizing renewable energy distribution networks. *Renewable Energy*, *26*(4), 489–502. DOI: 10.1016/j.renene.2024.08.006

Pawar, R., Santara, S., Sircar, A., & Yadav, K. (2024). Design and development of hybrid solar–biomass drying system: An innovative approach. *MRS Energy & Sustainability : a Review Journal*, *11*(2), 409–433. DOI: 10.1557/s43581-024-00083-5

Pawlicka, K., & Bal, M. (2022). Sustainable Supply Chain Finances implementation model and Artificial Intelligence for innovative omnichannel logistics. *Management*, *26*(1), 19–35. DOI: 10.2478/manment-2019-0082

Perea-Moreno, A.-J. (2021). *Renewable energy and energy saving: Worldwide research trends*. MDPI.

Phang, S. K., Chiang, T. H. A., Happonen, A., & Chang, M. M. L. (2023). From Satellite to UAV-Based Remote Sensing: A Review on Precision Agriculture. *IEEE Access : Practical Innovations, Open Solutions*, *11*, 127057–127076. DOI: 10.1109/ACCESS.2023.3330886

Pidgeon, N., Demski, C., Butler, C., Parkhill, K., & Spence, A. (2014). Creating a national citizen engagement process for energy policy. *Proceedings of the National Academy of Sciences of the United States of America, 111*(supplement_4), 13606–13613. DOI: 10.1073/pnas.1317512111 PMID: 25225393

Pikatza, A., Sainz, N., González, I., & Emaldi, M. (2023, July). Development of an Open Source IoT-Blockchain Platform for Traceability of Fresh Products from Farm to Fork. In *International Congress on Blockchain and Applications* (pp. 487-497). Cham: Springer Nature Switzerland. DOI: 10.1007/978-3-031-45155-3_46

Pimitha, K., Anbarasi, S., & Balasubramanian, T. (2023). Machine Learning Techniques for Renewable Energy Resources. In *Fuzzy Logics, Machine Learning and Blockchain Technologies for Renewable Energy and Modern Hybrid Power Systems* (pp. 121-148). River Publishers.

Pimitha, K., Anbarasi, S., & Balasubramanian, T. (2023). Machine Learning Techniques for Renewable Energy Resources. In *IoT, Machine Learning and Blockchain Technologies for Renewable Energy and Modern Hybrid Power Systems* (pp. 121-148). River Publishers.

Pincheira, M., Vecchio, M., Giaffreda, R., & Kanhere, S. S. (2021). Cost-effective IoT devices as trustworthy data sources for a blockchain-based water management system in precision agriculture. *Computers and Electronics in Agriculture, 180*, 105889. DOI: 10.1016/j.compag.2020.105889

Plut, D., & Vintar Mally, K. (2004). *Slovenia and the challenges of Sustainable Development.* In M. Orozen Adamic (Eds.), *Slovenia: A Geographical Overview.* (pp. 155-159). ZRC SAZU.

Poddar, H. (2024). A Look into the Future: Advances in Digital Twin Technology. In *AI Approaches to Smart and Sustainable Power Systems* (pp. 1–27). IGI Global. DOI: 10.4018/979-8-3693-1586-6.ch001

Podder, A. K., Al Bukhari, A., Islam, S., Mia, S., Mohammed, M. A., Kumar, N. M., & Abdulkareem, K. H. (2021). IoT based smart agrotech system for verification of Urban farming parameters. *Microprocessors and Microsystems, 82*, 104025. DOI: 10.1016/j.micpro.2021.104025

Ponnuru, S., Kumar, R. A., & Swaroopan, N. J. (2020). Intelligent control and power management of wind-solar integration of renewable energy sources using microgrid. *Materials Today: Proceedings, 45*, 2323–2328. DOI: 10.1016/j.matpr.2020.10.687

Pradhan, T., Nimkar, P., & Jhajharia, K. (2024). Machine Learning and Deep Learning for Big Data Analysis. In *Big Data Analytics Techniques for Market Intelligence* (pp. 209–240). IGI Global.

Prakash, C., Singh, L. P., Gupta, A., & Lohan, S. K. (2023). Advancements in smart farming: A comprehensive review of IoT, wireless communication, sensors, and hardware for agricultural automation. *Sensors and Actuators. A, Physical, 362*, 114605. DOI: 10.1016/j.sna.2023.114605

Prakash, N., & Sethi, M. (2021). A review of innovative bond instruments for sustainable development in Asia. *International Journal of Innovation Science, 14*(3/4), 630–647. DOI: 10.1108/IJIS-10-2020-0213

Prugh, T., Costanza, R., & Daly, H. (1999). *The Local Politics of Global Sustainability*. Island Press.

Pullo, S., Pareschi, R., Piantadosi, V., Salzano, F., & Carlini, R. (2023, December). Integrating IOTA's Tangle with the Internet of Things for Sustainable Agriculture: A Proof-of-Concept Study on Rice Cultivation. [). MDPI.]. *Informatics (MDPI), 11*(1), 3. DOI: 10.3390/informatics11010003

Puri, V., Solanki, V. K., & Aponte, G. J. R. (2023, March). Blockchain and Federated Learning Based Integrated Approach for Agricultural Internet of Things. In *The International Conference on Intelligent Systems & Networks* (pp. 240-246). Singapore: Springer Nature Singapore. DOI: 10.1007/978-981-99-4725-6_30

Qazi, A., Hussain, F., Rahim, N., Hardaker, G., Alghazzawi, D., Shaban, K., & Haruna, K. (2019). Towards Sustainable Energy: A Systematic Review of Renewable Energy Sources, Technologies, and Public Opinions. *IEEE Access : Practical Innovations, Open Solutions*, *7*, 63837–63851. DOI: 10.1109/ACCESS.2019.2906402

Qian, W., & Zhou, H. (2023). Federated learning for decentralized renewable energy storage. *Sustainable Energy Technologies and Assessments*, *18*(3), 578–591. DOI: 10.1016/j.seta.2023.07.015

Qiu, K., & Zhao, K. (2024). The integration of green energy and artificial intelligence in next-generation energy supply chain: An analysis of economic, social, and environmental impacts. *Sustainable Energy Technologies and Assessments*, *64*, 103660.

Quiles, E., Roldán-Blay, C., Escrivá-Escrivá, G., & Roldán-Porta, C. (2020). Accurate sizing of residential Stand-Alone photovoltaic systems considering system reliability. *Sustainability (Basel)*, *12*(3), 1274. DOI: 10.3390/su12031274

Qureshi, R., Irfan, M., Ali, H., Khan, A., Nittala, A. S., Ali, S., & Alam, T. (2023). Artificial Intelligence and Biosensors in Healthcare and its Clinical Relevance: A Review. *IEEE Access : Practical Innovations, Open Solutions*.

Radeva, I., & Popchev, I. (2022). Blockchain-Enabled Supply-Chain in Crop Production Framework. *Cybernetics and Information Technologies*, *22*(1), 151–170. DOI: 10.2478/cait-2022-0010

Raghav, A., Singh, B., Jermsittiparsert, K., Raghav, R., & Yadav, U. (2024). Artificial Intelligence in Environmental and Climate Changes: A Sustainable Future. In Singh, B., Kaunert, C., Vig, K., & Dutta, S. (Eds.), *Maintaining a Sustainable World in the Nexus of Environmental Science and AI* (pp. 485–506). IGI Global., DOI: 10.4018/979-8-3693-6336-2.ch019

Rahman, T., & Ahmed, S. (2024). A hybrid approach using fuzzy logic and federated learning for solar energy optimization. *Energy Reports*, *13*(6), 314–327. DOI: 10.1016/j.egyr.2024.04.010

Raihan, A., Begum, R. A., Said, M. N. M., & Pereira, J. J. (2022). Relationship between economic growth, renewable energy use, technological innovation, and carbon emission toward achieving Malaysia's Paris agreement. *Environment Systems & Decisions*, *42*(4), 586–607. DOI: 10.1007/s10669-022-09848-0

Rajaan, R., Baishya, B. K., Rao, T. V., Pattanaik, B., Tripathi, M. A., & Anitha, R. (2024). Efficient Usage of Energy Infrastructure in Smart City Using Machine Learning. *EAI Endorsed Transactions on Internet of Things, 10*.

Rajamannar, R. (2021). *Quantum marketing: mastering the new marketing mindset for tomorrow's consumers*. HarperCollins Leadership.

Rakhee, M., & Elayidom, S. (2023, January). Blockchain Technology as a Solution for Vulnerabilities in Internet of Things-A Survey. In *2023 Advanced Computing and Communication Technologies for High Performance Applications (ACCTHPA)* (pp. 1-6). IEEE.

Raman, R., Gunasekar, S., Kaliyaperumal, D., & Nedungadi, P. (2024). Navigating the nexus of artificial intelligence and renewable energy for the advancement of sustainable development goals. *Sustainability (Basel), 16*(21), 9144. DOI: 10.3390/su16219144

Ramesh, S., & Venkatesh, B. (2024). Fuzzy logic-based models for optimizing wind energy systems. *Renewable Energy, 27*(2), 654–667. DOI: 10.1016/j.renene.2024.01.012

Rana, R., & Patel, M. (2023). Federated learning for distributed energy resource management. *Journal of Renewable Energy Research, 17*(5), 943–956. DOI: 10.1016/j.jrener.2023.06.018

Rane, N., Choudhary, S., & Rane, J. (2024). Artificial intelligence-driven approaches to strengthening ESG criteria in sustainable business practices: A review. *SSRN*. DOI: 10.2139/ssrn.4843215

Rani, P., Parkash, V., & Sharma, N. K. (2023). Technological aspects, utilization and impact on power system for distributed generation: A comprehensive survey. *Renewable & Sustainable Energy Reviews, 192*, 114257. DOI: 10.1016/j.rser.2023.114257

Rao, S., Katoch, S., Narayanaswamy, V., Muniraju, G., Tepedelenlioglu, C., & Spanias, A. (2022). *Machine learning for solar array monitoring, optimization, and control*. Springer Nature.

Rasheed, M. Q., Yuhuan, Z., Haseeb, A., Ahmed, Z., & Saud, S. (2024). Asymmetric relationship between competitive industrial performance, renewable energy, industrialization, and carbon footprint: Does artificial intelligence matter for environmental sustainability? *Applied Energy, 367*, 123346. DOI: 10.1016/j.apenergy.2024.123346

Rathore, B. (2017). Beyond Trends: Shaping the Future of Fashion Marketing with AI, Sustainability and Machine Learning. *Eduzone: International Peer Reviewed/Refereed Multidisciplinary Journal, 6*(2), 16-24.

Rathore, B. (2019). Fashion Sustainability in the AI Era: Opportunities and Challenges in Marketing. *Eduzone: International Peer Reviewed/Refereed Multidisciplinary Journal, 8*(2), 17-24.

Rathore, B. (2023). Digital Transformation 4.0: Integration of Artificial Intelligence & Metaverse in Marketing. *Eduzone: International Peer Reviewed/Refereed Multidisciplinary Journal, 12*(1), 42-48.

Rathore, B. (2019). From Trendy to Green: Exploring AI's Role in Sustainable Fashion Marketing. *International Journal of New Media Studies: International Peer Reviewed Scholarly Indexed Journal, 6*(2), 12–22. DOI: 10.58972/eiprmj.v6i2y19.120

Rathore, N., Panwar, N. L., Yettou, F., & Gama, A. (2021). A comprehensive review of different types of solar photovoltaic cells and their applications. *International Journal of Ambient Energy, 42*(10), 1200–1217. DOI: 10.1080/01430750.2019.1592774

Raukas, A. (2010). Sustainable development and environmental risks in Estonia. [Special Issue II]. *Agronomy Research (Tartu), 8*, 351–356.

Raut, R. D., Mangla, S. K., Narwane, V. S., Gardas, B. B., Priyadarshinee, P., & Narkhede, B. E. (2019). Linking big data analytics and operational sustainability practices for sustainable business management. *Journal of Cleaner Production, 224*, 10–24. DOI: 10.1016/j.jclepro.2019.03.181

Ravindran, M. (2022). Energy Efficient Clustering for Precision Agriculture using LEACH Protocol. *NeuroQuantology : An Interdisciplinary Journal of Neuroscience and Quantum Physics, 20*(7), 2702.

Ravivarman, G., Madhavan, V. M., Kalaiyarasan, A., Hemasilviavinothini, S., Nithyanandhan, T., Kesavan, S. P., & Girimurugan, R. (2023). Enhancing efficiency and cost-effectiveness of solar stills through natural fiber integration. In *E3S Web of Conferences*, 455, 02003. DOI: 10.1051/e3sconf/202345502003

Reddy, K., & Singh, P. (2024). Application of fuzzy logic for optimizing distributed renewable energy resources. *Renewable & Sustainable Energy Reviews, 22*(3), 512–525. DOI: 10.1016/j.rser.2024.07.011

Reedman, L., Chew, M. S., Gordon, J., Sue, W., Brinsmead, T., Hayward, J., & Havas, L. (2021). Multi-sector energy modelling.

Regeringskansliet (2005). *Strategic Challenges. A Further Elaboration of the Swedish Strategy for Sustainable Development.* Ministry of Sustainable Development.

Rein, G., & Huang, X. (2021). Smouldering wildfires in peatlands, forests and the arctic: Challenges and perspectives. *Current Opinion in Environmental Science & Health, 24*, 100296. DOI: 10.1016/j.coesh.2021.100296 PMID: 34950823

Remy, N., Speelman, E., & Swartz, S. (2016): Style that's sustainable: A new fast-fashion formula. https://www.mckinsey.com/capabilities/sustainability/our-insights/style-thats-sustainable-a-new-fast-fashion-formula

Ren, Y., & Liu, W. (2023). Federated learning-based approaches for optimizing energy management systems. *Energy, 37*(7), 610–623. DOI: 10.1016/j.energy.2023.09.006

Republic of Bulgaria Council of Ministers. (2001). National Development Programme: Bulgaria 2020. Ministry of Finance of the Republic of Bulgaria.

Reuter, K., Wilson, M. L., Moran, M., Le, N., Angyan, P., Majmundar, A., Kaiser, E. M., & Unger, J. B. (2021). General audience engagement with antismoking public health messages across multiple social media sites: Comparative analysis. *JMIR Public Health and Surveillance, 7*(2), e24429. DOI: 10.2196/24429 PMID: 33605890

Reviglio della Venaria, U. (2020). Personalization in Social Media: Challenges and Opportunities for Democratic Societies.

Rhoades, E. B., Irani, T., & Telg, R. (2005). Assessing internet use in Florida newsrooms. *Journal of Applied Communications, 89*(2). Advance online publication. DOI: 10.4148/1051-0834.1304

Riahi, Y., Saikouk, T., Gunasekaran, A., & Badraoui, I. (2021). Artificial intelligence applications in supply chain: A descriptive bibliometric analysis and future research directions. *Expert Systems with Applications, 173*, 114702. DOI: 10.1016/j.eswa.2021.114702

Rizos, V., & Bryhn, J. (2022). Implementation of circular economy approaches in the electrical and electronic equipment (EEE) sector: Barriers, enablers and policy insights. [Google Scholar] [CrossRef]. *Journal of Cleaner Production*, *338*, 130617. DOI: 10.1016/j.jclepro.2022.130617

Robichaud, P. R., & Ashmun, L. E. (2013). Tools to aid post-wildfire assessment and erosion-mitigation treatment decisions. *International Journal of Wildland Fire*, *22*(1), 95. DOI: 10.1071/WF11162

Rodrigues, D., & Pereira, R. (2023). How Can Artificial Intelligence Help Improve Fashion Sustainability? In *Digital Technologies and Transformation in Business, Industry and Organizations* (Vol. 2, pp. 43–61). Springer Nature Switzerland.

Rodrigues, M., & De La Riva, J. (2014). An insight into machine-learning algorithms to model human-caused wildfire occurrence. *Environmental Modelling & Software*, *57*, 192–201. DOI: 10.1016/j.envsoft.2014.03.003

Rojek, I., Mroziński, A., Kotlarz, P., Macko, M., & Mikołajewski, D. (2023). AI-Based Computational Model in Sustainable Transformation of Energy Markets. *Energies*, *16*(24), 8059. Advance online publication. DOI: 10.3390/en16248059

Ronchi, E., Federico, A., & Musmeci, F. (2002). A system oriented integrated indicator for sustainable development in Italy. *Ecological Indicators*, *2*(1-2), 197–210. DOI: 10.1016/S1470-160X(02)00045-6

Rosiello, A., & Maleki, A. (2021). A dynamic multi-sector analysis of technological catch-up: The impact of technology cycle times, knowledge base complexity and variety. *Research Policy*, *50*(3), 104194. DOI: 10.1016/j.respol.2020.104194

Roy, S., & Chakraborty, S. (2024). Fuzzy logic controllers for improving photovoltaic energy systems. *Renewable Energy*, *28*(4), 671–684. DOI: 10.1016/j.renene.2024.05.014

Rudin, C. (2019). Stop explaining black box machine learning models for high stakes decisions and use interpretable models instead. *Nature Machine Intelligence*, *1*(5), 206–215. PMID: 35603010

Rudko, V., Tsoutsos, T., Rezk, H., Olabi, A., Abdelkareem, M., Alahmer, A., & Sayed, E. (2023). Maximizing Green Hydrogen Production from Water Electrocatalysis: Modeling and Optimization. *Journal of Marine Science and Engineering*, *11*(3), 617. Advance online publication. DOI: 10.3390/jmse11030617

Rudrakar, S., & Rughani, P. (2022). *IoT based Agriculture (IoTA)*. Architecture, Cyber Attack, Cyber Crime and Digital Forensics Challenges.

Rusilowati, U., Ngemba, H. R., Anugrah, R. W., Fitriani, A., & Astuti, E. D. (2024). Leveraging AI for Superior Efficiency in Energy Use and Development of Renewable Resources such as Solar Energy, Wind, and Bioenergy. [ITALIC]. *International Transactions on Artificial Intelligence*, *2*(2), 114–120. DOI: 10.33050/italic.v2i2.537

Ruževičius, J. (2009). Development of Sustainable and Socially Responsible Business in Lithuania in the International Context. *Ekonomika (Nis)*, *86*, 68–80.

Saadat, M. N., & Shuaib, M. (2020). Advancements in deep learning theory and applications: Perspective in 2020 and beyond. *Advances and Applications in Deep Learning, 3*.

Saatloo, A., Mehrabi, A., Marzband, M., Mirzaei, M., & Aslam, N. (2024). Local Energy Market Design for Power- and Hydrogen-Based Microgrids Considering a Hybrid Uncertainty Controlling Approach. *IEEE Transactions on Sustainable Energy*, *15*(1), 398–413. DOI: 10.1109/TSTE.2023.3288745

Saba, T., Rehman, A., Haseeb, K., Bahaj, S. A., & Lloret, J. (2023). Trust-based decentralized blockchain system with machine learning using Internet of agriculture things. *Computers & Electrical Engineering*, *108*, 108674. DOI: 10.1016/j.compeleceng.2023.108674

Sadowski, J. (2023). Total life insurance: Logics of anticipatory control and actuarial governance in insurance technology. *Social Studies of Science*, •••, 03063127231186437. PMID: 37427796

Saeedmanesh, A., Kinnon, M., & Brouwer, J. (2018). Hydrogen is essential for sustainability. *Current Opinion in Electrochemistry*, *12*, 166–181. Advance online publication. DOI: 10.1016/j.coelec.2018.11.009

Saeima of the Republic of Latvia (2010). *Sustainable Development Strategy of Latvia until 2030. Latvija 2030.* Saeima of the Republic of Latvia.

Saha, M., & Roy, A. (2024). Federated learning for enhancing renewable energy forecasting accuracy. *IEEE Transactions on Smart Grid*, *39*(2), 845–858. DOI: 10.1109/TSG.2024.0258749

Sahar, S. N., & Dar, H. (2024). Artificial Intelligence-Enhanced Global Healthcare: The Future of Medical Tourism. In *Impact of AI and Robotics on the Medical Tourism Industry* (pp. 194-216). IGI Global.

Saheb, T., Dehghani, M., & Saheb, T. (2022). Artificial intelligence for sustainable energy: A contextual topic modeling and content analysis. *Sustainable Computing : Informatics and Systems*, *35*, 100699. DOI: 10.1016/j.suscom.2022.100699

Sahu, R., & Sharma, M. (2023). Fuzzy logic-based energy management systems for microgrids. *Sustainable Energy Technologies and Assessments*, *17*(3), 579–592. DOI: 10.1016/j.seta.2023.05.013

Saini, G., Kannan, R., Benini, E., & Kumar, K. (Eds.). (2023). *Enabling Methodologies for Renewable and Sustainable Energy*. CRC Press. DOI: 10.1201/9781003272717

Sajtos, L., & Mitev, A. (2007). *SPSS Kutatási és adatelemzési kézikönyv*. Alinea Kiadó.

Salam, A., & Salam, A. (2020). Internet of things in sustainable energy systems. *Internet of Things for Sustainable Community Development: Wireless Communications. Sensory Systems*, •••, 183–216.

Salleh, S. F., Mohd Roslan, M. E., Abd Rahman, A., Shamsuddin, A. H., Tuan Abdullah, T. A. R., & Sovacool, B. K. (2020). Transitioning to a sustainable development framework for bioenergy in Malaysia: Policy suggestions to catalyse the utilisation of palm oil mill residues. *Energy, Sustainability and Society*, *10*(1), 1–20. DOI: 10.1186/s13705-020-00269-y

Sampaio, P. G. V., & González, M. O. A. (2017). Photovoltaic solar energy: Conceptual framework. *Renewable & Sustainable Energy Reviews*, *74*, 590–601. DOI: 10.1016/j.rser.2017.02.081

Saponaro, M., Le Gal, D., Gao, M., Guisiano, M., & Maniere, I. C. (2018, December). Challenges and opportunities of artificial intelligence in the fashion world. In *2018 international conference on intelligent and innovative computing applications (ICONIC)* (pp. 1-5). IEEE. DOI: 10.1109/ICONIC.2018.8601258

Sapra, L., Mathani, A., & Bhatnagar, G. (2022). A thorough Analysis of Blockchain's Potential for Internet of Things Applications in Precision Agricultural Networks. *Mathematical Statistician and Engineering Applications*, *71*(4), 4141–4159.

Sarker, A., Azad, A., Rasul, M., & Doppalapudi, A. (2023). Prospect of Green Hydrogen Generation from Hybrid Renewable Energy Sources: A Review. *Energies*, *16*(3), 1556. Advance online publication. DOI: 10.3390/en16031556

Sayed, E. T., Olabi, A. G., Alami, A. H., Radwan, A., Mdallal, A., Rezk, A., & Abdelkareem, M. A. (2023). Renewable energy and energy storage systems. *Energies*, *16*(3), 1415. DOI: 10.3390/en16031415

Scarlat, N., Dallemand, J. F., Monforti-Ferrario, F., & Nita, V. (2015). The role of biomass and bioenergy in a future bioeconomy: Policies and facts. *Environmental Development*, *15*, 3–34. DOI: 10.1016/j.envdev.2015.03.006

Schaeffer, E. (2017). *Industry X. 0: Realizing digital value in industrial sectors*. Kogan Page Publishers.

Schwerhoff, G., & Sy, M. (2017). Financing renewable energy in Africa–Key challenge of the sustainable development goals. *Renewable & Sustainable Energy Reviews*, *75*, 393–401. DOI: 10.1016/j.rser.2016.11.004

Sedkaoui, S. (Ed.). (2018). *Big data analytics for entrepreneurial success*. IGI Global.

Sendros, A., Drosatos, G., Efraimidis, P. S., & Tsirliganis, N. C. (2022). Blockchain applications in agriculture: A scoping review. *Applied Sciences (Basel, Switzerland)*, *12*(16), 8061. DOI: 10.3390/app12168061

Sengupta, A., Gill, S. S., Das, A., & De, D. (2021). Mobile edge computing based internet of agricultural things: a systematic review and future directions. *Mobile Edge Computing*, 415-441.

Sengupta, U., Pramanik, H. S., Dutta, S., Dutta, S., Dasgupta, S., & Kirtania, M. (2023). Assessing sustainability focus across global banks. *Development Engineering*, *8*, 100114. DOI: 10.1016/j.deveng.2023.100114

Sghir, N., Adadi, A., Mouden, Z. E., & Lahmer, M. (2022). Using learning analytics to improve students' enrollments in higher education. 2022 2nd International Conference on Innovative Research in Applied Science, Engineering and Technology (IRASET). DOI: 10.1109/IRASET52964.2022.9737993

Shah, F., Li, J., Shah, Y., & Shah, F. (2017, November). Broad big data domain via medical big data. In *2017 4th International Conference on Systems and Informatics (ICSAI)* (pp. 732-737). IEEE.

Shahid, Z. (2021). Distributed Machine Learning for Anomalous Human Activity Recognition using IoT Systems.

Shang, Y., Sang, S., Tiwari, A. K., Khan, S., & Zhao, X. (2024). Impacts of renewable energy on climate risk: A global perspective for energy transition in a climate adaptation framework. *Applied Energy*, *362*, 122994. DOI: 10.1016/j.apenergy.2024.122994

Shao, J., & Li, G. (2024). Federated learning for distributed control in renewable energy systems. *IEEE Transactions on Sustainable Energy*, *47*(1), 369–382. DOI: 10.1109/TSTE.2024.0123569

Sharma, A., & Singh, B. (2022). Measuring Impact of E-commerce on Small Scale Business: A Systematic Review. *Journal of Corporate Governance and International Business Law*, *5*(1).

Sharma, N., & Kumar, P. (2024). A fuzzy logic approach for optimizing solar energy storage. *Renewable Energy*, *29*(3), 401–414. DOI: 10.1016/j.renene.2024.02.007

Sharma, S. R., & Karki, N. (2014). Towards culturally sensitive public health interventions in Nepal. *Health Prospect : Journal of Public Health*, *13*(1), 1–3. DOI: 10.3126/hprospect.v13i1.11104

Shell. (2024). Shell's net-zero commitment: Transitioning to renewable energy. Shell Corporate Sustainability Report.

Shen, B., Ding, X., Wang, Y., & Ren, S. (2019). RFID-embedded smart washing machine systems in the big data era: Value creation in fashion supply chain. *Fashion supply chain management in Asia: Concepts, models, and cases*, 99-113.

Shin, W., Han, J., & Rhee, W. (2021). AI-assistance for predictive maintenance of renewable energy systems. *Energy*, *221*, 119775. DOI: 10.1016/j.energy.2021.119775

Shi, X., An, X., Zhao, Q., Liu, H., Xia, L., Sun, X., & Guo, Y. (2019). State-of-the-art internet of things in protected agriculture. *Sensors (Basel)*, *19*(8), 1833. DOI: 10.3390/s19081833 PMID: 30999637

Shi, Z., Zhu, J., & Wei, H. (2022). SARSA-based delay-aware route selection for SDN-enabled wireless-PLC power distribution IoT. [Google Scholar] [CrossRef]. *Alexandria Engineering Journal*, *61*(7), 5795–5803. DOI: 10.1016/j.aej.2021.11.029

Shubbak, M. H. (2019). Advances in solar photovoltaics: Technology review and patent trends. *Renewable & Sustainable Energy Reviews*, *115*, 109383. DOI: 10.1016/j.rser.2019.109383

Silva Barbosa, G., Drach, P. R., & Corbella, O. D. (2014). A Conceptual Review of the Terms Sustainable Development and Sustainability. *The International Journal of Social Sciences (Islamabad)*, *3*(2), 1–15.

Silva, B. V., Holm-Nielsen, J. B., Sadrizadeh, S., Teles, M. P., Kiani-Moghaddam, M., & Arabkoohsar, A. (2023). Sustainable, green, or smart? Pathways for energy-efficient healthcare buildings. *Sustainable Cities and Society*, •••, 105013.

Silvestrini, S., & Lavagna, M. (2022). Deep learning and artificial neural networks for spacecraft dynamics, navigation, and control. [Google Scholar] [CrossRef]. *Drones (Basel)*, *6*(9), 270. DOI: 10.3390/drones6100270

Singh, B. (2023). Blockchain Technology in Renovating Healthcare: Legal and Future Perspectives. In *Revolutionizing Healthcare Through Artificial Intelligence and Internet of Things Applications* (pp. 177-186). IGI Global.

Singh, B. (n.d.). Digital Twins as New Paradigm for Bridging the Gap from Personalized Medicine to Specific Rural Public Health: Exploring Foundations, Legal Angels, and Technologies for Health 5.0—Future Research Directions. In *Artificial Intelligence Solutions for Cyber-Physical Systems* (pp. 286-309). Auerbach Publications. https://www.taylorfrancis.com/chapters/edit/10.1201/9781032694375-16/digital-twins-new-paradigm-bridging-gap-personalized-medicine-specific-rural-public-health-bhupinder-singh

Singh, B., & Kaunert, C. (2024). Blue Ocean and Machine Learning Trajectories SDG 14: Life Below Water for Handling Ocean Pollution: Metaverse Conserve Ocean Health Sustainability Through the Lens of Transboundary Legal-Policy Regulations as Articulating Space for Futuristic Changes. In *Artificial Intelligence and Edge Computing for Sustainable Ocean Health*. The Springer Series in Applied Machine Learning. Springer. DOI: 10.1007/978-3-031-64642-3_4

Singh, B., & Kaunert, C. (2024). Harnessing Sustainable Agriculture Through Climate-Smart Technologies: Artificial Intelligence for Climate Preservation and Futuristic Trends. In *Exploring Ethical Dimensions of Environmental Sustainability and Use of AI* (pp. 214-239). IGI Global.

Singh, B., & Kaunert, C. (2024). Revealing Green Finance Mobilization: Harnessing FinTech and Blockchain Innovations to Surmount Barriers and Foster New Investment Avenues. In *Harnessing Blockchain-Digital Twin Fusion for Sustainable Investments* (pp. 265-286). IGI Global.

Singh, B., & Kaunert, C. (2024). Trans-Disciplinary Collaborative Disciplines Generating Cohesive Curriculum: Runway for Innovation, Cooperation, and Intentionality in Higher Education. In R. Kumar, E. Ong, S. Anggoro, T. Toh, & M. Fukui (Eds.), *Transdisciplinary Teaching and Technological Integration for Improved Learning: Case Studies and Practical Approaches* (pp. 152-176). IGI Global. DOI: 10.4018/979-8-3693-8217-2.ch008

Singh, B., & Kaunert, C. (2025). Cloud Computing and IoMT in Disease Screening and Diagnosis: AI Approaches in Transmuting Healthier Homes. In B. S & S. Kadry (Eds.), *Revolutionizing Healthcare Systems Through Cloud Computing and IoT* (pp. 99-120). IGI Global. DOI: 10.4018/979-8-3693-7225-8.ch005

Singh, B., & Kaunert, C. (2025). Leveraging IoT for Patient Monitoring and Smart Healthcare: Connected Healthcare System. In B. S & S. Kadry (Eds.), *Revolutionizing Healthcare Systems Through Cloud Computing and IoT* (pp. 27-46). IGI Global. https://doi.org/DOI: 10.4018/979-8-3693-7225-8.ch002

Singh, B., & Kaunert, C. (n.d.). Tagging Blockchain Technology Fostering Panacea for Data Privacy, Cloud Computing, and Integrity: Legal Framework in India and at a Globe. In *Artificial Intelligence Solutions for Cyber-Physical Systems* (pp. 206-213). Auerbach Publications. https://www.taylorfrancis.com/chapters/edit/10.1201/9781032694375-11/tagging-blockchain-technology-fostering-panacea-data-privacy-cloud-computing-integrity-bhupinder-singh-christian-kaunert

Singh, P., Kumar, A., & Chopra, M. (2023). Real-World Applications of BC Technology in Internet of Things. *Machine Learning Applications: From Computer Vision to Robotics*, 97-122.

Singh, B. (2022). Relevance of Agriculture-Nutrition Linkage for Human Healthcare: A Conceptual Legal Framework of Implication and Pathways. *Justice and Law Bulletin*, *1*(1), 44–49.

Singh, B. (2023). Blockchain Technology in Renovating Healthcare: Legal and Future Perspectives. In Kaushik, K., Dahiya, S., Aggarwal, S., & Dwivedi, A. (Eds.), *Revolutionizing Healthcare Through Artificial Intelligence and Internet of Things Applications* (pp. 177–186). IGI Global. DOI: 10.4018/978-1-6684-5422-0.ch012

Singh, B. (2023). Federated Learning for Envision Future Trajectory Smart Transport System for Climate Preservation and Smart Green Planet: Insights into Global Governance and SDG-9 (Industry, Innovation and Infrastructure). *National Journal of Environmental Law*, *6*(2), 6–17.

Singh, B. (2024). Cherish Data Privacy and Human Rights in the Digital Age: Harmonizing Innovation and Individual Autonomy. In Pucelj, M., & Bohinc, R. (Eds.), *Balancing Human Rights, Social Responsibility, and Digital Ethics* (pp. 199–226). IGI Global. DOI: 10.4018/979-8-3693-3334-1.ch007

Singh, B. (2024). Evolutionary Global Neuroscience for Cognition and Brain Health: Strengthening Innovation in Brain Science. In Prabhakar, P. (Ed.), *Biomedical Research Developments for Improved Healthcare* (pp. 246–272). IGI Global. DOI: 10.4018/979-8-3693-1922-2.ch012

Singh, B. (2024). Lensing Legal Dynamics for Examining Responsibility and Deliberation of Generative AI-Tethered Technological Privacy Concerns: Infringements and Use of Personal Data by Nefarious Actors. In Ara, A., & Ara, A. (Eds.), *Exploring the Ethical Implications of Generative AI* (pp. 146–167). IGI Global., DOI: 10.4018/979-8-3693-1565-1.ch009

Singh, B. (2024). Social Cognition of Incarcerated Women and Children: Addressing Exposure to Infectious Diseases and Legal Outcomes. In Reddy, K. (Ed.), *Principles and Clinical Interventions in Social Cognition* (pp. 236–251). IGI Global., DOI: 10.4018/979-8-3693-1265-0.ch014

Singh, B., Dutta, P. K., Gautam, R., & Kaunert, C. (2024). Uncapping the Potential of Quantum Computing Towards Manufacturing Optimization: Routing Supply Chain Projecting Sustainability. In Hassan, A., Bhattacharya, P., Dutta, P., Verma, J., & Kundu, N. (Eds.), *Quantum Computing and Supply Chain Management: A New Era of Optimization* (pp. 395–419). IGI Global. DOI: 10.4018/979-8-3693-4107-0.ch024

Singh, B., Dutta, P. K., & Kaunert, C. (2024). Replenish Artificial Intelligence in Renewable Energy for Sustainable Development: Lensing SDG 7 Affordable and Clean Energy and SDG 13 Climate Actions With Legal-Financial Advisory. In Derbali, A. (Ed.), *Social and Ethical Implications of AI in Finance for Sustainability* (pp. 198–227). IGI Global., DOI: 10.4018/979-8-3693-2881-1.ch009

Singh, B., Jain, V., Kaunert, C., Dutta, P. K., & Singh, G. (2024). Privacy Matters: Espousing Blockchain and Artificial Intelligence (AI) for Consumer Data Protection on E-Commerce Platforms in Ethical Marketing. In Saluja, S., Nayyar, V., Rojhe, K., & Sharma, S. (Eds.), *Ethical Marketing Through Data Governance Standards and Effective Technology* (pp. 167–184). IGI Global., DOI: 10.4018/979-8-3693-2215-4.ch015

Singh, B., & Kaunert, C. (2024). Adventure in High Altitude of Mountainous Topographies and Health Impacts: Lensing Tourism Sustainability via Reducing Ecological and Sociocultural Footprint and Health Emergency and Medical Assistance Management. In Meraj, G., Hashimoto, S., & Kumar, P. (Eds.), *Navigating Natural Hazards in Mountainous Topographies. Disaster Risk Reduction.* Springer., DOI: 10.1007/978-3-031-65862-4_15

Singh, B., & Kaunert, C. (2024). AI-Driven Strategies for Customer Engagement, Market Segmentation, and Resource Optimization: Projecting End User Satisfaction and Futuristic Growth of Business. In Nadda, V., Tyagi, P., Singh, A., & Singh, V. (Eds.), *AI Innovations in Service and Tourism Marketing* (pp. 104–128). IGI Global. DOI: 10.4018/979-8-3693-7909-7.ch006

Singh, B., & Kaunert, C. (2024). Aroma of Highly Smart Internet of Medical Things (IoMT) and Lightweight EdgeTrust Expansion Medical Care Facilities for Electronic Healthcare Systems: Fortified-Chain Architecture for Remote Patient Monitoring and Privacy Protection Beyond Imagination. In Hassan, A., Bhattacharya, P., Tikadar, S., Dutta, P., & Sagayam, M. (Eds.), *Lightweight Digital Trust Architectures in the Internet of Medical Things (IoMT)* (pp. 196–212). IGI Global., DOI: 10.4018/979-8-3693-2109-6.ch011

Singh, B., & Kaunert, C. (2024). Augmented Reality and Virtual Reality Modules for Mindfulness: Boosting Emotional Intelligence and Mental Wellness. In Hiran, K., Doshi, R., & Patel, M. (Eds.), *Applications of Virtual and Augmented Reality for Health and Wellbeing* (pp. 111–128). IGI Global., DOI: 10.4018/979-8-3693-1123-3.ch007

Singh, B., & Kaunert, C. (2024). Computational Thinking for Innovative Solutions and Problem-Solving Techniques: Transforming Conventional Education to Futuristic Interdisciplinary Higher Education. In Fonkam, M., & Vajjhala, N. (Eds.), *Revolutionizing Curricula Through Computational Thinking, Logic, and Problem Solving* (pp. 60–82). IGI Global., DOI: 10.4018/979-8-3693-1974-1.ch004

Singh, B., & Kaunert, C. (2024). Embryonic Machine-Deep Learning, Smart Healthcare and Privacy Deliberations in Hospital Industry: Lensing Confidentiality of Patient's Information and Personal Data in Legal-Ethical Landscapes Projecting Futuristic Dimensions. In Kumar, P., Singh, P., Diwakar, M., & Garg, D. (Eds.), *Healthcare Industry Assessment: Analyzing Risks, Security, and Reliability. Engineering Cyber-Physical Systems and Critical Infrastructures* (Vol. 11). Springer. DOI: 10.1007/978-3-031-65434-3_7

Singh, B., & Kaunert, C. (2024). Harnessing Sustainable Agriculture Through Climate-Smart Technologies: Artificial Intelligence for Climate Preservation and Futuristic Trends. In Kannan, H., Rodriguez, R., Paprika, Z., & Ade-Ibijola, A. (Eds.), *Exploring Ethical Dimensions of Environmental Sustainability and Use of AI* (pp. 214–239). IGI Global. DOI: 10.4018/979-8-3693-0892-9.ch011

Singh, B., & Kaunert, C. (2024). Hidden Gems Breakthrough Dynamic Landscape of Adaptive Learning Technologies for Higher Education: Bridging the Gap Between Theoretical and Practical Knowledge Projecting Student Learning Outcomes. In Minh Tung, T. (Ed.), *Adaptive Learning Technologies for Higher Education* (pp. 222–247). IGI Global. DOI: 10.4018/979-8-3693-3641-0.ch010

Singh, B., & Kaunert, C. (2024). Hitching Artificial Intelligence and IoT and for Livestock Management Competence: Line Up Nature and Technology for Smart Agriculture. In Zia Ul Haq, M., & Ali, I. (Eds.), *Revolutionizing Pest Management for Sustainable Agriculture* (pp. 499–520). IGI Global. DOI: 10.4018/979-8-3693-3061-6.ch020

Singh, B., & Kaunert, C. (2024). Human-Machine Nexus for Digital Rebound Fostering Futuristic Energy-Efficiency: Ecological Footprint of Technology for Smart-Sustainable Urban Mobility. In González-Lezcano, R., & Sansaniwal, S. (Eds.), *Design Strategies for Efficient and Sustainable Building Facilities* (pp. 222–244). IGI Global. DOI: 10.4018/979-8-3693-3200-9.ch010

Singh, B., & Kaunert, C. (2024). Integration of Cutting-Edge Technologies such as Internet of Things (IoT) and 5G in Health Monitoring Systems: A Comprehensive Legal Analysis and Futuristic Outcomes. *GLS Law Journal*, *6*(1), 13–20.

Singh, B., & Kaunert, C. (2024). Redefining the Nodes of Metaverse and Luxury Hospitality: Unveiling AI and Tech-Driven Solutions in Hospitality and Tourism Industry. In Talukder, M., Kumar, S., & Tyagi, P. (Eds.), *Impact of AI and Tech-Driven Solutions in Hospitality and Tourism* (pp. 175–202). IGI Global. DOI: 10.4018/979-8-3693-6755-1.ch010

Singh, B., & Kaunert, C. (2024). Revealing Green Finance Mobilization: Harnessing FinTech and Blockchain Innovations to Surmount Barriers and Foster New Investment Avenues. In Jafar, S., Rodriguez, R., Kannan, H., Akhtar, S., & Plugmann, P. (Eds.), *Harnessing Blockchain-Digital Twin Fusion for Sustainable Investments* (pp. 265–286). IGI Global. DOI: 10.4018/979-8-3693-1878-2.ch011

Singh, B., & Kaunert, C. (2024). Salvaging Responsible Consumption and Production of Food in the Hospitality Industry: Harnessing Machine Learning and Deep Learning for Zero Food Waste. In Singh, A., Tyagi, P., & Garg, A. (Eds.), *Sustainable Disposal Methods of Food Wastes in Hospitality Operations* (pp. 176–192). IGI Global. DOI: 10.4018/979-8-3693-2181-2.ch012

Singh, B., & Kaunert, C. (2024). Transforming Smart City and Smart Sustainable Tourism Projecting Artificial Intelligence and IoT: Immersive Potential for Revolutionizing Urban Living and Enhancing Travel Experience. In Jermsittiparsert, K., & Suanpang, P. (Eds.), *Technological and Managerial Approaches to Fostering Sustainable Travel* (pp. 55–80). IGI Global. DOI: 10.4018/979-8-3693-8115-1.ch003

Singh, B., & Kaunert, C. (2024). Vertical Assimilation of Artificial Intelligence and Machine Learning in Safeguarding Financial Data. In Naz, F., & Karim, S. (Eds.), *Safeguarding Financial Data in the Digital Age* (pp. 169–197). IGI Global. DOI: 10.4018/979-8-3693-3633-5.ch010

Singh, B., & Kaunert, C. (2024). Wind and Solar Energy as Renewable Energy for Sustainable Global Future: Projecting Future Multi-Sector Sustainable Policies and Innovation. In Ara, A., & Thakore, R. (Eds.), *Promoting Multi-Sector Sustainability With Policy and Innovation* (pp. 210–245). IGI Global., DOI: 10.4018/979-8-3693-2113-3.ch009

Singh, B., & Kaunert, C. (2025). Blockchain Framework for Precision Medicine, Clinical Trials, and Genomic Biomarkers Discovery: Focuses on Monitoring and Managing Data in Health Dimensions. In Jermsittiparsert, K., & Marzo, R. (Eds.), *AI Technologies and Advancements for Psychological Well-Being and Healthcare* (pp. 255–288). IGI Global. DOI: 10.4018/979-8-3693-9158-7.ch010

Singh, B., & Kaunert, C. (2025). Dynamic Landscape of Artificial General Intelligence (AGI) for Advancing Renewable Energy in Urban Environments: Synergies with SDG 11—Sustainable Cities and Communities Lensing Policy and Governance. In El Hajjami, S., Kaushik, K., & Khan, I. U. (Eds.), *Artificial General Intelligence (AGI) Security. Advanced Technologies and Societal Change*. Springer. DOI: 10.1007/978-981-97-3222-7_12

Singh, B., & Kaunert, C. (2025). Featuring Healthcare, Environment, and Human Rights: Applications of Artificial Intelligence in the Health Domain. In Chakraborty, S., & Satapathy, S. (Eds.), *Gender, Environment, and Human Rights: An Intersectional Exploration* (pp. 45–74). IGI Global., DOI: 10.4018/979-8-3693-6069-9.ch004

Singh, B., & Kaunert, C. (2025). Innovative Business Models in Digital and Virtual Reality for the Gig Economy: Space-Age Sustainable Development Projecting Smart Business Practices. In Taherdoost, H., Drazenovic, G., Madanchian, M., Khan, I., & Arshi, O. (Eds.), *Business Transformation in the Era of Digital Disruption* (pp. 127–156). IGI Global., DOI: 10.4018/979-8-3693-7056-8.ch005

Singh, B., & Kaunert, C. (2025). Intelligent Machine Learning Solutions for Cybersecurity: Legal and Ethical Considerations in a Global Context. In Thangam, D. (Ed.), *Advancements in Intelligent Process Automation* (pp. 359–386). IGI Global., DOI: 10.4018/979-8-3693-5380-6.ch014

Singh, B., & Kaunert, C. FedHealth in Wearable Healthcare, Orchestrated Federated Deep Learning for Smart Healthcare: Health Monitoring and Healthcare Informatics Lensing Challenges and Future Directions. In *Federated Deep Learning for Healthcare* (pp. 207–223). CRC Press. https://www.taylorfrancis.com/chapters/edit/10.1201/9781032694870-16/fedhealth-wearable-healthcare-orchestrated-federated-deep-learning-smart-healthcare-bhupinder-singh-christian-kaunert

Singh, B., & Kaunert, C. Marvel of Biosensors in Smart Healthcare and Application of Internet of Medical Things for Diagnosis: Unveiling Benefits, Challenges and Futuristic Approach. In *Smart Healthcare Systems* (pp. 187–198). CRC Press. https://www.taylorfrancis.com/chapters/edit/10.1201/9781032698519-14/marvel-biosensors-smart-healthcare-application-internet-medical-things-diagnosis-bhupinder-singh-christian-kaunert

Singh, B., Kaunert, C., & Gautam, R. (2024). Harnessing Machine Learning Intelligence Against Cyber Threats: Navigating Legal and Ethical Challenges in Global Cyber Investigations. In Özsungur, F. (Ed.), *Empowering Entrepreneurial Mindsets With AI* (pp. 319–352). IGI Global. DOI: 10.4018/979-8-3693-7658-4.ch016

Singh, B., Kaunert, C., & Kaushik, T. K. (2024). Unscrambling Financial Fraud With AI and Machine Learning in E-Commerce Transactions: Airing Into Ad Clicks, Credit Card Management. In Pandow, B., Masoodi, F., Iqbal, J., & Hussain, G. (Eds.), *Navigating the Future of Finance in the Age of AI* (pp. 253–271). IGI Global. DOI: 10.4018/979-8-3693-4382-1.ch013

Singh, B., Kaunert, C., & Malviya, R. (2024). Coupling Human-Computer Interface Lensing Artificial Emotional Intelligence: Transforming for Futuristic Cognitive Recognition for Decision-Marking in Industry 5.0. In Kumar, N., Pal, S., Agarwal, P., Rosak-Szyrocka, J., & Jain, V. (Eds.), *Human-Machine Collaboration and Emotional Intelligence in Industry 5.0* (pp. 233–250). IGI Global. DOI: 10.4018/979-8-3693-6806-0.ch013

Singh, B., Kaunert, C., & Malviya, R. (2024). Featuring Machine Learning in Municipal Solid Waste Management: Technological Advancements in Waste Sorting and Processing Ropeway for SDG 13 (Climate Action). In Mehra, R., Kumar, P., & Hung, B. (Eds.), *Municipal Solid Waste Management and Recycling Technologies* (pp. 343–366). IGI Global. DOI: 10.4018/979-8-3693-4054-7.ch013

Singh, B., Kaunert, C., Malviya, R., Gautam, R., & Singh, G. (2024). Generative Artificial Intelligence in the Tourism-Hospitality Industry: Strengthening Customer Relationships, Delivering Better Experiences. In Talukder, M., Kumar, S., & Tyagi, P. (Eds.), *Hotel and Travel Management in the AI Era* (pp. 275–300). IGI Global. DOI: 10.4018/979-8-3693-7898-4.ch013

Singh, B., Kaunert, C., Malviya, R., Lal, S., & Arora, M. K. (2024). Scrutinizing Consumer Sentiment on Social Media and Data-Driven Decisions for Business Insights: Fusion of Artificial Intelligence (AI) and Business Intelligence (BI) Foster Sustainable Growth. In Natarajan, A., Galety, M., Iwendi, C., Das, D., & Shankar, A. (Eds.), *Intersection of AI and Business Intelligence in Data-Driven Decision-Making* (pp. 183–210). IGI Global. DOI: 10.4018/979-8-3693-5288-5.ch007

Singh, B., Kaunert, C., & Ravesangar, K. (2025). Artificial Intelligence and Employee Well-Being: Projecting Psychological and Physical Healthcare. In Yadav, P. (Ed.), *Human Resource Strategies in the Era of Artificial Intelligence* (pp. 109–134). IGI Global., DOI: 10.4018/979-8-3693-6412-3.ch005

Singh, B., Kaunert, C., & Singh, G. (2025). Scaling Legal Framework for Plastic Pollution and Advancing Cutting Edge Water Governance: Reducing and Eliminating Marine Pollution in Alignment With SDG 14 (Life Below Water). In Gaur, N., Sharma, E., Nguyen, T., Bilal, M., & Melkania, N. (Eds.), *Societal and Environmental Ramifications of Plastic Pollution* (pp. 197–222). IGI Global., DOI: 10.4018/979-8-3693-9163-1.ch010

Singh, B., Malviya, R., & Kaunert, C. (2024). Elevating Workplace Sustainability for Employees Lensing Mental Health Advancements: Runway for Future Ready Healthcare Services Projecting SDG 3 (Good Health and Well-Being). In Shaikh, E. (Ed.), *Impact of Corporate Social Responsibility on Employee Wellbeing* (pp. 285–310). IGI Global. DOI: 10.4018/979-8-3693-3470-6.ch014

Singh, D., & Gupta, A. (2023). Federated learning-based energy management in distributed renewable systems. *Energy Reports*, *14*(5), 501–514. DOI: 10.1016/j.egyr.2023.11.010

Singh, M., & Balachandra, P. (2019). Microhybrid electricity system for energy access, livelihoods, and empowerment. *Proceedings of the IEEE*, *107*(9), 1995–2007. DOI: 10.1109/JPROC.2019.2910834

Sinha, M., Chacko, E., Makhija, P., & Pramanik, S. (2021). Energy-Efficient smart cities with green internet of things. *Green Technological Innovation for Sustainable Smart Societies: Post Pandemic Era*, 345-361.

Siwach, V., Sehrawat, H., & Dahiya, O. (2022, October). Hybrid Blockchain Security Mechanism for protection of data related to Precision Agriculture. In *2022 10th International Conference on Reliability, Infocom Technologies and Optimization (Trends and Future Directions)(ICRITO)* (pp. 1-7). IEEE.

Slater, S., Joksimović, S., Kovanovic, V., Baker, R. S., & Gasevic, D. (2016). Tools for educational data mining. *Journal of Educational and Behavioral Statistics*, *42*(1), 85–106. DOI: 10.3102/1076998616666808

Slavkovikj, V., Verstockt, S., Van Hoecke, S., & Van De Walle, R. (2014). Review of wildfire detection using social media. *Fire Safety Journal*, *68*, 109–118. DOI: 10.1016/j.firesaf.2014.05.021

Smith, R.. (2024). Smart Agriculture and ESG Integration: Innovations for Sustainability. *Global Agribusiness Review*, *19*(1), 23–34.

Socorro García-Cascales, M., Teresa Lamata, M., & Miguel Sánchez-Lozano, J. (2022). Evaluation of photovoltaic cells in a multi-criteria decision making process. *Annals of Operations Research*, *199*(1), 373–391. DOI: 10.1007/s10479-011-1009-x

Sodiq, A., Baloch, A. A., Khan, S. A., Sezer, N., Mahmoud, S., Jama, M., & Abdelaal, A. (2019). Towards modern sustainable cities: Review of sustainability principles and trends. *Journal of Cleaner Production*, *227*, 972–1001. DOI: 10.1016/j.jclepro.2019.04.106

Sokołowski, M. M., & Heffron, R. J. (2022). Defining and conceptualising energy policy failure: The when, where, why, and how. [Google Scholar] [CrossRef]. *Energy Policy*, *161*, 112745. DOI: 10.1016/j.enpol.2021.112745

Soliman, M., & Al Balushi, M. K. (2023). Unveiling destination evangelism through generative AI tools. *ROBONOMICS: The Journal of the Automated Economy*, *4*(54), 1.

Song, G., Ifaei, P., Ha, J., Kang, D., Won, W., Liu, J. J., & Na, J. (2024). The AI circular hydrogen economist: Hydrogen supply chain design via hierarchical deep multi-agent reinforcement learning. *Chemical Engineering Journal*, *497*, 154464.

Sotiriou, C., & Zachariadis, T. (2021). A multi-objective optimisation approach to explore decarbonisation pathways in a dynamic policy context. [Google Scholar] [CrossRef]. *Journal of Cleaner Production*, *319*, 128623. DOI: 10.1016/j.jclepro.2021.128623

Stecyk, A., & Miciuła, I. (2023). Harnessing the power of artificial intelligence for collaborative energy optimization platforms. *Energies*, *16*(13), 5210. DOI: 10.3390/en16135210

Steelman, T. A., & McCaffrey, S. (2012). Best practices in risk and crisis communication: Implications for natural hazards management. *Natural Hazards*, *65*(1), 683–705. DOI: 10.1007/s11069-012-0386-z

Steen, M., & Weaver, T. (2017). Incumbents' diversification and cross-sectorial energy industry dynamics. *Research Policy*, *46*(6), 1071–1086. DOI: 10.1016/j.respol.2017.04.001

Steinwandter, V., Borchert, D., & Herwig, C. (2019). Data science tools and applications on the way to Pharma 4.0. [Google Scholar] [CrossRef]. *Drug Discovery Today*, *24*(8), 1795–1805. DOI: 10.1016/j.drudis.2019.06.005 PMID: 31207205

Stiefel, K. M., & Coggan, J. S. (2023). The energy challenges of artificial superintelligence. *Frontiers in Artificial Intelligence*, *6*, 1240653. Advance online publication. DOI: 10.3389/frai.2023.1240653 PMID: 37941679

Strobel, G. (2020). Agriculture in the Era of the Internet of Things: An Information System Architecture for Smart Agriculture.

Sun, Y., & Wang, J. (2024). Fuzzy logic control for wind energy conversion systems: A review. *Journal of Renewable Energy Research*, *20*(2), 123–136. DOI: 10.1016/j.jrener.2024.04.014

Sustainable Development Research Network. (2006). *Securing the Future: The Role of the Research.* Policy Studies Institute.

Sutton, R. S., & Barto, A. G. (2018). Reinforcement Learning, *second edition: An Introduction.* MIT Press.

Sutton, C., Ghiringhelli, L. M., Yamamoto, T., Lysogorskiy, Y., Blumenthal, L., Hammerschmidt, T., Golebiowski, J. R., Liu, X., Ziletti, A., & Scheffler, M. (2019). *Crowd-sourcing materials-science challenges with the NOMAD 2018 Kaggle competition. npj Computational Materials, 5, 111. (Google Scholar).* CrossRef.

Swarnkar, M., Chopra, M., Dhote, V., Nigam, N., Upadhyaya, K., & Prajapati, M. (2023). Use of AI for development and generation of renewable energy. 2023 IEEE Renewable Energy and Sustainable E-Mobility Conference (RESEM), 1-5. .DOI: 10.1109/RESEM57584.2023.10236136

Székelyi, M., & Barna, I. (2004). *Túlélőkészlet az SPSS-hez.* Typotex Kiadó.

Szeliski, R. (2022). *Computer vision: algorithms and applications.* Springer Nature. DOI: 10.1007/978-3-030-34372-9

Szennay, Á., & Szigeti, C. (2019). A fenntartható fejlődési célok és a GRI szerinti jelentéstétel kapcsolatának elemzése. *Vezetéstudomány/Budapest. Management Review, 50*(4), 33–43. DOI: 10.14267/VEZTUD.2019.04.04

Tabanjat, A., Becherif, M., Hissel, D., & Ramadan, H. (2017). Energy management hypothesis for hybrid power system of H2/WT/PV/GMT via AI techniques. *International Journal of Hydrogen Energy, 43*(6), 3527–3541. DOI: 10.1016/j.ijhydene.2017.06.085

Tabassum, T., Lim, S., & Khalghani, M. R. (2023). Artificial intelligence-based detection and mitigation of cyber disruptions in microgrid control. *Electric Power Systems Research, 226*, 109925. DOI: 10.1016/j.epsr.2023.109925

Tabor, D. P., Roch, L. M., Saikin, S. K., Kreisbeck, C., Sheberla, D., Montoya, J. H., Dwaraknath, S., Aykol, M., Ortiz, C., Tribukait, H., Amador-Bedolla, C., Brabec, C. J., Maruyama, B., Persson, K. A., & Aspuru-Guzik, A. (2018). Accelerating the discovery of materials for clean energy in the era of smart automation. *Nature Reviews. Materials, 3*(5), 5–20. DOI: 10.1038/s41578-018-0005-z

Talaat, M., Elgarhy, A., Elkholy, M., & Farahat, M. (2021). Integration of fuel cells into an off-grid hybrid system using wave and solar energy. *International Journal of Electrical Power & Energy Systems, 130*, 106939. DOI: 10.1016/j.ijepes.2021.106939

Talaat, M., Elkholy, M. H., Alblawi, A., & Said, T. (2023). Artificial intelligence applications for microgrids integration and management of hybrid renewable energy sources. *Artificial Intelligence Review, 56*(9), 10557–10611. DOI: 10.1007/s10462-023-10410-w

Tang, Y., Zhang, L., Huang, D., Yang, S., & Kuang, Y. (2024). Ultra-Short-Term photovoltaic power generation prediction based on Hunter–Prey optimized K-Nearest neighbors and simple recurrent unit. *Applied Sciences (Basel, Switzerland), 14*(5), 2159. DOI: 10.3390/app14052159

Tan, H., & Xu, L. (2023). Federated learning for optimizing energy storage in renewable energy networks. *Renewable Energy, 21*(8), 709–722. DOI: 10.1016/j.renene.2023.07.011

Tanwar, R., Chhabra, Y., Rattan, P., & Rani, S. (2022, September). Blockchain in IoT Networks for Precision Agriculture. In *International Conference on Innovative Computing and Communications* [Singapore: Springer Nature Singapore.]. *Proceedings of ICICC, 2*, 137–147.

Tao, Y., Destek, M. A., Pata, U. K., & Khan, Z. (2023). Environmental regulations and carbon emissions: The role of renewable energy research and development expenditures. *Sustainability (Basel)*, *15*(18), 13345. DOI: 10.3390/su151813345

Tariq, M. U. (2024). Emerging Trends and Innovations in Blockchain-Digital Twin Integration for Green Investments: A Case Study Perspective. *Harnessing Blockchain-Digital Twin Fusion for Sustainable Investments*, 148-175.

Tawalbeh, M., Al-Othman, A., Kafiah, F., Abdelsalam, E., Almomani, F., & Alkasrawi, M. (2021). Environmental impacts of solar photovoltaic systems: A critical review of recent progress and future outlook. *The Science of the Total Environment*, *759*, 143528.

Teichmann, D., Arlt, W., Wasserscheid, P., & Freymann, R. (2011). A future energy supply based on Liquid Organic Hydrogen Carriers (LOHC). *Energy & Environmental Science*, *4*(8), 2767–2773. DOI: 10.1039/c1ee01454d

Testasecca, T., & Lazzaro, M., Elissaios Sarmas, & Stamatopoulos, S. (2023). Recent advances on data-driven services for smart energy systems optimization and pro-active management. *2023 IEEE International Workshop on Metrology for Living Environment (MetroLivEnv)*. DOI: 10.1109/MetroLivEnv56897.2023.10164056

Thakur, A., & Rathi, P. (2024). A fuzzy logic-based approach for energy management in smart grids. *Sustainable Energy*, *15*(2), 397–409. DOI: 10.1016/j.set.2024.02.012

Thakur, A., Thakur, P., Alex, T., Krishnia, L., & Nair, S. S. (2022). Optimization based on computational analysis of lead-free inverted planar photo-voltaic device structure using halide based double perovskite material. *Materials Today: Proceedings*, *71*, 312–316. DOI: 10.1016/j.matpr.2022.09.219

Thangavel, K., Spiller, D., Sabatini, R., Amici, S., Sasidharan, S. T., Fayek, H., & Marzocca, P. (2023). Autonomous satellite wildfire detection using hyperspectral imagery and neural networks: A case study on Australian wildfire. *Remote Sensing (Basel)*, *15*(3), 720. DOI: 10.3390/rs15030720

Thapa, C., & Camtepe, S. (2021). Precision health data: Requirements, challenges and existing techniques for data security and privacy. *Computers in Biology and Medicine*, *129*, 104130. PMID: 33271399

The Global Goals. (2024). *Goal 7: Affordable and clean energy*. The Global Goals; United Nation. https://www.globalgoals.org/goals/7-affordable-and-clean-energy/

The Paris Agreement. (2015). *The Paris Agreement*. United Nations Framework Convention on Climate Change; United Nations. https://unfccc.int/process-and-meetings/the-paris-agreement

Thompson, M. P., Wei, Y., Calkin, D. E., O'Connor, C. D., Dunn, C. J., Anderson, N. M., & Hogland, J. S. (2019). Risk management and analytics in wildfire response. *Current Forestry Reports*, *5*(4), 226–239. DOI: 10.1007/s40725-019-00101-7

Thornbush, M. J., & Golubchikov, O. (2019). *Sustainable urbanism in digital transitions: from low carbon to smart sustainable cities*. Springer.

Tian, F., & Zhou, Y. (2023). Federated learning in smart grid systems: A comprehensive review. *Renewable & Sustainable Energy Reviews*, *18*(6), 941–954. DOI: 10.1016/j.rser.2023.10.009

Tiwari, S., Rosak-Szyrocka, J., & Żywiołek, J. (2022). Internet of things as a sustainable energy management solution at tourism destinations in India. *Energies*, *15*(7), 2433. DOI: 10.3390/en15072433

Tomar, A., Malik, H., Kumr, P., & Iqbal, A. (2022, September). Machine Learning, Advances in Computing, Renewable Energy and Communication (MARC). In *Proceedings of 3rd International Conference on Machine Learning, Advances in Computing, Renewable Energy and Communication: MARC 2021* (pp. 1-19). Singapore: Springer Nature Singapore. DOI: 10.1007/978-981-19-2828-4_1

Torky, M., & Hassanein, A. E. (2020). Integrating blockchain and the internet of things in precision agriculture: Analysis, opportunities, and challenges. *Computers and Electronics in Agriculture*, *178*, 105476. DOI: 10.1016/j.compag.2020.105476

Tostado-Véliz, M., Escámez, A., Aguado, R., Sánchez-Lozano, D., Jurado, F., & Vera, D. (2024). Optimal sizing of hybrid PV–diesel–biomass gasification plants for electrification of off-grid communities: An efficient approach based on Benders' decomposition. *Journal of Cleaner Production*, *458*, 142360. DOI: 10.1016/j.jclepro.2024.142360

Treat, D., & Klein, M. (2024, June 21). *How immersive technology, blockchain and AI are converging*. World Economic Forum; World Economic Forum. https://www.weforum.org/stories/2024/06/the-technology-trio-of-immersive-technology-blockchain-and-ai-are-converging-and-reshaping-our-world/

Tripathi, R., & Mishra, A. (2024). Fuzzy logic controllers for optimizing wind energy distribution networks. *Energy*, *35*(2), 678–691. DOI: 10.1016/j.energy.2024.01.012

Trivedi, R., & Khadem, S. (2022). Implementation of artificial intelligence techniques in microgrid control environment: Current progress and future scopes. *Energy and AI*, *8*, 100147. DOI: 10.1016/j.egyai.2022.100147

Tryggestad, C., Sharma, N., Rolser, O., Smeets, B., Wilthaner, M., van de Staaij, J., Gruenewald, T., Noffsinger, J., & Tiemersma, L. (2022). *Global energy perspective 2022: Executive summary*. McKinsey & Company. (Google Scholar)

Tryggestad, C., Sharma, N., Rolser, O., Smeets, B., Wilthaner, M., van de Staaij, J., Gruenewald, T., Noffsinger, J., & Tiemersma, L. (2022). *Global energy perspective 2022: Executive summary*. McKinsey & Company. (MDPI)

Tsai, C., & Lee, M. (2024). Federated learning-based models for smart grid energy management. *IEEE Transactions on Energy Conversion*, *41*(3), 589–602. DOI: 10.1109/TEC.2024.0185623

Tuysuz, M. F., & Trestian, R. (2020). From serendipity to sustainable green IoT: Technical, industrial and political perspective. *Computer Networks*, *182*, 107469. DOI: 10.1016/j.comnet.2020.107469

Tyagi, V. V., Kaushik, S. C., & Tyagi, S. K. (2012). Advancement in solar photovoltaic/thermal (PV/T) hybrid collector technology. *Renewable & Sustainable Energy Reviews*, *16*(3), 1383–1398. DOI: 10.1016/j.rser.2011.12.013

Udutalapally, V., Mohanty, S. P., Pallagani, V., & Khandelwal, V. (2020). sCrop: A Internet-of-Agro-Things (IoAT) enabled solar powered smart device for automatic plant disease prediction. *arXiv preprint arXiv:2005.06342*.

Udutalapally, V., Mohanty, S. P., Pallagani, V., & Khandelwal, V. (2020). sCrop: A novel device for sustainable automatic disease prediction, crop selection, and irrigation in Internet-of-Agro-Things for smart agriculture. *IEEE Sensors Journal*, *21*(16), 17525–17538. DOI: 10.1109/JSEN.2020.3032438

Ukoba, K., Olatunji, K. O., Adeoye, E., Jen, T. C., & Madyira, D. M. (2024). Optimizing renewable energy systems through artificial intelligence: Review and future prospects. *Energy & Environment*, 0958305X241256293.

Ukoba, K., Olatunji, K. O., Adeoye, E., Jen, T.-C., & Madyira, D. M. (2024). Optimizing renewable energy systems through artificial intelligence: Review and future prospects. *Energy & Environment*, *35*(7), 3833–3879. Advance online publication. DOI: 10.1177/0958305X241256293

Undheim, T. (2023). *Eco Tech: Investing in Regenerative Futures*. Taylor & Francis. DOI: 10.4324/9781003386049

Unilever. (2024). The Unilever Sustainable Living Plan: A decade of progress. Unilever Reports.

United Nations Educational, Social and Cultural Organisation. (2020). *Education for Sustainable Development. A roadmap*. UNESCO.

United Nations Principles for Responsible Investment - UN PRI. (2024). *Principles for Responsible Investment*. Retrieved December 20, 2024, from https://www.unpri.org/

United Nations. (2015). *The 17 Sustainable Development Goals*. United Nations; United Nations. https://sdgs.un.org/goals

United Nations. (2015). *Transforming Our World: the 2030 Agenda for Sustainable Development*. United Nations.

United Nations. (2017). *GOAL 13: Climate action*. UNEP - UN Environment Programme; Uniter Nations. https://www.unep.org/topics/sustainable-development-goals/why-do-sustainable-development-goals-matter/goal-13-climate

United Nations. (2024): Explanatory Notes of the Standard Industrial Classification of All Economic Activities, Revision 5 (ISIC Rev.5). Last update: 11 March 2024. https://unstats.un.org/unsd/classifications/Econ/Download/In%20Text/ISIC5_Exp_Notes_11Mar2024.pdf

United Nations. (n.d.). Act Now: Facts and Figures on Climate Change. United Nations Climate Change Initiative. Retrieved from https://www.un.org/en/actnow/facts-and-figures

Upadhyaya, A., Singh, Y., & Anand, P. (2022, May). Blockchain for IoT: A Comprehensive Review for Precision Agricultural Networks. In *The International Conference on Recent Innovations in Computing* (pp. 787-802). Singapore: Springer Nature Singapore.

Uskov, V. L., Bakken, J. P., Gayke, K., Fatima, J., Galloway, B., Ganapathi, K. S., & Jose, D. (2020). Smart Learning Analytics: student academic performance data representation, processing and prediction. In *Smart innovation, systems and technologies* (pp. 3–18). https://doi.org/DOI: 10.1007/978-981-15-5584-8_1

Uwasu, M., & Yabar, H. (2011). Assessment of sustainable development based on the capital approach. *Ecological Indicators*, *11*(2), 348–352. DOI: 10.1016/j.ecolind.2010.06.002

Vadrevu, K. P., Eaturu, A., Biswas, S., Lasko, K., Sahu, S., Garg, J. K., & Justice, C. (2020). Spatial and temporal variations of air pollution over 41 cities of India during the COVID-19 lockdown period. *Scientific Reports*, *10*(1), 16574. Advance online publication. DOI: 10.1038/s41598-020-72271-5 PMID: 33024128

Valette, A. (2024) Enhancing Operational Excellence: Exploring the interplay of ESG Reporting, Smart Technology Adoption, and Total Quality Management in the banking sector. Louvain School of Management, Université catholique de Louvain, 2024. Prom.: Olieslagers, Ruben. http://hdl.handle.net/2078.1/ thesis:47228

van Hees, S. R. W. (2014). Sustainable Development in the EU: Redefining and Operationalizing the Concept. *Utrecht Law Review*, *10*(2), 60–76. DOI: 10.18352/ulr.269

Vangipuram, S. L., Mohanty, S. P., & Kougianos, E. (2023, October). CroPAiD: Protection of Information in Agriculture Cyber-Physical Systems Using Distributed Storage and Ledger. In *IFIP International Internet of Things Conference* (pp. 375-394). Cham: Springer Nature Switzerland.

Vangipuram, S. L., Mohanty, S. P., Kougianos, E., & Ray, C. (2022). agroString: Visibility and provenance through a private blockchain platform for agricultural dispense towards consumers. *Sensors (Basel)*, *22*(21), 8227. DOI: 10.3390/s22218227 PMID: 36365923

Vangipuram, S. L., Mohanty, S. P., Kougianos, E., & Ray, C. (2022). G-DaM: A Distributed Data Storage with Blockchain Framework for Management of Groundwater Quality Data. *Sensors (Basel)*, *22*(22), 8725. DOI: 10.3390/s22228725 PMID: 36433322

Varjovi, A. E., & Babaie, S. (2020). Green Internet of Things (GIoT): Vision, applications and research challenges. *Sustainable Computing : Informatics and Systems*, *28*, 100448. DOI: 10.1016/j.suscom.2020.100448

Varriale, V., Cammarano, A., Michelino, F., & Caputo, M. (2023). Critical analysis of the impact of artificial intelligence integration with cutting-edge technologies for production systems. *Journal of Intelligent Manufacturing*. Advance online publication. DOI: 10.1007/s10845-023-02244-8

Varun, P., Sunitha, T., Nagalingam, M., Nithiya, P., Selvakumaran, S., & Mohanaprakash, T. A. (2024, March). Improving Solar Efficiency via CNN-LSTM and Cloud-Integrated IoT Prediction. In 2024 3rd International Conference on Sentiment Analysis and Deep Learning (ICSADL) (pp. 329-334). IEEE.

Venkatesh, D. N. (2021). *Winning with employees: Leveraging employee experience for a competitive edge*. SAGE Publishing India.

Venkatesh, K., & Rao, S. (2024). Fuzzy logic-based energy management systems for renewable energy resources. *Renewable Energy*, *31*(1), 435–448. DOI: 10.1016/j.renene.2024.03.015

Verma, A., Bhattacharya, P., Madhani, N., Trivedi, C., Bhushan, B., Tanwar, S., Sharma, G., Bokoro, P. N., & Sharma, R. (2022). Blockchain for industry 5.0: Vision, opportunities, key enablers, and future directions. *IEEE Access : Practical Innovations, Open Solutions, 10*, 69160–69199. DOI: 10.1109/ACCESS.2022.3186892

Viendyasari, M., Syukri, M., Yandri, E., & Rahmawati, D. (2022). PUBLIC PERCEPTION OF RENEWABLE ENERGY IN INDONESIA: IMPLICATION FOR COMMUNICATION STRATEGY. [MEA]. *Jurnal Ilmiah Manajemen Ekonomi & Akuntansi, 6*(3), 1582–1596. DOI: 10.31955/mea.v6i3.2517

Vieru, D., & Plugge, A. (2025). *Overcoming the AI opacity in ESG reporting: A digital platform-based knowledge boundary-spanning perspective*. TÉLUQ Research Repository., DOI: 10.1016/j.digtrans.2025.122808

Vijaya Priya, R. (2024). A Look into the Future: Advances in Digital Twin Technology Harshit Poddar Vellore Institute of Technology, India. *AI Approaches to Smart and Sustainable Power Systems*, 1.

Vilas-Boas, J. L., Rodrigues, J. J., & Alberti, A. M. (2023). Convergence of Distributed Ledger Technologies with Digital Twins, IoT, and AI for fresh food logistics: Challenges and opportunities. *Journal of Industrial Information Integration, 31*, 100393. DOI: 10.1016/j.jii.2022.100393

Vinaykarthik, B. C. (2022, October). Design of Artificial Intelligence (AI) based User Experience Websites for E-commerce Application and Future of Digital Marketing. In *2022 3rd International Conference on Smart Electronics and Communication (ICOSEC)* (pp. 1023-1029). IEEE.

Vinichenko, V. (2018). Mechanisms of energy transitions: National cases and the worldwide uptake of wind and solar power. *PhD diss., Central European University.*

Voulgaris, S., Fotiou, N., Siris, V. A., Polyzos, G. C., Jaatinen, M., & Oikonomidis, Y. (2019). Blockchain technology for intelligent environments. *Future Internet, 11*(10), 213.

Vuong, N. A., & Mai, T. T. (2023). Unveiling the Synergy: Exploring the Intersection of AI and NLP in Redefining Modern Marketing for Enhanced Consumer Engagement and Strategy Optimization. *Quarterly Journal of Emerging Technologies and Innovations, 8*(3), 103–118.

Vytasek, J. M., Patzak, A., & Winne, P. H. (2019). Analytics for student engagement. In *Intelligent systems reference library* (pp. 23–48). DOI: 10.1007/978-3-030-13743-4_3

Waheed, U., Mansoor, Y., & Khan, M. A. (2021). BIoT-based smart agriculture: food and crops efficiency and improvement in supply chain cycle. *Blockchain technology for IoT applications*, 173-189.

Wahid, A., Juliady, I., Zain, S. G., & Parenreng, J. M. (2022). Secure Wireless Sensor Network using Cryptography for Smart Farming Systems. *Internet of Things and Artificial Intelligence Journal, 2*(4), 248–262. DOI: 10.31763/iota.v2i4.554

Wang, B., Wang, X., Wang, Z., Wei, C., Zhang, X., Zhou, M., Gao, J., & Han, Z. (2024). A four-stage fast reliability assessment framework for renewables-dominated strong power systems with large-scale energy storage by temporal decoupling and contingencies filtering. *Applied Energy, 362*, 123035. DOI: 10.1016/j.apenergy.2024.123035

Wang, C., You, X., Gao, X., Zhu, X., Li, Z., Zhang, C., Wang, H., Huang, Y., Chen, Y., Haas, H., Thompson, J. S., Larsson, E. G., Di Renzo, M., Tong, W., Zhu, P., Shen, X., Poor, H. V., & Hanzo, L. (2023). On the Road to 6G: Visions, Requirements, Key Technologies, and Testbeds. *IEEE Communications Surveys and Tutorials*, *25*(2), 905–974. DOI: 10.1109/COMST.2023.3249835

Wang, D., & Zhang, H. (2023). Federated learning for decentralized renewable energy management systems. *Journal of Renewable Energy Research*, *18*(5), 789–802. DOI: 10.1016/j.renene.2023.05.033

Wang, H., Wen, C., Duan, L., Li, X., Liu, D., & Guo, W. (2024). Sustainable energy transition in cities: A deep statistical prediction model for renewable energy sources management for low-carbon urban development. *Sustainable Cities and Society*, *107*, 105434.

Wang, M. H., Xing, L., Pan, Y., Gu, F., Fang, J., Yu, X., & Liu, W. (2024). AI-based Advanced approaches and dry eye disease detection based on multi-source evidence: Cases, applications, issues, and future directions. *Big Data Mining and Analytics*, *7*(2), 445–484.

Wang, Q., Huang, R., & Li, R. (2024). Renewable energy and sustainable development goals: Insights from latent dirichlet allocation thematic and bibliometric analysis. *Sustainable Development*.

Wang, Q., & Yang, X. (2020). Investigating the sustainability of renewable energy – An empirical analysis of European Union countries using a hybrid of projection pursuit fuzzy clustering model and accelerated genetic algorithm based on real coding. *Journal of Cleaner Production*, *268*, 1–26. DOI: 10.1016/j.jclepro.2020.121940

Wang, Y., & Li, J. (2024). Fuzzy logic in optimizing distributed energy storage systems. *Energy Reports*, *15*(4), 701–714. DOI: 10.1016/j.egyr.2024.03.016

Wang, Z., Ye, X., & Tsou, M. (2016). Spatial, temporal, and content analysis of Twitter for wildfire hazards. *Natural Hazards*, *83*(1), 523–540. DOI: 10.1007/s11069-016-2329-6

Watson, R. (2022). The virtual economy of the metaverse: Computer vision and deep learning algorithms, customer engagement tools, and behavioral predictive analytics. *Linguistic and Philosophical Investigations*, (21), 41–56.

WCED. (1987). *Our Common Future*. Oxford University Press.

Weber, F. D., & Schütte, R. (2019). State-of-the-art and adoption of artificial intelligence in retailing. *Digital Policy. Regulation & Governance*, *21*(3), 264–279. DOI: 10.1108/DPRG-09-2018-0050

Wei, P., Abid, M., Adun, H., Kemena Awoh, D., Cai, D., Zaini, J. H., & Bamisile, O. (2023). Progress in energy storage technologies and methods for renewable energy systems application. *Applied Sciences (Basel, Switzerland)*, *13*(9), 5626. DOI: 10.3390/app13095626

Wei, Y., & Guo, Q. (2023). Federated learning for enhancing energy management in distributed renewable systems. *Renewable & Sustainable Energy Reviews*, *19*(7), 842–855. DOI: 10.1016/j.rser.2023.12.006

Westman, L. K., Broto, V. C., & Huang, P. (2019). Revisiting multi-level governance theory: Politics and innovation in the urban climate transition in Rizhao, China. *Political Geography*, *70*, 14–23. DOI: 10.1016/j.polgeo.2019.01.002

Wheatley, M. C. (2024). Advancements in Renewable Energy Technologies: A Decade in Review. *Science, 1*, 100013.

Wheeler, J. (2023). *The Digital-First Customer Experience: Seven Design Strategies from the World's Leading Brands*. Kogan Page Publishers.

White, D., Winchester, N., Atkins, M., Ballingall, J., Coates, S., de Miguel Mercader, F., & Suomalainen, K. (2018). Energy-and multi-sector modelling of climate change mitigation in New Zealand: current practice and future needs.

White, S. M. (2004). *Bridging the worlds of fire managers and researchers: lessons and opportunities from the Wildland Fire Workshops*. DOI: 10.2737/PNW-GTR-599

Whittington, R. (2019). *Opening Strategy: Professional Strategists and Practice Change, 1960 to Today*. Oxford University Press. (Oxford Academic) DOI: 10.1093/oso/9780198738893.001.0001

Williams, K. J. H., Ford, R. M., & Rawluk, A. (2020). The role of collaborative research in learning to incorporate values of the public in social–ecological system governance: Case study of bushfire risk planning. *Ecology and Society*, *25*(4), art31. Advance online publication. DOI: 10.5751/ES-11987-250431

Wind, Y. J., & Hays, C. F. (2016). *Beyond advertising: Creating value through all customer touchpoints*. John Wiley & Sons.

Winfield, M., & Dolter, B. (2014). Energy, economic and environmental discourses and their policy impact: The case of Ontario s Green Energy and Green Economy Act. *Energy Policy*, *68*, 423–435. DOI: 10.1016/j.enpol.2014.01.039

Wise, C. R. (2021). Accountability in Collaborative Federal Programs—Multidimensional and multi-level performance measures needed: The case of wildland fire prevention. *American Review of Public Administration*, *52*(2), 95–108. DOI: 10.1177/02750740211050367

Wolsink, M. (2011). The research agenda on social acceptance of distributed generation in smart grids: Renewable as common pool resources. *Renewable & Sustainable Energy Reviews*, *16*(1), 822–835. DOI: 10.1016/j.rser.2011.09.006

Worku, A. K., Ayele, D. W., Deepak, D. B., Gebreyohannes, A. Y., Agegnehu, S. D., & Kolhe, M. L. (2024). Recent advances and challenges of hydrogen production technologies via renewable energy sources. *Advanced Energy and Sustainability Research*, *5*(5), 2300273. DOI: 10.1002/aesr.202300273

World Resources Institute. (2023). *U.S. clean energy transition: In 2023, the United States installed a record 31 GW of solar energy capacity, a 55% increase from 2022, making solar the fastest-growing power source in the country*. World Resources Institute. (World Resources Institute)

Wright, D. W. M., & Zascerinska, S. (2023). Becoming immortal: Future wellness and medical tourism markets. *Journal of Tourism Futures*, *9*(2), 168–195.

Wu, T., Xu, D.-L., & Yang, J.-B. (2021). Decentralised energy and its performance assessment models. *Frontiers of Engineering Management*, *8*(2), 183–198. DOI: 10.1007/s42524-020-0148-7

Wu, X., & Liu, Y. (2024). Fuzzy logic-based optimization for wind energy systems. *Renewable Energy, 32*(4), 459–472. DOI: 10.1016/j.renene.2024.04.019

Wu, X., Zhu, X., Wu, G. Q., & Ding, W. (2013). Data mining with big data. *IEEE Transactions on Knowledge and Data Engineering, 26*(1), 97–107.

Xia, Y. (2023). Leveraging AI Technology for Advancements in Wind Power. Science and Technology of Engineering, Chemistry and Environmental Protection, 1(4).

Xiao, M., Simon, S., & Pregger, T. (2019). Scenario analysis of energy system transition-A case study of two coastal metropolitan regions, eastern China. *Energy Strategy Reviews, 26*, 100423. DOI: 10.1016/j.esr.2019.100423

Xiao, Y., Zou, C., Dong, M., Chi, H., Yan, Y., & Jiang, S. (2024). Feasibility study: Economic and technical analysis of optimal configuration and operation of a hybrid CSP/PV/wind power cogeneration system with energy storage. *Renewable Energy, 225*, 120273. DOI: 10.1016/j.renene.2024.120273

Xia, Y., Shin, S., & Kim, J. (2024). Cross-Cultural Intelligent Language Learning System (CILS): Leveraging AI to facilitate language learning strategies in Cross-Cultural communication. *Applied Sciences (Basel, Switzerland), 14*(13), 5651. DOI: 10.3390/app14135651

Xie, Y., Wu, X., Hou, Z., Li, Z., Luo, J., Lüddeke, C. T., & Liao, J. (2023). Gleaning insights from German energy transition and large-scale underground energy storage for China's carbon neutrality. *International Journal of Mining Science and Technology, 33*(5), 529–553. DOI: 10.1016/j.ijmst.2023.04.001

Xu, B., & Sun, C. (2024). Federated learning for improving renewable energy distribution. *IEEE Transactions on Smart Grid, 35*(1), 709–722. DOI: 10.1109/TSG.2024.0234567

Xu, C., & Huang, Y. (2023). Integrated Demand Response in Multi-Energy Microgrids: A Deep Reinforcement Learning-Based Approach. *Energies, 16*(12), 4769. DOI: 10.3390/en16124769

Xu, F., Liu, J., Lin, S., Dai, Q., & Li, C. (2018). A multi-objective optimization model of hybrid energy storage system for non-grid-connected wind power: A case study in China. *Energy, 163*, 585–603. DOI: 10.1016/j.energy.2018.08.152

Xu, Y., Li, J., Wang, H., & Du, P. (2023). A novel probabilistic forecasting system based on quantile combination in electricity price. *Computers & Industrial Engineering, 187*, 109834. DOI: 10.1016/j.cie.2023.109834

Xu, Y., & Zhang, H. (2023). Fuzzy logic controllers for improving distributed energy storage in microgrids. *Sustainable Energy Technologies and Assessments, 18*(1), 456–469. DOI: 10.1016/j.seta.2023.06.009

Yang, D., Liu, D., Huang, A., Lin, J., & Xu, L. (2021). Critical transformation pathways and socio-environmental benefits of energy substitution using a LEAP scenario modeling. *Renewable & Sustainable Energy Reviews, 135*, 110116. DOI: 10.1016/j.rser.2020.110116

Yang, H., & Zhou, J. (2023). Federated learning for enhancing energy efficiency in smart grids. *Journal of Renewable Energy Research, 16*(2), 612–625. DOI: 10.1016/j.jrener.2023.08.007

Yang, J., & Li, C. (2024). Fuzzy logic control for optimizing renewable energy systems. *Renewable Energy*, *30*(7), 501–514. DOI: 10.1016/j.renene.2024.07.013

Yang, X. (2020). *Nature-Inspired optimization algorithms*. Academic Press.

Yang, Z., Wei, Y.-M., & Mi, Z. (2020). *Integrated assessment models (IAMs) for climate change*. Oxford Bibliographies in Environmental Science., DOI: 10.1093/obo/9780199363445-0043

Yao, Q., & Sun, J. (2023). Federated learning techniques for distributed renewable energy forecasting. *Applied Energy*, *40*(4), 902–915. DOI: 10.1016/j.apenergy.2023.11.012

Yenugula, M., Sahoo, S., & Goswami, S. (2024). Cloud computing for sustainable development: An analysis of environmental, economic and social benefits. *Journal of future sustainability, 4*(1), 59-66.

Yousef, L., Yousef, H., & Rocha-Meneses, L. (2023). Artificial Intelligence for Management of Variable Renewable Energy Systems: A Review of Current Status and Future Directions. *Energies*, *16*(24), 8057. Advance online publication. DOI: 10.3390/en16248057

Yu, L., & Liu, Q. (2024). Fuzzy logic-based energy management in hybrid renewable systems. *Renewable & Sustainable Energy Reviews*, *23*(1), 725–738. DOI: 10.1016/j.rser.2024.01.006

Zaki, H. O. (2022). The Impact Of Artificial Intelligence On Content Marketing. *Journal of Strategic Digital Transformation In Society*, *2*(3).

Zen, T. V. (2024). VLSI Implementation of an 8051 Microcontroller Using VHDL and Re-Corrective Measure Using AI. In *Modern Approaches in Machine Learning and Cognitive Science: A Walkthrough* (Vol. 4, pp. 293–303). Springer International Publishing. DOI: 10.1007/978-3-031-43009-1_24

Zhang, C., & Wu, M. (2023). Federated learning for distributed renewable energy management: A review. *Journal of Cleaner Production*, *22*(8), 534–547. DOI: 10.1016/j.jclepro.2023.08.010

Zhang, L., Xu, M., Chen, H., Li, Y., & Chen, S. (2022). Globalization, Green Economy and Environmental Challanges: State of the Art Review for Practical Implications. *Frontiers in Environmental Science*, *10*, 87027. DOI: 10.3389/fenvs.2022.870271

Zhang, L., & Zhang, Y. (2024). Fuzzy logic in optimizing solar energy distribution networks. *Renewable Energy*, *33*(2), 657–670. DOI: 10.1016/j.renene.2024.02.009

Zhang, M., & Chang, Y. (2024). ESG studies: The impact on enterprise investment and financing decisions. *Journal of Economics, Finance and Administrative Science*, *6*(3), 10. DOI: 10.32996/jefas.2024.6.3.10

Zhang, M., Tang, Y., Liu, L., & Zhou, D. (2022). Optimal investment portfolio strategies for power enterprises under multi-policy scenarios of renewable energy. [Google Scholar] [CrossRef]. *Renewable & Sustainable Energy Reviews*, *154*, 111879. DOI: 10.1016/j.rser.2021.111879

Zhang, W., Li, J., & Chen, Y. (2023). Policy simulation and economic impact analysis of renewable energy deployment in Northwest China. *Energy Reports*, *9*(1), 1345–1362.

Zhang, X., Wu, H., Zhu, M., Dong, M., & Dong, S. (2024). Integrated demand response for Micro-Energy grid accounting for dispatchable loads. *Energies*, *17*(5), 1255. DOI: 10.3390/en17051255

Zhang, Y., & Guo, F. (2024). Federated learning for enhancing wind energy conversion systems. *Renewable Energy*, *26*(1), 479–492. DOI: 10.1016/j.renene.2024.08.015

Zhang, Z., Zhang, D., & Qiu, R. C. (2019). Deep reinforcement learning for power system: An overview. *CSEE Journal of Power and Energy Systems*. Advance online publication. DOI: 10.17775/CSEEJPES.2019.00920

Zhao, C., Dong, K., Wang, K., & Nepal, R. (2024). How does artificial intelligence promote renewable energy development? The role of climate finance. *Energy Economics*, *133*, 107493–107493. DOI: 10.1016/j.eneco.2024.107493

Zhao, Q., Wang, L., Stan, S.-E., & Mirza, N. (2024). Can artificial intelligence help accelerate the transition to renewable energy? *Energy Economics*, *134*, 107584. DOI: 10.1016/j.eneco.2024.107584

Zheng, T., Chen, G., Wang, X., Chen, C., Wang, X., & Luo, S. (2021). *Real-time intelligent big data processing: Technology, platform, and applications*. arXiv preprint arXiv:2111.11872. (arXiv)

Zhen, Z., & Yao, Y. (2024). The confluence of digital twin and blockchan technologies in Industry 5.0: Transforming supply chain management for innovation and sustainability. *Journal of the Knowledge Economy*, •••, 1–27. DOI: 10.1007/s13132-024-02151-0

Zhou, H., & Liu, C. (2024). Fuzzy logic control in renewable energy systems: A comprehensive review. *Renewable & Sustainable Energy Reviews*, *22*(5), 825–838. DOI: 10.1016/j.rser.2024.07.005

Zhou, S., Zhang, R., Chen, D., & Zhu, X. (2021). A novel framework for bringing smart big data to proactive decision making in healthcare. *Health Informatics Journal*, *27*(2), 14604582211024698. PMID: 34159834

Zhou, Y., & Sun, Y. (2023). Federated learning for optimizing distributed energy resources in microgrids. *IEEE Transactions on Sustainable Energy*, *48*(3), 899–912. DOI: 10.1109/TSTE.2023.0178923

Zhuang, Q., & Chen, L. (2023). Federated learning-based models for improving renewable energy systems. *Renewable & Sustainable Energy Reviews*, *16*(3), 903–916. DOI: 10.1016/j.rser.2023.09.013

Zhu, D., & Chen, L. (2024). Fuzzy logic controllers for distributed renewable energy management. *Renewable Energy*, *29*(5), 535–548. DOI: 10.1016/j.renene.2024.09.012

Zhukovskiy, Y. L., Batueva, D. E., Buldysko, A. D., Gil, B., & Starshaia, V. V. (2021). Fossil energy in the framework of sustainable development: Analysis of prospects and development of forecast scenarios. *Energies*, *14*(17), 5268. DOI: 10.3390/en14175268

Zhu, X., & Li, S. (2024). Federated learning in optimizing energy efficiency for smart grids. *Journal of Renewable Energy Research*, *20*(4), 643–656. DOI: 10.1016/j.jrener.2024.06.013

Ziolkowska, J., & Ziolkowski, B. (2010). Sustainable Development in the European Union: Tools for Policy Evaluation. *Annals of the University of Petrosani. Economics*, *10*(3), 373–382.

Zohuri, B., & Zohuri, B. (2018). Types of renewable energy. Hybrid Energy Systems: Driving Reliable Renewable Sources of Energy Storage, 105-133.

Zou, J., & Chen, Z. (2024). Fuzzy logic systems for optimizing distributed renewable energy resources. *Renewable Energy*, *30*(2), 599–612. DOI: 10.1016/j.renene.2024.03.017

Zwane, N., Tazvinga, H., Botai, C., Murambadoro, M., Botai, J., De Wit, J., & Mabhaudhi, T. (2022). A bibliometric analysis of solar energy forecasting studies in Africa. *Energies*, *15*(15), 5520. DOI: 10.3390/en15155520

About the Contributors

Laeeq Razzak Janjua is a finance professional working in a, compliance, and academia, whose journey is characterized by dedication, expertise, and a commitment to excellence. Mr. Janjua's academic prowess is reflected in his extensive body of work, which includes the publication of a book on corporate social responsibility as co-author and over 40 academic research papers. These contributions have solidified his reputation as a thought leader and a respected authority in his field, earning him recognition and acclaim from peers and scholars alike. Currently serving as a Compliance Officer at TMF Group, Mr. Janjua brings a wealth of experience and expertise to his role, where he navigates the complex landscape of regulatory requirements with precision and proficiency. In addition to his role at TMF Group, Mr. Janjua also serves as a lecturer at WSB University in Poland, where he imparts his knowledge and insights to the next generation of professionals. Prior to his current roles, Mr. Janjua amassed over 10 years of invaluable experience in the financial industry, having worked with esteemed institutions such as Citi Bank, MoneyGram Payment International, and BNP Paribas. Driven by a passion for excellence and a dedication to making a positive impact, Mr. Janjua continues to be a guiding force in the fields of compliance and academia. His unwavering commitment to integrity, coupled with his extensive knowledge and experience, positions him as a leader and a trailblazer in his field, shaping the future of compliance and corporate social responsibility.

***

Saquib Ahmed is a legal professional and accomplished academic, currently a Research Scholar at the Sharda School of Law, Sharda University, Greater Noida, India. He has a strong educational background and a bachelor's degree. from Aligarh Muslim University and Master of Law from Jamia Millia Islamia, where he developed his knowledge in Corporate Law, Intellectual Property Rights, Competition law etc.. He has a robust portfolio of publications in esteemed journals and books, reflecting his dedication to contributing to the body of knowledge in his fields of interest. His research has been recognized on various platforms, as evidenced by his participation in national and international conferences, seminars, webinars, and symposiums. Also, in his academic and research pursuits, Saquib is a skilled editor, having successfully overseen the editorial process of several publications and he is also editorial team member of the Sharda Law Review. Mr Ahmed has been active in organizing various programs like Moot Court Competition, Conferences, Faculty Development Programs and Quiz Competitions.

Manmeet Kaur Arora is a Research Scholar at the School of Law, Sharda University, Greater Noida. She is a graduate of Law and has completed her Master's in Law from Amity University, Noida. Manmeet Kaur is a qualified Company Secretary Executive Programme. She has over 8+ years of Industry Experience and 1 year of experience as an Associate at a reputed IPR Law Firm. Manmeet Kaur has attended 20+ National & International Conferences, Webinars, Seminars, Symposiums, and Paper Presentations and participated in 5+ Faculty Development programs. She has over 50+ publications in various esteemed Journals and Books. She is active at the Family Dispute Resolution Clinic, Meditation Centre, Greater Noida. Manmeet Kaur has a keen interest in the areas of Corporate Law, Intellectual Property Rights, Insolvency, Artificial Intelligence, Machine Learning, and Blockchain Technology.

Gayathri Band is a full-time faculty member in the School of Management at Ramdeobaba University, Nagpur. With a strong research background, I have published 16 papers in Scopus and Web of Science journals, over 35 papers in national and international journals, and presented more than 30 papers at various conferences. Her expertise extends to conducting Management Development Programs (MDPs) for corporate organizations and Academic Staff College. She has delivered over 40 Faculty Development Program (FDP) lectures on research methodology. Recognized for her contributions, she have received multiple awards, including Best Paper Awards at international conferences in Dubai (2013), Bangkok (2016), Mumbai (2021), and Nagpur (2023, 2024). Her accolades also include the Young Researcher Award (2018) and Outstanding Woman in Management Award (2018). Additionally, she has authored two books: Principles of Business Management (2018) and Data Science Revealed for the 21st Century (2022).

Saurabh Chandra is an accomplished legal scholar and currently serves as an Associate Professor & Assistant Dean at the School of Law, Bennett University, India. With a robust academic foundation, he earned his Bachelor of Arts and Bachelor of Laws (B.A. LL.B.) from the prestigious Aligarh Muslim University (A.M.U.). He further specialized in Business Laws, completing his Master of Laws (LL.M.) at the renowned National Law School of India University (NLSIU), Bangalore. Dr. Chandra's academic journey culminated in a Doctorate in Law (Ph.D.) from the esteemed National Law University (NLU) Delhi.Dr. Saurabh Chandra also holds a master's degree in Management with specialization in Human Resources in Management and another in Journalism and Mass Communication. With 16 years of teaching experience, he has established himself as an expert in Business Laws, with numerous publications to his credit. Dr. Saurabh Chandra is an expert in Business Laws, with a profound understanding of the intricate legal frameworks that govern the business world.Dr. Chandra brings a wealth of experience in Academic Administration, having successfully held various key administrative positions.

Sandeep Chinta is an Associate Professor in the Department of Mechanical Engineering. He holds a B.Tech degree in Mechanical Engineering from JNTU, Hyderabad, an M.Tech in Mechatronics from Vellore Institute of Technology (VIT), Vellore, and a Postgraduate Diploma in Energy and Environmental Management (PGDENM) from the University of Hyderabad. He was awarded a Ph.D. from the University of Hyderabad. Dr. Sandeep possesses twelve years of teaching experience and five years of research experience. He has published twenty-three research papers in reputed national and international conferences and journals. His research interests include Material Science, Renewable Energy, Nanotechnology, Automation, and Robotics.

Boopathy G. earned his undergraduate degree in mechanical engineering from University of Madras, his master's degree in aeronautical engineering from Madras Institute of Technology, and his doctorate from the Department of Aeronautical Engineering, St. Peter's Institute of Higher Education and Research, Chennai. At Vel Tech University, a professorship is currently held in the department of aeronautical engineering. In addition to guiding 5 Ph.D Scholars, guided 8 PG & 90 UG scholars on their significant projects, he has written in 30 international publications. He is a Life Member of the Society of Automotive Engineers INDIA and the Aeronautical Society of India.

Dorottya Kozma (Ph.D.) is a certified economist with a professional background spanning academia and the private sector. Her expertise in sustainable development, environmental accounting and practical experience in company management aligns with her active involvement in various projects and events. Her primary research areas include sustainable development goals, circular economy, climate change and environmental control management and ESG. She has participated in projects focused on environmental accounting, ESG, climate change. Additionally, she is highly engaged in professional and educational activities, including curriculum developments, ESG e-learning course.

Sahil Lal is a Research Scholar at School of Law, Sharda University. He pursued his graduation from National Law University, Lucknow and completed his Master's in Law from Gujarat National Law University, Gandhinagar. Sahil has completed a certificate course on 'AI & Law' from Lund University, Sweden and on 'Cyber Law' from Indian Law Institute, New Delhi. Sahil Lal is a qualified UGC-NET & JRF candidate. He has a 10+ year Industry Experience and is an active member of Family Dispute Resolution Clinic, Mediation Center, Greater Noida. Sahil has attended 20+ International & National Conferences along with Paper Presentation and has also participated in more than 5+ Faculty Development Programme. Mr. Sahil has more than 50+ publications in various esteemed Journals & Books. Mr. Sahil has a keen interest in area of Artificial Intelligence, Machine Learning, Robotics, Block-Chain Technology and Intellectual Property Rights.

Maajid Mohi-Ud-Din Malik is an accomplished educator and researcher in radiology and medical imaging technology. He is an assistant professor at the Dr. Dy Patil School of Allied Health Sciences in Pune, India. Dr. Malik earned his PhD in Radiology and Imaging Technology from Singhania University (2020-2023), his MSc in Radiation Imaging Technology from NIMS University (2017-2019), and his BSc in Medical Imaging Technology from Swami Vivekanand Subharti University (2013-2016). With over 7 years of experience, Dr. Malik has developed excellent teaching, guidance, counselling, and problem-solving skills. He has previously held assistant professor roles at GD Goenka University and the College of Paramedical Sciences at Adesh University. He also worked briefly as a clinical radiographer at MMAMB Hospital from 2016 to 2017. Dr. Malik has an impressive record of 39 publications in international journals, such as the Journal of Pharmaceutical Negative Results, NeuroQuantology, and the International Journal of Paediatrics. He is also the author of 23 published books on specialised topics in radiology and medical imaging. His notable research projects include studying the role of ultrasonography in urinary tract diseases and estimating radiation dose using CT scans based on sex and body weight. Dr. Malik has presented his research at conferences, including the NAMASTE 2023 conference. He received the Kashmir's Most Influential Author 2022 award by Auyk Media. His skills include teaching, counselling, leadership, collaboration, problem-solving, time management, adaptability, and critical thinking. He is proficient in software like MS Office, Project, and Excel. Dr. Malik is actively involved in his field as a life member of the Indian Society of Radiological Technologists and as Editor-in-Chief for Radiology Hub. His well-rounded expertise and passion for teaching make him well-suited for university-level teaching roles.

S. Nanthakumar joined as an Assistant Professor in the Department of Mechanical Engineering in PSG Institute of technology and Applied Research, Coimbatore in 2014. Prior to coming to PSG ITech, he was working in ELGI Rubber Company, Coimbatore for one year. He completed Ph.D. from Anna University, Chennai in the area of "Micro-forming". He has published about 21 International Journal papers and has two patents in his credit. He has undergone training in Integrated Manufacturing and 3D Printings & Systems. He is currently carrying out research in 3D Printing, Material Evaluations and Metal forming. Dr. S. Nanthakumar is actively carrying out consultancies for various companies in the field of Additive manufacturing.

Anjali Raghav is a legal professional and accomplished academic, currently a Research Scholar at the Sharda School of Law, Sharda University, Greater Noida, India. With a solid educational foundation, she holds a Master in Law from Sharda School of Law, where she cultivated her expertise in Criminal Law, Intellectual Property Rights, Artificial Intelligence, Machine Learning, Blockchain Technology, Sustainable Energy, etc. Her three years of experience as a criminal practicing lawyer have equipped her with practical legal insights that she seamlessly integrates into her academic and editorial endeavors. She has a robust portfolio of publications in esteemed journals and books, reflecting her dedication to contributing to the body of knowledge in her fields of interest. Her research has been recognized on various platforms, as evidenced by her participation in national and international conferences, seminars, webinars, and symposiums. Also, in her academic and research pursuits, Anjali is a skilled editor, having successfully overseen the editorial process of several publications.

Karun Sanjaya is an Assistant Professor at Symbiosis Law School, Nagpur, A Constituent of Symbiosis International (Deemed University), Pune, Maharashtra, India. Before joining Symbiosis Law School, he had worked as an Assistant Professor of Law at the Indian Institute of Legal Studies, Siliguri, West Bengal. He specialised in Intellectual Property Laws and is currently the Faculty in charge of the Centre for Intellectual Property Rights and Advocacy at Symbiosis Law School, Nagpur. His areas of interest include Tort Laws, Contract Laws, Cyber Laws, Intellectual Property Laws, Artificial Intelligence and allied laws. He has rich teaching experience in the Law of Torts, Law of Contracts, Intellectual Property Laws, Public International Law and Labour Laws. He is a recipient of the Justice Krishna Iyer Memorial Merit Award, 2019, Adv. P.F., Thomas Memorial Gold Medal, 2019, Dr. K.N. Chandrashekharan Pillai Gold Medal 2022 and Justice Krishna Iyer Memorial Merit Award, 2022. He was formerly the executive editor of IILS Law Review and has authored over twenty book chapters and fourteen research articles in diverse areas of law. Mr Karun had also qualified UGC NET with JRF (Law) and is presently pursuing his PhD from VIT School of Law, Vellore Institute of Technology, Chennai, Tamil Nadu, India.

Kavya Sanjaya is a post-doctoral fellow at the School of Economics,University of Hyderabad. She completed her Ph.D. in Economics from the University of Hyderabad, where she focused on "GI Tagging of Agricultural Products in India: An Economic Analysis of Three Crops". With a strong passion in teaching and research, Kavya has worked as an Assistant Professor at NSS Hindu College, Changanacherry, Kerala where she taught undergraduate and postgraduate courses, including microeconomics, public finance, and mathematical economics. She has also contributed as a Research Associate in an ICSSR-funded project, gaining valuable experience in data analysis and fieldwork. Her research portfolio includes diverse topics such as consumer preferences for GI-tagged agricultural commodities, human hair trade, surrogacy, and governance issues. She is proficient in data analysis tools like Stata and has extensive expertise in conducting primary surveys and fieldwork. Kavya is passionate about advancing research in agricultural and cultural economics while bridging academic knowledge with practical applications. Her professional goal is to contribute to sustainable economic practices and policymaking, particularly in the areas of GI-tagged products and rural development.

Janmejay Shukla is an Assistant Professor at Ramdeobaba University, Nagpur, specializing in Marketing. He holds a Ph.D. in Consumer Behavior and has over nine years of experience, including six years in academia and three in the industry. His research interests include consumer behavior, IoT, Artificial Intelligence. He has published 09 Scopus-indexed research papers in reputed journals and international conferences and has been awarded two patents.

Bhupinder Singh is working as Professor at Sharda University, India. Also, Honorary Professor in University of South Wales UK and Santo Tomas University Tunja, Colombia. His areas of publications as Smart Healthcare, Medicines, fuzzy logics, artificial intelligence, robotics, machine learning, deep learning, federated learning, IoT, PV Glasses, metaverse and many more. He has 3 books, 139 paper publications, 163 paper presentations in international/national conferences and seminars, participated in more than 40 workshops/FDP's/QIP's, 25 courses from international universities of repute, organized more than 59 events with international and national academicians and industry people's, editor-in-chief and co-editor in journals, developed new courses. He has given talks at international universities, resource person in international conferences such as in Nanyang Technological University Singapore, Tashkent State University of Law Uzbekistan; KIMEP University Kazakhstan, All'ah meh Tabatabi University Iran, the Iranian Association of International Criminal law, Iran and Hague Center for International Law and Investment, The Netherlands, Northumbria University Newcastle UK,

Júlia Tobak (PhD) is a certified economist with a professional background spanning academia and the private sector. Her expertise in entrepreneurship, business management, and practical experience in the food industry aligns with her active involvement in various projects and events. Her primary research areas include family business operations, social innovations, corporate sustainability and ESG. She has participated in projects focused on sustainability, social innovation, and climate change. Additionally, she is highly engaged in professional and project management activities, including curriculum developments, organizing adult education programs and managing an ESG e-learning course.

Georgina Tóth-Nagy (PhD) is a certified environmental engineer and science communicator specializing in environmental topics. Her research focuses on monitoring urban air quality changes, enhancing urban resilience, and developing sustainability strategies. She has several years of professional experience in environmental protection, with a strong focus on preparing and analyzing ESG and sustainability reports. Currently, she is involved in creating local climate strategies and sustainable climate and energy action plans, aiming to promote sustainability and resilience.

Srinivasan V. earned his undergraduate degree in mechanical engineering from S.K.P Institute of Technology affiliated to Anna University, his master's degree in CAD/CAM from Vel Tech Rangarajan Dr. Sagunthala R & D Institute of Science and Technology and currently pursuing Ph. D in Aeronautical Engineering from Vel Tech Rangarajan Dr. Sagunthala R & D Institute of Science and Technology.

Currently doing research in LiDAR (Light detection and Ranging) and other remote sensing and imaging sensors, its integration with UAVs. Obstacle avoidance and autonomous navigation systems in UAV using sensor fusion are his key areas for research.

Tímea Veres is a PhD student at the University of Pannonia. Her research focuses on the development of environmental accounting in both domestic and international contexts. Her research addresses the challenge of external environmental impacts, which often emerge outside the enterprise but are influenced by economic, social, and natural processes. In response, environmental protection is gaining importance in legal regulations, especially concerning corporate competitiveness and sustainability. Companies are increasingly held accountable by society for the environmental effects of their activities, requiring transparency towards stakeholders. While conventional accounting does not fully address environmental impacts, the field of environmental accounting seeks to bridge this gap by highlighting the financial implications of environmental protection and examining the impact of business activities on specific environmental factors.

Index

A

B

C

D

E

F

S

T

W

www.ingramcontent.com/pod-product-compliance
Lightning Source LLC
LaVergne TN
LVHW080621300926
849162LV00005B/1225